W9-BNY-721

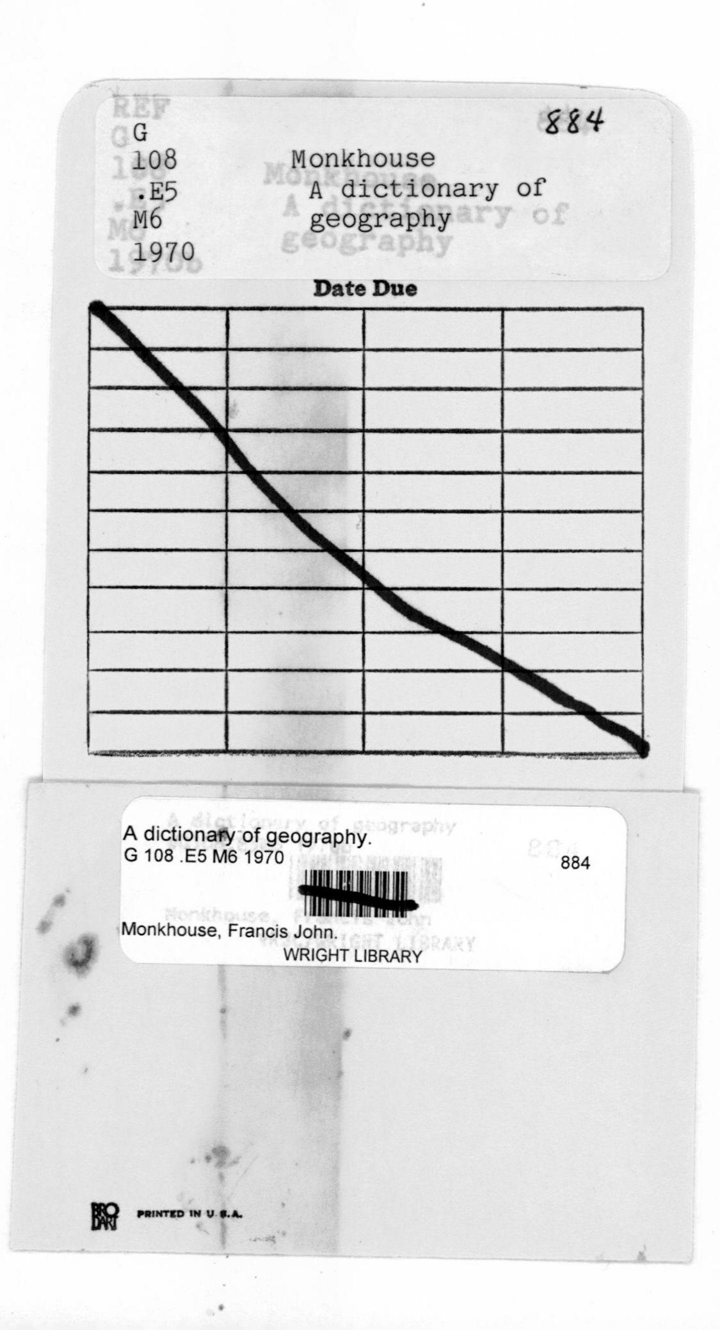

G
108
.E5
M6
1970
884
Monkhouse
A dictionary of
geography
Date Due
A dictionary of geography.
G 108 .E5 M6 1970
884
Monkhouse, Francis John.
WRIGHT LIBRARY
PRINTED IN U.S.A.

A Dictionary of Geography

Second Edition

BY

F. J. MONKHOUSE, M.A., D.SC.

Formerly Professor of Geography in the University of Southampton

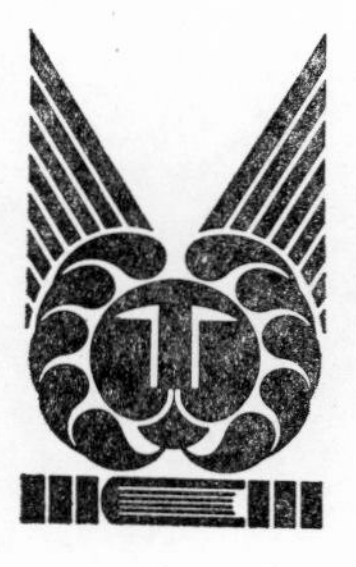

CHICAGO

ALDINE PUBLISHING COMPANY

First published in the United States 1965 by
ALDINE PUBLISHING COMPANY
529 South Wabash Avenue
Chicago, Illinois 60605

Second Edition 1970

SBN 202 10052
Library of Congress Catalog Card Number: 76 115941

Set in Monotype Imprint, and printed and bound in Great Britain by The Camelot Press Ltd., London and Southampton

PREFACE

The geographer seeks to describe the diverse features of the earth's surface, to explain if possible how these features have come to be what they are, and to discuss how they influence the distribution of man with his multifarious activities. Geography therefore stands transitionally yet centrally between the natural sciences, the social studies and the humanities. While in its concept and content it is an integrated whole, of necessity it impinges on the associated disciplines, and inevitably makes use of a wide range of kindred terminology. In compiling the 3400 entries for this Dictionary, the main criterion for inclusion has been usage. Current geographical textbooks and periodicals have been systematically combed, and where a term has been used in a specific geographical context, or in a specialist sense which differs from general practice or popular usage, it has been included. Examination papers at advanced school and university level have also been closely analysed. Foreign words are listed where they have been accepted into English geographical literature, especially where no satisfactory translation exists (KARST, CIRQUE, FIORD, FIRN, INSELBERG). Cross-references are freely given, printed in SMALL CAPITALS, where it is necessary to assist the user in tracing cognate and supplementary entries, or where the meaning of the word thus shown is essential to the understanding of the entry. The emphasis throughout is on specific factual information, conveniently accessible on a strict alphabetical basis, rather than a bare definition. Statistical material and formulae are appended, where it would seem helpful, in the form of tables under the relevant entries. Thus under RIVER, the user can expect to find listed the longest in the world, with their lengths in miles; under CONTINENT and OCEAN, the name and area of each; and under MOUNTAIN, the altitudes of some of the highest in the world, in each continent, and in Great Britain.

The main categories of entry are as follows:

(a) *Landforms*, a category which in view of the modern development of Geomorphology is of necessity lengthy.

(1) General characters and processes
(2) Names of features
(3) Rocks and minerals
   (i) Names (ii) Character (iii) Geological time and divisions
(4) Earth-movements and their results
(5) Other geophysical terms
(6) Vulcanicity
(7) Weathering and mass-movement
(8) Underground drainage and water supply
(9) Rivers and river basins
(10) Lakes
(11) Glaciation and periglaciation
(12) Arid and semi-arid landscapes
(13) Coastal features

In the case of (3), (4) and (5), it is emphasized that though this volume is not a dictionary of geology, the geographer as a student of landforms would expect to find the more important terms, of which he inevitably makes considerable use.

(b) *Oceanography*

(1) Physical oceanography
(2) Biological oceanography

This is considered in so far as it concerns the geographer, partly because the pattern of the ocean is the complement of that of the continents, partly because the oceans contribute to the climates of the land-masses, partly because of their economic value.

(c) *Climate and weather*

(1) General features
(2) Temperature
(3) Pressure and winds
(4) Humidity and precipitation
(5) Climatic types and regions

(d) *Cartography and surveying*

(1) General
(2) Map projections

The ability to illustrate his work with a wide variety of maps and diagrams is an integral part of the geographer's skill, and he is intimately concerned with the interpretation of published maps which form basic sources of information.

(e) *The earth as a spheroid*

(f) *Scales, units and dimensions*

Conversions are appended to and from the metric scales.

(g) *The soil*

(h) *Vegetation and biogeography*

(i) *Economic geography*

(1) General economic activities
(2) Products and power resources
(3) Transport and communications

(j) *Human geography*

(k) *Population and settlement*

(l) *Town- and country-planning*

(m) *Political and historical geography*

(n) *Archaeology*

(o) *Geography and methodology, geographical thought*

Since this dictionary is neither a gazetteer nor a compendium of current affairs, lists of countries and capitals, regional names and international groupings are not included, since these can be found conveniently in *The Stateman's Year Book*, *Whitaker's Almanack*, and specialist dictionaries of current affairs and politics. To have listed these terms would have inflated this dictionary to uneconomic proportions, and would indeed have departed from its primary purpose.

An indication is given of the origin of foreign words by the inclusion of (Fr.), (Sp.), etc., after the entry. The derivation and etymology of a word is not analysed in detail; this is done fully in the invaluable *A Glossary of Geographical Terms* (Longmans, 1961), edited by L. D. Stamp. Occasionally where a term was specifically coined by some authority, which happens with increasing frequency, the name of the authority and the date of the original usage are given. It is appreciated that there is still debate about the exact meaning of some terms, notably those which have apparently undergone some evolution of meaning; this dictionary, guided by advice from students and teachers, takes a firm line and proposes in a number of cases a definition which seems to be accepted by reputable authorities.

I expect, anticipate, and indeed welcome criticisms of both omission and commission, since my chosen entries are, after all, only a judicious selection of the vast number possible, and in places I may seem to have been unjustifiably arbitrary. Special attention is given to American terms which differ in meaning from the English usage. Some of the material is derived from textbooks for which I have been responsible; the rest has been garnered from far and wide. Should I have quoted inadvertently from other works without acknowledgement, which despite every care may have occurred (since my card-index has been built up over a long period of time from a multiplicity of sources), I trust that the author concerned will accept my apologies and thanks.

While working on this dictionary, I contributed a series of entries to the *Great Twentieth Century Dictionary*, published by Reader's Digest Limited. Though this is on a much less detailed scale, some of the factual material used is necessarily similar to the comparable entries in this dictionary. I am most grateful to both Reader's Digest Limited and Messrs. Arnold for permission to use some material in each context.

Mr. M. J. Clark, B.A., and Mr. J. Lewin, B.A., research students in the Department of Geography, University of Southampton, have played a major part in the compilation of this dictionary. An essential feature is the inclusion of 224 diagrams, compiled and mainly drawn by Mr. Clark, with the assistance of Mr. P. Hurn, Cartographer in the Department of Geography, University of Southampton. These illustrations must be regarded not as mere adornments to the text, but as an integral part of it. Each is inserted under what seems to be the most relevant entry. Where another entry is illustrated on the same diagram, this is indicated by the cross-reference [*f*] and the name under which the diagram appears. Mr. Lewin has patiently worked through the standard textbooks and reference volumes, checking and rechecking, entering and deleting, expanding and condensing. The immense labour of typing and re-typing the manuscript from the card-index was undertaken by Mrs. G. A. Trevett, B.A., Secretary to the Department of Geography, University of Southampton. Finally, I am most grateful to my colleagues and to the research students in this Department for constant help and advice, and for readily putting their particular expertise at my disposal.

It is my hope that this volume will be of practical assistance not only to a wide range of students in schools, colleges and universities, but also to the general public, for Geography in one form or another is the concern of us all.

F. J. M.

Southampton, December 1964

PREFACE TO THE SECOND EDITION

The five hundred and seventy-two additional entries to this *Dictionary of Geography*, together with a few minor modifications to the existing material, are the result of extensive correspondence and discussion since the appearance of the first edition in 1965. I convey my grateful thanks to reviewers, students and others from many parts of the world who have troubled to write to me; to my former colleagues at the University of Southampton; and to temporary colleagues at the Universities of Southern Illinois and Maryland, who contributed stimulating trans-Atlantic points of view. I am particularly grateful to Dr. John Leighley of the University of California (Berkeley), Dr. R. A. Harper of the University of Maryland, Dr. M. J. Clark of the University of Southampton, Dr. W. K. D. Davis of the University of Wales (Swansea), and my son, Mr. L. J. Monkhouse, and his colleagues in the planning department of the East Suffolk County Council.

In this edition I have paid additional attention to some of the many terms used in quantitative developments in geography, and to various aspects of planning and urban studies. In accordance with the official British policy towards metrication, quantities are given in S.I. terms, though Imperial equivalents are retained where necessary.

Ennerdale, September, 1969 F. J. M.

ABBREVIATIONS USED IN THE TEXT

Note. Where the listed word is repeated in the text entry, it is indicated by its initial.

abbr.	abbreviation
adj.	adjective
alt.	alternative
app.	appertaining
approx.	approximate(ly)
avge.	average
c.	about, with numbers such as dates
C.	Centigrade, Celsius
cf.	compare
cm.	centimetre
c.p.	chief producer (of commodities)
ct.	contrast
cu. cm.	cubic centimetre
cu. m.	cubic metre
cu. ft.	cubic foot, feet
E.	east
e.g.	for example
=	equals
esp.	especially
estim.	estimated
excl.	excluding
[*f*]	figure attached
[*f* word]	figure attached to word quoted, with some reference to present entry
F.	Fahrenheit
Fr.	French
ft.	foot, feet
fthm.(s)	fathom, fathoms
gen.	generally, general
Germ.	German
Gk.	Greek
gm.	gram(me)
Gt.	Great
I.	island, isle in proper name
i.e.	that is
incl.	including
in.(s)	inch, inches
It.	Italian
km.	kilometre
L.	Lake
Lat.	Latin
l.c.	lower case
lb.(s)	pound(s)
lit.	literally
m.	metre
max.	maximum
mb.	millibar
mi.	mile
min.	minimum
mm.	millimetre
mtn.	mountain
N.	north
N.E.	northeast
N.W.	northwest
occas.	occasionally
opp.	opposite
orig.	originally
oz.(s)	ounce(s)
partic.	particularly
p.h.	per hour
pl.	plural
pron.	pronounced
R.	River
ref.	reference
rel.	relative, relatively
resp.	respectively
S.	south
S.E.	southeast
S.I.	*Système Internationale (d'Unités)*
sing.	singular
Sp.	Spanish
specif.	specifically
sq.	square
S.W.	southwest
syn.	synonymous
vb.	verb
W.	west
yd.(s)	yard(s)

A DICTIONARY OF GEOGRAPHY

aa (Hawaiian) (pron. *ah-ah*) A LAVA-flow which has solidified into irregular block-like masses of a jagged, clinkerous, angular appearance, the result of gases escaping violently from within the lava and the effects of the drag of the still molten material under the hardening surface-crust. E.g. on the slopes of Mauna Loa (4171 m., 13,680 ft.), Hawaiian National Park. Ct. PAHOEHOE, PILLOW LAVA.

abime (Gk.) A deep vertical shaft in limestone country (KARST), opening at the bottom into an underground passage. In its lit. sense, it means 'bottomless'.

ablation The wasting or consuming of snow and ice from the surface of an ice-sheet or glacier. It involves: (i) *melting*, caused by solar radiation acting esp. by conduction by way of solid debris on the surface or of neighbouring rock walls, by relatively warm rainfall, and by melt-water streams; (ii) *sublimation*, the direct transference of water from the solid to the gaseous state, depending on the wind, temperature and humidity; (iii) ABRASION, caused by powerful winds blowing hard ice-particles along the surface (esp. in Polar regions); (iv) the calving of icebergs (by some authorities), where the ice-margins reach tide-water. *A. factor*: the rate at which the snow- or ice-surface wastes; *a. moraine*: uneven piles of rock material, formerly contained in the ice, left after its a. Ct. ALIMENTATION.

Abney level A surveying instrument, comprising a spirit-level mounted above a sighting-tube, the bubble being reflected in the eye-piece. It is used to measure the angle of inclination of a line joining an observer to another point. If the linear distance between the points is measured, their difference in height can be calculated using the tangent ratio. This affords a rapid and convenient field-survey method when an accuracy of ± a half-degree is acceptable.

aboriginal, -gines The original human inhabitants (sometimes also applied to flora and fauna), which are true natives of a country; thus named esp. after its discovery by Europeans. The term is applied partic. to natives of Australia (familiarly *Abo*).

abrasion The mechanical or frictional wearing-down, specif. of a rock by material (e.g. quartz-sand) which forms the abrasive medium, transported by running water, moving ice, wind and waves (which supply the energy or momentum) (cf. sand-blast). A. is the result of the process of CORRASION of rock. A. of an ice-surface can be caused by wind-blown ice-particles.

abrasion platform A nearly smooth rock p. which has been worn by the forces of ABRASION, as along the coast (a wave-cut p.).

absolute age In GEOCHRONOLOGY, the dating of rocks in actual terms of years; ct. RELATIVE A. Various tables of dating exist; e.g. by A. Holmes:

	(*beginning millions of years ago*)
Quaternary	1
Tertiary	70
Cretaceous	135
Jurassic	180
Triassic	225
Permian	270
Carboniferous	350
(in U.S.A.) Pennsylvanian / Mississippian	
Devonian	400
Silurian	440
Ordovician	500
Cambrian	600

Note: Other estimates date the beginning of the Cambrian at between 500 and 550 million years ago.

absolute drought A period of at least 15 consecutive days, each with less than 0·25 mm. (0·01 in.) of rainfall (British climatology). The record duration of a d. in Britain is 60 days during the spring of 1893, in Sussex.
Note: In U.S.A. a criterion of 14 days without measurable rain is called a DRY SPELL (which itself has a different definition in Gt. Britain).

absolute flatland map A m. in which all areas with slopes below a selected critical value are outlined and distinctively shaded.

absolute humidity The mass of water-vapour per unit volume of air, expressed in gm. per cu. m., or in grains (=0·0648 gm.) per cu. ft. A body of air of a given temperature and pressure can hold water-vapour up to a limited amount, when it becomes saturated (at the DEW-POINT). Cold air has a low a. h., warmer air has a higher figure; e.g. air at 10°C. (50°F.) can contain 9·41 gm. per cu. m.; at 20°C. (68°F.), 17·117 gm.; at 30°C. (86°F.), 30·036 gm. A. h. over the land is highest near the Equator, lowest in central Asia in winter. Ct. RELATIVE H., SPECIFIC H., MIXING RATIO.

absolute instability The state of a column of air with a LAPSE-RATE greater than the DRY ADIABATIC LAPSE-RATE of the atmosphere around, and which is therefore unstable. A. i. is regardless of moisture content, in ct. to CONDITIONAL I., where there is a dependence on moisture content.

absolute stability The state of a column of air where the LAPSE-RATE is less than the SATURATED ADIABATIC LAPSE-RATE of the atmosphere around, and which is therefore stable.

absolute temperature A scale of t. based on Absolute Zero (= −273·16°C.), the point at which thermal molecular motion ceases; i.e. 0° Absolute or KELVIN (K). The Kelvin degree has the same value as the CENTIGRADE degree; it is sufficiently accurate to use a scale obtained by adding 273°C. to the observed t. e.g. −10°C. = 263°K. This scale is esp. valuable in that there are no negative quantities, and it is sometimes used in meteorology in expressing upper air t.'s.

absorption In meteorology, the physical process by which a substance retains radiant energy (heat- and light-waves) in an irreversible form of some other kind of energy, as opposed to reflecting, refracting or transmitting it; ct. a dull black surface which absorbs a high proportion (e.g. a BLACK BULB THERMOMETER) with ice and snow (which absorb little), and burnished silver (which absorbs only 5%).

abstraction Used by some geomorphologists as syn. with river CAPTURE. Strictly a. involves lateral widening of the master-stream; capture involves HEADWARD EROSION by the master-stream.

abyss (Lat.) Lit. bottomless (cf. ABIME), hence indicating something of very great, almost unfathomable, depth (a chasm, ocean DEEP).

abyssal App. to ocean depths between 2200 m. and 5500 m. (1200 and 3000 fthms.); some authorities use the term loosely for depths of only 1800 m. (1000 fthms.) or even 900 m. (500 fthms.) and even gen. as the ocean-floor. *A. plain:* an undulating deep-sea plain. *A. zone:* an area of accumulation of PELAGIC marine deposits, notably ooze. See SUBMARINE RIDGE [*f*]. [*f*]

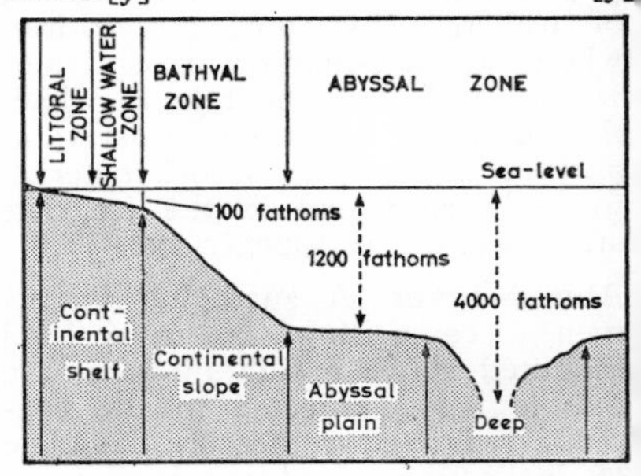

accessibility Used specif. in urban and transport geography to indicate ease or difficulty of movement between points; e.g. between a city centre and its urban periphery, between a capital city and the country it administers. It also implies the ability of vehicles to move to a destination and to stop (park) on arrival.

accessibility isopleth An i. indicating time and/or distance between a central point (e.g. a city) and the area around, on any scale: suburban or local, country, continent, or world.

accessory minerals A large number of varied m.'s widely distributed in relatively small quantities in an igneous rock, whose absence would not alter its essential nature.

access road A r. affording direct a. to premises or land on either or both sides of its line.

accident, climatic A term used by W. M. Davis and adopted partic. by C. A. Cotton to indicate interruptions to the 'normal' CYCLE OF EROSION which are the result of marked changes in climate. The 'normal' cycle was conceived as taking place under humid temperate conditions, from which aridity or glaciation might provide accidental departures. The term has become virtually obsolete since the idea of a humid temperate climate being 'normal' no longer holds credence, in view of increased knowledge of geomorphological processes elsewhere in the world.

accidented relief A rugged, 'broken' or highly dissected physical landscape.

acclimatize (vb.) To become inured to an unaccustomed climate (e.g. a European living in tropical latitudes), or to high altitudes where there is a shortage of oxygen in the atmosphere (e.g. on a Himalayan mountaineering expedition). *Acclimation* is sometimes used in U.S.A.

accommodation unit Used by town planners for the a. of a single household, whether a house, flat (apartment) or single room.

accordant drainage A systematic relationship apparent between rock-type and structure on the one hand, and surface d. pattern on the other. [*f*]

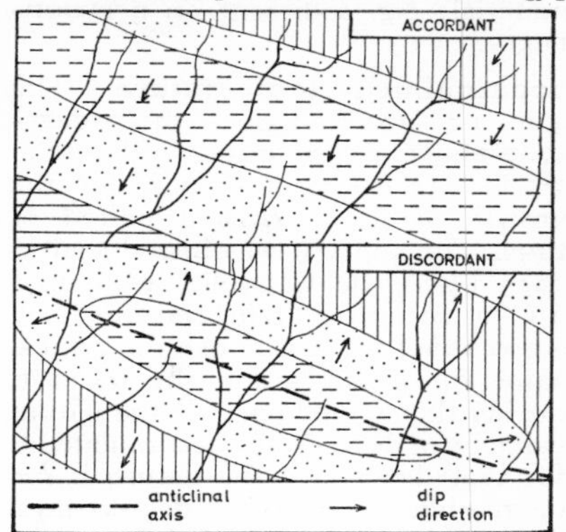

accordant junction (of rivers) A tributary which joins a main river along a course leading normally to a junction at that level. The principle was first enunciated in 1802 by J. Playfair as PLAYFAIR'S LAW

accordant (or **concordant**) **summit levels** Where the s.'s of hills or mountains rise to approx. the same elevation; this accordance may be explained either: (i) by assuming the existence of a former upland with a plane surface, which has been much dissected; or (ii) as the result of the uniform denudation of such an area of evenly spaced valleys, where stream erosion and weathering reduce the hilltops uniformly in height. A. summits may be plotted by drawing a SUPERIMPOSED PROFILE from a contour map. [*f*]

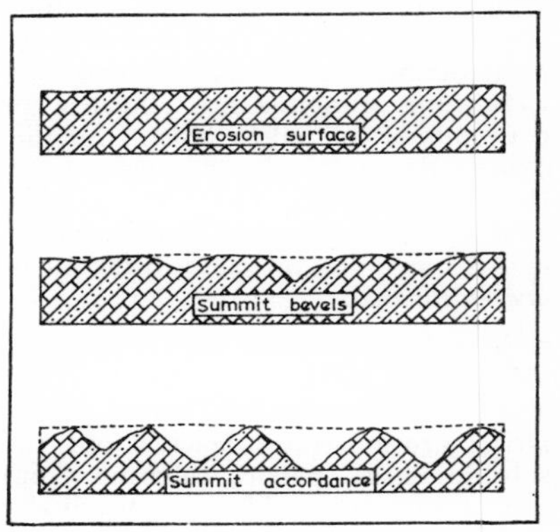

accretion The accumulation of material; e.g. sediment on a FLOOD-PLAIN.

accumulated temperature The sum or 'accumulation' of DEGREE-DAYS above a basic critical value (e.g. 5·5°C. (42°F.) as a base t. for the growth of grass) over a period of time. The concept was introduced in 1855 by A. de Candolle, who used 6°C. as his critical temperature. If on a given day the t. is above the critical value for *h* hours and the mean t. during that period exceeds the datum value by *m* degrees, the accumulated t. for that day above the datum is *hm* degree-hours or *hm*/24 degree-days. To cut down the calculations necessitated by using daily figures, estimates based on monthly means may be made. If 5·5°C. (42°F.) is the datum, and 7°C. (45°F). is the mean for a specific month, it will count (in a 31-day month) as 1·5 × 31 = 46·5 degree-days (C.) or 3 × 31 = 93 degree-days (F.) towards the final total.

acid lava A mass of molten igneous material, flowing slowly from a volcanic vent, stiff and viscous, rich in silica, and with a high melting-point (about 850°C.). Hence it solidifies rapidly and does not flow far, forming a steep-sided dome; e.g. Mount Lassen, Cascades, California. Some a. l.'s solidify in a fine-crystalled state as rhyolite or dacite, others in a glassy form as obsidian.

acid rock An igneous r. with a high percentage of free or combined silica (over 65%), or consisting of minerals rich in silica. Many common r.-forming minerals are silicates, or compounds of silica with metallic oxides. Examples of a. r.'s are granite, rhyolite, obsidian. The term a. r. is, however, considered to be misleading, and is becoming obsolete in favour of *over-saturated*.

acid soil A s. with a hydrogen-ion value (pH) below 7·2, base-deficient and sour. In cool, moist areas, percolating ground-water leaches out the soluble bases, partic. calcium, from the A-HORIZON. The s. gradually becomes lime-deficient; i.e. increasingly acid. See PODZOL.

acre A unit of English measure of area.

1 a. = 4840 sq. yds.
640 a.'s = 1 sq. mi.
1 a. = 0·4047 hectares (ha.)
2·4711 a.'s = 1 ha.
247·11 a.'s = 1 sq. km.

acre-foot Used in U.S.A., esp. in irrigation engineering, signifying the amount of water required to cover an a. of land to the depth of 1 ft. This amounts to 43,560 cu. ft.

actinometer A gen. term for any instrument for measuring the intensity of RADIATION, esp. from the sun.

action area Under the 1968 Town and Country Planning Act, an area indicated in the structure plan of a local planning authority which has been selected for early DEVELOPMENT, redevelopment or improvement.

active layer, in the soil The thickness of the soil which is frozen in winter, but thaws out in summer. Syn. with *mollisol*. Ct. PERMAFROST.

activity rate The proportion of the population in the working age-group (males, 15–64, females 15–59) who are registered as employed or who are unemployed but are seeking work. The geographical variation in this factor is most critical in the female sector; a low rate may indicate poor opportunities for female employment.

actual isotherm An i. for which the values plotted are actual means, not reduced to sea-level by adding a correction for the altitude of the station; the pattern of i.'s therefore closely resembles that of a contour map.

adiabatic App. to the change in temperature of a mass of gas (notably air), which is undergoing expansion (cooling) or compression (heating) without actual loss or gain of heat from outside. This commonly occurs within an ascending or descending air-mass, since on expansion individual molecules are more widely diffused, while on compression they are more closely packed. On expansion the lowering temperature may reach DEW-POINT, causing condensation and precipitation. Hence *adiabatically*. See DRY and SATURATED A. LAPSE-RATE.

adit A nearly horizontal passage by which a mine is worked or drained from a hillside. A.-mines are approached in this way; e.g. coal-mines in the valleys of S. Wales; iron-mines in the Orne and Fentsch valleys in French Lorraine.

adobe (i) Orig. unburnt, sun-dried bricks, and the buildings made from these, used in Mexico, S.W. U.S.A., and Argentina, later extended to the fine-grained, clayey, usually calcareous, deposits from which the bricks were made (cf. LOESS). (ii) In a more limited sense, the term is restricted to a hard-baked clayey deposit in basins of the deserts of U.S.A. It is probably of windblown origin, but was possibly reworked and re-deposited by running water.

adolescence A stage in the CYCLE OF EROSION, following youth and preceding maturity; the features characteristic of the latter are as yet only slightly developed. There has been much recent criticism of the use of such evocative language in descriptive analysis.

adret (Fr.) A hill-slope, esp. in the French Alps, which faces S. and so receives the max. available amount of sunshine and warmth, esp. in ct. to the N.-facing shady side (UBAC). In Germ. *Sonnenseite*, in It. *adretto* or *adritto*, in Turkish *günvey*.

adsorption The concentration and physical adhesion of particles of one substance on the surface of another, as of molecules of gases or of a material in solution held on the surface of a solid, though not involved in chemical combination. In sediments, it includes films of water surrounding individual particles. In soils, it includes base-salts in colloidal form surrounding mineral particles.

advanced-dune A sand-d. formed ahead of a larger d. accumulating round some obstacle, kept distinct from the main d. by eddy motion of the wind. [*f* DUNE]

advection The movement of air, water and other fluids in a horizontal direction; ct. CONVECTION, where movement is in a vertical direction. In the case of air, it may result in the transfer of heat and of water-vapour, as from lower to higher latitudes, or from a warm sea to a cooler land (and vice-versa).

advection fog A f. formed when a warm, moist air-stream moves horizontally over a cooler land or sea surface, thus reducing the temperature of the lower layers of the air below the DEW-POINT; e.g. in the neighbourhood of the Grand Banks of Newfoundland, where warm air from over the Gulf Stream, moving in a N. direction from the Florida Channel, passes over the waters of the Labrador Current; this is 8° to 11°C. (15° to 20°F.) cooler, since it brings melt-water from disintegrating PACK-ICE further N., thus forming dense f. on 70 to 100 days in the year. Also a. f. occurs on an average of 40 days in the year off the Golden Gate, San Francisco, caused by warm air moving E. over cold offshore currents.

adventive cone A parasitic or subsidiary c., which may break out on the flanks of an existing volcano; e.g. on Mt. Etna.

aegre See BORE.

aeolian App. to the effects of wind, esp. on relief; e.g. in deserts where the unconsolidated surface is unprotected by vegetation, or along a sandy sea-coast. *A. erosion* (akin to sand-blast) may produce ZEUGEN, YARDANGS, DEFLATION hollows; *a. transport* and *deposition* may produce DUNES, LOESS. Often spelt eolian, esp. in U.S.A. The term is derived from Aeolus, the Gk. god of the winds.

aerial photograph A p., vertical or oblique, of the earth's surface from an aircraft. Used for mapping

(PHOTOGRAMMETRY) and for gen. study, esp. of landforms and archaeology. A. p.'s are taken in strips (*sorties*) of overlapping prints, and may be used to make a *mosaic* (in U.S.A. a '*print lay-down*'). The scale of an a. p. is the relation between the height of the aircraft and the focal length of the camera-lens; e.g. with a 100 cm. camera at 10,000 m. height, R.F. = 1/10,000.

aerobic In the biological sense, referring to organisms living in the presence of free oxygen, specif. with ref. to those in the soil. Ct. ANAEROBIC.

aerological App. to the atmosphere; hence the U.K. Meteorological Office's *Daily A. Record.*

aeroplankton Minute organisms (spores, etc.) which float freely in the atmosphere.

aerosphere A gen. term for the entire gaseous envelope surrounding the earth, including the TROPOSPHERE and the STRATOSPHERE. The term was used because some authorities limited ATMOSPHERE to the lowest layer only of the gaseous envelope, but the latter is usually the more accepted term for the whole.

affluent An obsolescent term for the tributary of a river.

afforestation (i) Formerly, in a legal sense, the placing of an area under forest law (e.g. by the Norman kings) as a royal hunting-ground; e.g. the New Forest. (ii) Esp. in Scotland, the clearing of land of sheep and cattle to create a deer-forest. (iii) The deliberate planting of trees, usually where none grew previously or recently; e.g. by the Forestry Commission on heathlands and moorlands. Ct. REFORESTATION.

after-glow A faint and diffuse arch of radiance occasionally visible in the W. sky after sunset, when the sun is 3° or 4° below the horizon, probably caused by the scattering effect on light of dust-particles in the atmosphere. Ct. ALPINE GLOW.

aftershock Vibrations of the earth's crust after the main earthquake waves have passed, originating at or near the same SEISMIC FOCUS, caused by minor adjustments of the rocks after their main rupture. These shocks may go on for hours, days or even months.

age-pyramid, age/sex pyramid A frequency distribution HISTOGRAM of the population of a specific area, built up in 1-, 5- or 10-year age-groups, with males on one side, females on the other. This usually takes the form of a p., with the base representing the youngest group, the apex the oldest. The horizontal bars are drawn proportional in length either to the percentages or to the actual numbers in each group. [*f*]

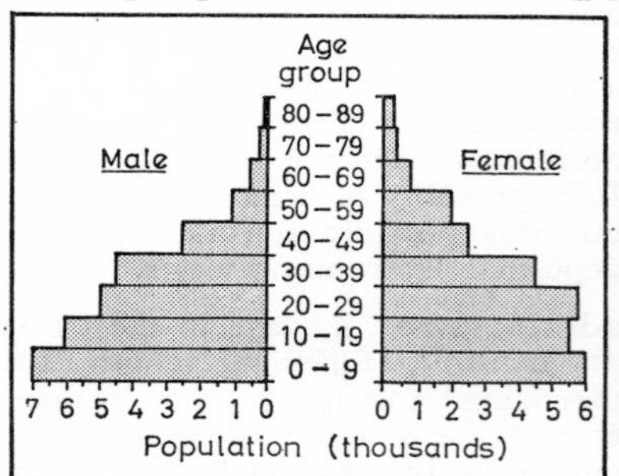

agglomerate A mass of angular fragmental material ejected by the action of a volcano, and cemented in an ash or tuff matrix. Ct. BRECCIA.

agglomeration (i) A group of houses, specif. where several villages, suburban or urban areas have expanded and coalesced in a somewhat formless and usually unplanned manner. (ii) The growth of water particles by the collision of minute droplets in the process of cloud formation, resulting in either *coalescence* (where both droplets are water), or *accretion* (if one is an ice crystal).

aggradation The building-up of the land-surface by the deposition and accumulation of solid material derived from denudation by a river. It is also applied to the development of a marine beach. Vb., to *aggrade.*

aggregate (i) Used of a structural unit in the soil, in which the individual particles are held together. (ii) An a. of mineral particles is one definition of a rock.

agonic line A l. on a map joining the earth's magnetic poles, along which the MAGNETIC DECLINATION is zero; i.e. a magnetic needle points to True North. This l. seems to be moving slowly in a W. direction. [*f*]

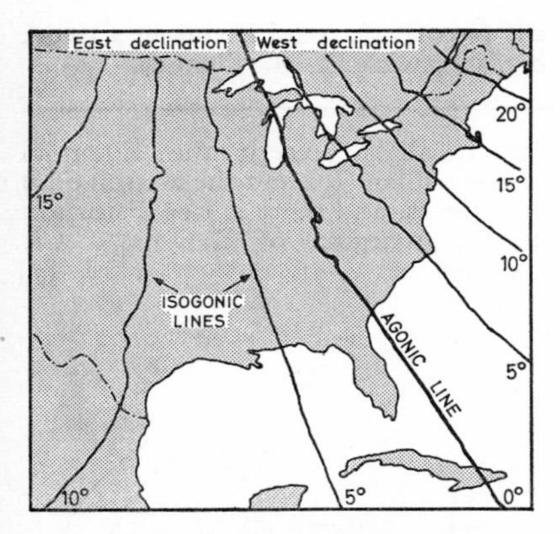

agricultural climatology Gen., c. in its application to agriculture. See MICROCLIMATE.

agricultural region An extensive area of land with broadly similar conditions and patterns of a. practice, distinctive from those of adjacent areas.

agriculture Used in a wide sense as the growing of crops and the rearing of livestock; the whole science and practice of farming. However, some writers restrict the term to the growing of crops alone.

agronomy The theory and practice of agricultural economics. In the U.S.A., the application of scientific principles to the production of crops.

A-horizon The top zone in the SOIL PROFILE [*f*], immediately below the surface, from which the soluble salts (esp. bases) and colloids have been leached. It is usually black or grey in colour, and contains some HUMUS.

aiguille (Fr.) A prominent needle-shaped rock-peak, esp. in the Mont Blanc massif; e.g. the A. du Midi, Charmoz, Grépon, Verte.

air See ATMOSPHERE.

airfield A generic term which should be used in place of the obsolescent *aerodrome*; it has certain technical installations, and excludes a mere landing-ground. A *service a.* is virtually coterminous with a military a., but this also includes experimental a.'s. An *airstrip* is not classed as an a. by the Board of Trade (formerly the Ministry of Aviation), but affords merely a landing-ground, sometimes solely of an emergency nature. See also AIRPORT.

airglow The faint light of the night sky, even when moonless.

air-mass A largely homogeneous mass of air, sometimes extending over hundreds of miles (though the term can be applied to more limited local phenomena), with marked characteristics of temperature and humidity, bounded by FRONTS, and originating in a specif. source-region. An a.-m. may travel a great distance, transporting its orig. characteristics, though becoming gradually modified. On a basis of temperature, an a.-m. is known as *Polar* or *Tropical*; on a basis of humidity as *Maritime* (having crossed the oceans and so moist) or *Continental* (originating over continents and so dry). In combination, the a.-m.'s are Polar Maritime (*Pm* or in U.S.A. *mP*), Polar Continental (*Pc*), Tropical Maritime (*Tm*), and Tropical Continental (*Tc*). Other classifications include a.-m.'s from over the Arctic Ocean (*A*), the Antarctic continent (*AA*), and the equatorial oceans (*E*). An indication of warming through equatorward movement is given by adding *W* (warm) (e.g. *TcW*); of cooling through polarward movement by adding *K* (Germ. *kalt*) (e.g. *TcK*). If it is monsoonal in character, the suffix *M* is added (e.g. *PcM* and *TmW*(*M*)). Considerable modification in temperature or humidity since an a.-m. left its source-region is denoted by the prefix *N* (e.g. *NTm*); stable or unstable conditions by the suffix *S* or *U* respectively; and its source-region by suffix initials, *NP* (N. Pacific Ocean), *SI* (S. Indian Ocean). A mixed a.-m. may be indicated by *X*.

air-meter A simple form of ANEMOMETER, consisting of a wheel (with a calibrated dial) which rotates in the wind. It is useful for measuring winds of low velocity, but tends to continue rotating by its own momentum for some time after a gust ceases, thus giving inaccurate readings.

airport An airfield used by civil, rather than military, aircraft and personnel, with Customs facilities; e.g. Heathrow (London); Schiphol (Amsterdam); Kennedy and La Guardia (New York); Orly and Le Bourget (Paris); Tempelhof (West Berlin).

air-stream A moving current of air; a wind.

ait (eyot) A small island in a river; e.g. Chiswick E., R. Thames.

Aitoff's Projection A p. based on the ZENITHAL EQUIDISTANT P., in which the horizontal distances from the central meridian are doubled; it resembles the MOLLWEIDE P., but the parallels (except the Equator) and meridians (except the central one) are curves, and there is less distortion at the margins of the map. Cf. HAMMER P. [*f*]

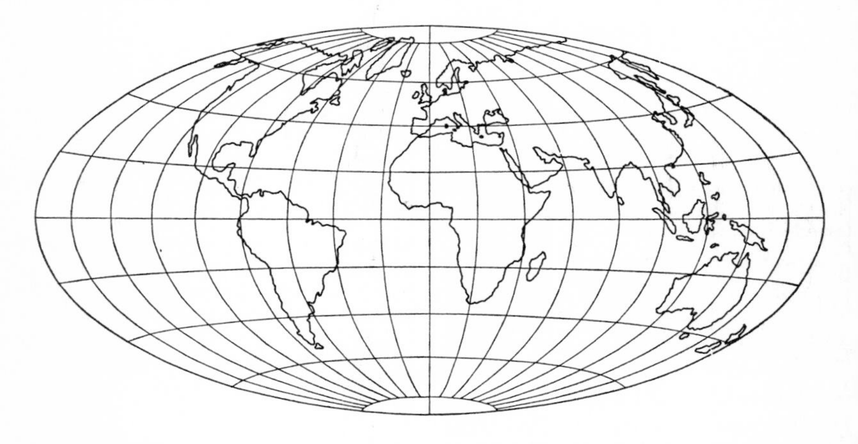

albedo The reflection coefficient, or ratio between the total solar RADIATION falling (incident) upon a surface and the amount reflected, expressed as a decimal or as a percentage. The earth's average a. is about 0·4 (40%); i.e. 4/10 of the solar radiation is reflected back into space. It varies from 0·03 for dark soil to 0·85 for a snow-field. Water has a low a. (0·02) with near-vertical rays, but a high a. for low-angle slanting rays. The figure for grass is about 0·25.

Albers' Projection A CONICAL p. of equal area properties, with 2 standard parallels along which the scale is correct. Parallels are concentric circles spaced more closely to N. and S. of the standard parallels; meridians are radiating straight lines, spaced equally, with true scale on the standard parallels, the scale too great between them and progressively smaller outside them. Meridian and parallel scales are constructed in inverse proportion to each other to obtain the equal area property. There is little distortion of shape on a continental scale. It is used for maps of U.S.A., since the greatest scale error for that area is only 1·25%. [f]

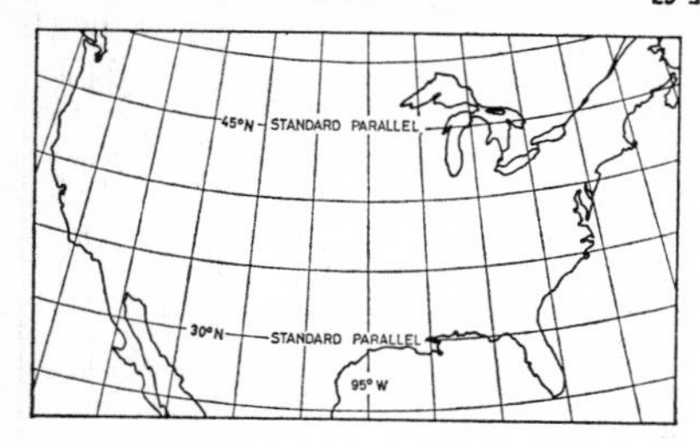

Aleutian 'low' The mean sub-polar atmospheric low pressure area in the N. Pacific Ocean, most marked in winter. It is not a very intense stationary 'low', but an area of rapidly moving individual 'lows', interrupted by occasional anticyclones. Cf. ICELANDIC LOW.

alfalfa A perennial leguminous fodder-plant (*Medicago sativa*), sometimes known as *lucerne*. It has very long tap-roots, a mass of branching stems, and purple flowers which develop into small pods. It can withstand drought, heat and cold, and with its capacity for dense growth it can produce very high yields per acre, as much as 10 tons under irrigation; used as green fodder, silage and hay. About half the world's total is grown in U.S.A., and a further 40% in Argentina.

Algonkian A term derived from the Algonkian Indians, between the Great Lakes and Hudson Bay, denoting Late Pre-Cambrian sedimentary rocks.

alidade (i) A rule with peep-sights at either end, or (on more elaborate instruments) with a telescope mounted exactly parallel to the rule. It is used in conjunction with a PLANE-TABLE to draw lines of sight on a distant object when surveying. (ii) The index of any graduated survey instrument, such as a SEXTANT.

alimentation In glaciology, the accumulation of snow on a FIRN-field, through direct snowfall, the contribution of avalanches, and the refreezing of melt-water. This may nurture an outflowing glacier. When a. near the source of a glacier exceeds wastage (ABLATION) at its end, it will 'advance'; when these are the same it will be stationary; when a. is less than wastage the glacier will shrink, 'recede' or 'retreat'.

alkali In chemistry, the soluble hydroxide of metal, esp. of sodium, potassium and calcium, which reacts with an acid to form a salt and water. It is specif. applied to soils which give a pH reaction of above 7·2, found usually in dry areas where the soluble salts have not been washed or leached away. In a SOIL PROFILE the alkaline earths (calcium carbonate, magnesium carbonate, calcium sulphate or gypsum, sodium chloride, potassium carbonate) accumulate in the B-HORIZON, having been washed down from the A-HORIZON. *A.-flat:* a level plain of sediment with a high proportion of a. salts, often crusted, formed by the evaporation of a former lake; e.g. in the Jordan Valley (Israel); near the Great Salt Lake (U.S.A.). Ct. PLAYA.

alkaline rocks Igneous rocks rich in sodium, potassium, occas. lithium; e.g. the riebeckite-microgranite of Ailsa Craig, W. Scotland.

allocthon, -ous Something which has been transported; ct. AUTOCTHON, -OUS. Used specif. in connection with an overthrust and far-travelled rock-mass, i.e. a NAPPE. It also refers to COAL MEASURES formed from transported vegetation.

allogenic Having an origin elsewhere. The term may be applied to: (i) streams which derive their water-supply from outside the immediate area, such as those crossing a desert or an area of limestone country; (ii) the constituents of certain sedimentary rocks which were orig. part of other rocks; and which have been transported, redeposited and compacted; e.g. the pebbles of CONGLOMERATES.

allotment A small piece of land rented cheaply to an individual to cultivate; usually owned by a Local Authority and located away from the cultivator's house. The word is derived from a piece of land allotted to or set apart for landless villagers during the time of the Enclosure Awards in England and Wales.

alluvial App. to ALLUVIUM.

alluvial cone A form of an A. FAN, though with a higher angle of slope, in which the mass of material is thick and coarse, and its surface steep, as in semi-arid areas, where deposits are carried by short-lived torrents. It is sometimes termed a *dejection c.* E.g. in Arizona, S. Utah and S. California. [*f*]

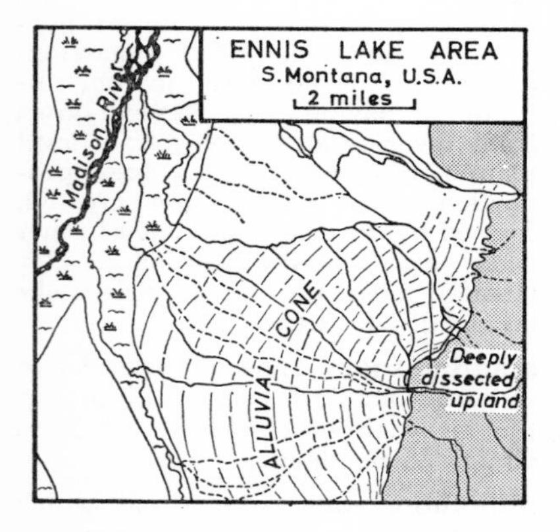

alluvial fan A fan-shaped mass of sand and gravel, with its apex pointing upstream and with a convex slope, deposited by a stream where it suddenly leaves a constricted course for a main valley or an open plain. Recent research seems to show that f.'s are produced where the constriction of a valley abruptly ceases, not solely where there is a change of gradient, as was previously suggested. E.g. in the upper Rhône valley in Switzerland, where torrents from the Bernese and Pennine Alps leave the gorges for the open floor of the main valley.

alluvial flat A near-horizontal area near a river, on which ALLUVIUM is deposited in time of flood; on a larger scale it becomes an *a. plain.* [*f*]

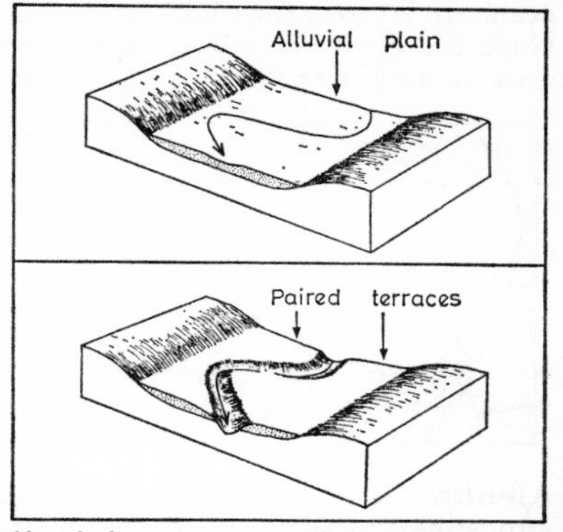

alluvial terrace Following REJUVENATION a river cuts down its channel, leaving at a higher level portions of an A. FLAT, which may be paired on either side as terraces. Some authorities regard a. t.'s as syn. with river t.'s, but this is only possible if the term 'a.' is taken to include coarse sands and gravels as well as fine-grained deposits. [*f* A. FLAT]

alluvium, adj. **alluvial** (i) In a broad sense, it denotes all unconsolidated fragmental material laid down by a stream as a cone or fan, in its bed, on its flood-plain, and in lakes, deltas and estuaries; comprising silt, sand, gravel. (ii) In a restricted sense, it denotes fine-grained silt and silt-clay, as on maps of the Geological Survey of G.B. *A. soils*, consisting of fine, well-mixed rock waste, frequently replenished in flood, with a high mineral content, and near the river responsible for deposition (therefore irrigation is possible), are often of great agricultural value; e.g. the valleys of the Nile, Indus, Ganges, Mekong, Yangtse, Hwangho. The chief disadvantage of an alluvial plain is its liability to destructive flooding.

almwind A local name for a wind of a FÖHN type which blows from the S. across the Tatra Mtns., descending into the foreland of S. Poland. It may be strong and blustery, and raises the temperature very rapidly, causing avalanches in late winter and spring.

alp A high-lying gentle slope, bench or 'shoulder' in the mountains, esp. in Switzerland, commonly above a U-shaped glaciated valley, at the level where a marked change of slope occurs. Though snow-covered in winter, it provides rich summer pasture to which animals are driven (TRANSHUMANCE). It may be the site of a small permanent settlement (e.g. Belalp, Fafleralp in the Bernese Oberland); sometimes a winter-sports resort (e.g. Wengernalp); or merely a summer residence for herdsmen. [*f*]

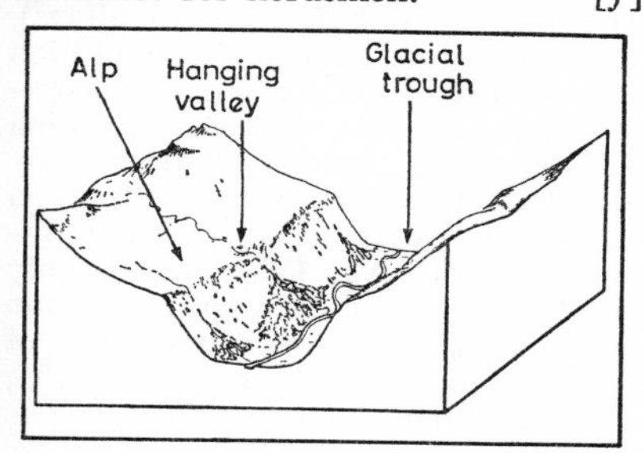

Alpine App. to the Alps; also used more widely (with l.c.) of any high mountains, their characteristic relief (esp. glacial features), climate and flora. Specif. the term refers to the last series of major mountain-building movements (the *A. Orogeny*) in mid-Tertiary times, responsible for the main ranges of Europe and Asia (sometimes known as the *Alpides*).

alpine glacier See GLACIER.

alpine glow (Germ. *Alpenglühen*) A short-lived pinkish tint on mountain peaks, esp. when snow-covered, visible just after sunset and just before sunrise. The g. may start when the rim of the sun is about 2° above the horizon. The colours in the morning have a purplish tint, in the evening an orange tint. Ct. AFTER-GLOW.

Altaides Ranges of mountains uplifted during the *Altaid Orogeny* of Upper Carboniferous-Permian times, of which the Altaid range in Asia is the type-example. See ARMORICAN, HERCYNIAN, VARISCAN.

altimeter An instrument used in aircraft or by surveyors to show height above sea-level, based on the fall in atmospheric pressure with height, which averages 34 mb. (1 in. of mercury) for each 300 m.; i.e. an ANEROID BAROMETER calibrated for height. In surveying, an a. should be used only to determine heights relative to a near-by BENCH-MARK or SPOT-HEIGHT, since absolute determinations may give a large error. For accurate work, tables are available which make allowances for latitude and for air temperature at each station. Some modern types of *radio a.* use electronic techniques. An *altigraph* is a self-recording a.

altimetric frequency graph A method of analysing relief by graphing the f. of occurrence of specif. heights above sea-level, esp. useful when seeking to recognize and correlate PLANATION surfaces. The f. of occurrence may be obtained by: (i) counting the summit spot-heights; (ii) covering the area with a grid of small squares, and in each square noting the highest point, or the mean of the highest and lowest points, or the

$$\frac{\textit{height of each corner} + 4\ (\textit{height at centre})}{8}$$

On the graph, plot heights above sea-level in feet on the horizontal scale, the % frequency on the vertical scale. In the construction, the heights are grouped in altitude class intervals; e.g. 25-ft. groups (0–25, 26–50, 51–75). [*f*]

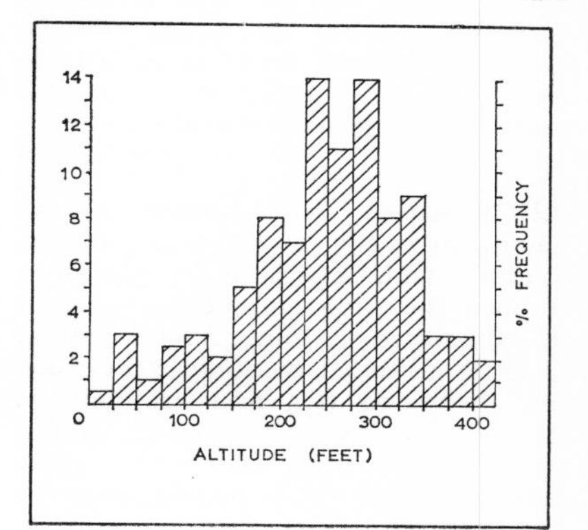

altiplanation A levelling process caused by MASS-WASTING, which may under certain conditions produce terrace-like forms and flattened summits. Some of these are accumulation features of rock-material, but it is also suggested that planation of solid rock may be caused by FREEZE-THAW processes under PERIGLACIAL conditions. The recognition of a. terraces of this kind in S.W. England (e.g. on Dartmoor) has been claimed.

altiplano A high-lying plateau-basin in the Andes, specif. in Bolivia, between the E. and W. Cordillera, at about 4000–4500 m. (13–15,000 ft.) above sea-level. It is covered with sheets of alluvium and glacial drift. L. Titicaca (fresh water) and L. Poopo (saline) are the foci of an area of inland drainage on this plateau.

altithermal phase, of post-glacial climate Used esp. by American workers to denote the post-glacial thermal maximum; in Alaska and the Yukon this has been dated as *c.* 3500 B.C.

altitude (i) The height above a chosen datum surface (mean sea-level) in any topographical survey system. (ii) In surveying, the vertical angle between the horizontal plane of the observer and any higher point; e.g. the summit of a peak. (iii) In astronomy, the a. of a heavenly body is the angle of elevation between the plane of the horizon and the body, measured along a GREAT CIRCLE through the body and the ZENITH. A. is used in conjunction with AZIMUTH to fix the position of a heavenly body. [*f* AZIMUTH]

altocumulus (*Ac.*) A fleecy cellular cloud, in bands or waves of globular masses, gen. separated by blue sky, though they may be so close together that their edges join. They occur at middle altitudes, at about 2400–6000 m. (8–20,000 ft.), and are usually, though not invariably, a sign of fair weather. The sun and moon, when seen through these clouds, may be surrounded by a CORONA

altostratus (*As.*) A greyish uniform sheet-cloud, usually of wide extent, through which it is possible to see the sun 'as through ground-glass', with a 'watery look'. It usually heralds rain, since it is associated with the approach of a WARM FRONT, and is often formed from the thickening of CIRROSTRATUS. It occurs at middle altitudes, at about 2400–6000 m. (8–20,000 ft.).

aluminium (in U.S.A., **aluminum**) A metal of remarkable lightness, high strength cf. weight (esp. in alloys), great resistance to corrosion, and high electrical conductivity. The ore, BAUXITE, is first roasted to produce alumina (Al_2O_3). The metal is extracted by the electrolysis of a mixture of alumina and molten cryolite (sodium aluminium fluoride, Na_3AlF_6), mined only in W. Greenland. It requires 25,000 KWh. of electricity to produce 1 ton of metal. C.p. of refined metal = U.S.A., Canada, France, Japan, Norway.

amber A yellowish translucent substance, a fossilized resin from coniferous trees, occurring partic. in the Oligocene strata along the Baltic coast of Samland (formerly East Prussia, now U.S.S.R.); it is of value for ornaments and jewellery.

amenity That element of the physical environment which has a beneficial physical or psychological personal effect, though it cannot be measured in economic terms.

amorphous The state of a mineral which has no definite crystalline structure; e.g. limonite (Fe_2O_3).

amphibole A group of ferromagnesian silicate minerals, of which the most important is hornblende.

amphidromic system A unit area in the sea, within which the surface of the water is set oscillating by the tide-producing forces, with a period related to the dimensions of the unit. In addition, a gyratory movement is produced by the earth's rotation (CORIOLIS FORCE), so that high water rotates around the *nodal* (or *amphidromic*) *points* in an anti-clockwise direction in the N. hemisphere. At these points the water level remains at approx. the same level (zero range), while CO-TIDAL LINES radiate outwards, along which the tidal range increases. In the English Channel a KELVIN WAVE is set up, and the a. point is theoretically located on the land (actually in Wiltshire), when it is known as a *degenerate a. point*. [*f*]

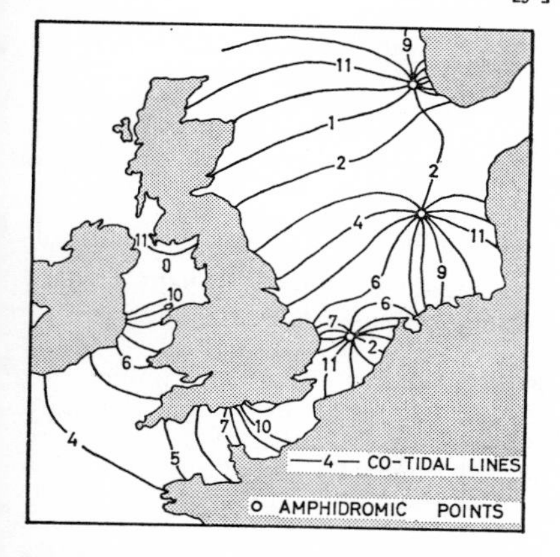

amygdale A VESICULE, a gas- or steam-escape bubble cavity in a volcanic rock, spherical or oval, filled with a secondary mineral. Hence *amygdaloidal*. E.g. some of the rocks of the Borrowdale Volcanic Series, English Lake District.

anabatic wind A local w. blowing up-valley during the afternoon, esp. in summer. The air on the mountain slopes is heated by conduction to a greater extent than air at the same level above a valley floor. This causes convectional rising of air from above the slopes, hence air moves up from the valley to take its place.

anaclinal Opposed to the DIP of the surface rocks; the term is applied partic. to streams and valleys trending against the dip of the beds which they cross. The term was introduced by J. W. Powell in 1875, together with CATACLINAL, but is now rarely used. Ct. OBSEQUENT STREAM.

anaerobic In the biological sense, the term refers to organisms living in the absence of free oxygen. In soil science, an *a. soil* is in an airless state, notably when it is waterlogged.

anafront A COLD FRONT in which a warm AIR-MASS is for the most part rising over a wedge of cold air. Ct. KATAFRONT.

anaglyph A method of visualizing relief in three dimensions. Two photographs printed side by side in red and green are viewed through lenses tinted green and red respectively. This is used effectively in some American textbooks on landforms.

analemma A graph which plots the sun's DECLINATION for each day of the year on a vertical scale, and the EQUATION OF TIME plotted on a horizontal scale, with sun times fast (in ct. to MEAN SOLAR TIME) to the left of the centre axis, sun times slow to the right. An a. is sometimes printed on a globe. [*f*]

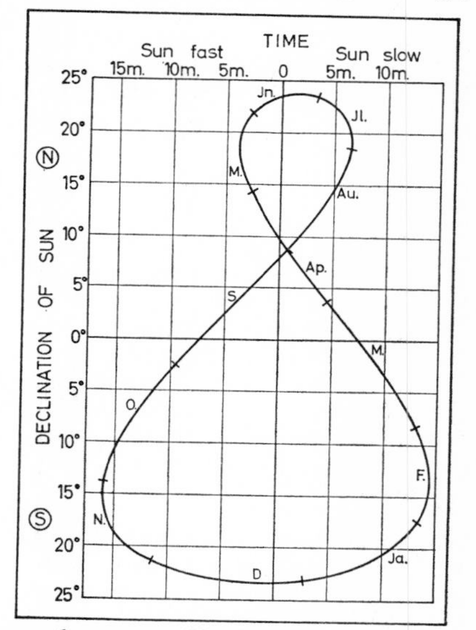

analogue Used specif. in long-range meteorological forecasting, involving the deduction of a series of future repeating patterns of weather situations by analysing and comparing sequences of previous years. See also MODEL.

analogy A method of research in which hypotheses or ideas can be tested in terms of equivalence or likeness of relations between two apparently dissimilar features or processes (*positive a.*). E.g. the 'SHIFTING RULE'; the application of the concept of gravitational fields between heavenly bodies to the behaviour of population groups. An a. may also be *negative* (with unlike features) or *neutral* (with irrelevant features). Cf. ANALOGUE.

anastomosis (In physiology, the cross-connections between arteries and veins in the body.) In geomorphology, applied to streams which are braided into a network of small water-courses (*anastomosing streams*), often the result of extensive deposition in the stream-bed.

anchor ice I. formed on and frozen to the bed of a stream or other moving water-body, when the rest of the water is not frozen. This may be in the form of FRAZIL I. on submerged objects around which the cold water is flowing. A. i. is not GROUND I., according to some authorities, though others regard it as syn.

andesite A fine-crystalled, light grey, igneous rock of intermediate composition, with 52–65% silica. It is in the form of tiny crystals embedded in a glassy ground-mass. While gen. extrusive in origin, it may be found also in minor intrusions. It was first studied in the Andes of S. America, hence the name. It forms massive crags and buttresses in the English Lake District, Snowdonia and the Ben Nevis group.

Andesite Line A petrological boundary in the Pacific Ocean, which can be traced from Alaska, Japan, the Marianas, the Bismarck Archipelago, Fiji and Tonga to the E. of New Zealand. To the W. of the A. L. the rocks are intermediate, with 52–65% silica (andesite, dacite, rhyolite); to the E. they are basic, with less than 52% silica (basalt, trachyte, olivine). On the E. of the Pacific the A. L. runs close to the coast of the Americas. Its significance is that inside the A. L. is the true ocean-basin, floored with basic rocks; outside it are continental types of rock. [*f*]

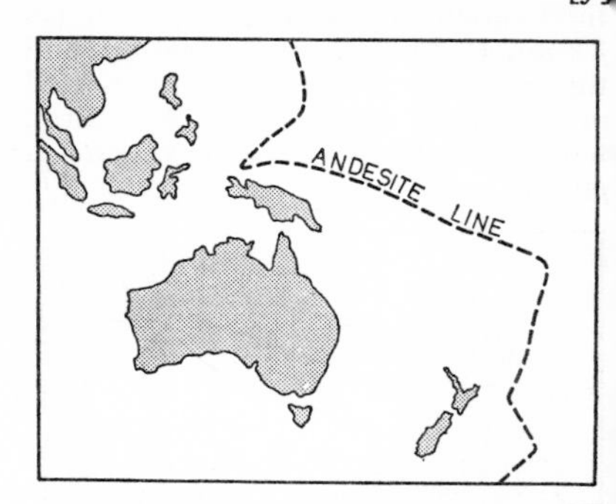

anecumene The uninhabited area of the earth's surface. Ct. ECUMENE.

anemometer An instrument for indicating, and, in more elaborate forms, for automatically recording (*anemograph*) the direction and velocity of the wind. The types used are: (i) *pressure-plate a.*, a simple device with a wooden base and an upright support holding a metal plate suspended from a knife-edge, placed at right-angles to the wind; the angle to which the plate blows is observed, and the velocity is read off from tables; (ii) *cup-a.*, with cups mounted on horizontal cross-arms attached to a vertical rotating spindle; (iii) *Dines Tube*, which depends on the difference in pressure between 2 pipes, one facing the wind, the other connected to a system of suction holes. The pressure difference is communicated to a scribing device on a revolving drum, while the vane on the top of the tube indicates changes of direction, which are also recorded.

aneroid barometer An instrument for measuring atmospheric pressure, consisting of a metallic box, almost exhausted of air, whose flexible sides expand and contract with the changing air pressure; the magnified movements are communicated by a spring to a needle on a calibrated circular dial (ct. BAROGRAPH). It was invented by Lucien Vidie in about 1843. While the instrument is light, portable and convenient, it is subject to slight errors. It has gained use in geomorphology for the field determination of heights, giving readings accurate to ± 1·5 m. under favourable conditions. Very accurate 'precision models' are produced for meteorological and aircraft use.

angel An unexplained radar echo, possibly produced by sharp temperature or humidity changes in the lower atmosphere.

Angerdorf (Germ.) A long GREEN VILLAGE.

angle of rest The max. slope at which a mass of moving unconsolidated rock (e.g. SCREE) becomes stable. This may be upset by the lubricating effect of heavy rain, occas. by earth-tremors, and by the passage of animals or man.

angular unconformity An u. in which the older underlying rock strata dip at a different angle, usually steeper, from that of the younger overlying ones. Ct. DISCONFORMITY, NON-SEQUENCE, UNCONFORMITY.

angulate pattern (of drainage) A modified form of TRELLIS DRAINAGE, in which tributaries join the main streams at acute or obtuse angles. It reflects the influence of major jointing in the rocks, sometimes of faulting.

anhydrite Anhydrous calcium sulphate, i.e. without its water of crystallisation. Specif. gravity 2·89–2·98. Derived from Gk. *anhydros*, waterless. See GYPSUM.

annular drainage A pattern of SUBSEQUENT d. in which streams follow arcuate courses as discontinuous portions of what may appear to be concentric circles. This occurs partic. around dissected domes, where subsequent streams are eroding valleys in the less resistant strata; e.g. the Black Hills of S. Dakota, with a central crystalline dome rising to 2207 m. (7242 ft.) (Harney Peak), round which the Cheyenne and its tributaries have developed a. patterns, eroded in weak shales. [*f*]

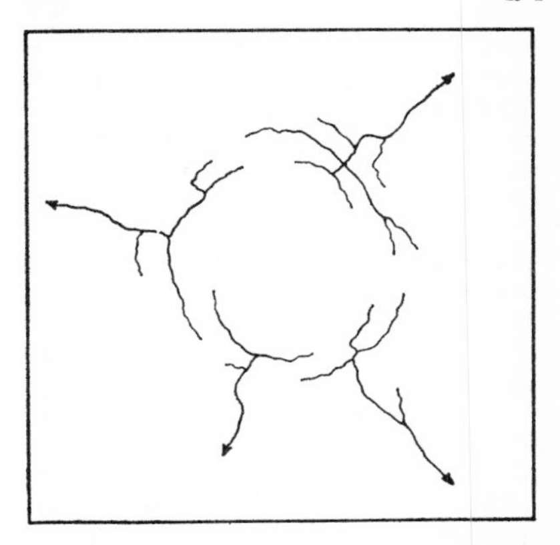

anomaly The departure of any element or feature from uniformity, or from a normal state, used partic. in meteorology in connection with temperature, in oceanography, and in connection with gravity. (i) A *temperature a.* is the difference in degrees between the mean temperature (reduced to sea-level) of a station, and the mean temperature for all stations in that latitude. The result is either a *positive* (higher than average) or a *negative* (lower than average) *a.* Points with equal a. may be plotted and joined up with ISANOMALOUS LINES. (ii) A *gravity a.* is the difference between the observed gravity and that computed for an idealized globe. If allowance is made for height above sea-level (and thus distance from the earth's centre), a *free-air a.* is obtained. If the attraction of other masses is allowed for, the *Bouguer a.* is obtained. The *isostatic a.* allows for disturbances due to ISOSTATIC movements. (iii) A *salinity a.* is the difference between the observed salinity at any point and the mean salinity for the entire oceans.

Antarctic App. to the S. Polar regions, opp. to ARCTIC. Strictly it is an adjective, but the word is also used as a noun to describe that portion of the earth's surface lying within the A. CIRCLE (66° 32′ S.), the *A. land-mass* (*Antarctica*), and the *A. ice-sheet*. The area of the last is about 13 million sq. km. (5 million sq. mi.) and its max. thickness is about 2300 m. (7500 ft.), though near the coast ranges of mountains with high rock-peaks (NUNATAKS) project above the ice. The term is also loosely used to describe features or conditions of a type similar to those found in the A. region.

Antarctic Circle The parallel of latitude 66° 32′ S., along which on about 22 Dec. (summer SOLSTICE in the S. Hemisphere) the sun does not sink below the horizon, and on about 21 June (winter solstice) it does not rise above it. S. of this line the number of days without sun in winter increases until at the S. Pole there are six months of darkness succeeded by six months of daylight. Ct. ARCTIC C.

antecedent boundary A political b. defined before any distinctions of culture, language, settlement type, etc., had developed, often in a virtually unpopulated area. Thus later developed patterns of culture may now be defined by such a b. Ct. *subsequent* and *superimposed* b.'s, drawn after the cultural pattern of the landscape has developed, which may cause conflicts and result in MINORITIES.

antecedent drainage, antecedence A river system originating before a period of uplift and folding of the land as a result of earth-movements. The river continues to cut down its valley at approx. the same rate as the uplift, and so maintains its gen. pattern and direction. This is one type of INCONSEQUENT D. E.g. the Indus in Kashmir, the Brahmaputra where it crosses from Tibet into Assam, and the Ganges and its headstreams; in each case deep gorges have been cut across lofty mountains. The R. Colorado has eroded a canyon across the plateaus of S.W. U.S.A.; the term a. d. was coined by J. W. Powell (1875) in specif. connection with this river. [*f*]

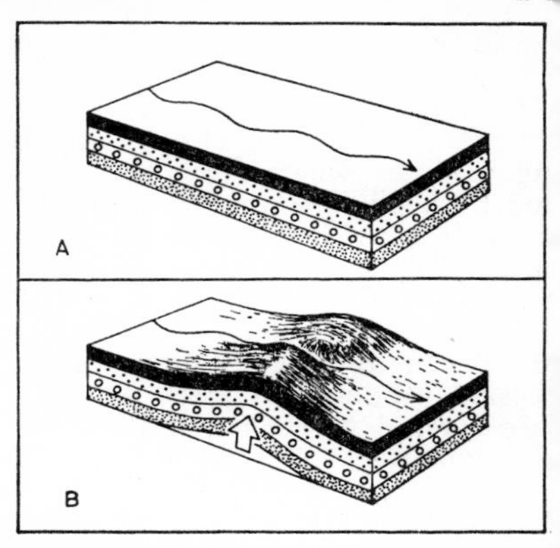

anteposition Used to denote a combination of ANTECEDENT and SUPERIMPOSED DRAINAGE; e.g. the R. Colorado.

anthodite A delicate, flower-like formation of white CALCITE, found on the roof of a limestone cave; e.g. in the Skyline Caverns, Virginia, U.S.A.

anthracite A type of hard, shiny coal, which contains (according to various classifications) 85–95% + of carbon, and 2–8% volatile matter, and which burns smokelessly with great heat. It is formed by the large-scale decomposition and compaction of the buried deposits of the coal-forests and swamps of Carboniferous times. Theophrastus, a philosopher-pupil of Aristotle, knew of coal and called it *anthrax*, from which the word a. is derived. Large a. fields are the E. Appalachian (in Pennsylvania. U.S.A.) and the S. Wales coalfields.

anthropogeography A somewhat misleading term, meaning broadly the geography of Man. It now implies the geography of Man on an anthropological basis, whereas it was initially intended to cover the relation of the earth to its inhabitants on a wide social basis as well. Sometimes (undesirably) it is interpreted as syn. with HUMAN GEOGRAPHY. The term is derived from a Germ. cognate, *Anthropogeographie*, used by F. Ratzel in 1882.

anthropology The scientific study of man, including not only physical a. (evolution of human physical features, their measurement and classification), but also in its broad sense ETHNOGRAPHY and ETHNOLOGY, and archaeology (which studies the remains of the cultures of early man, i.e. prehistoric antiquities).

anticentre The point at the ANTIPODES of the EPICENTRE of an earthquake.

anticline An upfold caused by compressive forces in the earth's crust; the strata dip outwards, forming limbs on either side of the AXIS or central line. The a. may be overfolded so that it appears to lie on its side. If the axis of an a. begins to dip, the fold is said to PITCH in that direction. [*f*A]

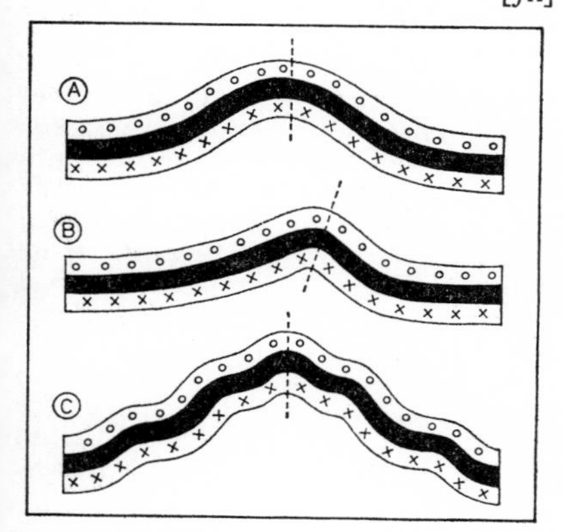

anticlinorium A complex ANTICLINE, on which minor upfolds and downfolds are superimposed; e.g. the Weald of S.E. England. [*f* ANTICLINE C]

anticorona See BROCKENSPECTRE.

anticyclone An area of atmospheric pressure high in relation to its surroundings, diminishing outwards from the centre, indicated on a synoptic chart by a series of roughly concentric, usually widely spaced, closed isobars. In the N. hemisphere air moves clockwise around an a., in the S. hemisphere it moves anti-clockwise; see BUYS BALLOT'S LAW. An a. system is slow-moving, winds are light, variable or absent near the centre, and the weather is usually settled: dry, warm and sunny in summer, and either cold, frosty and clear, or foggy ('anticyclonic gloom') in winter. The air tends to subside within it. The term was first introduced by Francis Galton in 1861. The word '*high*' is now commonly used as a noun. [*f*]

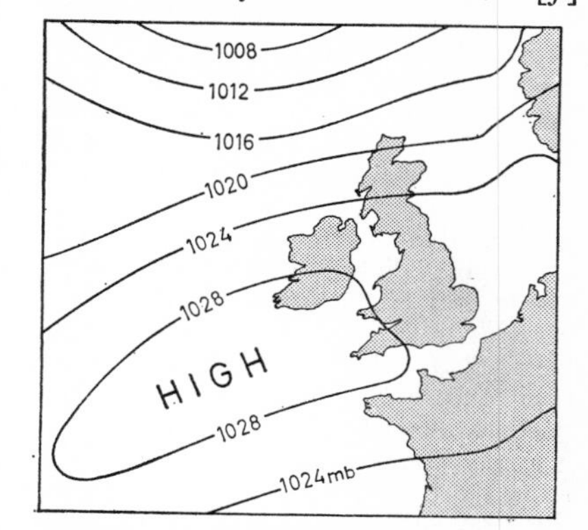

antimony A metal, derived from the ore *stibnite* (Sb_2S_3), used as an alloy-metal; e.g. in printer's type, bullets, accumulator plates, to make 'hard lead', 'Babbitt metal' (with tin and copper), 'white metal', and pewter (with lead and tin). A. oxide is employed as a paint pigment. C.p.= China, Rep. of S. Africa, Bolivia, U.S.S.R., Mexico, Yugoslavia.

antipleion Syn. with *meion*, indicating a climatological station with a high negative temperature ANOMALY. If ISANOMALS are drawn, areas of high negative anomaly (conventionally tinted blue) stand out; e.g. the continental interiors and the E. coasts of continents in high latitudes.

antipodal bulge The tidal effects at a point a. to the 'tidal bulge' on the side of the earth nearest to the moon (where the lunar attraction is greatest); at the a. b., 13,000 km. (8000 mi.) further from the moon, lunar attraction is weakest and in effect the crust is 'pulled away', to produce an apparent a. b. in the ocean waters.

antipodes, adj. **antipodal** Points at either end of a diameter of the earth, i.e. 180° of longitude apart, one being as many degrees of latitude S. of the Equator as the other is N. Antipodes Island is the nearest point which is a. to Britain (actually to the Channel Islands), situated S.E. of New Zealand at 49° 42′ S., 178° 50′ E. Note the broadly a. arrangement of the world's land-masses and oceans, with a few exceptions (Patagonia is a. to N. China, New Zealand is a. to Spain). As Australia and New Zealand lie on the other side of the world from Britain, they are sometimes loosely referred to as 'the A.'.

anti-trades Winds in the upper atmosphere (above 1800 m., 6000 ft.) above the surface TRADE WIND belt, blowing in a gen. W. direction; they were also called the *Counter trades* (now obsolete). The term was formerly used to describe surface W. winds in higher latitudes, so that it is liable to some confusion. In modern usage only the first meaning is employed. They are not high altitude 'return currents', transporting to higher latitudes rising air at the INTER-TROPICAL CONVERGENCE ZONE, but represent part of the troposphere Westerlies.

anvil cloud The flattening of the tops of large convective clouds when they reach the base of the STRATOSPHERE, spreading in the direction of high-level winds. Ice and snow crystals falling below the spreading layer produce the characteristic wedge-shaped cloud.

apartheid The policy of racial segregation, esp. as practised in the Republic of South Africa. It lit. means 'separation'; in a South African context it implies segregation of the Coloured and White populations.

aphelion The furthest point in the orbit of a heavenly body from the sun; the earth at a. on 4 July is 152 million km. (94·5 million mi.) away. The velocity of the earth in its orbit is least at a. Ct. PERIHELION.

apogean tide When the moon is at its furthest distance (APOGEE) from the earth, its gravitational attraction is less; high t.'s are lower and low t.'s are higher than usual, and the tidal range is less. See NEAP T.

apogee (i) The point in the orbit of a planet when at its max. distance from the earth; ct. PERIGEE. In a. the moon is 407,000 km. (253,000 mi.) from the earth. Clearly the term had far greater significance when the earth was believed to be the centre of the universe. It is now used with specific reference to the moon. (ii) The meridional altitude of the sun on the longest day of the year, when it reaches its greatest altitude (at midday).

apparent dip The D. of a rock stratum in a section which is not at right-angles to the STRIKE; i.e. the angle of a. d. is between 0° and that of TRUE D.

apparent time (or local solar time) The t. of day indicated by the apparent movement of the sun, as shown on a sun-dial; *a. noon* is when the sun's centre crosses in transit the local meridian (i.e. the highest point of its apparent diurnal course). The interval between two successive diurnal transits is not the same, because the sun's orbit is an ellipse and is inclined to the Equator; hence the need for MEAN SOLAR TIME, as calculated from the EQUATION OF T. See also ANALEMMA.

Appleton Layer A l. in the IONOSPHERE occurring at about 240 km. (150 mi.) above the earth's surface, which reflects short radio-waves (which have penetrated the HEAVISIDE-KENNELLY LAYER) back to earth.

applied geography The application of geographical methods of survey and analysis to world problems such as under-development, overpopulation, urban and rural land-use, and town-planning.

apse line A l. joining the points of PERIHELION and APHELION, which moves slowly around the earth's orbit in the same direction as its orbital movement. The times of perihelion and aphelion are progressively later, about 1¼ seconds per annum.

aquiclude In a strict sense, a rock stratum which is porous and may hold much water (e.g. clay), but the pores are filled with water held by surface tension, so sealing the rock against the downward movement and free passage of water. Some writers use the term more widely as syn. with AQUIFUGE. Neither is much employed in Britain, though more commonly in the U.S.A.

aquifer (alt. **aquafer**) A stratum of PERMEABLE rock, such as chalk and sandstone, which can hold water in its mass and will allow it to pass through. If underlain by an IMPERMEABLE stratum, it will act as a storage reservoir for GROUND WATER. The chief a.'s in S.E. England are the Chalk and Lower Greensand, in S.W. Lancashire the Triassic sandstones underlie more than 1·2 million sq. km. (0·5 million sq. mi.). The Dakota Sandstone on the E. flanks of the Rockies in U.S.A. allows water to move underground from the mountain slopes in an E. direction into the Great Plains.

[*f* ARTESIAN BASIN]

aquifuge A little-used term in Gt. Britain, more popular in U.S.A., for an IMPERMEABLE rock stratum, which will neither hold water in its mass nor allow it to pass through; e.g. shale, granite, quartzite. Ct. AQUICLUDE.

[*f* ARTESIAN BASIN]

arable land Farm-land which is ploughed and cultivated, not necessarily annually, including ploughland, fallow, short-ley grassland and market-gardens, though not usually domestic gardens, allotments or vineyards. It may also include land suitable for such purposes.

aragonite A crystalline form of CALCIUM CARBONATE.

arboriculture The cultivation of trees.

arc A line of islands in the form of a curve; e.g. the Aleutians; the island a.'s of E. Asia. Usually it lies along a line of folding in the earth's crust, with distinct crustal weakness, and is associated with earthquakes and volcanoes. Ocean deeps commonly lie close and parallel to an a. on its outer (oceanic) side.

arc, of the meridian A geodetic measurement along a meridian, made in order to determine accurately the shape and size of the earth, esp. with ref. to its oblateness. The first a.'s measured were by French expeditions in Peru (1735–43) and Sweden (1736–7).

arch A natural opening through a rock mass, caused in several ways: (i) by the collapse of a limestone cavern, leaving a portion of the 'roof'; e.g. at Gordale Scar, Malham (Yorkshire); the Marble Arch, the valley of the R. Cladagh, near Enniskillen (N. Ireland); (ii) on the coast, where caves are worn by marine erosion penetrating a rock projection, e.g. Durdle Door (Dorset); Needle Eye, near Wick (N. Scotland); see STACK [*f*]; (iii) where a river forms a very acute meander, and then breaks through the narrow 'neck', thus abandoning the meander and flowing through the a.; e.g. Rainbow Bridge (Utah); (iv) by the weathering of the weak part of a mass of rock, finally forming a hole right through; e.g. Arches National Monument and Natural Bridge, Bryce Canyon (both in Utah).

Archaean (i) In gen. usage, syn. with PRE-CAMBRIAN, i.e. referring to the oldest rocks. (ii) By some authorities, the rocks of the ARCHAEOZOIC era, the 2nd of the 3 eras of Pre-Cambrian time. (iii) In U.S.A. and Canada, the earlier of the 2 eras of Pre-Cambrian time.

Archaeozoic By some authorities, the middle of the 3 eras into which Pre-Cambrian time is divided, and the associated groups of rocks. The preceding era is the Eozoic, the following one is the Proterozoic. By others, it signifies all time before the beginning of the Palaeozoic era.

archipelago (It.) A group of islands scattered in near proximity about a sea. This word orig. meant 'chief gulfs' (e.g. the Aegean Sea), but this meaning is now obsolete. The term was later applied to a sea over which numerous islands are scattered, and now simply to the islands themselves; e.g. the Sporades and Cyclades in the Aegean Sea. Some groups are actually called A.; e.g. the Tuamotu A. in the S. Pacific Ocean.

Arctic Of or pertaining to N. Polar regions, strictly that part of the earth's surface lying within the A. CIRCLE (66° 32′ N.). The word is also used to describe features or conditions of climate, landscape, animals and plants characteristic of A. regions. Though strictly an adjective, A. is also used as a substantive (the A.), esp. in Canada and U.S.A. Some climatologists define an A. climate as one where the mean temperature for the coldest month is below 0°C. (32°F.), for the warmest month below 10°C. (50°F.).

Arctic air-mass A very cold AIR-MASS originating over the A. Ocean, denoted by the symbol *A*. Note the possible confusion with POLAR A.-M., a term which was coined orig. merely to denote a temperature contrast with a TROPICAL A.-M.; the term A. was introduced later, after Polar had become established, and not all authorities distinguish between them.

Arctic Circle The parallel of latitude 66° 32′ N., along which about 21 June (the summer solstice in the N. hemisphere) the sun does not sink below the horizon, and about 22 Dec. (winter solstice) it does not rise above it. N. of this line the number of days without the sun in winter increases until at the N. Pole there are 6 months of darkness succeeded by 6 months of daylight. Ct. ANTARCTIC C.

Arctic Front A semi-permanent frontal zone (see FRONTOGENESIS) along which (in the N. hemisphere) cold air from the A. region meets moderately cool air in latitudes 50°N. to 60°N. Though shortage of observations prevents any detailed know ledge of its location, it probabl extends from N. of Iceland along th N. coast of Eurasia, across the N Pacific and N. Canada; i.e. it lies N of the POLAR FRONT in the Atlanti and Pacific Oceans. It is not a ver active frontal zone, as temperatur contrasts between A. and POLAR AIR MASSES are not very marked.

'Arctic smoke' A type of fo formed in high latitudes, when col air passes over a warmer water surfac and moisture condenses, so that th water appears to 'smoke'. The fog i usually shallow, and is typically founc in bays in the Arctic Ocean. Som regard A. s. as syn. with STEAM-FOG but strictly the former develops ove salt water, the latter over fresh.

arcuate delta A DELTA with rounded arcuate, convex-outward margin; e.g. the Nile, Danube.

[*f* DELTA

area The extent of a surface measured in square units:

144 sq. ins.	= 1 sq. ft.
9 sq. ft.	= 1 sq. yd.
30¼ sq. yds.	= 1 sq. pole
40 sq. poles	= 1 rood
4 roods	= 4840 sq. yds.
	= 1 acre
640 acres	= 1 sq. mi.

English	*Metric*
1 sq. mi.	= 2·58999 sq. km.
0·386103 sq. mi.	= 1 sq. km. (100 hectares) (ha.)
1 acre	= 0·40468 ha. = 405c sq. m.
2·47106 acres	= 1 ha. (100 ares)
1 sq. in.	= 645·16 sq. mm.
1 sq. ft.	= 0·0929 sq. m.
1 sq. yd.	= 0·83613 sq. m.
1·19599 sq. yds.	= 1 sq. m.

To convert units per sq. mi. to units per sq. km., multiply by 0·3861; to convert units per sq. km. to units per sq. mi., multiply by 2·58998.

area-height diagram A d. indicating the relationship between area and altitude, comprising a graph with a vertical scale of heights in ft.

and horizontal scale of either: (i) areas in sq. mi. between each pair of selected contours; or (ii) the percentage of the total area occupied by the area between each pair.

areal differentiation The varied nature of the earth's surface, as shown by the character, pattern and inter-relationship of relief, climate, soil, vegetation, population, political units, etc., thus producing a mosaic of unique and dissimilar units. This is one definition of geography, partic. as propounded by R. Hartshorne.

area measurement grid A g. ruled on transparent paper or plastic, covered with mathematically spaced dots which can be used for rapid calculation of individual land-parcels or for sampling the incidence of such features as land utilization, field size, types of terrain, etc. Known as the *Blakerage G.* after its inventor, R. Blake.

arenaceous App. to a rock composed largely of cemented quartz-grains and other small particles; e.g. sandstone, gritstone.

arête (Fr.) (i) A steep-sided rocky ridge, specif. the crest between two adjacent CIRQUES; e.g. Striding and Swirral Edges on Helvellyn, English Lake District. It is also known, esp. in U.S.A., as a *combe-ridge.* (ii) Applied more gen. to any clean-cut ridge in high mountains; e.g. the Brouillard and Peuteret A.'s on the S. face of Mont Blanc; the Zmutt, Hörnli, Fürggen and Italian A.'s on the Matterhorn; the A. du Diable on Mont Blanc du Tacul. In Germ., *Grat*; e.g. the Viereselsgrat on the Dent Blanche in the Pennine Alps.
[*f* PYRAMIDAL PEAK]

aretic drainage Syn. with INTERNAL or inland D.

argillaceous App. to a rock largely composed of clay minerals; e.g. mud, clay, mudstone, shale.

arid, noun **aridity** Lit. dry, parched or deficient in moisture. Various definitions include: (i) less than 250 mm. (10 ins.) of rainfall per annum (U.S.A.); (ii) insufficient rainfall to support vegetation in any quantity; (iii) insufficient rainfall to support agriculture without irrigation; (iv) where the total EVAPORATION potential exceeds the actual precipitation. Various formulae have been devised to define the boundary of an arid climate, mostly empirical, involving relationships between temperature, precipitation and/or evaporation; values are plotted and isopleths are drawn. Formulae include: (i) $I = \frac{P}{T+10}$; where I is the index of aridity, T is the mean annual temperature in °C., and P is the mean annual rainfall in mm. (E. de Martonne).

(ii) Rain (or moisture) factor $= P/T$ (R. Lang)

(iii) Rain mainly in winter: $R = t$
Rain evenly distributed through the year: $R = t + 7$
Rain mainly in summer: $R = t + 14$

where t is the mean annual temperature in °C.; if the annual rainfall (R) in cm. is less than $2t$ and greater than t, the climate is regarded as a. (W. Köppen and R. Geiger). (iv) The PRECIPITATION-EFFICIENCY INDEX of C. W. Thornthwaite (1931), on which an a. climate has an index of less than 16.

arkose A coarse-grained sandstone or grit containing a high percentage of fragments of FELSPAR, usually with a siliceous cement. The felspar fragments are usually little altered by weathering, which suggests a rapid disintegration of the parent granite or gneiss.

Armorican orogeny Applied by E. Suess (1888) to the Late Palaeozoic (Carbo-Permian) mountain-building period of W. Europe, corresponding to the VARISCAN of central Europe. It was named after Armorica, that part of N.W. France more usually known as Brittany. The term is also applied to features of similar nature; e.g. A. times, A. trend-lines. The resultant mountains are sometimes regarded as part of the ALTAIDES. The gen. usage is to regard the A. mountains as the W. representatives of the whole HERCYNIAN system, the Variscan as the E. representatives.

array (of goods) A gen. term used to distinguish any grouping of particular types of CENTRAL G.'s within an establishment. It introduces another stage in the generalization of the data between individual g.'s and individual shop types; e.g. within a newsagent's establishment there are many types of individual g.'s, but stationery, newspapers and confectionery form distinct a.'s of g.'s.

arroyo (Sp.) A stream-bed, usually dry, in an arid area, but occasionally carrying a short-lived torrent after intensive rain; the term is used esp. in Latin America and the S.W. U.S.A. Cf. WADI, NULLAH.

artesian basin A b. in the earth's crust, sometimes of great extent, in which one or more AQUIFERS are enclosed above and below by impermeable strata; e.g. the London B. Formerly water here saturated both the Chalk and the layer of Lower Eocene sandstones, lying in a syncline below the London Clay and above the Gault Clay; when wells were sunk the water rose to the surface. But the great withdrawals (at present about 1200 million litres, or 260 million gallons a day) have so reduced the orig. supply that consumption now exceeds intake on the N. Downs' and Chilterns' CATCHMENT areas. The WATER-TABLE and the hydrostatic pressure are falling, and the water has to be pumped up. The greatest a. b.'s in the world are in Australia, where Jurassic sandstone aquifers underlie 1·3 million sq. km. (0·5 million sq. mi.), deriving their intake of water from rain falling on the E. Highlands. Many of the 9000 wells are deep, descending to as much as 1·6 km. (1 mi.), but much of the water is slightly saline and is used for watering stock rather than for irrigation. Some experts consider that the amount withdrawn each year exceeds the annual intake from rainfall; i.e. part of the water is derived from accumulation during the past, which is not being replaced. [f

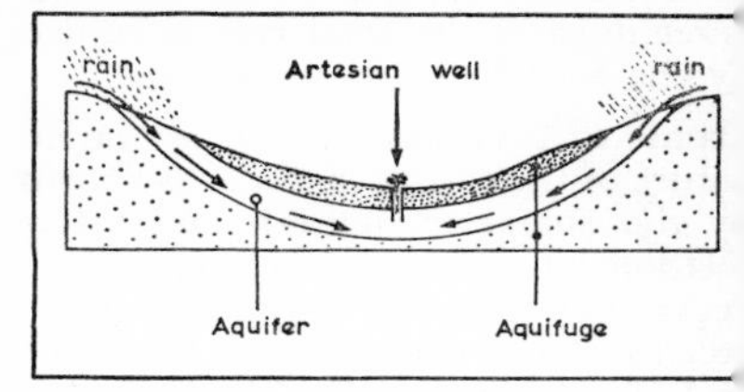

artesian well A boring put down into an AQUIFER in an A. BASIN; if the outlet of a w. in the centre of the basin is at a lower level than the WATER TABLE within the aquifer around the edges of the basin, water will rise in the well under hydrostatic pressure. It is named from w.'s of this type in Artois, N.E. France. The term is also loosely applied to any w. in which water rises some distance under pressure, though not to the surface; sometimes called '*sub-artesian*'; e.g. a deep w. was sunk near Wool in Dorset through the Tertiary beds into the Chalk to 221 m. (726 ft.) depth and the water rose to within 28 m. (9 ft.) of the surface. Pumping is required in a sub-a. w. [f ARTESIAN BASIN

artifact A man-made object, app. specif. to primitive or prehistoric weapons and tools, notably of FLINT.

asbestos A highly fibrous group of minerals, the most important being *chrysotile*, which occurs attached to the walls of veins traversing such metamorphic rocks as serpentine. The fibres can be spun and woven into a textile. A. has a high resistance to electricity and a very low conductivity; it is heat-resistant, will not melt, corrode or decay, and is used for fire-proof clothing. The fibres are also processed into sheets. C.p. = Canada (mostly near Thetford), Brazil, S. Africa, Rhodesia, U.S.A.

ash Fine material ejected from the crater of a volcano during an eruption; the term is really a misnomer, for the a. is not a product of combustion, but consists of finely comminuted particles of lava. Specif., coarse a. has grains 0·25–4·0 mm. in diameter, fine a. has grains less than 0·25 mm. The a. is so fine that it may be carried very great distances. After the eruption of Krakatoa in the Soenda Straits in 1883, a. was carried twice round the world. When Aniakchak in Alaska blew up in 1912, 0·3 m. thickness of a. fell on the town of Kodiak, 80 km. away. In March 1963, the dormant volcano Irazu in Costa Rica began to pour out a. across 650 sq. km. (250 sq. mi.) of country; this lay 'like a deep fall of black snow'.

ash cone A small volcanic c. of a.; its shape depends on the nature of the material, but is usually concave due to the spreading outwards of material near the base, and is less steep than a CINDER C. There are many a. c.'s in Iceland; e.g. a group of ninety, each 35–45 m. (120–150 ft.) high, at Rauholar, near Reykjavik. Monte Nuovo, W. of Naples in Italy, grew in three days to a height of over 140 m. (460 ft.) as the result of a single eruption. In 1937 an a. c., Vulcan, accumulated rapidly at Blanche Bay near Rabaul in the Bismarck Archipelago, growing to 180 m. (600 ft.) during the first day and 226 m. (740 ft.) in three days. A few major c.'s are made entirely of a.; e.g. Volcano de Fuego in Guatemala, 3350 m. (11,000 ft.) in height.

ash flow See NUÉE ARDENTE.

aspect The direction in which a slope faces, partic. with reference to possible amounts of sunshine and shadow; see ADRET, UBAC. A. has marked effects on the siting of settlements, vegetation and cultivation. Ct. the S.-facing sides (with villages, farms, orchards) and the N.-facing sides (with coniferous forests) of Alpine valleys which trend W. to E.

association A common word given a specialized ecological meaning, as adopted by botanists and geographers. It is one of the hierarchy of plant-groupings, an assemblage of plants living in close inter-dependence, with similar growth and habitat requirements, and with one or more DOMINANT species which may be used to denote it; e.g. oak forest. Rather more complex definitions of an a. are to be found in botanical literature, including its relationship to a CLIMAX COMMUNITY.

Asteroids A belt of about 1500 small heavenly bodies (*planetoids*), each less than *c.* 800 km. (500 mi.) in diameter (the largest is called *Ceres*), revolving in the Solar System between Mars and Jupiter. Possibly they are the result of the break-up of a single former planet.

astrolabe An instrument formerly used: (i) in the fixing of latitude by observing the apparent transit of the sun across the meridian at midday; (ii) for measuring the altitude of any heavenly body.

asymmetrical fold An ANTICLINE [*f*B] or SYNCLINE wherein one limb dips more steeply than the other; e.g. the Hampshire Basin syncline. In its extreme form it becomes an OVERFOLD.

Atlantic Polar Front The POLAR FRONT between POLAR MARITIME and TROPICAL MARITIME AIR-MASSES in the N. Atlantic Ocean.

Atlantic stage, of climate A former climatic phase, *c.* 5500 to 3000 B.C., when the climate of W. Europe (and probably elsewhere) was milder, cloudier and damper, with temperatures about 2° to 3°C. above those of the present; hence the phrase 'climatic optimum'. In Gt. Britain there was extensive growth of mixed oak forest, widespread peat formation, and a gen. rise of sea-level of about 3 m. (10 ft.) as a result of the return of water from the melting ice-sheets. The formation of the Straits of Dover occurred *c.* 5000 B.C. (the accepted start of the Upper Holocene). It is sometimes called the *Megathermal Period*.

Atlantic type, of coastline A type of coastline which develops where the trend of mountain ridges and the 'grain' of the relief gen. are at right-angles or oblique to the coastline; e.g. the coast of S.W. Ireland, N.W.

France, N.W. Spain, Morocco. Ct. Pacific (CONCORDANT) type of coastline.

atlas A uniform collection of maps in a bound volume. Many early collections (e.g. Ptolemy, Ortelius) did not use the actual name. The *Mercator A.* (published after G. Mercator's death) used the name *A. sive Cosmographicae Meditationes de Fabrica Mundi et Fabricati Figura* (1595), consisting of 107 maps, with a title-page on which was depicted the mythical character Atlas supporting the earth. This was followed by the larger *Mercator-Hondius A.* (1606). An a. may include other than terrestrial maps; e.g. the Dutch sea-a.'s of the 17th century, the *Palomar Observatory Sky A.* (1956), the *Hubble A. of Galaxies* (1961). Some countries have produced National A.'s; e.g. France, Belgium, Canada.

atmosphere A thin layer of odourless, colourless, tasteless gases surrounding the earth. It consists of nitrogen (78·08%), oxygen (20·95%), argon (0·93%), carbon dioxide (0·03%), and very small proportions of neon, krypton, helium, methane, xenon, hydrogen, etc., with amounts of water-vapour varying from 0 to 4·0%. The a. is held to the earth by gravitational attraction. A very small amount of ozone (O_3), the allotropic form of oxygen, is present, with its max. density at about 30 to 50 km. (20 to 30 mi.). Half the a. lies within 5·6 km. (3·5 mi.) of the earth's surface, 75% within 11 km. (7 mi.), 90% within 16 km. (10 mi.), and 97% within 27 km. (17 mi.). The 'weather-making' layers are limited to a height of a few km., esp. as more than half the water-vapour is below 2300 m. (7500 ft.). The a. is divided into a number of 'layers' or 'shells', commonly by temperature characteristics, hence the TROPOSPHERE and STRATOSPHERE. Other layers are defined on their physico-chemical properties (ozonosphere, IONOSPHERE). The total weight of the a. is $5{\cdot}9 \times 10^{15}$ tons. In the atmosphere, the properties typical of a gas really cease to exist at 600 km. (370 mi.).

atmospheric circulation The ger c. and movement of air, in the form pressure-cells and wind-systems, bot near sea-level and also in the upper See PLANETARY WINDS. [f

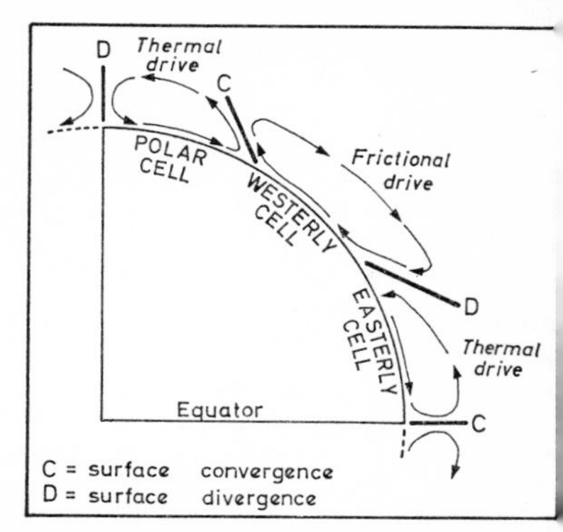

atmospheric pressure The p. exerted by the atmosphere as a result of it weight on the surface of the earth expressed in millibars (1000 mb. = bar = 1 million dynes per sq. cm.) The average p. over the earth' surface at sea-level is 1013·25 mb. equivalent to the weight of a colum of 760 mm. (29·92 ins.) of mercur at 0°C., or to a weight of air of 1033· gm. per sq.cm. (14·66 lb. per sq. in.) See also BAROMETER, BAROGRAPH.

atoll A CORAL REEF of circular elliptical or horse-shoe shape, enclosing a lagoon. These are found mos commonly in the W. and centra Pacific; e.g. the Gilbert and Ellic I.'s, Cook I.'s, Marshall I.'s. These reefs pose problems both of explaining their shape and also the fact that the coral extends downward to depths at which, under present conditions it is unable to grow. Charles Darwin's subsidence theory postulated tha rings of coral grew around islands subjected to slow gradual submergence These have sunk gradually beneath the sea (or the sea has risen), while the reef has maintained itself through upward growth. However, some reefs occur where insufficient submergence is evident. R. A. Daly suggested tha coral formation could be related to a rise in sea-level associated with the return of melt-water to the oceans after the Quaternary glaciation. Sir J. Murray put forward the idea that coral reefs could grow up on banks of their own debris, where growth would be most vigorous on the outside, thus causing a coral ring to grow outwards. [*f*]

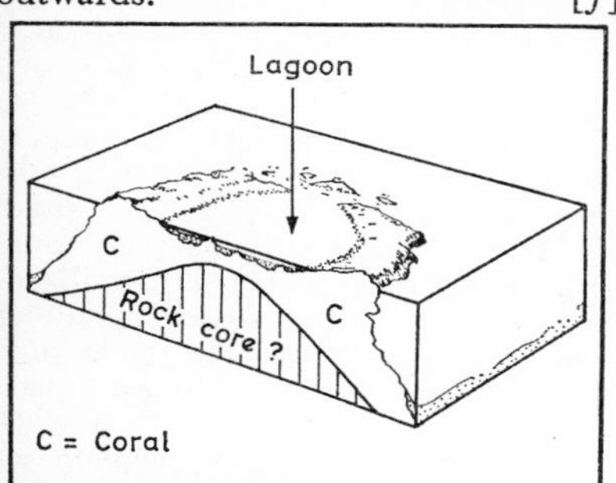

attached-dune A sand-DUNE, either a TAIL-DUNE or a HEAD-DUNE, accumulating as a sand-drift around a rock or other obstacle in the path of wind-blown sand in desert lands. [*f* DUNE]

attitude App. to the disposition of a rock-stratum, which may be horizontal or tilted. The term includes the relationship of a BEDDING-PLANE to the horizontal plane, given in terms of its DIP and STRIKE.

attrition The constant wearing-down into progressively more rounded, finer particles of a load of rock-material through frictional rubbing and grinding during transport by running water, the wind and waves. A. does not include the wearing away of river-beds and banks and of sea-cliffs; this is ABRASION or CORRASION.

aureole A zone of alteration (METAMORPHISM) of rocks in contact with an igneous intrusion, so that complex mineralogical changes were caused by the high temperatures. It can be traced around the margins of a BATHOLITH, e.g. around Dartmoor, and is usually associated with the presence of metal ores. [*f* BATHOLITH]

aurora The luminous effect of electromagnetic phenomena in the IONOSPHERE, visible in high latitudes as red, green and white arcs, 'draperies', streamers, rays and sheets in the night sky, best developed at a height of about 100 km. (60 mi.). Probably it is the result of magnetic storms and of electrical discharges from the sun during periods of sun-spot activity, mainly of electron particles, funnelled into the earth's magnetic field and accelerated to the high energies necessary, causing ionization of gases, though this is still a matter of research. It is called the *A. Borealis* (or 'N. Lights') in the N. hemisphere, the *A. Australis* in the S. Occasionally the A. Borealis is seen in England, but it is more common in N. Scotland, the Orkneys and Shetlands, and presents a magnificent spectacle in N. Scandinavia and N. Canada.

australite A small 'button' of silica found in the interior of Australia, believed to be debris formed *c.* 5000 years ago when some large meteorites came into collision with the moon.

autarchy Absolute sovereignty, autocratic rule, despotism.

autarky Though frequently used as an alt. spelling for AUTARCHY, this is incorrect, and the term should be used only to denote economic self-sufficiency within a national state, striving to produce its requirements internally and so diminish its dependence on imports, for national prestige, as a preparation for time of war and possible blockade, and because of inadequate foreign credits to buy goods abroad.

autobahn (Germ.) A road in Germany with separate carriageways in each direction, with limited access, used by high speed motor-traffic. The term was employed briefly in Gt. Britain, but is now replaced by *motorway*.

autochthon, -ous (i) Strata that have been shifted little by earth-movements from their orig. sites, though often strongly faulted or folded; e.g. the Helvetides on the N. side of the Alpine ranges in Switzerland and

Austria. Crystalline massifs, fragments of the old Hercynian continent which have not moved horizontally, though involved within the folding (e.g. Pelvoux, Mont Blanc, Aar-Gotthard), are also a. (ii) Descriptive of the formation of coal from *in situ* vegetation. Ct. ALLOCTHON, -OUS.

automated cartography Techniques utilizing lines and points to enable mappable features to be automatically plotted. This information can be stored in a computer, the lines as strings of co-ordinates on magnetic tape, the points as pairs of co-ordinates either on tape or punched cards. See also COMPUTERGRAPHICS, SYMAP, DATA BANK.

autonomy Derived from the Gk. *autonomia*, independence, and indicating the right of self-government, the attribute of an independent state. Hence *autonomous*.

autumn Astronomically, the transitional period between the autumnal EQUINOX and the winter SOLSTICE; in the N. hemisphere this extends from *c.* 21 Sept. to 22 Dec., in the S. hemisphere from *c.* 21 March to 21 June. Popularly, a. in Gt. Britain comprises the months of Sept. and Oct.; in America ('the Fall') it comprises Sept., Oct. and Nov.

available relief The vertical distance between the height of a land-surface undergoing dissection and the height of the valley-floors of adjacent dissecting streams. This figure may be used in quantitative statements about landscape geometry and in morphometric calculations.

avalanche The fall by gravity of a mass of material down a mountain-side; it can be used for rock (better LANDSLIDE or ROCK-FALL), but it is usually restricted to masses of snow and ice. On steep slopes a.'s may occur both in winter, when newly fallen non-coherent snow slides off the older snow surface, and in spring when wet, partially thawed, masses of enormous size fall down the valley slopes, and where a crust compacted by the wind (WIND-SLAB) breaks aw suddenly, often through the passa of a skier. Other falls may come fro the margins of a HANGING GLACIE An a. can be very destructive; duri the spring of 1951 a sudden tha after heavy winter snowfall cause widespread a.'s in Switzerland an Austria, with much loss of life an destruction of property. In suc ranges as the Alps, the probable tracks are known, and partic. dange ous areas are avoided. Villages, road and railways are carefully site natural a. breaks such as rock spu and thick pine-woods are utilize and steel a.-sheds are erected critical points to protect roads (e. the new Great St. Bernard road tunnel) and approaches to railway The recently completed section the Trans-Canada Highway throug the Rogers Pass in British Columbi is protected by many miles of a.-shed

avalanche wind The rush of win in front of an AVALANCHE produce by the large falling mass; this may b very destructive, and can cut swathe through a forest or a village.

aven (Fr.) (i) A deep shaft-like hol in limestone country (syn. wit PONOR), orig. derived from a *patoi* word in the Causses region of th Central Massif of France; e.g. the A Armand in the Causse Méjan, 198 m (650 ft.) deep. leading to a vast series o caves. (ii) The term is used mor specif. in England to describe th enlarged vertical joints in the roof of cave which narrow upward (the resul of carbonation-solution), and some times, though not always, open ou into depressions on the surface.

axial plane A p. bisecting th upper (or lower) angle between th limbs of a fold on each side o the crest-line of an anticline, or th trough-line of a syncline. Inclination of this plane from the vertical characterize different types of fold.

[*f* AXIS, OF FOLD

axis, of earth The diameter betwee the N. and S. Poles, about which th earth rotates in 24 hours, tilted at a angle of about 66½° to the plane of the ECLIPTIC, i.e. at an angle of about 23½° from a line perpendicular to that plane. The length of the axis is 12,714 km. (7900 mi.). [*f*]

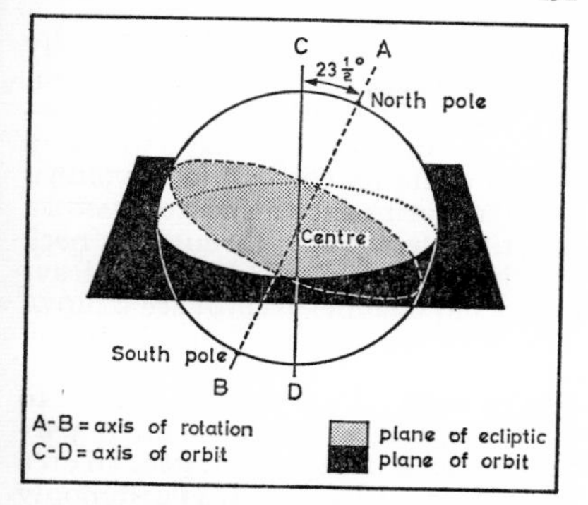

axis, of fold The central line of a FOLD (crest or trough) from which the strata dip away (as in an ANTICLINE: the crest) or rise (as in a SYNCLINE: the trough) in opposing directions. See AXIAL PLANE. [*f*]

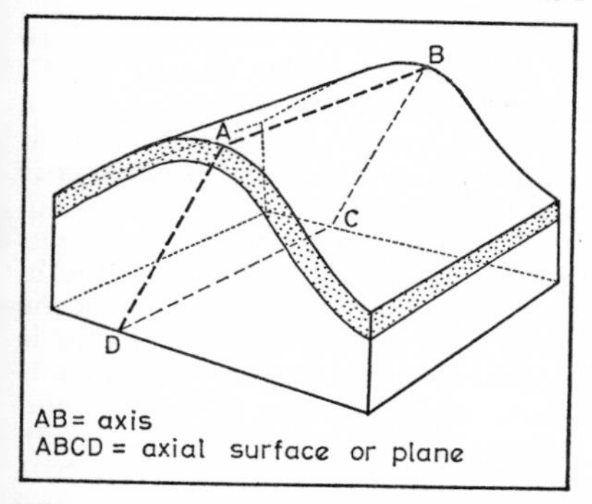

ayala A local wind in the Central Massif of France, strong, sometimes violent, and very warm; similar to the MARIN.

azimuth (i) In astronomy, the angle intercepted between the meridian plane of an observer and the vertical plane passing through a heavenly body, measured the nearest way from the meridian plane. Ct. ALTITUDE. (ii) In surveying, a bearing read clockwise from true N.; i.e. the a. of a point due E. is 90°, of W. is 270°. By the U.S. Coast and Geodetic Survey, a. is measured clockwise from zero = S.; i.e. an a. of 90° = W., 270° = E. It may be either the *magnetic a.* (measured from the MAGNETIC N.), or the *true a.* (measured from the TRUE N.). In a geodetic survey, the a. is used to fix the orientation of the system of TRIANGULATION. [*f*]

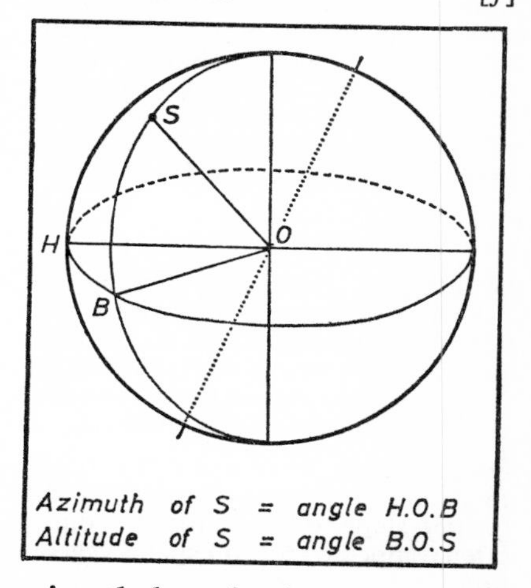

azimuthal projection Syn. with ZENITHAL P.

Azoic An obsolescent term derived from the Gk., 'without life', the meaning of which has undergone change. It has been used: (i) for all PRE-CAMBRIAN stratified rocks; (ii) for the earliest part of Pre-Cambrian times, irrespective of the nature of the rocks.

azonal soil A 'young' s. which has not been sufficiently long under the effects of the soil-forming processes, agencies and influences to develop mature characteristics. A. s.'s include: (i) *mountain* or SCREE s.'s, on unstable slopes of rock fragments; (ii) *alluvial* s.'s, derived from ALLUVIUM deposited by running water, which provides the material for a good agricultural s. (with its well-mixed, fine texture, good mineral content, and its replenishment in times of flood); (iii) *marine* s.'s derived from mud-banks, sand-banks and dunes, often saline;

(iv) *glacial* s.'s, developed on DRIFT; (v) *wind-blown* s.'s, including sand and esp. LOESS (LIMON) as parent-materials; (vi) *volcanic* s.'s, derived from recent volcanic parent-materials (LAVA, ASH, CINDER, PUMICE), which because of their high mineral content form good soils when weathered.

Azores 'high' The subtropical anticyclone in the atmosphere, situated gen. over the E. side of the N. Atlantic Ocean. It is more extensive and continuous during the N. hemisphere summer, and may extend N.E. to affect W. Europe and Gt. Britain.

backing A change of direction of the wind in an anti-clockwise direction; e.g. from S.W. to S., to S.E. In U.S.A. the term is thus used for the N. Hemisphere, but in the reverse (i.e. clockwise) for the S. Ct. VEERING.

backset bed Used mainly in U.S.A. to denote a deposit of sand on the gentler slope to windward of a dune, often trapped by tufts of sparse vegetation.

backshore The area extending from the average high-water line to the coastline. [*f* COASTLINE]

back-slope The gentler slope of a CUESTA, which may be compared with the steeper ESCARPMENT. Most authorities use DIP-SLOPE in connection with a cuesta, but as the slope is seldom the same as the dip of the rock, this can be misleading, and b.-s. is strictly more correct. [*f* CUESTA]

back-swamp Marshy areas in a low-lying FLOOD-PLAIN behind the LEVEE bordering a river; used specif. in U.S.A. in respect of the Mississippi flood-plain.

back-wall The steep rock wall at the back of a CIRQUE, rising to an ARÊTE enclosing it. In cirques in the British mountains, the b.-w. may be from 60 to 180 m. (200 to 600 ft.) high. The huge Walcott Cirque in Antarctica has a b.-w. 3000 m. (10,000 ft.) high.

backwash A mass of water running back down the slope under the influence of gravity, after a wave breaks on a beach; the receding movement of a wave. See also LONGSHORE DRIFT [*f*].

backwater An area of virtually stagnant water, still joined to a stream, but relatively unaffected by its current. It develops readily where a stream splits its channel (see BRAIDED RIVER) or forms such an acute MEANDER that the current cuts through its neck and flows on a more direct course, leaving its old channel as a b.; see OXBOW. The term is also used figuratively.

backwoods First used in U.S.A. to denote sparsely settled, though partially cleared, forest-land; gen. an area of pioneer settlement. It is commonly applied to any sparsely settled area away from an urban centre.

Bad (Germ. = *bath*) A common prefix attached to names of German towns which are health-resorts.

badlands A name orig. applied to an area of semi-arid climate in S. Dakota (called '*les terres mauvaises à traverser*' by the French), because it was so difficult to cross. This area is now a National Monument 5200 sq. km. (2000 sq. mi.), with a spectacular landscape of gullies and 'saw-tooth' ridges, cut in vari-coloured Oligocene shales and limestones. The name is applicable to any such erosional landscape, where a maze of ravines and valleys dissects plateau-surfaces. B. formation may result from severe SOIL EROSION, esp. in an area of sparse and intermittent rainfall; the first stage is commonly caused by the destruction of a protective turf mat, sometimes by over-grazing.

baguio A tropical storm in the Philippine I.'s, experienced esp. from July to November, derived from the name of a city in the islands. During July 1911, one of these storms resulted in a rainfall of 1170 mm. (46 ins.), one of the wettest days (24 hours) ever recorded on the earth's surface.

Bai-u The season of heaviest rain in parts of China and Japan, occurring in late spring and summer. The B. rains are sometimes called the '*plum rains*', since they occur during the plum-ripening season.

bajada (Sp.), anglicized to *bahada* A continuous, gently sloping fringe of angular scree, gravel and coarse sand around the margins of an inland basin, or along the base of a mountain range, in a semi-arid region. It has been formed by the coalescence of a series of adjacent ALLUVIAL CONES, each deposited by a torrential, usually intermittent, stream where it leaves a constricted valley; e.g. in Arizona, Nevada, Mexico and the Atacama Desert of central Chile. [*f* PLAYA]

balance of trade The relation between the value of the imports and exports (incl. INVISIBLE EXPORTS) of any country; this may be *favourable* (exports higher) or *adverse* (imports higher). In a strict sense, the term should be limited to visible t.; if invisible imports are included, a more suitable term is *balance of payments*, which includes also the bullion account, in fact, all economic transactions between 2 countries.

balk A section of unploughed land between ploughed areas. In open fields with a strip system, holdings were divided by such areas, often under grass, while larger b.'s separated groups of strips.

ball lightning A rare form of L., in which a luminous ball, sometimes moving, sometimes apparently stationary, is seen. The cause or substance of this effect is not known.

balloon-sonde, ballon-sonde (Fr.) A balloon, hydrogen-filled, which is liberated into the atmosphere, carrying self-recording meteorological instruments. It is thus possible to obtain data at great altitudes. The b. rises, expanding as it goes, and ultimately bursts. The remains of the b. and the carefully protected instruments return to earth by parachute. From the recorded measurements, a profile of changes of temperature and pressure with altitude is obtained. Ct. RADIOSONDE.

bamboo A genus of giant grasses (*Bambusoidea*), growing widely within tropical latitudes (esp. S.E. Asia), and in favoured areas even well outside the tropics (as in S.W. Ireland and Cornwall). It grows profusely and extremely rapidly, ranging from a few cm. to 36 m. (120 ft.) high, and up to 20 cm. (8 ins.) in diameter. Some b.'s are climbers, over 60 m. (200 ft.) in length. It is so plentiful that it is one of the commonest and cheapest media for housing, scaffolding and construction work in S.E. Asia.

band A long ridge-like hill or SPUR, esp. in the English Lake District; e.g. the Band, Bowfell.

band-graph A graph with years plotted on the horizontal axis and quantities on the vertical axis, showing trends over a period of time in both the total commodity involved and its constituents; the 'bands' between successive graphed lines may be shaded distinctively. This is sometimes known as a *compound line-graph* or an *aggregate line-graph*.

bank (i) A vague term for the margin of a river. In the cases of a mountain torrent or of a river in a CANYON, which usually occupies the entire valley-floor, or of some wide FLOOD-PLAINS, b.'s are hardly distinguishable. It is best marked in cases where a channel with distinct sides has been cut in the floor of a valley by a stream with powers of vertical erosion. At high-water the space between the b.'s is full (BANK-FULL stage) and water then overflows on to the FLOOD-PLAIN; at low-water beds of gravel and alluvium appear. (ii) An area of mud, sand or shells (not rock) in the sea, usually covered with fairly shallow water, though deep enough not to be a hazard to navigation. The waters above a b. are usually good fishing-grounds; e.g. the Grand B.'s of Newfoundland, the Dogger B. (iii) In the N. of England and Scotland a hill-side; e.g. 'Ye b.'s and braes o' bonnie Doon'.

bank-caving On the outside of a curve in a river's course, the force of the current impinges against the bank. Undercutting and undermining may result, causing the slumping of masses of clay, sand and gravel into the river, where it is swept away as part of its load.

banket (Afrikaans) A pebble-conglomerate containing gold, found in the Witwatersrand of the Republic of S. Africa.

bankfull The state of river flow when the space from bank to bank is completely filled with water; beyond this, FLOOD-STAGE is reached. At b. stage the velocity of the river is thought to be constant along its whole length.

banner-cloud A cloud which streams out on the lee-side of a peak in a clear sky, the result of condensation within a rising air-current produced as a forced up-draught by the mountain. Descent and re-warming of the air will occur down-wind, and here the 'banner' ends as the water droplets evaporate. While this flow of air continues, the position of the cloud itself remains fixed. E.g. the 'Table Cloth' over Table Mountain (Republic of S. Africa). A b.-c. may be partic. pronounced in the case of a prominent isolated peak; e.g. the Matterhorn. See LEE-WAVE.

bar (i) A unit of atmospheric pressure, equivalent to 1 million dynes per sq. cm. At 45°N. latitude, at sea-level, and at a temperature of 0°C., 1 b. = 29·5306 ins. (750·076 mm.) of mercury. It is divided into 1000 millibars (mb.). On the S.I., 1 b. = 10 NEWTONS. (ii) A bank of mud, sand and shingle, deposited in the water offshore parallel to the coast (*offshore b.*), across the mouth of a river, across the exit to a harbour (*harbour b.*), across a bay between 2 headlands (BAY-B.), and between an island and the mainland (TOMBOLO). A b. may be either an upstanding and exposed feature of sand, or a submerged ridge covered at least at high tide. More specif. the terms BARRIER BEACH or BARRIER ISLAND should be used for the former to avoid confusion. (iii) In American literature, a b. denotes deposits of alluvium, sand and gravel found in stream-channels, at the mouth of a river, and even in a lake, causing navigational obstruction (e.g. to Mississippi steam-boats). [*f*]

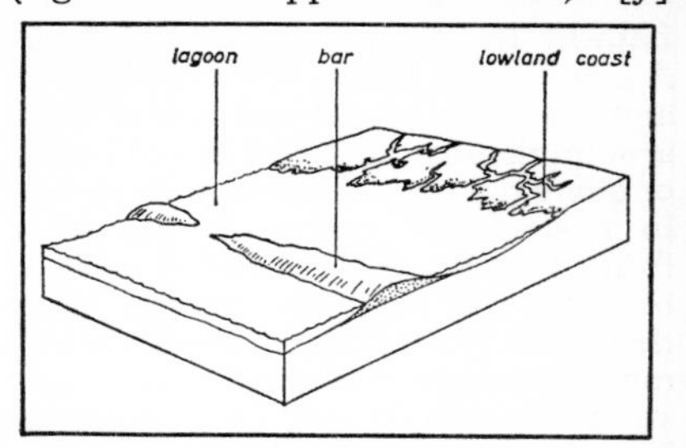

barbed drainage pattern A p. of d. in which tributaries form obtuse angles with the main stream; they appear to point up-river. This results from CAPTURE, which has caused the main stream to flow in a direction opposite to its original one. [*f*]

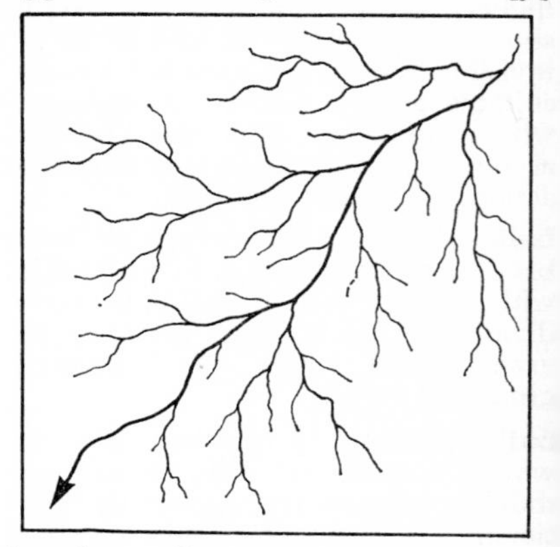

barchan See BARKHAN.

bar-graph A diagram consisting of a series of bars or columns proportional in length to the quantities they represent. They may be: (i) *simple*, where each bar shows a total value; (ii) *compound*, where each bar is divided to show constituents as well as the total value. The bars may be placed either vertically, horizontally, or in pyramidal form; the first is usually most satisfactory when a time-scale is involved. After the scale of values has been determined, the exact length of each bar is computed and parallel lines are drawn. It is preferable to leave a space slightly less wide than the bar itself between each, although in some cases, particularly for rainfall diagrams, there need not be any intervening space; these are often descriptively known as 'battleship-diagrams' from their profiles. When the outline of each bar has been drawn, it can be filled either in solid black or with diagonal shading. In a compound b.-g., several shadings and stipples may be used to distinguish each constituent.

[*f* ALTIMETRIC FREQUENCY GRAPH]

barkhan (barchane, barchan) (Turki) A crescentic sand-dune, lying transversely to the wind direction, with 'horns' trailed out downwind. It forms when the wind direction remains constant. It varies in height to over 30 m., with a gentle slope on the windward side, a steeper one on the 'slip-face' or lee side where eddy motion seems to assist in maintaining a concave profile. If the supply of sand is continuous, the dune may advance as a result of the endless movement of sand up the windward slope and over the crest, sometimes constituting a threat to oases. A dune is sometimes found as an individual hill, but usually they occur in groups, a chaotic ever-changing 'sand sea'. The main problem is how it originates, since it appears to occur readily on fairly level open surfaces; possibly patches of pebbles, a shrub, even a dead animal, or a low bump in the ground may cause the accumulation of a heap of sand, from which the features of a characteristic b. can develop. [*f*]

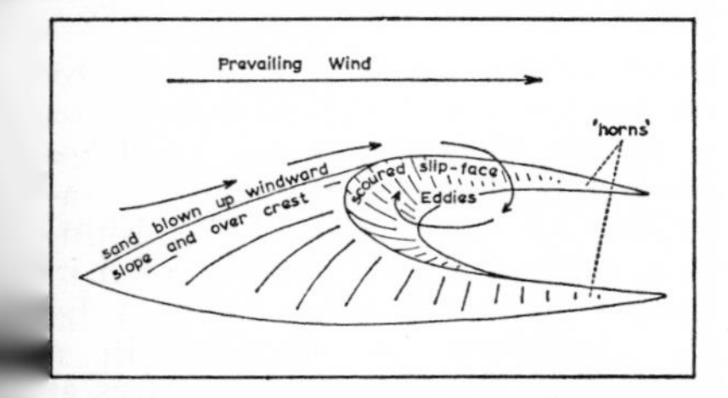

barley A cereal grown in middle and high latitudes, but which is more tolerant than wheat, since it requires a shorter growing season (e.g. it can be grown beyond the Arctic Circle), and can stand greater summer drought and lighter sandier soils. Much is fed to stock, and certain high quality varieties are malted for brewing and distilling. C.p. = U.S.S.R., Gt. Britain, U.S.A., France, W. Germany, Turkey.

baroclinic The state of a fluid or gas in which surfaces of constant pressure intersect surfaces of constant temperature; a measure of *baroclinicity* (sometimes called *baroclinity*) may be derived for the relationship between these. This implies in meteorology a condition of large-scale atmospheric INSTABILITY, and the presence of a frontal zone. [*f*]

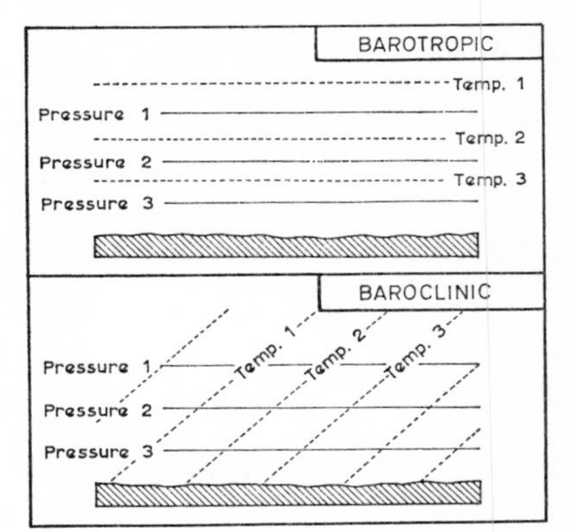

barograph A self-recording ANEROID BAROMETER; differences of atmospheric pressure are transmitted to a needle, which traces an inked line on a moving drum, thus giving a continuous record (*barogram*). In another form, the position of a mercury meniscus is recorded photographically.

barometer An instrument invented by E. Torricelli (1643) for measuring ATMOSPHERIC PRESSURE, by balancing the weight of a column of mercury against the weight of a column of the atmosphere. Modern b.'s (e.g. the KEW or FORTIN patterns) have various adjustments for applying corrections, and vernier-scales for exact readings. Corrections are applied for: (i) *latitude*, standardized to 45°N.; (ii) *temperature*, standardized to 12°C., 53·6°F.; (iii) *altitude*, an average decrease of 33·9 mb. (1 in. of mercury) for each 274 m. (900 ft.) for the lowest 1000 m., above which the rate of decrease becomes progressively less; see ALTIMETER; (iv) *instrumental individuality*, as compared with a standard instrument. Another type is the ANEROID B. See also BAROGRAPH and BAR.

barometric gradient The amount of change in atmospheric pressure between 2 points, as indicated by the distance apart on a level surface of the ISOBARS on a synoptic chart, i.e. a 'steep' g. involves considerable difference in pressure, with closely spaced isobars, while a 'gentle' g. has only slight difference, with widely spaced isobars. The direction of max. g. is at right-angles to the isobars. A steep g. is associated with strong winds. In a tropical storm (e.g. a HURRICANE or TYPHOON), there may be a difference of 40 mb. over only 30 km. (20 mi.). It is also known as *pressure g*.
[*f* GEOSTROPHIC FLOW]

barometric tendency The *nature* (increasing or decreasing) and the *amount* of change of atmospheric pressure during a specified period, gen. 3 hours.

barotropic In a fluid or gas, where surfaces of constant pressure are parallel to surfaces of constant temperature; this b. state is equivalent to zero BAROCLINICITY. The noun is *barotropy*. [*f* BAROCLINIC]

barrage A large structure, usually of concrete, sometimes of earth, across a river, impounding a body of water much deeper than the orig. river; e.g. the Lloyd B. in Pakistan, the Asyut B. in Egypt. By some users, b. is used when no power-station for electricity is included, DAM if there is, but this is not universal. Another distinction regards a b. as being concerned with annual water storage, a dam with perennial storage, the latte impounding more than 1 year's flow.

barrel A measure of petroleum 1 b. = 32 Imperial gallons, 42 American gallons, 159 litres. 1 ton o petroleum is equivalent to 7–8 barrels depending on its specific gravity.

barren lands, barrens A rathe gen. term applied to much of N Canada, characterized by much bar rock, swamp, TUNDRA vegetation permanently frozen subsoil (PERMAFROST) and a short intense summer The term has fallen into disuse, partl because the areas are not as barren a once believed (esp. in respect of minera wealth).

barrier-beach, -island A sand bar above high tide, parallel to th coastline, and separated from it by a lagoon; it may be sufficiently above high tide, with dunes lying on it, to be called a *b.-island*. From New Jerse along the E. coast of U.S.A. to Florida, and along the shores of the Gulf of Mexico as far as the Mexica border, extends a string of b.-b.'s where a flat coastal plain is bordered by shallow water. Some probabl started as offshore bars of sand far ou from the mainland, but have bee pushed gradually shoreward by wave action, forming lagoons known a SOUNDS; e.g. Albemarle and Pamlic Sounds behind Cape Hatteras (N Carolina). Some b.-islands represen portions of former SPITS. Others ar fragments of a breached dune-line; e.g the Frisian I.'s off the mainland coas of the Netherlands, Germany an Denmark. Many b.-islands are now seaside resorts; e.g. Miami Beach, Palm Beach (Florida); the Lido at Venice.

barrier-lake A lake produced by the formation of a natural dam across a valley. Types include: (i) lakes ponded behind a land-slide or avalanche; e.g. Earthquake Lake, Montana, formed after a vast rock-fall in August 1959; (ii) lakes enclosed by the deposition of alluvium in a DELTA; e.g. the Étang de Vaccarès a the mouth of the R. Rhône; (iii) lakes formed by the deposition of a TERMINAL MORAINE across a glaciated valley; e.g. the lakes of the English Lake District (in part), the Finger Lakes of New York State; (iv) lakes formed by ice-dams; e.g. Vatnsdalur in Iceland; (v) lakes ponded by vegetation dams; e.g. in the heathlands of W. Europe; (vi) lakes ponded behind a dam of calcium carbonate; e.g. L. Plitvicka, on the Crna R. in Yugoslavia; (vii) lakes ponded behind a lava dam; e.g. Lac d'Aydat in the Central Massif of France.

barrier reef A CORAL r. parallel to the coast, but separated from it by a lagoon of considerable depth and width. The Great Barrier R. of Australia extends from the Torres Strait at 9°S. to about 22°S., over 1900 km. (1200 mi.) long, and varying in width from 30 to 50 km. (20 to 30 mi.) off Cairns to 160 km. (100 mi.) in the S. Numerous individual r.'s, crowned with islands, are spread over an irregular platform covered with shallow water, possibly the result of the depression of a denuded surface below sea-level to a depth which allowed reef-building organisms to flourish. Many smaller examples of b. r.'s are found around Pacific islands (e.g. Aitutaki in the Cook group). For theories of the development of coral reefs, see ATOLL. [*f*]

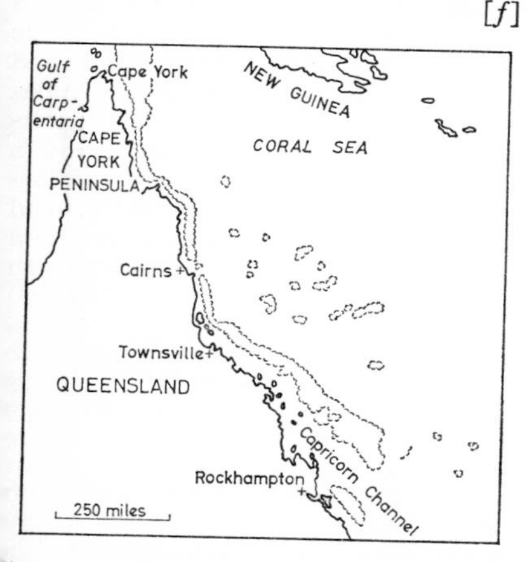

barrow (tumulus) A mound of earth and stones covering a prehistoric burial place, in plan either elongated ('long barrow') or circular ('round barrow'), partic. common on the chalklands of S. England, such as Salisbury Plain (there are estimated to be over 6000 b.'s in Wessex). The earliest b.'s in Gt. Britain were constructed *c.* 2500–1800 B.C. The last b. burials were as late as Saxon times.

barysphere Used rather loosely with several meanings. (i) The whole of the earth's interior beneath the CRUST (or *lithosphere*), i.e. including both the MANTLE and the CORE; it is syn. with *centrosphere*, and sometimes (incorrectly) is called the *bathysphere*. (ii) In a more limited sense, the core only, of 3476 km. (2160 mi.) radius, consisting of nickel-iron under great pressure. (iii) By some writers, b. is used as syn. with mantle and in contrast to both the surface crust and the inner core. The consensus of opinion favours (i).

basal complex The ancient rocks in the great 'shield' areas of the earth's crust.

basal conglomerate A CONGLOMERATE found in the lowest part of a series of strata resting unconformably on older rocks. It may appear above any erosional break, typically as a beach deposit spread over a former land-surface.

basal sapping The disintegration of rocks along the back-wall of a CIRQUE, thought to be the result of melt-water making its way down the BERGSCHRUND, and thus enabling alternate freezing (by night) and thawing (by day) to shatter the rocks.

basal-slip A sliding movement of the basal ice in a glacier over the underlying rock-floor, caused by the solid character of the ice, its weight higher up, and the gradient of its floor.

basalt A fine-grained, usually dark-coloured, igneous rock, belonging to the basic group, with 45–52% silica (over half plagioclase felspars and the rest ferro-magnesian silicates), extruded from a fissure in the earth's crust, hence a VOLCANIC ROCK. It flows readily, forming extensive sheets, and may solidify in hexagonal columns; e.g. the Giant's Causeway, N. Ireland, and Fingal's Cave, Staffa. With andesite, it makes up 98% of the extrusive varieties. Most of the large SHIELD VOLCANOES (Mauna Loa, Hawaii) and extrusive plateaus (Columbia-Snake in U.S.A., N.W. Deccan in India, Antrim in N. Ireland) consist of b.

basal wreck An alt. name for a CALDERA.

base-flow, of a river Used, esp. in U.S.A., to indicate that portion of a river's volume contributed by groundwater inflow, as distinct from direct surface- or overland-flow following periods of rain. The b.-f. fluctuates during the year, corresponding to the height of the WATER-TABLE; in mid-latitudes it usually is at its lowest in late autumn following high EVAPOTRANSPIRATION during summer.

base-level The lowest level to which a river can erode its bed (or to which a land-surface can be reduced by running water), normally assumed to be sea-level (*sensu lato*), though there may be such local temporary b.-l.'s as a lake or a resistant stratum of rock. While b.-l. is usually thought of as a horizontal surface, some writers, recognizing that streams require a gradient in order to flow, regard b.-l. as an inclined or curved surface which is the theoretical limit to stream erosion. Such a surface is practically impossible to define.
[*f* REJUVENATION]

base-line (i) A carefully measured line, the initial stage in a triangulation survey. From its ends a system of triangles is projected by angular measurement as an accurate framework for detailed topographical surveying. The British Ordnance Survey has measured numerous bases in Britain (e.g. on Hounslow Heath and Romney Marsh); the current triangulation depends on: (*a*) the Ridgeway base, 11,260·1931 m. in length, from White Horse Hill in Berkshire t Liddington Castle in Wiltshire; an (*b*) the Lossiemouth base (7170·723 m.) near the shores of the Moray Firth A b.-l. is measured with every refinement, using an INVAR tape, an various corrections, as for temperature, are applied, so that an accurac of 1 in 300,000 is attained. (ii) Se LAND SURVEY SYSTEM (U.S.A.).

base map A m. on which furthe information may be plotted. A series o related distributions may be depicte on a common b. m.

basement complex A mass of ver ancient igneous and metamorphi rocks, usually (though not always Pre-Cambrian in age and very complex in structure, which underlies th stratified sedimentary rocks.

basic function Used in urban geography to distinguish those activitie that form the economic support of a town. In ct., those activities that serv only the local community are known a NON-B. or service f.'s.

basic grassland A type of g occurring mainly on chalk and limestone terrain, the dominant grasse being sheep's fescue (*Festuca ovina* and red fescue (*F. rubra*).

basic lava A mass of molte igneous material on the surface of th earth, rich in iron, magnesium an other metallic elements, and relatively poor in silica. It has a low melting-point and will flow readily for a considerable distance before solidifying. Usually its flow from the vent o a VOLCANO is unchecked and free fro much explosive activity. The b. l. ma form a large flattish cone (a SHIEL VOLCANO) or, if it flows from a number of fissures, an extensiv plateau. B. l. congeals usually a basalt, sometimes known as *floo basalt*.

basic–non-basic component (sometimes expressed as a ratio) In urba geography the b. c. represents th income earned by firms and individual from outside a defined region (e.g. a city), while the n.–b. c. is the income earned inside the region. The terms are also used for actual employment and activities outside and inside the region resp., and for goods and services sold outside (exports) and inside (local) resp.

basic rock A rather loosely used term for an igneous r. with a silica content of less than about 52%, and with more than 45% of basic oxides (of aluminium, iron, calcium, sodium, magnesium and potassium); e.g. basalt, dolerite, gabbro. Sometimes it is used still more loosely to indicate an igneous r. composed of dark-coloured minerals. The term is apt to be misleading, even inaccurate, and is obsolescent. In U.S.A. it has been suggested that the term should be replaced by either *subsilicic* (in respect of its silica content), *mafic* (in terms of its base content), or *melanocratic* (in respect of the dark-coloured minerals).

basic slag The waste from a blast-furnace in which iron ore with a high phosphorus content is smelted; b. s. contains tetracalcium phosphate, among other compounds. When crushed, it makes a valuable fertilizer.

basin (i) A large-scale depression on the earth's surface, occupied by an ocean, hence *ocean b.* (ii) The area drained by a single river-system, hence *river b.* (iii) A shallow structural downfold in the earth's crust; e.g. the Hampshire, London, Paris B.'s. (iv) A shallow downfold containing *Coal Measures*, hence a *coal b.*; e.g. Saar B. (v) An area simply enclosed by higher land, with or without an outlet to the sea; e.g. the Great B. of U.S.A., the Tarim B. (central Asia). (vi) Syn. with CIRQUE in parts of W. America. (vii) A depression resulting from the settlement of the surface through the removal in solution of underground deposits of salt or gypsum, naturally or by man's action. (viii) Part of a dock-system filled at high tide (*tidal b.*), or a part of a canal or navigable river with facilities for handling barges.

basin-and-range A series of tilted FAULT-BLOCKS, forming asymmetrical ridges alternating with basins. It is applied as a proper name ('Basin and Range country') to part of S.W. U.S.A. between the Sierra Nevada and the Wasatch Mountains, where the ranges have steep E. faces and more gentle W. ones.

basin cultivation A method of c. used to avoid soil erosion resulting from torrential rainfall. A field is divided into many small b.'s surrounded by low earth ridges, so as to hold the water and allow it to sink in, rather than run off rapidly. E.g. on the margins of the tropics, esp. in Nigeria, Sudan and Ghana.

basin irrigation A type of i. whereby flood-waters are directed into 'basins' surrounded by earth banks; e.g. along the Nile valley in Egypt.

'basket-of-eggs' relief See DRUMLIN.

batholith (bathylith) A large mass of igneous rock, usually granite, formed by the deep-seated intrusion of MAGMA on a large scale. The dome-like upper surface may be ultimately exposed by prolonged denudation to form uplands; e.g. Dartmoor and Bodmin Moor in S.W. England; the Wicklow and Mourne Mountains in Ireland; and the British Columbia Coast Range b., 2400 km. (1500 mi.) long by 160 km. (100 mi.) wide (the last is now known to be a series of separate b.'s, intruded at different times). Many b.'s are elongated along the axes of fold-ranges; e.g. the Aar-Gotthard, Mont Blanc and Pelvoux massifs along the axes of the Alps; the Montagnes Noires and the M. d'Arrée along the Armorican axes of Brittany. The mechanism by which b.'s are emplaced is not wholly understood. One theory is of EMPLACEMENT, by which a vast block of country-rock foundered and sank into the underlying magma; another is that the country-rock has been absorbed by, or transformed into, magma by granitization. A b. may be surrounded by a mineralized AUREOLE. [*f*]

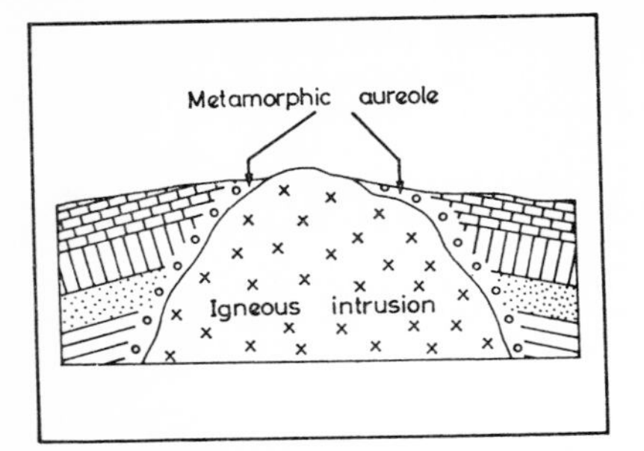

bathyal App. somewhat loosely to that part of the ocean between 180 m. and 1800 m. (100 and 1000 fthms.), broadly the CONTINENTAL SLOPE, partic. to the benthonic life (BENTHOS) and the types of deposit found there. Hence b. deposits of Blue, Red, Green and Coral muds. [*f* ABYSSAL ZONE]

bathymetric Relating to the measurement of the depth of a water-body; a b. map of an ocean, sea or lake, with contours (known as *isobaths*) related to the datum of the mean water-surface, depicting the relief of its floor; e.g. the *Carte générale bathymétrique des Océans*, published by the International Bureau of Hydrographics in Monaco. Hence *bathymetry*.

bathy-orographical An adj. applied to a map depicting both the altitude of the land and the depths of the sea, usually by layer-colouring, with water conventionally shown in shades of blue, land in green, yellow, brown.

bauxite An amorphous mass of clay with a high content of aluminium hydroxide ($Al_2O_3.2H_2O$), usually with ferruginous and other matter; the chief ore of ALUMINIUM, the third most important element in the earth's crust (after oxygen and silicon), comprising about 15% by volume or 8% by weight. It occurs widely in felspars and other silicates. Under tropical weathering, felspars break down into clay-minerals, leaving b., varying in colour from off-white to red and brown. C.p. of crude bauxite in terms of metal content = Jamaica, Surinam, U.S.S.R., Guyana, France.

bay (i) A wide, open, curving indentation of the sea or a lake into the land; e.g. Hudson B., B. of Fundy, B. of Biscay, Georgian B. (L. Huron). Some authorities use b. in a hierarchy of coastal openings, larger than a COVE, smaller than a GULF; cf. BIGHT, EMBAYMENT. But in gen. b. is used rather loosely. However, a b. has to be defined precisely in delimiting TERRITORIAL WATERS. A straight line is drawn between the natural promontories on either side of the indentation, and the water area so delimited is considered to be a b. if its area is as large as, or larger than, a semicircle whose diameter is equal to the delimiting line. (ii) (Special case.) A shallow elongated depression in the coastal plain of E. U.S.A. (the Carolina B.'s). (iii) Used as a translation of the Germ. *Bucht*, indicating an extension of lowland into an upland area along a river valley; e.g. *Kölnische Bucht* or Cologne 'Bay', the Leipzig 'Bay'. [*f*]

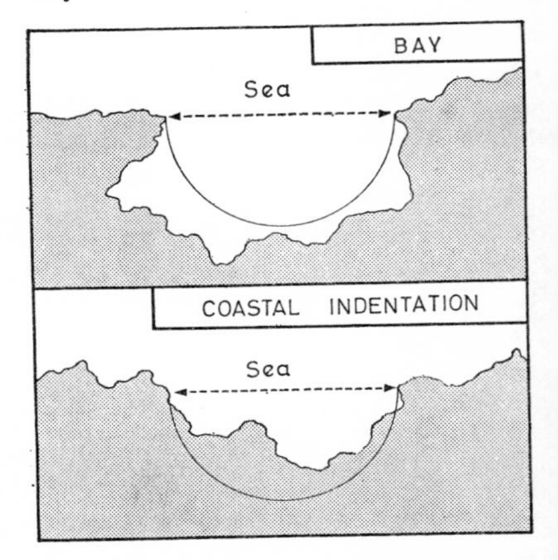

bay-bar (in U.S.A., a **baymouth-bar**) A bank of sand, mud or shingle, extending wholly across a bay, usually linking two headlands, and thus straightening off the coast. Where only one end of such a feature is attached to the coast, it is a SPIT. A bar linking the mainland with an island is a TOMBOLO. B.-b.'s are produced: (i) by the convergence of two spits from opposite sides of a bay; (ii) by a single spit extending in a constant direction across the bay; (iii) as a development from an offshore BAR driven onshore. E.g. Loe B. (Cornwall); the b.-b. across Corpus Christi Bay (Texas); the NEHRUNGEN on the E. Baltic coast.

bay-head beach, pocket beach A crescent of sand and shingle at the head of a small cove between 2 headlands; e.g. along the Pembroke and Gower coasts of S. Wales; Cornwall; along the Maine and Oregon coasts of U.S.A.

bay-head delta A d. formed by a heavily laden stream depositing much of its load at the head of a bay.

bay-ice Sea ice, formed in a bay during a series of winters, thick enough to impede navigation.

bayou A marshy creek or sluggish, swampy backwater, specif. along the lower Mississippi and its delta, and along the Gulf coast of U.S.A.

beach The gently sloping accumulation of material (shingle, sand) along the coast between the low-water spring-tide line and the highest point reached by storm-waves. The most typical b. is one with a gently concave profile; the landward side is backed by sand-dunes, succeeded by shingle, then with an area of sand, and rocks covered with seaweed at or about the low-tide mark. Some b.'s may comprise an extensive area of sand uncovered at low tide; e.g. in Morecambe Bay (Lancashire); Long B. (Washington State); the Florida b.'s; the coast of New Jersey (Atlantic City, Ocean City). See also RAISED B.

beach cusp A projection of sand and shingle, alternating with rounded depressions, along a b. The projections are frequently of open cone-shape, apexes pointing seawards, giving a kind of scalloped pattern. They are the result of a powerful SWASH and BACKWASH, esp. when waves break at or near right-angles to the coast. It is difficult to explain their initiation, though once started the eddying swash scours the depression, moving coarser material on either side on to the cusps, and progressively emphasizing it.

beach profile A p. transverse (at right-angles) to the coastline, usually concave upwards, since it is steeper above high water and gentler below. The b. p. is in equilibrium when the amount of material accumulated is more or less balanced by the amount removed, representing the net result of the average set of conditions obtaining along that stretch of coast. This balance is easily disturbed by strong onshore winds, esp. with exceptionally high tides, producing destructive storm-waves which will comb down and remove material. The b. p. of equilibrium is part of the overall shore p.

beaded esker A winding ridge of gravel and sand of fluvioglacial origin, with larger humps of material strung out along its length, the results of periods of pause in the retreat of the glacier nurturing the stream which deposited the e. material. E.g. behind Flamborough Head, Yorkshire.

[*f* ESKER]

beaded valley A v. in which a narrow section alternates with a wider, more open one; e.g. the upper Yosemite V., California.

bearing The horizontal angular difference between the meridian and a point viewed by an observer, measured in degrees clockwise from the meridian (*true b.*); e.g. a true b. of 270° is due W., of 180° is due S. If it is measured from magnetic N., it is known as a *compass* or *magnetic b.* In this context it is syn. with AZIMUTH.

Beaufort Notation A code of letters devised by Admiral Sir Francis Beaufort, F.R.S., in the early 19th century, to indicate the state of the weather: *b.* blue sky; *c.* cloudy; *o.* overcast; *g.* gloom; *u.* ugly threatening sky; *q.* squalls; *kq.* line squall; *r.* rain; *p.* passing showers; *d.* drizzle; *s.* snow; *rs.* sleet; *h.* hail; *t.* thunder; *l.* lightning; *tl.* thunderstorm; *f.* fog; *fe.* wet fog; *z.* haze; *m.* mist; *v.* unusual distant visibility; *e.* wet air, but no rain falling; *y.* dry air; *w.* dew; *x.* hoar-frost. Refinements include capital letters to indicate intensity (*R* = heavy rain); double capitals for continuity (*RR* = continued heavy rain); suffix $_o$ for slightness (r_o = slight rain); *i* for intermittence (*ir* = intermittent rain). On weather maps, this code is now replaced by symbols; e.g. ● rain, ❟ drizzle, ▲ hail, ✳ snow, ⋖ lightning, ⌓ dew.

Beaufort Wind Scale A scale of wind force, devised by Beaufort in 1805, modified in 1926; it is related to the descriptions of wind effects and estimated velocity at 10 m. above the ground.

Scale No.	*Wind*	*Force (m.p.h.)*	*Observed effects*
0	calm	0	Smoke rises vertically
1	light air	1-3	Wind direction shown by smoke drift but not by vane
2	light breeze	4-7	Wind felt on face; leaves rustle; vane moves
3	gentle breeze	8-12	Leaves and small twigs in motion; a flag is extended
4	moderate breeze	13-18	Raises dust; small branches move
5	fresh breeze	19-24	Small trees sway; small crests on waves on lakes
6	strong breeze	25-31	Large branches in motion; wind whistles in telephone wires
7	moderate gale	32-38	Whole trees in motion
8	fresh gale	39-46	Breaks twigs off trees
9	strong gale	47-54	Slight structural damage to houses
10	whole gale	55-63	Trees uprooted. Considerable structural damage
11	storm	64-75	Widespread damage
12	hurricane	above 75	Devastation

beck A small stream in N. England, with a rapid flow, a winding course, and an irregular bed.

bed (i) A layer or stratum of rock, usually a feature of the deposition of sedimentary rocks, divided from the layers above and below by well-defined BEDDING-PLANES. The term is used in 2 ways: (*a*) in a gen. context, as simply syn. with layer; (*b*) in a hierarchical sense, within an order of increasing magnitude of LAMINATION, layer, b., STRATUM, FORMATION. By contrast, some writers consider that a b. is larger than a stratum, consisting of several strata. (ii) Applied to a layer of PYROCLASTS, (e.g. an ash-b.) and to individuals in a sequence of lava flows; e.g. in N. Skye, where 25 distinct b.'s of basalt have an overall thickness of 300 m.; in the N.W. Deccan where one boring revealed 29 b.'s. (iii) The floor of an area covered with water (river, lake, sea), usually permanently, though it may dry out temporarily.

bedding The pattern of layers of sedimentary rocks, sometimes of distinctive banding, colour, thickness, composition and texture, separated above and below by well-defined planes. When the beds are very clearly defined, the rock may be described as *well-bedded*. The term is also sometimes applied to the tendency of certain metamorphic and igneous rocks to split in well-defined planes. See also CURRENT B. [*f*].

bedding-plane The surface or plane separating distinctive layers of sedimentary rock, indicating the end of one phase of deposition and the beginning of another. Sometimes it is referred to as part of the STRATIFICATION. [*f* STRIKE]

bed-load, of river The solid material carried along the bed of a river by SALTATION and by gravity; i.e. as a *traction load*, the particles being pushed and rolled along the stream bed. It contrasts with the load carried in suspension; the b.-l. consists of coarser, larger material. See SIXTH-POWER LAW.

bed-rock (i) The solid unweathered rock underlying the superficial layer of top-soil, sub-soil and unconsolidated material (the weathered or residual mantle). (ii) Sometimes used specif. of solid rock underlying PLACER deposits of gold or tin.

beet, sugar A yellowish-white root grown for its sucrose content, which averages 15–20% of its total bulk. It requires a fine, rich, loamy soil, such as derived from loess, limon or silt, and needs a large labour supply for hoeing, weeding and topping, despite developments in mechanization. B. s. is more expensive than cane s., but it has a higher s. content bulk for bulk, and transport costs are lower in W. Europe since production is closer to areas of consumption than in the case of cane s. About 1/3rd of the world's s. now comes from beet, mainly grown in the U.S.S.R., U.S.A., France, Poland, W. Germany, E. Germany, Czechoslovakia and Gt. Britain. The crushed pulp is used as cattle food, and by-products include molasses and alcohol.

beheading, of river See CAPTURE, RIVER.

belt (i) A zone of distinctive character, sometimes elongated, sometimes concentric (as around a city); e.g. of rock outcrops (the B. Series of metamorphic rocks in N.W. U.S.A.); of climate (TRADE-WIND b.); of vegetation (CONIFEROUS FOREST b.); of agriculture (*cotton, corn b.*'s of U.S.A.); of land use (GREEN B. around cities); of a residential area (commuter-b.). (ii) A narrow strip of water; e.g. the Great B., Little B., in the Baltic Sea. (iii) An elongated area of PACK-ICE.

belted outcrop plain An EROSION SURFACE upon which a series of rocks outcrop in parallel strips. Parallel CUESTAS and lowlands may develop from such areas; e.g. the coastal plain of S.E. U.S.A., lying E. of the FALL-LINE.

ben A Scottish peak, derived from the Gaelic *beinn* or *beann*. The term is used as a prefix to many hundreds of Scottish mountains; e.g. B. Nevis, Lomond, More, Cruachan.

bench A terrace, step or ledge, usually narrow, backed and fronted by a perceptible steepening of slope, produced by natural denudational processes (e.g. a WAVE-CUT B.), structural movements (STEP-FAULT), or artificial excavation (e.g. the result of quarrying or open-cast mining).

bench-mark (B.M.) A surveyor's defined and located point of reference. The British Ordnance Survey uses a broad arrow with a bar across its apex cut in solid rock, walls and buildings, with a specific height related to ORDNANCE DATUM at Newlyn, Cornwall. The U.S. Geological Survey, U.S. Coast and Geodetic Survey and U.S. Corps of Engineers use an engraved brass disc, fixed in solid rock or concrete, related to the Sea-Level Datum of 1929 (there is a $250 fine for disturbing a B.M.).

benefit–cost analysis See COST-BENEFIT ANALYSIS. It is commonly written this way round in the U.S.A.

benthos Plant and animal organisms living on the sea-floor, in ct. to free-floating life (PLANKTON) and swimming life (NEKTON). B. may be divided into: (i) *littoral* (between high-water spring tide mark and about 200 m. depth); and (ii) *deep-sea* groups. They may live either in a fixed position, or be capable of crawling or burrowing. See also NERITIC and PELAGIC.

berg A Germ. and Afrikaans word used as an element in proper names: (i) as a single hill or mountain in Germany (ct. *Gebirge*, a range); e.g. Feldberg, Hesselberg; (ii) as a range in S. Africa; e.g. Drakensberg, Groot Zwartberg. The element b. is used in numerous combinations for physical features; e.g. BERGSCHRUND, BERG WIND, ICEBERG.

bergschrund (Germ.) (in Fr. *rimaye*) A crack around the head of a FIRN-filled CIRQUE, separating the steep ice-slope on the BACK-WALL above from the main snow-field. It represents the point where the moving ice-mass which will become a glacier is drawing away from the enclosing walls of the cirque. Ct. RANDKLUFT. [*f* CIRQUE]

berg wind Lit. a 'mountain wind', but specif. a warm, dry, sometimes gusty, w. blowing mainly in winter down from the plateau of S. Africa towards the coast, thus warming ADIABATICALLY.

berm (i) A narrow terrace, shelf or ledge of shingle thrown up on the beach by storm-waves. (ii) A remnant of a partially eroded valley-terrace, indicating an interruption of the CYCLE OF EROSION of a river, leaving portions of the earlier valley-floor above the present river-level.

Bernhard's Index of Concentration A formula devised by J. Bernhard (1931) for indicating the degree of c. of settlement. He argued that c. of settlement is a function of the number of houses in each settlement, $\frac{H}{S}$, and in the number of settlements in a given area, $\frac{A}{S}$. Considering each function, he obtained the formula $C=\frac{H}{S}\times\frac{A}{S}=\frac{HA}{S^2}$. Ct. DEMANGEON'S COEFFICIENT OF DISPERSION.

Bessemer process of steel-making Molten iron from a blast-furnace is put into the B. converter, a large egg-shaped container, through which a blast of hot air (in the latest plant oxygen) is blown to oxidize the carbon present. A certain amount of spiegeleisen (an alloy of iron, manganese and carbon) is added. In a *Basic B.* the converter is lined with dolomite or magnesite, in an *Acid B.* it is lined with silica. It was invented by Sir Henry Bessemer (1813–98).

bevel (i) Used in geomorphology to describe a surface that has been planed-off, such as the crest of a hill-top or a CUESTA, forming a flat surface. (ii) The inclined plane of the junction between a horizontal and vertical surface in a sea-cliff which has been reduced by subaerial action; e.g. the bevelled cliffs in N. Cornwall. [*f* ACCORDANT SUMMIT LEVEL]

B-horizon The layer in the soil beneath the A-HORIZON, where much of the material (colloids, bases and mineral particles) washed down is deposited or precipitated; it is sometimes referred to as the zone of ILLUVIATION, or the *illuvial horizon*. It may be marked by the presence of a HARDPAN. It is usually sub-divided in detailed soil-studies into B_1, B_2, B_3, of which the 1st and 3rd are transitional to the horizons above and below.

bield (Anglo-Saxon) A rough enclosure for sheep, surrounded by dry-stone walls, esp. among the mountains of the English Lake District. Commonly used in place-names; e.g. Nan B. (head of Mardale), Cat B.'s (Wasdale).

bifurcation ratio The proportion between the number of streams of one order of magnitude to the number of larger streams of the next order of magnitude which they create through their confluence. In hydrological work streams are divided into orders, the highest being the main river or 'trunk stream'; see STREAM ORDER. If in a particular drainage basin 150 rivulets of order 1 unite to form 50 streams of order 2, the b. r. is 3·0. The term is somewhat of a misnomer, since b. would imply division, not uniting.

bight (i) A wide gentle curve or indentation of the coast, commonly between two headlands; e.g. B. of Benin, B. of Biafra, Great Australian B. (ii) Sometimes used specif. of a curved re-entry in an ice-edge in Polar regions.

bill A long narrow promontory, usually ending in a prominent 'beak' or spur; e.g. Portland B. The term has been extended to include other coastal projections; e.g. Selsey B.

billabong (Austr. aboriginal) A stagnant backwater in a stream that flows only temporarily.

binodal tidal unit An AMPHIDROMIC tidal system with two nodes, rather than the usual one.

biochore (i) A region or unit-area with a distinctive plant or animal life. (ii) The part of the earth's surface which can support life (W. Köppen).

bioclimatology The study of climate in relation to organic life, incl. human beings, other animals and plants. This may involve esp. questions of human habitability: housing, clothing and other health requirements depending on climatic conditions.

biogeography The study of geographical aspects of plant and animal life, esp. in terms of reasoned distributions.

bioherm (i) An old CORAL REEF (ct. BIOSTROME). (ii) The sedentary organisms (corals, gasteropods, foraminifera) which have contributed to its formation. Sometimes a distinction is drawn between the mound-like form of a b. and the more horizontally bedded BIOSTROME.

biosphere The surface zone of the earth and its adjacent atmosphere, in which organic life exists.

biostrome A coral reef currently in course of formation, with a more or less horizontal structure.

biotite A common rock-forming mineral in both IGNEOUS and METAMORPHIC rocks, consisting of ferromagnesian silicates. It ranges in colour from brown to dark green, with a glassy or pearly lustre, and is transparent and translucent. One of the MICA group of minerals.

biotic Of or app. to life. B. factors reflect the influence of living organisms, esp. in the development of soil and vegetation (e.g. bacteria, earthworms, ants, termites, moles, field-mice; in U.S.A. also prairie-dogs, ground-squirrels and gophers) in ct. to climatic, EDAPHIC and physiographic factors.

biotic complex 'The interacting complex of soils, plants and animals which, in response to climatic and other environmental conditions, forms a varied covering over much of the earth' (S. R. Eyre).

bird's foot (or **birdfoot**) **delta** A d. which extends out into the sea, with DISTRIBUTARIES bordered by narrow strips of sediment; e.g. the Mississippi d. [*f*]

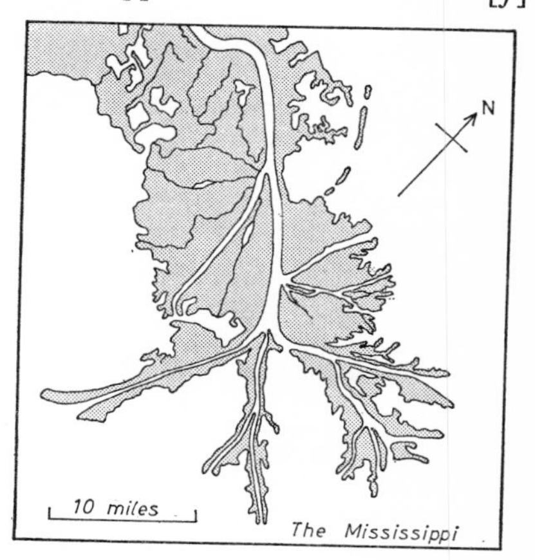

birth-rate A vital statistic of importance to the geography of population, since natural increase is a basic fact. The crude b.-r. is

$$\frac{\textit{number of births}}{\textit{total population}} \times 1000.$$

A more refined figure for studying fertility is the *standardized b.-r.*, in which age and sex anomalies are smoothed out by comparison with a hypothetical standard population; e.g. in U.S.A. in 1940, the crude b.-r.'s for urban and rural areas were 16·8 and 18·3, while standardized b.-r.'s were 15·8 and 19·5 respectively.

'biscuit-board' relief The result of CIRQUE erosion cutting into several sides of a mountain massif; e.g. the Snowdon massif.

bise, bize (Fr.) A dry, usually rather cold N. or N.E. wind, blowing across parts of Switzerland, N. Italy and S. France, similar in nature to the MISTRAL and the TRAMONTANA. It occurs most frequently in spring, when it may bring bright sunny weather, and is also common in winter when it may bring snow (the *b. noire*). It can cause harmful frosts in spring.

bitumen A group of solid or liquid hydrocarbons which are soluble in carbon disulphide. Applied generally to pitch, tar, asphalt; e.g. the substance obtained from the Pitch Lake, Trinidad.

bituminous A composite term applied to a free-burning coal containing 12–35% volatile matter; it includes house-coal and gas-coal. The name was applied in error long ago under the impression that these coals contained BITUMEN.

blackband A layer of bedded carbonaceous ironstone or *siderite* ($FeCO_3$) commonly occurring in the Coal Measures, containing about 30% of iron. These ores are so closely associated with coal, esp. in the Lanarkshire and N. Staffs. coalfields, that they are virtually self-smelting, i.e. require little fuel. Considerable reserves of b. ore still exist, but increasing cost and difficulty of extraction make them uneconomical. They were not discovered until 1801, but were very important in the 19th century. Ct. CLAYBAND.

black bulb thermometer A mercury t. with its bulb blackened, mounted horizontally in a glass tube from which air has been removed, and exposed to the sun's rays, thus giving 'sun temperatures'. This instrument is now rarely used.

black cotton soil See REGUR.

black earth See CHERNOZEM.

blanket-bog A type of BOG formed under conditions of high rainfall, swathing the whole land-surface where it is relatively horizontal; e.g. the Moor of Rannoch in Inverness- and Perthshire.

blast-furnace A furnace with a steel shell lined with refractory bricks, charged with iron ore, limestone and coke, in which the ore is smelted to produce molten iron; this may be cast in 'pigs', or passed directly to a steel-converter.

blende See ZINCBLENDE.

'blight' In planning terms, the effect on the value of property of planning proposals indicating that the life of the property will be limited. Even though compensation is payable on acquisition, such properties may become virtually unsaleable.

blind valley A v. in limestone, dry or occupied by a stream, enclosed at the lower end by a rock-wall, at the base of which the stream disappears underground. This may be the result of: (i) the collapse of the roof of an underground stream-course; (ii) the grading of a surface stream to a progressively falling BASE-LEVEL, forming a horizontal cave-passage through which the stream goes underground. E.g. the Cladagh R., Marble Arch Cave, near Enniskillen (N. Ireland); the Bonheur, a tributary of the Dourbie (thence the Tarn) in the Grands Causses in the Central Massif of France.

blizzard A very strong, bitterly cold, wind accompanied by masses of dry powdery snow or ice-crystals, with poor visibility (a WHITE-OUT), under polar or high-altitude conditions. The term was derived from N.W. gales experienced in U.S.A. in winter, but now it is used widely for all such phenomena, esp. in Antarctica.

block-diagram A drawing in either 1-point or 2-point (true) perspective, giving a 3-dimensional impression of a landform; in effect it is a sketch of a relief model. Geological sections can be appended to the sides of the diagram.

block disintegration The break-up of well-bedded and jointed rocks by mechanical means, notably frost action. When water freezes in rock interstices, its volume increases by about 10%, exerting great pressure and shattering the rocks along lines of weakness, producing rectangular blocks; e.g. the Cambrian quartzites of the Canadian Rockies in Banff National Park.

block-faulting The division of a section of the earth's crust into individual blocks by FAULTING, some of which may be raised, others depressed, others tilted. See FAULT-BLOCK, TILT-BLOCK. [*f*]

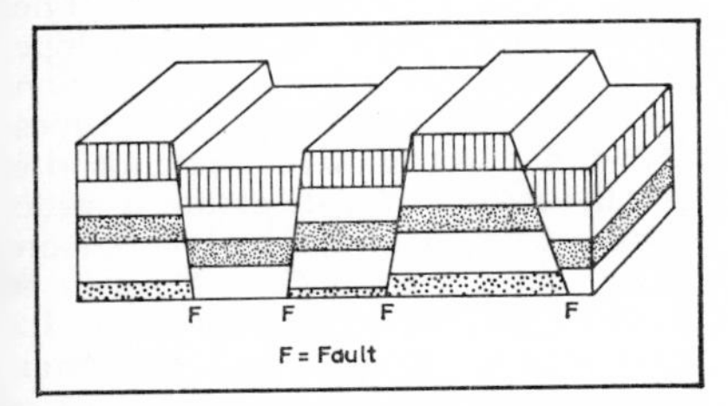

block-field, -spread See FELSENMEER.

'blocking high' An area of atmospheric high pressure (an ANTICYCLONE), which remains relatively stationary by comparison with approaching depressions, and which thus blocks their passage across its particular location.

block lava Usually syn. with AA, but sometimes restricted to truly angular, non-clinkerous masses of congealed acid lava.

block mountain A FAULT-BLOCK (i.e. a block bordered and outlined by faults), standing up prominently because of either: (i) its elevation by earth-movements; or (ii) the sinking of the surrounding area. E.g. the Harz Mtns., Vosges, Black Forest. See HORST, TILT-BLOCK.

block-slumping A type of MASS-MOVEMENT down the steep faces of escarpments and sea-cliffs, esp. where a stratum of clay underlies more massive but well-jointed rocks such as chalk and limestone. The action of water (esp. SPRINGS) will undermine the upper strata, and will lubricate the clay. This b.-s. involves a shearing of the rocks, a tearing away of a soggy mass of material, usually with a distinct rotational movement on a curved plane (ROTATIONAL SLIP), leaving a scar on the hillside, while the material slumps downwards; e.g. the Cotswold scarp between Gloucester and Cheltenham. [*f* ROTATIONAL SLIP]

blood-group index Differing proportions of *A*, *O*, *B* and *AB* elements are present in the blood of various population groups; random tests indicate that in the world's population gen. the groups are present in average proportions of *A:* 38%, *O:* 37%, *B:* 18% and *AB:* 7%. The racial affinities between one population and another can be expressed in the form of an index derived from comparing their respective blood-group ratios, as devised by N. Lahovary, and adding the differences.

Blood-group	*Population X*	*Population Y*	*Difference*
A	38	40	2
O	37	45	8
B	18	10	8
AB	7	5	2
	100	100	20

Thus the index of racial affinity between *X* and *Y* is 20.

blood rain Rain-drops containing fine particles of reddish dust, which have been transported by the wind from deserts, and washed down from the atmosphere during precipitation; e.g. in S. Italy, Sicily, Malta, even as far as the Canary I.'s, after which plants are often covered with a reddish powder.

blow-hole A near-vertical cleft linking a sea-cave with the surface inland of the cliff-edge; spray is sometimes thrown out of the hole by the compressional force of a wave surging into the cave below. It is formed by erosion concentrated along a well-marked JOINT or FAULT extending vertically from the surface into the roof of the cave. A b.-h. is usually formed in well-jointed hard rocks; e.g. the Old Red Sandstone of the Orkneys, the Isle of Soay (off Skye), and the coast of Caithness (Mermaid's Kirk); and in Carboniferous Limestone (Huntsman's Leap, near St. Gowan's Head, S. Pembrokeshire). It is sometimes called a *gloup* or *gloap* in Scotland. [*f* STACK]

blow-out A hollow in a sandy terrain, esp. among dunes in deserts, heathlands and along coasts, or in an arid plain, formed by wind-eddying (DEFLATION), partic. where the protective vegetation has been destroyed. In a rock-desert, a b.-o. may form from a break in a resistant surface-layer (e.g. by faulting or localized weathering), which the wind may enlarge by eddying into a deflation-hollow. The floor of an arid intermont basin (e.g. in Arizona, S. California and New Mexico) is partic. subject, esp. after a temporary salt-lake has dried out. A b.-o. can also occur in an area of peat; e.g. on the Millstone Grit plateaus of the Pennines where wind-erosion of peat is definitely occurring.

blow-well, blowing well An artificial w. or a natural spring, found esp. in E. Yorkshire and E. Lincolnshire, which, working on the ARTESIAN principle, creates fountains or flows of water above ground. The chalk AQUIFER is sometimes covered by a clay AQUICLUDE, which when penetrated leads to a water-flow of this kind.

blubber The subcutaneous fat of a whale, from which train-oil is obtained. Some 50–80% of b. by weight consists of oil, yielding about 10 tons per average sized whale. The oil is a raw material for margarine and cooking-fat.

blue-band A distinctive layer of blue ice in a GLACIER, free from air-bubbles, probably the result of frozen melt-water.

Blue Mud A widely occurring m. on the CONTINENTAL SLOPE, of terrigenous origin, earthy and plastic in character, deriving its colour from the presence of iron sulphide. The individual particles are fine (less than 0·03 mm. in diameter).

blue sky The apparent colour of the cloudless s. during the daytime, the result of the scattering of sunlight by molecules of air. The short waves at the blue-violet end of the solar spectrum are scattered more readily by the finer molecules at high altitudes, where the s. appears deep blue. The shade of blue can be measured on a scale devised by F. Linke, with 14 shades ranging from white to ultramarine.

bluff A steep prominent slope. The term is used specif. of a valley-slope cut by the lateral erosion of a river on the outside of a MEANDER (sometimes *river-cliff*). As a river extends its lateral erosion, it will create a wide FLOOD-PLAIN, bounded by low b.'s on either side, cut back wherever a meander impinges against them. In the case of a large river (e.g. the lower Mississippi), the b.'s may lie back a mile or more from the river channel. A continuous series of b.'s is sometimes referred to as the *bluffline*.

bocage (Fr.) A rural landscape in N.W. France, with small fields enclosed by low banks crowned with hedges, often containing pollarded oaks and ash. In W. Brittany the banks may be replaced by dry-stone walls. It is also used as a prefix in a PAYS-name; e.g. *B. normand*, *B. breton*, *B. angevin*, *B. vendéen*.

Bodden (Germ.) An irregularly shaped inlet along the S. Baltic coast produced by a rise of sea-level over a former uneven lowland surface. The island of Rügen (off the coast of E. Germany) consists of a few former irregular islands, now linked by sand-spits, and enclosing the Wieker B.

Jasmünder B., Kubitzer B., Greifswalder B. [*f*]

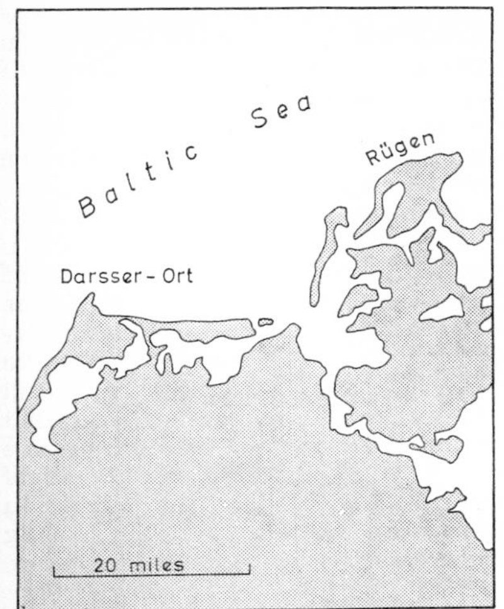

bog (i) Waterlogged, spongy ground, with a surface layer of decaying vegetation, esp. sphagnum and cotton-grass, ultimately producing a thickness of highly acid PEAT. (ii) The vegetation complex thus associated. E.g. on Dartmoor, the gritstone plateaus of the Pennines, W. Scotland (Moor of Rannoch), Offaly (B. of Allen), Connemara and Mayo (Ireland). Widespread b.'s occur in high latitudes in Canada (MUSKEG), Scandinavia, Finland and U.S.S.R., where an uneven surface of impermeable rocks with a partial boulder-clay cover causes waterlogging. See also BLANKET-BOG, RAISED BOG.

bogaz plur. **bogazi** (Serbo-Croat) An elongated trench-like chasm in limestone country, notably in the KARST of Yugoslavia. It has developed along a clearly defined joint through CARBONATION-SOLUTION.

boghead coal A close-grained black or dark brown type of c., containing algal remains, but with much less carbon than BITUMINOUS C. On distillation it may produce oil, though not as much as richer OIL-SHALES. So called after a locality W. of Edinburgh, where it was worked.

bog-ore A layer of hydrated iron oxide (usually *limonite*–$2Fe_2O_3.3H_2O$), found in peat bogs or shallow lakes, probably precipitated by the agency of bacterial organisms. It has long been worked, since it is fairly rich in iron (40–60%) and easy to obtain; e.g. in central Sweden, Finland and Luxembourg ('*le fer fort*' found as 'alluvial iron' in the valleys of the R. Alzette, Mamer, Attert and Eisch). The term also includes the Clinton ores near Birmingham (Alabama).

bolson A basin of inland drainage, occurring among the high plateaus of S.W. U.S.A. in Arizona, New Mexico and S. California, commonly containing a salt-lake, with sheets of rock-salt or gypsum, and usually rimmed by mountains. Should a permanent or temporary stream flow across the basin, it may be referred to as a *semi-b.*

bomb, volcanic A mass of lava thrown into the air during a volcanic eruption, solidifying in globular masses of rock before reaching the ground. This is included in the class of PYROCLASTS.

bonanza A Sp. term, applied in U.S.A. to something profitably productive; e.g. (i) a 'mine in b.', yielding rich ore; (ii) in the past, cereal growing in newly opened up temperate grasslands, with an accumulated high humus and mineral content in the soil, and resultant high yields if only for a short time ('b. farming').

bone bed A stratum containing fragments of fossil b.'s, teeth and scales of vertebrates, esp. of fishes. The concentration of this accumulation may be the result of some rapid catastrophe; e.g. a submarine earthquake which killed all life simultaneously. The b. is rich in calcium phosphate. E.g. the Ludlow B.B. (Silurian).

bonitative map A m. which shows land favourable or unfavourable to specif. types of economic development, esp. potentiality for improvement. It is usually constructed by means of a series of SIEVE-M.'s.

Bonne, Bonne's Projection An EQUAL AREA P., of modified CONICAL type. The chosen central meridian (*c.m.*) is divided truly (i.e. 10° lat. = $\frac{2\pi R}{36}$, where R is the radius of the globe to scale), and is crossed by a standard parallel (*s.p.*). Each *c.m.* and *s.p.* is selected as centrally as possible relative to the area to be mapped; e.g. for N. America, *c.m.* 90°W.; *s.p.* 40°N. The *s.p.* is drawn using the customary formula for a CONICAL P. (for which see table), $r = R \cot \theta$ (where θ is *s.p.*). All other parallels are concentric to the *s.p.*, whose function is merely to control their curvature. Each parallel is divided truly; e.g. 10° of meridian interval $= \frac{2\pi R . \cos lat}{36}$. The other meridians are curves drawn through the division points on each parallel. Each quadrangle formed by intersecting parallels and meridians has its base and height true to scale, and therefore the p. is equal area, but scale and shape distortion increase rapidly to W. and E. The p. is commonly used for compactly shaped countries in mid-latitudes (e.g. the official maps of France on a scale of 1/80,000), and for such continents as N. America and Eurasia. The limiting case of B., using the equator as the *s.p.*, is in fact the SANSON-FLAMSTEED (SINUSOIDAL). [*f*]

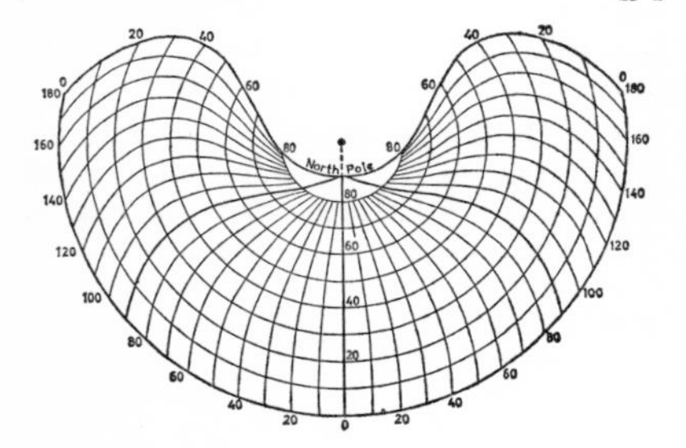

booley The Irish equivalent of the Scottish SHIELING.

bora (It.) A bitterly cold, squally wind, blowing violently down the mountains from N. or N.E. into the N. Adriatic Sea, from a high-pressure area over central Europe towards the rear of a depression over the Mediterranean, strengthened by gravity flow (cf. MISTRAL). It is most common in winter, but is known in a weaker form in summer as the *borino*. Wind velocities are at times very high, when it is sometimes called the *boraccia*; 137 km. (85 mi.) p.h. has been recorded at Trieste, with gusts up to 209 km. (130 mi.) p.h. Its strength is partic. marked when the trend of bays, islands and ridges coincides with the wind direction, thus funnelling it. It can be a great danger to navigation and to transport on coastal roads. The b. is sometimes associated with cloud and precipitation (rain and snow), when the cold air undercuts the moist warm air of the depression to the S. ('the rainy b.'), but usually the depression lies more distant, when the air is dry and the skies clear ('the dry b.').

border An area or zone lying along each side of the boundary between one state and another; usually syn. with FRONTIER. 'The B.' is specif. between England and Scotland.

bore A 'wall' of broken water moving upstream in a shallow-water estuary with an appreciable tidal range; it is caused by the progressive constriction of the advancing flood-tide at spring tide, retarded by friction at its base and by a powerful opposing river current. The b. gradually diminishes in height and ultimately dies out. E.g. the Severn b., sometimes a metre high; the *eagre* (also *aegir*) on the Trent, and the *mascaret* on the Seine (both much reduced by dredging for navigation); the Hooghly in the Bay of Bengal. On the Tsien-tang-kiang in N. China (near Hangchow), the front of the wave attains a height of 3–4·5 m. (10–15 ft.) and advances upstream at 16 km. (10 mi.) p.h. The tide rises so markedly in the Bay of Fundy that the term b. is commonly applied, though not in the strict sense.

Boreal (i) Gen. term for a climatic zone in which winters experience snow, and summers are short. (ii) App. to the N., specif. the northern coniferous forests. (iii) A climatic period from about 7500 to 5500 B.C., generally dry, with cold winters and warm summers; indicated by the development of a pine-hazel flora.

bornhardt (Germ.) A residual hill rising abruptly from the surface of a plain; cf. INSELBERG, of which term B. W. Bornhardt was the originator. Some writers use it with a precise meaning, implying a granite-gneiss inselberg, associated with the second cycle of erosion in a rejuvenated desert or semi-desert landscape (L. C. King); others regard it as gen. syn. with inselberg.

borough Originally a fortified place (Old English, *burh*), then a place with some type of municipal organization; e.g. in *Domesday Book* (1087), Worcestershire had three places (Worcester, Droitwich, Pershore) regarded as b.'s, termed either *civitas* or *burgus*, with burgesses listed. The term now has a distinct administrative significance. In England it denotes a town with a royal charter; there are two types: (i) a *County B.*, a town with the status of an administrative county; (ii) a *Municipal B.*, which is a dependent part of an administrative county. The term is also loosely used for other towns without the official status. There are various local definitions for a b. in parts of the U.S.A., Australia, New Zealand. Note also *Metropolitan B.* in the County of London. Cf. BURGH.

boss A small BATHOLITH with a more or less circular plan, gen. regarded as having a surface area (when exposed by denudation) of about 100 sq. km. (40 sq. mi.); e.g. in N. Arran (W. Scotland); Shap (Westmorland). Usually it is regarded as syn. with STOCK, though some regard a b. as circular, a stock as more irregular.

bottom (i) The low-lying alluvium-covered FLOOD-PLAIN along a river valley, esp. in U.S.A.; e.g. the Mississippi Bottomlands. (ii) The floor of a lake, sea or ocean. (iii) A dry valley in chalk country; e.g. Rake B., near Butser Hill, in Hampshire. (iv) The former head of a lake in a U-shaped valley, now infilled with sediment deposited by inflowing streams; e.g. Warnscale B. in the Buttermere valley, English Lake District.

bottomset beds In a DELTA, the fine material carried out to sea and deposited in advance of the main delta; over them in due course the FORESET BEDS are laid down as the delta advances seaward. [*f*]

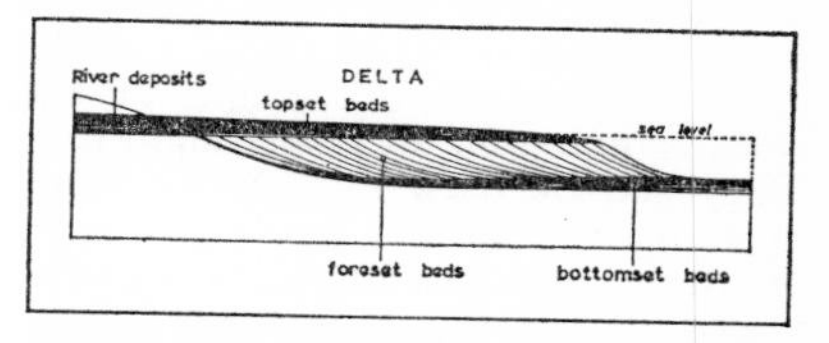

Bouguer Anomaly A phenomenon first observed by P. Bouguer (1735), who made gravitational observations near Chimborazo in the Andes, and discovered that the actual plumb-line deviation towards that peak was less than had been calculated. The same discrepancy was later found during surveys in the Indo-Gangetic Plain, where the estimated deviation resulting from the mass of the Himalayas was 15″ of arc, but the measured deflection of the plumb-line was only 5″. This a. results from the deficiency of mass in the crust underlying the mountain ranges. In calculating the a., the observed gravity is compared with a formula 'corresponding to a spheroidal surface with an average value of the ellipticity' (H. Jeffreys), corrected for height above sea-level.

boulder A large individual fragment of rock, exceeding 200 mm. (8 ins.) in diameter (*Brit. Stand. Inst.*), or 256 mm. (about 10 ins.) (*U.S. Wentworth Scale*). An example of a common word given a precise meaning.

boulder-clay An unstratified mass of clay, containing stones of all shapes and sizes, often striated, their nature depending on their source of origin. It is the product of glacial erosion and deposition, without any water transport involved. Many local names are applied; e.g. Chalky B.-c. and Cromer Drift of E. Anglia; Hessle and Purple B.-c. of Lincolnshire and the E. Riding of Yorkshire. The term TILL is now preferable, since it does not imply any definite or specif. constitution.

boulder-field See FELSENMEER.

boulder-train A series of ERRATICS deposited by a glacier; they are usually worn from the same identifiable bedrock source, carried forward by the moving ice, and deposited either in a broad fan-pattern, with the apex pointing to the origin of the rock, or in a more or less straight line. They can be used to map the exact movement of former ice; e.g. a train of dark Silurian boulders can be traced across the limestone Craven district in W. Yorkshire.

boundary The dividing line between one political state and another. Ct. the district or region adjacent on either side, which is the BORDER (or FRONTIER).

boundary current Part of the circulation of oceanic water at depth, along the W. side of the oceans: (i) flowing S. in the Atlantic from a high latitude source of deep cold water off S. Greenland; (ii) flowing N. in the Pacific from S. of New Zealand; and (iii) a less marked N. flow from S. of Africa along the E. coast of that continent into the Arabian Sea. It has been suggested by H. Stommel that (i) meets in latitude 35°S. another b. c. flowing N. from the Weddell Sea, where the water turns E. and produces a great flow around the Antarctic continent. The edges of the currents are strongly marked by sudden changes in temperature and salinity, hence the term 'boundary'.

Bourdon tube A type of thermometer or barometer consisting of a closed curved tube, elliptical in section, containing spirit; its curvature changes as the temperature or pressure changes.

bourne A temporary or intermitten stream which may flow in a DR VALLEY in chalk country after heav rainfall, partic. in winter, when th WATER-TABLE rises above the level the valley-floor. It forms a commo place-name element in chalk countr e.g. Ogbourne in Wiltshire. A co siderable time-lag occurs between th period of heavy rain and the b.-flov the water percolates slowly throug the closely jointed chalk towards th saturated layer. They are also know as 'winter bournes', 'woebournes 'nailbournes', 'gypseys' and 'lavants'

brachycephalic (sometimes abbr. *brakeph*) Broad-headed. This defined by anthropologists as a sk with a CEPHALIC INDEX of 83·1 or mor Hence *brachycephaly*.

brackish Slightly saline, with a s content less than that of sea-wate sometimes defined as containing 1 30 parts of salt per thousand.

brae (Scottish) A hillside; the BRO of a hill. See BANK.

braided river A r. whose cou consists of a tangled network of inte connected diverging and convergi shallow channels, with bars or ba of alluvial material and shingle between. These bars may be reveal at times of low water. They formed esp.: (i) in a heavily lad though shallow r.; (ii) on the surf of an ALLUVIAL FAN; and (iii) by melt-water stream as it flows acr FLUVIOGLACIAL deposits. The braidi is encouraged when the banks of main stream consist of easily erodi alluvium, sands and gravels. E.g. Spree near Berlin; the Rhône tween St. Maurice and L. Geneva

brake A dense growth of brushwo

brakeph See BRACHYCEPHALIC.

brash (i) Broken rubbly rock neath the soil and sub-soil, parti weathered *in situ*. The term appear Cornbrash, a Jurassic rock noted the 'brashy' soils it has produced, which were long known as good corn growing. (ii) Loose masses of small ice fragments floating in the sea.

Brave West Winds A nautical term for winds of the S. hemisphere, blowing from a W. or N.W. direction over the ocean between latitudes 40°S. and 65°S., moving N. and S. with the seasonal change of the world pressure belts. They are characterized by their strength and persistence, and by stormy seas, overcast skies, and damp, raw weather. Seamen call the latitudes in which they blow the 'ROARING FORTIES'.

breached anticline The development of a drainage pattern on an a. may result in a valley along its axis, so removing the overlying rocks and exposing older ones. The result is an anticlinal valley with infacing escarpments. This is a common phenomenon, since: (i) the axis of an a. is usually structurally weaker, having been subjected to tension, than a neighbouring syncline; and (ii) it is further above BASE-LEVEL and is thus more readily attacked. E.g. the breached Fernhurst a. in the Weald; the Vale of Pewsey. On a large scale, the Bow R. in W. Alberta has cut a deep valley along the crest of a major a. trending N.N.W. to S.S.E., leaving the remnants of the W. limb as the main crest of the Canadian Rockies. [*f*]

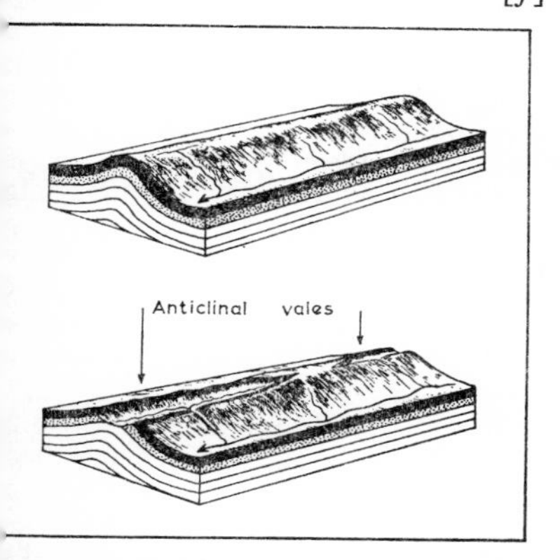

breadcrust bomb A volcanic b. with a glassy crust seamed with cracks, the result of shrinkage following cooling and congealing.

breaker A mass of turbulent broken water and foam rushing up a beach, formed when a wave travelling across deep water passes into shallow water; its crest steepens, curls over and breaks. Gen. this oversteepening occurs when the ratio of wave-height to wave-length exceeds 1/7. Waves can also break in deep water when generated by strong local winds (sometimes known as 'white horses' or 'whitecaps'), or if a mass of rock (e.g. a REEF) lies near the surface. Some authorities classify them into: (i) *spilling b.*'s; (ii) *plunging b.*'s; and (iii) *surging b.*'s. (i) and (iii) tend to be *constructive*, i.e. they move shingle and sand up the beach; (ii) is *destructive*, i.e. 'combs down' the beach and moves material seawards.

break of slope A marked or sudden change of inclination in a s. In the bed of a river, such a b. may represent a KNICKPOINT or a transverse resistant band of rock. In a U-shaped valley, the b. of s. in a cross-profile may represent the max. height of glacial erosion. Elsewhere it may indicate the edge of an EROSION SURFACE.

break-point bar An offshore BAR formed along the line where the waves first break.

breccia (It.) A rock composed of angular CLASTIC fragments mixed with finer material (ct. AGGLOMERATE and FAULT-B.). B. may be: (i) a SCREE deposit; (ii) a sedimentary rock cemented during consolidation; (iii) a 'volcanic b.', composed of PYROCLASTIC material; (iv) 'intrusion b.', formed of pieces of country rock embedded in material of magmatic origin; and (v) formed of some specific fragments, such as 'bone b.' found in caves from the remains of animals. The word is derived from It. *breccia*, meaning broken wall material. In English, a distinction is made between b. (angular constituents) and CONGLOMERATE (rounded constituents).

breckland (i) An area of heathland, with thickets and bracken. (ii) An area of land taken in from the heathland for temporary cultivation. (iii)

The proper name of an area on the borders of Norfolk and Suffolk, now largely under the coniferous plantations of the Forestry Commission.

breeze A wind between Force 2 (light breeze, 5 knots) and Force 6 (strong breeze, 28 knots) on the BEAUFORT WIND SCALE.

brickearth (i) Orig. any loamy clay used for brick-making. (ii) Specif., a fine-textured deposit found on river-terraces, resulting from wind-blown material which has been re-worked, re-sorted and re-deposited by water; e.g. on the Thames terraces. Soils derived from b. are usually fertile and easy to work.

brickfielder A dry, dusty, squally wind in Victoria, Australia, sometimes with temperatures exceeding 38C. (100°F.), blowing from the continental interior in a S. direction in front of a depression.

bridge A structure carrying a road or railway across a river, valley, entrance to a bay, or over another road or railway. (*a*) *Suspension:* Verrazano-Narrows, New York (1298 m., 4260 ft. span); Golden Gate, San Francisco (1280 m., 4200 ft. main span); Mackinac (Mich.) (1158 m., 3800 ft.); George Washington, New York (1067 m., 3500 ft.); new Forth Road B. (1006 m., 3300 ft.; 1·9 km., 1·25 mi. overall); (*b*) *steel arch:* Sydney Harbour (503 m., 1650 ft. span); Bayonne, New York (504 m., 1652 ft.); (*c*) *cantilever:* Forth railway b. (overall 1631 m., 5350 ft., main spans 521 m., 1710 ft.); Quebec (main span 549 m., 1800 ft.); (*d*) *movable:* bascule (Tower B., London, with 60 m., 200 ft. opening); vertical lift (Middlesbrough, 82 m., 270 ft.); smaller swing b.'s (over Manchester Ship Canal). Other types are of reinforced concrete (Gladesville, Sydney, N.S.W.), pre-stressed concrete (the Oosterschelde Bridge, Netherlands, 4·8 km., 3 mi. long, the longest in Europe, opened in 1965), cast-iron (Ironbridge on Severn), masonry, brick, wood, transporter, aqueducts carrying water, military (Bailey).

bridge-head In military operatior an outpost or footing on the banks o river, or on a coast, in territory mair held by an enemy.

bridge-point, bridging point p. at which a river is or could bridged; an important factor in t location of settlements, as indicat by numerous place-names involvi -*ford* or -*b*. elements in every langua (*pont, Brücke*). Of esp. importan are: (i) the *lowest* b.-p., which is n a fixed or specific p., but depen on technological ability and therefo the time-element; and (ii) the relatio ship between a b.-p. and the *lin of navigation* for sea-going ships.

bridle-path A p. which can accor modate pedestrians and equestria though not wheeled vehicles.

brigalow A type of SCRUB in sem arid areas of Australia, on the bord of the mulga; it consists mainly o species of *acacia* (*A. harpophylla*).

Brillouin Scale A logarithmic line s. produced by L. Brillouin (196 ranging from 10^{-50} to 10^{30}, and th including the largest (10^{27}) a smallest (10^{-13}) units yet measured man. The equatorial circumference the earth is taken as 0. Cf. G-SCALE.

brine A solution containing a mu higher proportion of common SA than sea-water, often occurring as spring.

British Summer Time See B.S.T

broad A sheet of reed-fringed, fre water, forming part of, or joined to slow-flowing river near its estua It is typically found in Norfolk a extends into Suffolk; here the te is a proper name. This area was shallow bay in Roman times, in whi peat accumulated. There has be much discussion about the origin of t B.'s, but the consensus of opinion that cutting and removal of peat h been a major factor. One category b. is separated from rivers by narr 'washlands' known as *ronds* (e. Salhouse), linked to the rivers artificial cuts; the other is an actual broadening of the river; e.g. Thurne B. Breydon Water is a portion of the old estuary, consisting of mud-flats crossed by creeks and the main river channel at low tide, a sheet of water 6 by 0·8 km. (4 by 0·5 mi.) at high tide.

Brockenspectre The apparent shadow of an observer standing on a mountain summit with the sun behind him, projected as a diffraction effect on to a cloud or fog-bank beyond. Round this shadow are rings of coloured light. The odd thing is that an observer sees only his own shadow, not those of his companions. It was so called after the Brocken, a summit in the Harz Mtns. (Germany). The terms *glory* and *anticorona* are sometimes used for this phenomenon.

Bronze Age The 3rd major phase in the development of Man's culture, succeeding the Palaeolithic and the Neolithic. In the early stages of the phase, copper was used in its pure form for objects of adornment (i.e. 'the Copper Age'), starting in Mesopotamia *c.* 4000 B.C. Its use spread during the next millennium. About 3000 B.C. the alloy of copper and tin (i.e. bronze) was discovered, and the B. A. reached its height in the 2nd half of the 2nd millennium. Towards the end of the last millennium, the gradual superseding of bronze by iron brought in the IRON AGE.

rook A small stream. Used loosely nd poetically in England (e.g. Lord Tennyson's 'Brook'). In U.S.A. it is egarded as smaller than a CREEK.

row The upper part of a hill. Cf. cottish *brae*.

rown-coal A brown fibrous deosit, intermediate in development, orm and character between peat nd coal proper, occurring in beds hich are worked by open-cast ethods. It is used for the generation f thermal electricity, the making of riquettes, and the extraction of gas nd oil. The chief producers are E. ermany, U.S.S.R., W. Germany nd Czechoslovakia. See also LIGNITE.

brown forest soil (brown earth) One of the main ZONAL SOIL-groups, characteristic of areas in middle latitudes formerly covered with deciduous woodland, rich in organic matter derived from the accumulation and decay of leaves. The surface A_1-horizon is usually a slightly acid humus layer, since the climate is humid and some leaching will have occurred. The A_2-horizon is a grey brown, somewhat leached, zone (but less so than in a PODZOL). The B-horizon is thick, dark brown, and contains colloids and bases carried down from the A-horizon. These soils are found notably in N.E. U.S.A., N. China, central Japan, and N.W. and central Europe. In the last, most of the s. has been cultivated for centuries, and the original forest has been long removed.

Brückner cycle A c. of climatic change, with an average (though irregular) periodicity of about 35 years. This was postulated by E. Brückner in 1890, who examined rainfall and temperature records going back to the 18th century. The actual amplitude of the oscillations is not great: for temperature less than 1°C., and for precipitation only from 9% above normal to 8% below. Individual c.'s may range from 25 to 50 years. Variation between cold and humid, and warm and dry, periods may be reflected in glacier fluctuation, the level of the Caspian and other inland seas, the date of grape harvests, patterns of tree-rings, and other phenomena.

B.S.T. British Summer Time (first introduced in 1916), during which clocks are advanced 1 hour in ct. to G.M.T. This is sometimes known as *Daylight Saving Time*. *Double B.S.T.* was instituted in 1941 (until 1947), whereby clocks were advanced 2 hours. In 1968 it was decided to retain B.S.T. throughout the year as *B. Standard T.* for an experimental period, in order to have the same time as W. Europe.

B.T.U. (British Thermal Unit) A unit of energy defined as the amount of heat which will raise the temperature of 1 lb. of water through 1°F., strictly from 63°F. to 64°F. at sea-level atmospheric pressure; or 1/180 of the heat required to raise 1 lb. of water from 0°C. to 100°C. It equals 252·1 calories. Now obsolescent.

buckwheat A species of herbaceous plant (not a grass) introduced into Europe from central Asia by the Turks. Its seed is actually a fruit, with a rind enclosing a kernel, borne on a plant about 0·6 m. high, with brown leaves and white flowers preceding the fruit. It grows well on poor soils and hilly slopes, matures quickly, and is a good CATCH-CROP between two others or if a main crop fails. C.p. = U.S.S.R., France, Poland, Canada and U.S.A. Its seed is used as animal fodder, and in some parts (e.g. U.S.A.) for b. ('griddle') cakes.

buffer state An independent s. situated between 2 powerful ones, thereby helping to prevent conflict between them, though it may be at times overrun itself; e.g. Belgium, between France and Germany, and Poland, between Germany and Russia. This does not mean that such s.'s exist only by virtue of being 'buffers', but it does mean that they have acquired and maintained their territory in the face of more powerful neighbours only because the latter allowed it, or other Great Powers enforced it. When a major war occurs, a b. s. is usually overrun and occupied, or it may form the zone of actual hostilities.

built-up area The a. in a town actually covered with buildings. It can be subject to planned development only by pulling down existing buildings (e.g. slum clearance), followed by rebuilding. The term is used as a category in planning.

bunch grass A coarse tufted g., separated by bare sandy or stony ground, which grows e.g. in the semi-arid parts of the American Midwest and on the intermont plateaus of the Western Cordillera. It provides extensive grazing, but for only a sm[illegible] number of head per unit of area.

bund (India) An embankment [illegible] dyke, used in irrigation, part[illegible] around rice-fields, mostly made [illegible] earth, sometimes of stone; the term [illegible] also applied to the enclosing walls [illegible] reservoirs and TANKS.

Bunter The lowest formation of t[illegible] Triassic system, consisting mainly [illegible] variegated red sandstones and pebb[illegible] beds. It was probably laid down un[illegible] arid or semi-arid conditions, so[illegible] times resting UNCONFORMABLY on [illegible] rocks below, sometimes on the P[illegible] mian with little indication of a bre[illegible] Hence the term New Red Roc[illegible] which may embrace the Permo-Tr[illegible] boundary.

buran A strong N.E. wind exp[illegible] enced in central Asia at all seaso[illegible] though occurring particularly as [illegible] fierce, bitterly cold blizzard ('wh[illegible] b.') in winter.

burgh Orig. equivalent to BOROU[illegible] the term is now limited to towns [illegible] Scotland which possess a char[illegible] with a council presided over b[illegible] provost, and with elected bail[illegible] Edinburgh, Glasgow, Dundee [illegible] Aberdeen are *city-burghs*.

burn (Scottish) A small stream [illegible] Scotland, N. England and N. Irela[illegible]

bush An area of uncleared shru[illegible] vegetation, sometimes with scatte[illegible] trees; in gen., wild and unsettled [illegible] opposed to cultivated, terrain. [illegible] S.W. U.S.A., the term is applie[illegible] areas of low bushes (e.g. creos[illegible] adapted to semi-arid conditions.

bushel A measure of volume, e[illegible] valent to 8 Imperial gallons (36[illegible] litres, 0·0476 cu. yd.). The weigh[illegible] threshed wheat required to fill [illegible] measure averages about 60 lb. (v[illegible] ing with the type of wheat); a b[illegible] oats weighs about 32 lb., of ba[illegible] 48 lb., and of maize 56 lb. Yield[illegible] cereals are often given in b. per a[illegible] e.g. average yield of wheat in [illegible] Britain is 50 b.p.a., in the Net[illegible] lands about 58, in Denmark 57

Canada only 23, U.S.A. 20, Australia 19, India 12; the world average is about 17. 1 British b. = 1·0321 U.S. b.

bushveld (Afrikaans, *bosveld*) A type of 'tall grass—low tree' SAVANNA, occurring in parts of subtropical and tropical Africa. The trees may be scattered so as to form open parkland, or be so dense as nearly to form forest.

butte A small, though prominent, flat-topped hill, usually capped with a resistant rock-stratum, remaining after the partial denudation of a plateau in semi-arid areas. It may be a small isolated portion of a MESA. E.g. in Arizona and New Mexico. It is usually flanked by slopes of angular SCREE, the result of desert-weathering. Isolated hills formed by other means are sometimes also regarded as b.'s; e.g. igneous intrusions which remain upstanding, or a hill protected by an extrusive cap of basalt; many examples of the last can be seen along the Front Range of the Rockies, esp. near Denver, Colorado. [*f* MESA]

butte témoin (Fr.) (in Germ. **zeugenberge**) A flat-topped OUTLIER, beyond the edge of a plateau or an ESCARPMENT of which it was once part, with its surface in broadly the same plane. Most true buttes are in fact b. t.'s. *t.* = witness, i.e. evidence of a former further extension of the plateau.

buttress A protruding or outstanding mass of rock on a mountainside; e.g. Kern Knotts B., Bowfell B., Gillercombe B. in the English Lake District; the Northeast, Tower and Observatory B.'s on Ben Nevis.

Buys Ballot's Law A law put forward in 1857 by C. H. D. Buys Ballot of Utrecht, a Dutch climatologist, that if an observer stands with his back to the wind in the N. hemisphere, pressure is lower on his left hand than on his right; in the S. hemisphere pressure is lower on his right. This is a result of the CORIOLIS FORCE (see FERREL'S LAW) on the movement of a body on the surface of the earth; air tends to move along the isobars of a pressure system, in the N. hemisphere clockwise round ANTICYCLONES, anti-clockwise round DEPRESSIONS. See GEOSTROPHIC FLOW.

bypass A road which skirts the margins of some locality, thus enabling through-traffic to avoid passing through the locality.

bysmalith A large IGNEOUS INTRUSION, in the rough shape of a cylinder, which has arched up the country rock above it. Its surface characteristics are similar to a LACCOLITH, but its sides plunge steeply down for a vast distance, and it is commonly associated with faulting round its margins. Sometimes it may be exposed as an upland mass by denudation; e.g. Mt. Hillers in the Henry Mtns. (Utah), Mt. Holmes in Yellowstone National Park (Wyoming).

caatinga (Portuguese) (i) A type of tropical woodland, growing in N.E. Brazil; XEROPHILOUS plants which can withstand long periods of aridity but can take advantage of seasonal rains. It forms a thick, virtually impenetrable mass of thorny jungle, from which rise occasional taller trees, such as giant cacti, wax-palms, acacias and euphorbias. For 7 months it is a grey, leafless, dead-looking tangle, but when the short rains occur the vegetation bursts into intensive life: trees develop brilliant blossom, herbs and bulbous plants flower. After 3 to 4 months, the vegetation reverts to its dormant state. (ii) Used in the Rio Negro valley, Brazil, for low evergreen forest, where there is no marked dry season.

cable The tenth part of a nautical MILE.

cacao A tree (*Theobroma cacao*) of tropical latitudes, on whose trunk grow large pods, containing 'beans'; these form the raw material of the cocoa and chocolate industry. C.p. = Ghana, Nigeria, Brazil, Ivory Coast, Cameroon, Ecuador.

cadastral From the Fr. word *cadastre*: a register of property, itself from the Lat. word *capitastrum*, a unit made for a Roman land-tax. Thus a *c. survey* is one in which property is surveyed. The word is also loosely used to describe large-scale mapping in which property divisions are shown.

Cainozoic (sometimes spelt **Cenozoic** or **Kainozoic**) Derived from Gk. 'recent life', indicating the 3rd of the main eras of geological time subsequent to the Pre-Cambrian (the last 70 million years), hence also called Tertiary, and the rock-groups deposited during that time. (Some authorities also include the Quaternary in the Cainozoic.) In A. Holmes' revised time-scale (adopted 1964), the Cainozoic is divided into 2 periods, TERTIARY and QUATERNARY; the former includes from the PALEOCENE to the PLIOCENE epochs (from 70 million to 2–3 million years ago), the latter the PLEISTOCENE and HOLOCENE epochs; i.e. the time-stages have been demoted 1 stage.

cairn, carn (Gaelic) Orig. a pile of stones raised as a monument; now a tall pile of stones on a mountain summit in N. England and Scotland. It forms part of the proper name of many mountains; e.g. Carn Dearg.

calamine An ore of ZINC, now commonly known in Britain as *smithsonite* ($ZnCO_3$).

calcareous Containing an appreciable proportion of calcium carbonate ($CaCO_3$); hence c. OOZE, c. TUFA (or travertine), c. rocks, c. soils. Derived from Latin, *calx*.

calcicole A plant which requires a soil rich in lime (adj. *calcicolous*); e.g. clematis, spindle-tree, rock-rose, traveller's joy, dogwood, viburnum, box.

calcification A soil-forming process in arid and semi-arid climates, so that calcium carbonate accumulates in the B-HORIZON.

calciphobe, calcifuge A lime-hating plant, restricted to soils with a marked acid character; e.g. azalea, rhododendron, *Calluna*, most *Ericaceae*, and other heath-plants. The two terms are virtually syn., thou calciphobe is stronger; a calcifuge c survive but not flourish in a chal soil, a calciphobe will quickly die.

calcite The crystalline form of c cium carbonate ($CaCO_3$), the ma constituent of all limestones, and common 'cement' in other coars grained sedimentary rocks. The c cium is derived from the weatheri of various igneous rocks, carried aw in solution as calcium bicarbon $Ca(HCO_3)_2$, and deposited as c. wh the bicarbonate decomposes. crystals have a hardness of 3 (on MOH SCALE), specific gravity 2·7, and a usually colourless unless stained impurities. C. is readily deposited the beds of streams, and as STALACTIT and STALAGMITES in caves. *Iceland sp* is a pure variety of c.

calcium carbonate $CaCO_3$, wi two crystalline forms, *calcite* a *aragonite*, and occurring commonly limestone, marble, chalk and coral. is insoluble in pure water, but d solves in water containing carb dioxide (e.g. rain-water) to form t readily soluble calcium bicarbona $Ca(HCO_3)_2$, which is present most surface waters of the ear When the calcium bicarbonate d composes, calcite is deposited temperatures below 30°C., aragon at those above.

calcrete A SEDIMENTARY depo derived from coarse fragments of oth rocks 'cemented' by calcium carbo ate.

caldera (also known as **basal wrec** The large shallow cavity which r mains: (i) when a PAROXYSMAL ERU TION removes the top of a former con (ii) as a result of ENGULFMENT. E. Askja in Iceland; Crater Lake, Or gon, lying in a c. nearly 10 km. (6 m across. There are numerous exampl in Alaska and the Aleutian I.'s, in Katmai, formed by an explosion 1912, when a peak rising to 2285 r (7496 ft.) was reduced to a c. 4 k (2·5 mi.) across, with a jagged summi rim now reaching to only 1370 r (4500 ft.). The largest c. is said to

Aso in Japan, 27 by 16 km. (17 by 10 mi.). [*f*]

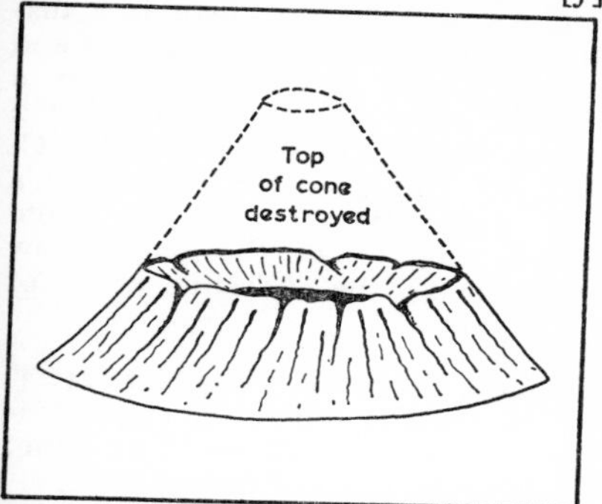

Caledonian First used by E. Suess (1888) to describe the ancient mountain ranges which can be traced through W. Scandinavia and Scotland into N.W. Ireland. These are sometimes called '*Caledonides*', the worn-down relics of the once lofty fold mountain systems. Folds of the same age are found in the S. Uplands of Scotland, the English Lake District and in N. Wales; in these cases the trend is more W. to E. The term C. is now mainly used to describe the period of mountain-building (late Silurian-Devonian) which led to the upfolding of these mountains, typically with a N.E. to S.W. trend in N.W. Europe.

calf (from the Norse *kalv*) (i) An islet adjacent to a larger island; e.g. the C. of Man. (ii) A fragment of floating ice; hence the term CALVING, to form an iceberg.

caliche (Sp.) (i) A crust-like accumulation of impure sodium nitrate ($NaNO_3$) in the soils of arid areas, esp. in the Atacama Desert of Chile and Peru, where it was first exploited in 1825. Its origin is not really understood; it may be formed by the leaching of bird guano, by the bacterial fixation of nitrogen, by leaching from volcanic tuffs, and by the drying-up of former shallow lakes. In the 19th century it was the world's chief source of fertilizers and of nitrogenous compounds; it is still important in the chemical industry, but nitrogen is now more easily obtained from coal and from the atmosphere. (ii) Applied in S.W. U.S.A. to a hard surface encrustation of calcium carbonate, as the result of solutions rising to the surface by CAPILLARITY, when the water evaporates; cf. DURICRUST.

calina (Sp.) A leaden dust-haze, occurring during summer in lands bordering the Mediterranean Sea.

calm Force 0 on the BEAUFORT WIND SCALE, with a velocity of less than 1 knot. A c. may occur in any latitude at any time, partic. under anticyclonic conditions. The chief areas of calms are within 5° latitude N. and S. of the equator in the DOLDRUMS, and in the 'HORSE LATITUDES'.

calorie The amount of heat-energy required to raise 1 c.c. (or 1 gm.) of water at sea-level pressure from 14·5°C. to 15·5°C. (technically known as the *15°C. calorie* or *gramme-calorie*).

calving The formation of an ICEBERG through the breaking-off of an ice-mass from the tongue of a glacier reaching the sea, or the edge of an ICE-BARRIER; e.g. the Ross Ice-Barrier, Antarctica.

camber An arching of rock strata caused by some superficial disturbance of the rock itself, rather than by earth-movements, esp. when it overlies plastic clay. Thus a cap-rock may be inclined towards the valleys, as in the Northamptonshire Ironstone fields and in the Cotswolds, where the oolites and sandstones are cambered over the Lias. [*f*]

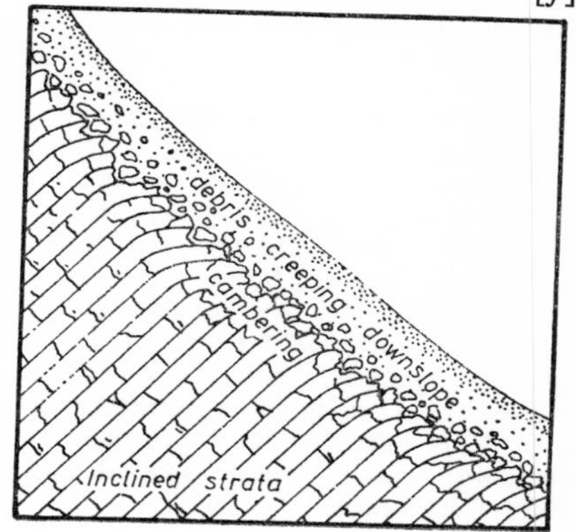

Cambrian The 1st of the geological periods of the Palaeozoic era, and the system of rocks laid down during that time, the first with clear evidence of abundant organic life, though only invertebrates. The name was given by A. Sedgwick in 1836 to rocks now divided into C. and Ordovician. The C. was gen. believed to have lasted from 500 to 410 million years ago, though recent research has dated its start at 600 ± 20 million years. Rocks of this age are found in Gt. Britain in N. Wales (e.g. the Harlech Series of grits and slates, Tremadoc Series, Llanberis Slates); S. Wales, the Welsh borders and the Midlands (quartzites, sandstones, limestones); and N.W. Scotland (white quartzite and limestone). There is sometimes a distinct break between the C. rocks and the earlier (PRE-CAMBRIAN) ones. The chief economic product is slate, quarried in N. Wales from Bethesda to Nantlle; some quartzites are used for road metal.

camera lucida A simple prismatic device for copying a map or diagram on to a sheet of paper.

Campbell-Stokes recorder A device for measuring and recording the duration of bright sunshine over a period of time. A fixed spherical lens burns an image of the sun on to a sensitized recording-card; as the sun moves, so does the position of the image, and a line will be burnt out by continuous sunlight.

campo (Portuguese) (i) SAVANNA country in central Brazil, probably not a natural CLIMAX vegetation, but the result of human interference. It is gen. woodland and parkland, which varies considerably in aspect. It covers an area in Brazil of some 2750 km. (1700 mi.) from N. to S. and 2400 km. (1500 mi.) from W. to E. It is usually classified as: (*a*) *c. cerrado* (with numerous scattered trees); (*b*) *c. limpo* (mostly treeless, with tall coarse grass). (ii) Locally in Argentina, the term c. is used for an intermont depression, usually trough-shaped and defined by faults.

canal (i) An artificial inland waterway, used for the cheap transport of bulky commodities. There is a distinction between small barge-c.'s and SHIP-C.'s. (ii) Occas. applied to a narrow piece of water connecting 2 larger stretches of sea; e.g. Lynn C. (Alaska). (iii) Used for irrigation c.'s and water supply c.'s. (iv) Note *canale*, plur. *canali*, the long narrow gulfs along the Adriatic coast of Yugoslavia.

Cancer, Tropic of The N. t., at 23° 32′ N. latitude, where the sun's midday rays are vertical about 21 June; the most N. point of the ECLIPTIC.

cane-brake A thick mass of reeds, growing in places up to 10 m. high, forming a dense swampy growth along the banks of the Mississippi, Red, Arkansas and other rivers.

cannel coal A dull, non-laminated coal, usually dark grey in colour. It has a high content of ash and volatile matter, is non-caking, and burns easily with a rather smoky yellow flame. It occurs in parts of the Lancashire coalfield, whence its name is derived from the dialect name for a candle.

canopy An almost continuous stra- tum of foliage formed by the crowns o tall trees, esp. as in the Tropical Rain- forest.

canyon (from Sp. **cañon**) (i) A deep, steep-sided gorge with a rive at the bottom; mainly found in ari or semi-arid areas, where a rapidl eroding river maintains its volum from snow-melt on distant mountains but where weathering is slight, s maintaining the steepness of the walls e.g. the Grand C., Zion C., C. d Chelly in S.W. U.S.A. The c. is mor pronounced if the land has bee slowly uplifted by earth-movement at more or less the same rate as th river has cut down, as in the case c the Colorado Plateau. (ii) A sub marine trough, a feature found be neath the sea on the continental she and beyond, which has all the appea ances of a steep-sided winding valley

. off the mouth of the Hudson R. l the Congo R.; the Fosse de Cap :ton off the Biscay coast of France. origin is uncertain, possibly the ult of: (*a*) faulting; (*b*) erosion by verful mud-laden submarine tur- ity currents or by springs bursting on the sea-floor; (*c*) former erosion rivers when the land was higher.

pacity, of a stream Introduced G. K. Gilbert in 1914 to indicate total load a s. can carry. Ct. MPETENCE, which is concerned with size of the largest particle carried. :h concepts were worked out in ory under laboratory conditions; it not always easy to apply them to ural s.'s, and they have met much icism.

pacity strain A state which is the ult of exceptional conditions (e.g. :r flooding, peak traffic), and which result requires physical expansion. is may well produce a 'shift' in the izontal dimension, and therefore a ocation (e.g. the river changes its nnel, the road is realigned or a ble-carriageway is constructed), ugh it will remain as near to the old ition as the area of INDUCED ANSION will allow.

pe A prominent headland or montory projecting into the sea; Cod, Finisterre, Comorin. 'The C.' cif. refers to the C. of Good Hope, ublic of S. Africa.

pe Doctor' A S.E. wind, blow- chiefly in summer, often very ongly, so called because it produces nulatingly fresh conditions in Cape wn, Republic of S. Africa.

pillarity The property of holding :er in the soil by surface tension as films around individual particles l in the capillary spaces (minute r-like tubes), above the height at ich it could be held by hydrostatic ssure. This water may be absorbed plant roots, or drawn to the sur- through the hair-tubes by evapor- n. The zone in which c. can cessfully act against gravity in respect of soil-water is known as the *capillary fringe.*

capital (i) The chief town of a country or province, normally the seat of government. It may also be used loosely as the 'commercial c.' (e.g. New York), and as the ecclesiastical c. (e.g. Canterbury). Amsterdam is the c. of the Netherlands, but not the seat of government, which is at The Hague. South Africa has 3 c.'s, in each of which an essential function of government is carried on. A federal c. (e.g. Canberra in Australia, Washington in U.S.A., Ottawa in Canada) may not be the most important city, except in that it is the seat of the federal government. (ii) One of the 3 elements in economic productivity: land, labour and c.; in this sense c. means the stock or source of money used to finance an enterprise.

capital goods Commodities (incl. primary raw materials, semi-manufactured and manufactured items, parts, machinery) which are to be used in the fabrication of other articles; ct. CONSUMER G.'s.

Capricorn, Tropic of The S. t., at 23° 32′ S. latitude, where the sun's midday rays are vertical about 21 Dec.; the most S. point of the ECLIPTIC.

cap-rock (i) A resistant stratum; it may be responsible for a waterfall (where a river spills over its edge, eroding the less resistant rock below; e.g. Niagara), or for a prominent hill, where the capping protects the rock below (see BUTTE). (ii) An impermeable stratum covering an AQUIFER, an oil-DOME, or a SALT-DOME. (iii) A mass of barren rock covering an ore-body. [*f* BUTTE]

capture, river The diversion of the headwaters of a river system into the basin of a neighbour with greater erosional activity which is flowing at a lower level; it is also known as *beheading, river piracy* and *abstraction.* The development of contiguous river systems must lead to one becoming more powerful; gradually it becomes the 'master-stream', more deeply entrenched, and so affording a lower BASE-LEVEL for its tributaries. These push back their watersheds with neighbouring streams by headward erosion and ultimately capture them, thus increasing their volume and that of the main river, and so making it still more powerful. At the point of diversion there is usually a marked bend, known as an *elbow of c.* Examples are widespread; e.g. t rivers of Northumberland and Yor shire; the c. of the upper Blackwat by the Way (Surray); of a form headstream of the Rhine by t Doubs; the many c.'s in the scar lands of the Paris Basin. 'It is normal incident in a veritable strugg for existence between rivers, and m occur in very varied circumstanc (S. W. Wooldridge). [

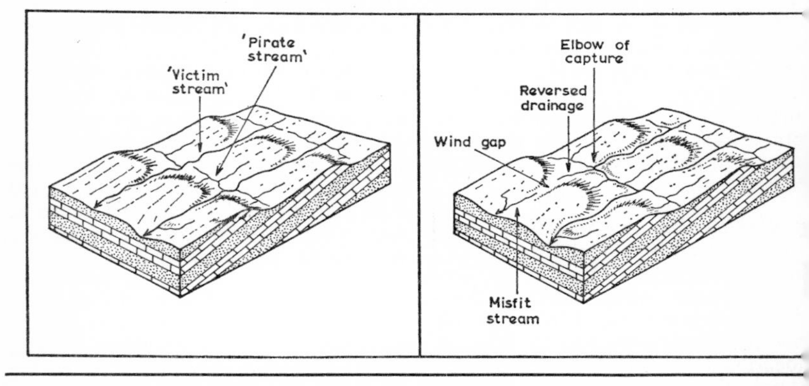

carat (i) A term indicating the degree of purity or 'fineness' of a GOLD alloy; 24 c. is pure gold, a sovereign is 22 c. (with 2 parts of copper), a fountain-pen nib is usually 14 c.; silver, palladium and nickel are also used as alloys. 'White gold' has a high proportion of nickel or palladium; 'rolled gold' is a thin film of 9 c. gold over brass. In continental Europe and U.S.A. the symbol *k* (from *karat*) is used; e.g. 22 *k.* (ii) A unit of measurement of the weight of diamonds and other precious stones. Formerly 1 c. = 0·2054 gm. The international metric c. has become standard for U.S.A., Gt. Britain, France, West Germany and the Netherlands = 0·20 gm. In Gt. Britain there are 142 c. to the ounce; a c. is divided into 100 points.

carbonaceous App. to sedimentary rocks containing an appreciable amount of organic material and its derivatives: peat, brown-coal, lignite, coal, c. shales. Carbon is widely distributed in the world, occurring in the free state as diamond and graphi more gen. in combination; e.g. co (avge. carbon content, 80%), petr leum (84%), timber (50%). It for only 0·2% of the earth's crust.

carbonation (c.-solution) A fo of chemical weathering of rocks rain-water containing carbon dioxi in solution (e.g. weak carbonic ac H_2CO_3), esp. on limestones and rocks containing other basic oxid The limestone is dissolved and r moved in solution in the form calcium bicarbonate. It is commo referred to simply, though not who accurately, as solution. It forms a ma agent of sculpture in limestone count both on the surface and undergrou

carbon dating, carbon 14 S RADIOCARBON DATING.

Carboniferous The 5th of t geological periods of the Palaeoz era, and the system of rocks laid do during that time (C. Limesto Millstone Grit, Coal Measures). Ge logists divide the C. into the *Avon*

.ower C.), *Namurian*, *Westphalian* ıd *Stephanian* (youngest). It occurred om *c.* 350 to 270 million years ago. ı U.S.A., the C. is divided into two: ıe lower or MISSISSIPPIAN, the upper : PENNSYLVANIAN.

arboniferous Limestone (i) :ratigraphically, the lowest of the rmations of the C. system, including so sandstones and shales (e.g. Yoreıle Shales), corresponding approx. to ıe Mississippian in N. America. In ngland it is divided into the Lower ıd Upper *Avonian*, corresponding in '. Europe to the *Tournaisian* and *iséan.* (ii) Lithologically, the most ıportant rock in the Lower C., also ıown as *Mountain Limestone*, consting of a hard, grey, crystalline, ell-jointed limestone, often markedly ssiliferous (containing crinoids, rals and brachiopods). It occurs in .W. Yorkshire, the Peak District, e Mendips, S. Wales, N. Wales, the ɔrest of Dean, Devonshire, on the W. and S. margins of the English ake District, in the Midland Valley Scotland, and over much of Irend. Many areas of C. L. form ateaus (e.g. the N. and S. Pennines, endips), with a marked absence of rface drainage, SWALLOW-HOLES, tricate cave-systems, dry valleys and rges, and pavements of bare rock ith CLINTS and GRIKES. Cf. KARST.

arbo-Permian The end of the arboniferous and beginning of the ermian periods; sometimes used to note the HERCYNIAN phase of ountain-building.

ırdinal points The 4 major points the compass: N., S., E. and W.

ırr A FEN containing, apart from eds (esp. *Carex paniculata*) and her aquatic plants, shrubs such as ler (which is dominant), sallow, ier and willow. It occurs on watergged terrain which is neither too id nor too poor in mineral elements; ;. in E. Norfolk, parts of Lincolnire. The term is cognate with the lancdic *kjörr* and the Swedish *kärr*.

carse (Scottish) Fertile alluvial lands along the estuary of a river; e.g. the C. of Gowrie.

cartogram A map on which statistical information is presented in diagrammatic form.

cartography In its widest sense, the whole series of processes of map-making, from an actual survey of the ground to printing the map; in a more limited sense, the drawing of a map.

cartouche A panel on a map, usually decorative, enclosing its title, scale and other information. Elaborately drawn c.'s were the delight of the Italian Renaissance, the later Flemish and Dutch cartographers, and the map-makers of Elizabethan England.

cascade A stepped series of small waterfalls, often of an artificial ornamental character.

Cascadian The name given in U.S.A. to the mountain-building movements at the end of the Tertiary period, *c.* 1 million years ago, when the Cascade Mtns. were uplifted, accompanied by widespread VULCANICITY, and the FAULT-BLOCK of the Sierra Nevada was also bodily uplifted and tilted. Following the 'C. Revolution' (the term *C. orogeny* is sometimes applied, though this implies folding which did not occur), there was a period of quiescence in the early Quaternary, followed by renewed volcanic activity when the present great Cascade peaks were built (Mt. Baker, Rainier, St. Helens, Adams, Hood, Jefferson, Washington, Shasta and Lassen Peak).

cash crop A c. grown for sale for consumption by others (ct. SUBSISTENCE C., consumed by the grower); e.g. cocoa in Ghana, rubber in Malaysia, sisal in Tanzania, large-scale cereal-farming in Canada, market-gardening.

cassiterite The main ore of TIN as an oxide (*SnO_2*).

caste A hereditary social group among the Hindus, into which a person is born and remains throughout his life. Once very rigid, numbering about 2400 different c.'s, the system is now less so, though it still forms a problem in the modern development of India.

castellated iceberg An i. with a towering pinnacled superstructure above the water, in contrast to horizontal or tabular masses.

castellatus (castellanus) clouds A mass of c.'s of turret shape when seen from the side.

cataclinal Applied by J. W. Powell (1875) to streams which flow in the direction of the DIP. A term now little used (see ANACLINAL and DIACLINAL).

cataclysmal Formerly applied to the processes and results of exceptional deluges, earthquakes and catastrophes in gen., which the *catastrophists* believed to be responsible for most of the relief features of the world, in ct. to the *uniformitarians* who did not.

cataract Orig. denoted a large waterfall; now it is used for a series of rapids, as on the Nile, with 5 main c.'s (numbered upstream from the 1st near Aswan to the 5th above the Atbara confluence), and several minor named ones. The Nile has cut its way down vertically through the Nubian Sandstone until in places it has reached ancient crystalline rocks. These offer greater resistance to erosion, so forming complicated rapids, divided channels and broken water.

catastrophism The concept that the earth's features are the result of sndden major catastrophic events, not of slow continuous processes operating with gradual inevitability. Ct. UNIFORMITARIANISM. Certain catastrophes may indeed occur (e.g. floods, eruptions, earthquakes), but these are essentially temporary and local.

catch crop A quick-growing c., grown between 2 main c.'s in a rotation, or between the rows of a main c., or in place of a failed c. ' principle is thus to 'snatch' a quic on land that would otherwise be te orarily unproductive. A market- dener finds this a valuable metho increasing his monetary yield per a

catchment (i) The area drainec a single river; a natural drainage which may coincide with a river ba in which the DIVIDES direct w derived from rainfall and PERCC TION into a river. However, wh underground flow is involved, th area may be larger or smaller than which may be apparent from sur relief. *C. Board:* a statutory autho in Gt. Britain, responsible for dr age and flood control within drainage basin of a river (e.g. Ous Board), though the boundaries some C. Boards extend beyon single river basin. *Note:* In U.S.A. term WATERSHED is applied to a c. The intake area of a single AQUIFE

catena A sequence of different which varies with relief and drain though normally derived from the s parent-material. Such a sequence be seen when following a line of file from a hill-top to a valley-bott

'cat's paw' A term used by yachts and others for a light breeze or pu wind affecting only a small area of w

cauldron subsidence The fou ering and collapse of a block of cou rock into the underlying MAC Glencoe in W. Scotland is an exar of a BATHOLITH mass which has lapsed, demarcated by *ring-faults* RING-DYKES.

causality An attempt to explain operation or relation of cause effect between 2 objects or eve these causal relationships (e.g. tween Man and his ENVIRONMENT) vital aspect of geography. See CH LOGY.

causse(s) (Fr.) Sometimes usec limestone country in general, sp for the Grands Causses in the S of the Central Massif of Fra Derived from a patois word meaning *chaux*, lime.

cave, cavern (usually syn., though the latter is sometimes regarded as larger) A subterranean chamber, usually natural, with an entrance from the surface. It is found partic. at the base of sea-cliffs where wave action has enlarged natural lines of weakness (*sea-c.*'s) (see *f* STACK) and in limestone country as a result of CARBONATION-SOLUTION, partic. along JOINT-PLANES. The term is used also as: (i) *caving*, to explore caves; (ii) to collapse (*c.-in*); (iii) adjectivally (*c.-art, c.-man*). The study and exploration of c.'s is known as *speleology*. The term c. is occasionally used for artificial hollows; e.g. the Chislehurst C.'s, the champagne c.'s near Reims, c.'s near Maastricht cut for lime-working, and the Tilly Whim C.'s near Swanage, Dorset, produced by quarrying.

cavitation The wearing of rocks by running water, when bubbles of air and water-vapour (formed by a rapid increase in the velocity of a stream) collapse, causing minute shock-waves against the bed and banks. This is partic. common in PLUNGE-POOLS and in TURBIDITY currents.

cay (key, kay) A low island of sand and CORAL fragments, built up by waves on a reef-flat, at or just above high tide, drying at low water; e.g. Florida Keys (terminated by Key West), Marquesas Keys, Grand C.'s Bahamas), the West Indies gen., Cayo Grande (off the coast of Venezuela). Similar features occur in Indonesia and on the Great Barrier Reef of Australia.

cedar-tree laccolith An intrusion consisting of a series of LACCOLITHS one above another, fed from a single vent or pipe from the MAGMA reservoir below.

'ceiling' (i) Some specif. level in the atmosphere, notably the lowest substantial layer of cloud; e.g. 'c. zero' means fog at ground level. (ii) The height to which a particular aircraft or balloon can climb. (iii) A physiological limit to which a man can climb without oxygen; this varies with individuals.

Celestial Sphere The imaginary 'bowl' of the heavens, representing a sphere with the earth at the centre, on the 'inner surface' of which the heavenly bodies appear to be placed. Distances are so vast (the nearest star is 4·29 light-years away) that the radius of the C. S. may be regarded as infinite. The plane of the earth's equator produced cuts the C. S. in the *C. Equator*, and the earth's axis produced meets the C. S. at the *C. N. Pole* and *C. S. Pole* respectively; the ZENITH is a point on the C. S. directly above the observer. The C. S. is used as the basis for many astronomical and navigational problems, solved by spherical trigonometry. [*f* RIGHT ASCENSION]

cell, atmospheric A large area of predominantly high or low a. pressure resulting from the interruption of the planetary pressure system, mainly because of unequal solar heating caused by the irregular distribution of continents and oceans. In the N. winter the Eurasiatic (Siberian) and American high pressure c.'s dominate the N. continents, while the Icelandic (N. Atlantic) and Aleutian (N. Pacific) c.'s are over the oceans. In the N. Summer the N. Atlantic and the N. Pacific are dominated by the Azorean and Hawaiian 'highs', with 'lows' over N. America and S. Asia. A c. is 3-dimensional, and the height to which the air-mass extends is of fundamental importance.

Celsius scale The internationally accepted name for the CENTIGRADE s. of temperature. A. Celsius, a Swedish astronomer (1701–44), developed a thermometer, with the division of the interval between the freezing and boiling points of water into 100, first described in a paper to the Swedish Academy of Science in 1742. Actually he used 0 for boiling and 100 for freezing point, but this was reversed in 1743. The name C. officially replaced Centigrade in 1948, but the latter is still widely used. For a conversion table from °C. to °F., see under FAHRENHEIT.

Celtic field A roughly square f. in use before the Saxons introduced the strip-f. system. It can be traced as a system of banks or earth-markings over wide areas, esp. on chalk down-land when viewed from the air. The term is loosely applied on some maps to the physical traces of any ancient f.-system.

'cement' Siliceous, calcareous or ferruginous material, deposited from circulating water, which has converted loose deposits, such as sand and gravel, into a hard compact rock. The nature of the c. has a marked influence on the subsequent weathering of the rock. A siliceous c. usually produces a hard, resistant rock; e.g. quartzite.

Centigrade scale A graduated s. of temperature on which 0° represents the melting point of ice, 100° the boiling point of water, at sea-level.

$$C=\frac{5}{9}(F-32), \text{ or } \frac{C}{100}=\frac{F-32}{180}.$$

Officially known as CELSIUS since 1948. For a conversion table from °C. to °F., see under FAHRENHEIT.

centimetre A unit of metric measurement. 1 cm. = 0·01 m. = 0·3937 in.; 1 in. = 2·54 cm.; 1 sq. cm. = 0·155 sq. in.; 1 cu. cm. = 0·06103 cu. in.; 1000 cu. cm. = 1 litre = 0·22 Imp. gallon.

centography Used by T. M. Poulson in 1959, involving the determination of the centres of distribution of any phenomenon (e.g. population), and their cartographical plotting.

Central Business District (C.B.D.) An American term indicating the heart of a city; commonly referred to as 'downtown'. Now it is used frequently in geographical literature for the focus of a town's activities. It can be defined by applying the C.B. Index Method, devised by R. E. Murphy and J. E. Vance.

central eruption An eruptive form of volcanic activity which proceeds from a single vent or a group of closely related vents, in ct. to a linear or FISSURE E. The product of a c. e. is a cone of some kind.

central good, service, function Any g. that is sold, or a s. or f. that is performed, at any C. PLACE. The term usually refers to the TERTIARY element of activity, excluding PRIMARY and SECONDARY productivity.

centrality The rel. importance of a place with regard to the region surrounding it, or the degree to which such a place exercises CENTRAL FUNCTIONS. The term was orig. used by W. Christaller in relation to the 'surplus importance' of a place, this place being defined as a town. Since these central functions are also 'central' to the town then restriction of the term to the area beyond the built-up area means that any comparison of c.'s will be biased by the size of the town.

centrality, index of Devised by W. Christaller in connection with settlement studies, based on telephone services: $C=(t-\frac{pT}{P})$, where C = index of centrality, t = number of telephones in a town, p = population of town, T = number of telephones in whole region, P = population of whole region. Sven Godlund, in his study of Swedish towns, suggested a formula based on the relationship between the total population of a town (F) and the number of people actively occupied in retail trade and services (D). Index = $\frac{D}{F}\times 100$. In a regional centre the index will exceed 6·5 (e.g. Leeds, 9·

centralization In an ecological sense, the drawing together of institutions and activities into an area of urban CONCENTRATION. By functioning in relation to the whole area, the institutions and activities are said to be centralized at their location. The opposite situation, the break-down of c., is referred to as *de-c.*

central place Any location providing goods and/or services for a surrounding tributary area. The c. p. possesses CENTRALITY, not because of any intrinsic qualities, but because of the localization of the functions. A c. p. is usually regarded as syn. with a to but in fact urban areas can possess several c.p.'s as distribution points, serving each respective adjacent area, whether the service is restricted to the actual built-up area, or to the country around, or to both.

central place hierarchy C.P. theory predicts that c.p.'s fall into a h., or series of discrete classes. Each class of p. performs all the functions of lower order centres, as well as a group of c. functions that differentiates it from, and sets it above, the lower-order p.'s.

central place system In a geographical sense, the c. p. s. refers to the spatial distribution of any set of C.P.'s. The pattern of this distribution is usually known as the 'network of c. p.'s'.

central tendency A form of average in a group of statistical data, expressed as (*a*) a *mean* (the arithmetic average of the sum of all the values); (*b*) a *median* (the actual middle value of a whole set of values if an odd series, or the mean of the two central values if an even series); and (*c*) the *mode*, in grouped data the class containing most values.

centripetal drainage A d. pattern in which numerous rivers converge from many directions on to a main stream. E.g. in the Katmandu valley of Nepal, many streams converge upon the Bagmati R., which drains S. through a gorge cutting across the surrounding mountain ranges. In extreme cases a basin of INLAND D. is formed.

centrocline See PERICLINE.

centrogram A diagram on which can be graphed regional trends in the distribution of population. A number of points which coincide as nearly as possible with the successive centres of gravity, or with the median points, of the population of a particular country as enumerated in successive census returns, are plotted on a map. These points are then joined by a line, which reflects any tendency of the centre of gravity, or median point, to shift in time.

centrosphere Syn. with BARYSPHERE.

cephalic index Used in anthropology to determine exactly the skull shape: max. breadth of head divided by max. length × 100. A skull of c. i. 75 or less is *dolichocephalic* (long-headed); of 83·1 or more is *brachycephalic* (broad-headed); in between it is *mesocephalic*.

cereal Applied to a species of grass grown for its grain, including wheat, barley, oats, rye and maize (corn) (temperate), and rice, millet and sorghum (tropical). They supply man's chief foodstuff.

cerrado (Portuguese) A type of SAVANNA in Brazil, a mixture of low contorted trees, 4 to 7 m. (12–24 ft.) high, and tall grass, with closely spaced tangled growth. The *cerradao* has similar species, but the trees are taller, 9 to 15 m. (30–50 ft.).

chain (i) Any linear sequence of related physical features; e.g. c. of lakes, islands, reefs. It is used esp. of mountains where there is a complex series of more or less parallel or *en échelon* ranges, as distinct from a single range. (ii) A unit of measurement = 66 ft. = 4 poles, derived from E. Gunter's surveying c.; 10 sq. c. = 1 acre. 1 c. = 100 links, each 7·92 in. (obsolete).

chalk Lithologically, a soft, amorphous whitish limestone, consisting almost entirely (*c.* 97%) of calcium carbonate ($CaCO_3$), up to 600 m. thick in England. Formerly it was thought to be wholly an organic deposit, consisting of the tests of foraminifera, coccoliths and other marine micro-organisms in a matrix of finely divided calcite; some writers even believed it to be a fossil abyssal ooze or calcareous mud, including the foraminifera *Globigerina*. Its origin is still problematical, but it is clear that while some chalk is undoubtedly of organic origin (laid down in shallow water, as indicated by the large shells of shallow-water life), with little terrigenous matter (possibly the result of low-lying adjacent land where little erosion was in progress), other c. may be due in part to chemical precipitation. It often contains nodules of FLINT, usually in bands. It forms characteristic relief features: rolling hills and undulating plateaus, with open expanses of downland, covered with short turf growing on a thin layer of soil, now mainly cultivated (esp. with cereals). Where the strata are tilted, a CUESTA may be formed. Surface drainage is slight, because of the numerous close joints, and DRY VALLEYS are common. C. occurs widespread in E. and S.E. England, incl. the Yorkshire Wolds, Lincoln Wolds, E. Anglian Heights, Chiltern Hills, Salisbury Plain, Dorset Downs, N. and S. Downs. In France c. outcrops extensively in the Paris Basin (Picardy, Artois, '*Champagne pouilleuse*').

Chalk, The A stratigraphical name, applied to the upper beds of the Upper Cretaceous system (which is usually considered also to include the Gault and the Upper Greensand). In S. England, this sequence can be recognized:

Upper Chalk	White chalk, with flints Chalk Rock
Middle Chalk	Soft white chalk, with some beds of marl, a few flints
	Melbourn Rock
Lower Chalk	Grey chalk Chalk-marl

Chalk-marl A stratigraphical horizon near the base of the Chalk, consisting of greyish calcareous material containing up to 30% argillaceous (clay) material; e.g. near Cambridge. See MARL. *Note:* If the rock is referred to, other than the horizon, the term used is chalky-marl.

Chalky Boulder-clay A type of BOULDER-CLAY found especially in East Anglia, greyish in colour, containing fragments of chalk and numerous FLINTS. Formerly it was classified into the Great C. B.-c. (now renamed the Lowestoft Till) and the Little C. B.-c. (now the Gipping Till). It is tentatively regarded as the product of the 2nd and 3rd glacial periods.

chalybeate An adj. describing water containing hydrated iron compounds, which may appear on the surface as a c. spring, of alleged medicinal value; e.g. at Harrogate.

chalybite Natural ferrous carbonate ($FeCO_3$), in its bedded form (SIDERITE) a major source of iron; e.g. in the Coal Measures and Jurassic limestones of England.

champagne (champaign, champain, campagne) (Fr.) An area of open hedgeless plains, as in N. France (ct. BOCAGE). Derived from *campania*, a Latin word meaning 'plain'. It was used as a proper name for a former French duchy, and now for the PAYS of an open, gently undulating landscape, incl. *C. pouilleuse* (the chalk-country to the E. of the Ile de France) and *C. humide* (the Gault Clay further E. still). It is also used for other smaller pays; e.g. *C. berrichonne, C. charentaise*). The name has been transferred to the sparkling wine produced in the neighbourhood of Reims.

chañaral (Sp.) An area of thorny scrub, largely composed of *chañar* bushes, with large thorns; found in N Argentina and central Chile.

channel (i) The deepest part of a river-bed, containing its main current naturally shaped by the force of water flowing in it. It can be measured in terms of: (*a*) depth, from the surface of water to the bed; (*b*) width; (*c*) cross-sectional area; (*d*) wetted perimeter (the length of the line of contact between the water and the bed); (*e*) HYDRAULIC RADIUS; (*f*) FORM RATIO; (*g*) GRADIENT or slope. (ii) A funnel-shaped estuary; e.g. the Bristol C. (iii) An irrigation ditch. (iv) A stretch of sea (wider than a strait) between 2 land-areas, linking more extensive seas; e.g. the English C., St. George's C. (v) The main shipping-lane or fairway, usually dredged, within a wide estuary, c between shoals. (vi) Used in urban geography to designate lines of movement: roads, canals, railways, sewers, gas- and water-mains, electricity and telephone cables etc., passages and lifts in buildings. An air-lane is an '*unbounded channel*'.

channel flow The RUNOFF of surface water in a more or less narrowly defined trough between banks, rather than spread out laterally over a wide area (SHEETFLOOD or OVERLAND FLOW).

channel storage The hold-up of flood-water within a section of river-c. which is receiving inflow more rapidly than can be passed on downstream. The max. storage before flooding occurs is *c. capacity*, or BANKFULL stage.

chapada (Portuguese) A tableland in Brazil, formed of an extensive, rel. horizontal, sheet of sedimentary rocks, such as sandstone, lying over the crystalline basement rocks of the Brazilian Plateau.

chaparral (Sp.) A type of evergreen scrub vegetation in California and N.W. Mexico, similar to the MAQUIS in S. Europe, the result of a long period of summer drought. It consists of tough, broad-leaf, evergreen scrub-oak, interlaced with vines and with patches of scanty grass.

Charnian An OROGENY which took place in late Pre-Cambrian times, the worn-down remnants of which can be seen in the Midlands of England (Charnwood Forest, the Wrekin, Malvern Hills, Caer Caradoc).

chart (i) A map of the sea and the coastline for the use of navigators, prepared by the Hydrographic Office of the Admiralty for British use, usually drawn on a MERCATOR PROJECTION, because this shows constant bearings as straight lines. (ii) A map prepared specially for aviation use, emphasizing aeronautical information (landing-fields, pylons, high towers, peaks and other dangers). (iii) A weather map, showing atmospheric pressure, winds and other information; e.g. Daily Weather C.

chart datum The plane from which soundings on a navigational chart are computed. The British Admiralty uses low-water springs; i.e. the worst possible navigational state of depth for a particular chart. A Port Authority (e.g. Southampton Harbour Board, Mersey Docks and Harbour Board) has its own c. d., so fixed as to avoid minus quantities in tidal data. C. d. at Southampton was 2·28 m. (7·48 ft.) below ORDNANCE DATUM until 1 Jan., 1965, when it was lowered to 2·74 m. (8·98 ft.) below O.D.

chase Orig. an area of unenclosed land used for hunting, similar to a FOREST (iii), but not necessarily a royal preserve. Now found commonly in place-names; e.g. Bringewood C. (near Ludlow, Shropshire), Cannock C., Chevy C.

chatter marks, chattermarks (i) A kind of 'bruising' or 'scarring' (as distinct from scratches), curved or crescentic in pattern, on the rock-floor formerly beneath a glacier. These were probably caused by a vibratory 'knocking' of loosely embedded boulders in the GROUND-MORAINE carried at the base of the glacier. (ii) Sometimes the term is applied to marks on wave-worn pebbles.

cheesewring The proper name of a striking granite TOR on Bodmin Moor (Cornwall), with a narrow stem and overhanging upper block. The term in gen. is now used by some writers to denote any similar features, including a GARA (plur. *gour*) in hot deserts.

chemical weathering See CORROSION.

chernozem (Russian) A 'black-earth', of loose crumbly texture, rich in humus and bases, covering large areas of land in temperate latitudes where the prevailing natural vegetation was grassland. The A-HORIZON is a black layer, 0·6 to 1 m. thick, grading into the B-HORIZON, brown or yellow-brown, in which colloids and bases have accumulated. C. extends across central Asia from Manchuria, through

S. Siberia and central U.S.S.R. into the Ukraine, and into Roumania and Hungary. Similar soils are found in central Canada, and in U.S.A. from N. Dakota to Texas. Parts of the Argentine Pampas and the S.E. Australian 'downs' have similar characteristics. These soils afford a striking example of an origin due to climate (warm summers, cold winters, some early summer rainfall but considerable periods of drought). Their texture and richness in plant-foods, together with the wide extent of gently rolling land and a climate favouring annual grasses, make them good cereal-growing soils.

chert A layer of irregular CONCRETIONS of an amorphous rock composed of hardened chalcedonic or opaline silica, which splinters easily, occurring in calcareous formations other than chalk; e.g. the Portland Stone Cherty Series.

chestnut soil A zonal s., characteristic of areas of sparse dry steppe with 200–250 mm. (8–10 ins.) of rain; it is loose, friable and dark brown in colour. It is actually a variety of CHERNOZEM, modified by greater aridity. It is little leached because of the low rainfall. The A-HORIZON is dark brown, becoming paler with depth. It has a wide distribution, covering the drier STEPPES to the S. of the chernozem in Russia, extending W. into Roumania and Hungary, in the High Plains of U.S.A., and in the drier parts of the Argentine PAMPA and the S. African VELD.

chevron crevasse A cr. near the margins of a glacier which has been rotated or twisted into the pattern of a chevron as a result of glacier motion.

chili A type of SIROCCO wind, experienced in Tunisia, dry and very warm.

chimney (i) A steep vertical cleft in a cliff, wider than a crack; e.g. Kern Knotts C. on Great Gable, English Lake District. (ii) Sometimes applied to the vent of a volcano. (iii) In U.S.A., a vertical pillar of rock.

china-clay See KAOLIN.

china-stone Granite which has been partially kaolinized and which is free from dark minerals. It is a hard rock which requires grinding for KAOLIN extraction. The term is also used for other whitish porcelainous rocks; e.g. those found in the Ordovician rocks of the Welsh Borderland.

chine A narrow cleft in soft earthy cliffs, through which a stream descends to the sea, esp. in the I. of Wight and Hampshire. It usually consists of a steep-sided 'inner' valley, contained within a wider 'outer' valley. Some were probably formed by small tributaries flowing to the former 'Solent River' system, while the inner valley was cut under present conditions. Marine erosion is cutting back the cliffs, and the small streams in the inner valley cut down more rapidly; e.g. Branksome C. (near Bournemouth), Blackgang C. (I. of Wight).

'Chinese wall' glacier The edge of an ice-sheet which reaches the sea as a vertical or even overhanging ice-cliff; e.g. along the coast of Greenland.

chinook A dry warm S.W. wind blowing down the E. slopes of the Rockies in Alberta, W. Saskatchewan and Montana, warmed ADIABATICALLY; in spring it causes a swift rise of temperature and rapidly melts snow (Indian name = 'snow-eater') Cf. FÖHN.

chi-square (χ^2) test A statistical significance test used to compare 2 frequency distributions. It can be used, e.g., to assess the probability that 2 distributions belong to a common parent population by examining the probability of differences between the 2 distributions being due simply to random (sampling) errors.

C-horizon In soil science, the layer of relatively little-altered material underlying the soil proper; the parent material from which the soil has been derived by the various soil-forming processes, loosely called the 'sub-soil'. [*f* SOIL PROFILE

chorochromatic map A m. in which broad distributions (e.g. lar use, soil, geological, agricultural, industrial) are shown non-quantitatively by areas of colour or tinting. It is known as a 'colour-patch' m. in U.S.A.

chorography A term widely used in the 17th and 18th centuries; e.g. by Bernard Varen, who implied thereby 'special' or 'regional' geography, as opposed to 'general' or 'systematic' geography. It is used today by some writers to denote: (i) the delineation or description of individual regions; (ii) a broad account of a large regional area, as distinct from the more detailed description of a small area (TOPOGRAPHY). Ct. CHOROLOGY. Some American writers refer to *chorographic* as a term relating to an area on a sub-continental scale, and to a c. map as one between 1:500,000 and 1:5,000,000 scale.

chorology The explanatory study of the causal connections and relations of phenomena within a particular region; 'the science of regions', as used by F. Marthe and F. von Richthofen in Germany. A. Hettner had a wider concept; he defined 3 groups of sciences: (i) *systematic* or material; (ii) *chronological* or historical; (iii) *chorological* or spatial, of which there are two branches: (*a*) the arrangement of things in the universe (astronomy), and (*b*) the arrangement of things on the surface of the earth (geography).

chorometrics The quantitative or statistical study of spatial distributions.

choromorphographic map A type of m. which delimits and classifies areas of land (Gk. *khora*) according to their surface configuration (Gk. *morphe*). G. M. Lewis's c.m. of W.-central U.S.A. shows 12 types; e.g. 'flat sand plains', 'closely dissected low plains', 'sand hills'.

choropleth (from Gk. *choros*, a place, and *plethos*, a measure) App. to maps drawn on a quantitative areal basis, calculated as average values per unit of area within specif. units, usually administrative (since statistics are conveniently available on this basis); e.g. the density of population per sq. mi., the percentage of land under cultivation, the yield per

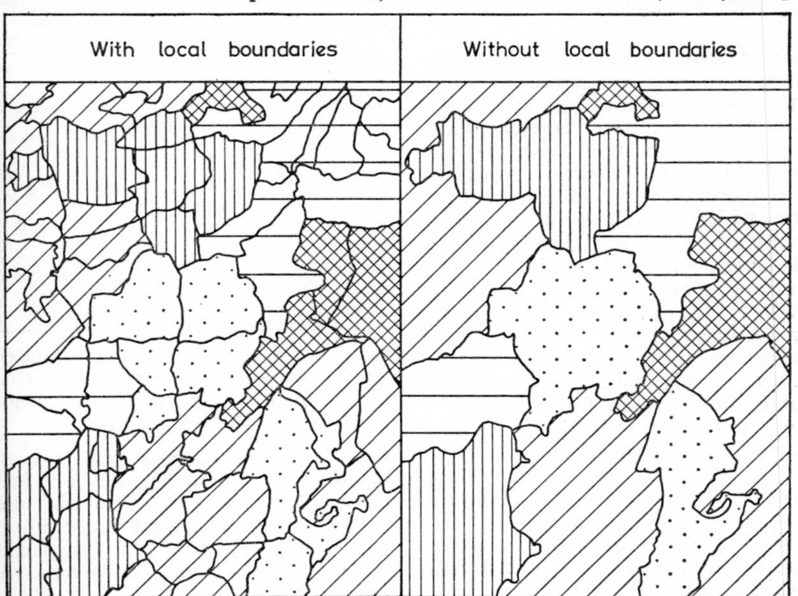

acre of arable land, calculated on a basis of parishes, *communes*, counties, *départements*, provinces, states or countries. It is also possible to produce c. maps by dividing an area into squares or hexagons, and finding a mean value for each; see also DASYMETRIC TECHNIQUE. A range of shading, stippling or colouring may be applied to show successive orders of density. These c. maps reveal only average values over a unit area and thus may mask a wide range of local variations. [*f*, *page 67*]

chott An alt. form of SHOTT.

chromium A hard metal derived from ferrous chromite ($FeCr_2O_4$), used as an alloy to produce 'stainless steel', and as a pure metal in 'electro-plate'. C.p. = S. Africa, Turkey, Rhodesia, Philippines.

chrono-isopleth diagram A graph in which the hourly values of pressure, temperature, etc. are plotted as abscissae and their times of occurrence in the month as ordinates. Similar values are joined by isopleths.

cinder cone A co. around a volcanic vent, composed exclusively of small fragmentary material, sometimes defined as 3–5 mm. in diameter. When lava is highly charged with gas, the rapid release of pressure causes explosive eruptions, which fragment the lava which solidifies while flying through the air. *Note:* The term ci. is a misnomer, originating when it was believed that it actually was the product of combustion. E.g. Sunset Crater, Arizona, a co. 300 m. (1000 ft.) high rising from the plateau at 2100 m. (7000 ft.), with an unbreached crater 120 m. (400 ft.) deep and a about 400 m. in diameter. It was the result of an eruption, *c.* A.D. 1064. Other examples are Cinder Cone (proper name) in Lassen Volcanic National Park (California); Parícutin n Mexico, which first erupted in 1943 and formed a 450 m. (1500 ft.) cone within a year.

circumference, of earth Along the equator = 40,076 km. (24,902 mi.); Polar c., 40,008 km. (24,860 mi.).

cirque (Fr.) A steep-walled rock-basin of glacial origin (also *corrie*, *coire*, *combe*, *cwm*, *kar*). A shallow pre-glacial hollow has been progressively enlarged, first by alternate thaw-freeze of a snow-patch within it, which causes the rocks to disintegrate (NIVATION). Melt-water helps to move the resulting debris. As the snow-patch grows, it develops into a FIRN-field, from which may issue a small C.-GLACIER. Freeze-thaw eats both into the back-wall of the c. (BASAL SAPPING), thus maintaining its steepness, and also into the floor, thus maintaining and emphasizing its basin shape; it also provides debris which freezes into the base of the ice and so acts as an abrasive. The out-moving ice in the c. appears to pivot about a central point (ROTATIONAL SLIP), which also emphasizes the basin shape. It is a common feature in glaciated mountain ranges, often containing a small lake. It varies from a tiny rock basin in N. Wales and the English Lake District to the Walcott C. in Antarctica with a BACK-WALL 3,000 m. high [*f*]

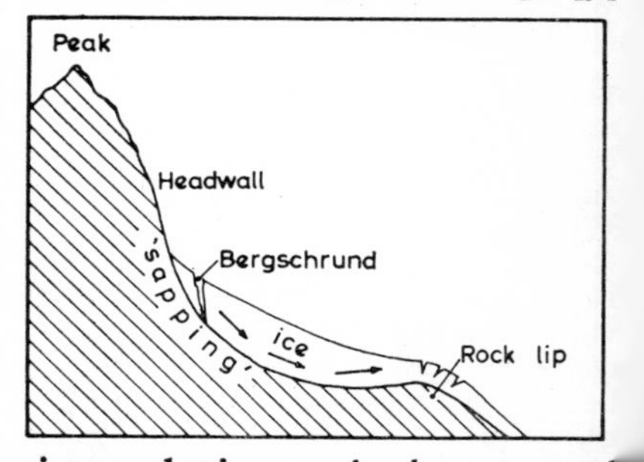

cirque-glacier A short-tongued glacier which barely protrudes from the basin in which the FIRN accumulates; e.g. in the Sierra Nevada.

cirrocumulus (*Cc*) A type of high cloud, usually consisting of ice-crystals, in lines of small globular masses with a rippled appearance with blue sky between ('mackerel sky'). It usually occurs above 6000 m (20,000 ft.).

cirrostratus (*Cs*) A uniform milky layer or veil of high sheet-cloud, above 6000 m. (20,000 ft.), through which the sun often shines with a distinct HALO. It affords a common indication of the approach of a WARM FRONT. It may thicken and develop into ALTOSTRATUS.

cirrus (*Ci*) A high-flying (6–12,000 m., 20–40,000 ft.), delicate, fibrous, wispy cloud, consisting of tiny ice-spicules, so light that it hardly interferes with sunlight or even moonlight. It is often a fair-weather cloud, but if it thickens to form CIRROSTRATUS it may be an indication of an approaching DEPRESSION. When the c. is drawn out as 'mare's tails' or 'stringers', it indicates strong winds in the upper atmosphere. Adj. *cirriform*.

citrus fruit A category of usually evergreen trees of the genus *Citrus*, including the lemon (orig. in India), orange (from China), and grape-fruit, lime, citron, tangerine. Now grown widely in lands with Mediterranean climates (Italy, S. Spain, Greece, Israel, California, S. Africa, S. Australia) and in subtropical or warm temperate climates (Florida, Texas, Mexico, S. China).

city (i) A loosely used term, implying a very large town. (ii) An old-established town, with a cathedral and bishop, not necessarily large; e.g. Winchester. (iii) In England and the Commonwealth, a title created by royal charter or letters patent, but with no specif. privileges other than ceremonial. Most, though not all, c.'s have a Lord Mayor, in contrast to the Mayor of a BOROUGH. (iv) In some states of the U.S.A., a collective body of people with a Mayor; an incorporated town; some c.'s in this sense have a population as low as 1500, though in some states 10,000 people are required.

city region The area round a large c. that is functionally bound to, and dominated by, this c.

civil day A mean solar d. of 24 hours, reckoned from midnight to midnight. See MEAN SOLAR TIME, EQUATION OF TIME, SOLAR DAY.

civil twilight See TWILIGHT.

clachan Derived from a Gaelic word meaning a circle of stones, now used in the Highlands of Scotland and N. Ireland for a small settlement or hamlet without any obvious plan.

clarain structure, in coal Very thin, finely banded structure in coal, usually parallel to the BEDDING-PLANE. Though lustrous, the bands are less glassy than VITRAIN, and are often embedded in the duller DURAIN.

'classical' geography In this sense, not the g. of the Ancient World, but denoting an approach to g. in the 19th century begun by A. von Humboldt and K. Ritter, continued by F. Ratzel and others (mainly German scholars). It sought to avoid purely descriptive writing, to organize factual information derived from both the physical and social sciences, and to investigate the relationship between Man and his environment. A reason for its decline was criticism of its apparent rigidity, esp. with regard to DETERMINISM; another was the development of new approaches, esp. in the Social Sciences. A reaction to c.g. was the development of REGIONAL G., later of REGIONAL SCIENCE, and of the various aspects of SYSTEMATIC G.

clastic (derived from the Gk. *clastos*, broken) App. to rock composed of broken particles from other rocks, usually transported from some distance, deposited, and converted by the process of LITHIFICATION into a consolidated coherent rock; this takes place by: (i) cementation; (ii) compaction; and (iii) desiccation. They include: (*a*) sedimentary rocks: conglomerate, sandstone, clay, shale; (*b*) pyroclastic rocks: tuff, volcanic ash and agglomerate. A consolidated mass of comminuted shell-fragments, though essentially organic in origin, could also be regarded as a c. rock.

clatter A Devonshire dialect-word for SCREE or individual boulders, as on Dartmoor; also called *clitter*.

clay (i) A fine-textured, plastic, sedimentary rock (*argillaceous*), derived from the compaction of mud, consisting mainly of hydrous aluminium silicates, derived from the weathering and resultant decomposition of various felspathic rocks. No structure is developed, and when it dries out c. is traversed by irregular cracks. When wet, it forms a virtually IMPERMEABLE rock, since the minute pore spaces between the fine particles are filled with water held by surface tension, so sealing the rock against the downward passage of water. Many geological varieties of diverse age are distinguished and named; e.g. greyish-blue Lias C., dark grey Kimmeridge C., bluish-grey Oxford C., yellow or brownish Wealden C., bluish-grey London C. (ii) In soil science, a soil with individual particles less than 0·002 mm. (in U.S.A., 0·005 mm.) in size; a c. soil has 30% or more of its bulk of c. (iii) A complex group of minerals, composed largely of aluminium and iron silicates. See also LATERITE, RED C., BOULDER-C., KAOLIN.

clayband A layer of clay-ironstone, consisting of ferrous carbonate or *siderite* ($FeCO_3$) mixed with earthy material, found within the Coal Measures. The iron occurs in nodular bands or thin seams, separated by shales and sandstones. It was extensively worked in Great Britain between about 1750 and 1870; large reserves still exist, but increasing cost and difficulty of extraction make them uneconomic and uncompetitive with cheaper Jurassic ores.

claypan A stratum of stiff compact clay, forming an impermeable layer below the surface of the soil, causing impeded drainage and waterlogging. Ct. HARDPAN.

clay-slate A category of slate derived from compacted clay (i.e. an argillaceous slate), in ct. to a slate compacted from volcanic ash (e.g. the green slate of Buttermere).

Clay-with-Flints A mass of reddish-brown clay containing flint fragments, lying unevenly on the surface of chalk country, and also in funnel-shaped PIPES which may penetrate to a considerable depth. Its age and origin are problematical and complex. In part it may represent the insoluble residue of the Chalk, and in part it may be derived from formerly overlying Tertiary rocks.

cleavage The natural tendency of a rock, such as slate or compacted volcanic ashes, to split into thin sheets along parallel planes, formed as the result of past metamorphic pressure. It may make any angle with the BEDDING-PLANES (ct. LAMINATION), and occurs particularly in fine-grained rocks. The geological interpretation of the causes of c. is a complex problem, however. There are 4 degrees of scale of thickness of layers. *Fracture c.* occurs when rocks respond to folding by slipping along shear-planes (see SHEARING), producing numerous small, closely-spaced joints. C. is of importance to quarrymen in the extraction and dressing of commercial slates for roofing purposes. *Note*: A crystallographer uses the term to indicate the tendency of a crystal to split along planes determined by its molecular structure. [*f*]

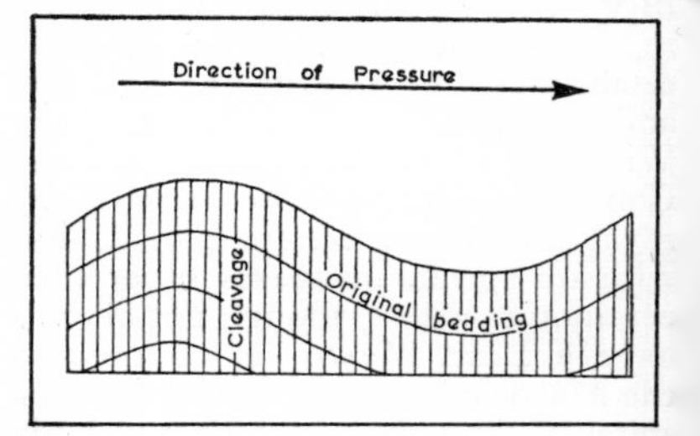

cliff A rock-face along the sea-coas[t] where marine denudation is active an[d] the land rises steeply and appreciabl[y] in height inland. The character of [a] *sea-c.* depends on the nature of th[e] rocks, their bedding and jointing[,] homogeneity or heterogeneity, th[e] presence of bands of weakness such a[s] the SHATTER-BELTS of faults, thei[r] gen. resistance both to wave attac[k] and weathering. C.'s are formed i[n] many rocks, esp. massive ones; e.g. granite (Cornwall), Old Red Sandstone (Orkneys, W. Scotland, Pembroke), Jurassic limestone (Durlston Head, Dorset), New Red Sandstone (St. Bees Head, Cumberland), Carboniferous Limestone (Pembroke). They also occur in softer rocks, where the rate of marine erosion at their base is rapid; e.g. chalk (Dorset, Isle of Wight, Sussex, Kent, E. Riding of Yorkshire), Tertiary clays and sandstones (Hants.), Eocene clays and sandstones (Alum Bay, Isle of Wight), boulder-clay (Norfolk; Holderness, Yorkshire). The term c. is also used for any high steep rock-face or precipice in the mountains, or rising above a lake shore, and for the side of a deeply incised river valley.

climate The total complex of weather conditions, its average characteristics and range of variation over an appreciable area of the earth's surface. Usually conditions over many years (e.g. 30 or 35) are taken into consideration. The term is derived from the Gk., meaning 'slope' (possibly the slope of the earth's axis), later being considered as a zone lying in a particular latitude, and then being applied to characteristic weather conditions. C. is studied in terms of the various climatic elements: TEMPERATURE, incl. RADIATION, ATMOSPHERIC PRESSURE, WIND, HUMIDITY (WATER-VAPOUR, CLOUDS, PRECIPITATION, EVAPORATION). These elements result from the interplay of factors: latitude, altitude, the distribution of land and sea, ocean currents, relief features, and the influence of soil and vegetation.

climatic optimum See ATLANTIC STAGE.

climatograph A circular graph for depicting seasonal temperature conditions. Mean monthly temperatures are plotted from a centre with the aid of a graduated table. The distance from the centre of the circle to 100°F. is taken as 10 times the distance from the centre to 0°F. If the latter distance is x, then the difference y for any temperature $t°$ is given by the formula $y = x\,\frac{(colog\ t)}{100}$. If the limiting temperatures of hot, warm, cool and cold seasons are assumed to be 68°, 50°, and 32°F., then the length and nature of such seasons at any place may be read from the graph by noting where the temperature curve cuts the lines representing the limiting temperatures. The slope of the temperature curve correctly represents the degree of change in temperature from month to month. [f]

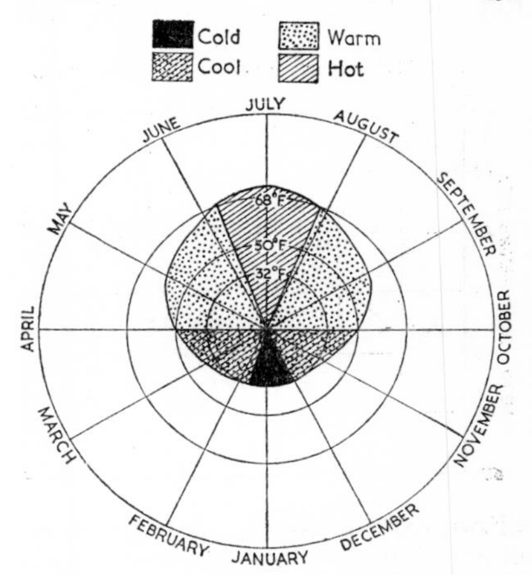

climatology The scientific study of climate; an expression, description and where possible explanation of: (i) its distribution and regional patterns; and (ii) its contribution to the environment of life. Ct. METEOROLOGY, MICROCLIMATOLOGY, LOCAL CLIMATE.

climax vegetation, c. community When the optimum vegetation of a particular long-term plant SUCCESSION, left undisturbed, has been established in relation to the particular physical environment (climate, relief, soil), a state of equilibrium is attained. Thus the c. v. of a hot, wet climate within a few degrees of latitude of the equator is 'Tropical Rain-forest'. *Climatic c. v.* is now preferred to the obsolete term 'Natural Vegetation'.

climograph, climogram A diagram in which the data for 2 climatic elements (e.g. WET-BULB and DRY-BULB temperatures; temperature and precipitation) at a particular place are plotted against each other on a graph as abscissae and ordinates respectively. The shape and position of the resultant graph provides an index of the general climatic character at that place. The term was first used by T. Griffith Taylor to correlate and depict climatic conditions in terms of human physiological comfort; see also HOMOCLIME and HYTHERGRAPH. E. Raisz uses the term *climatogram*. [*f*]

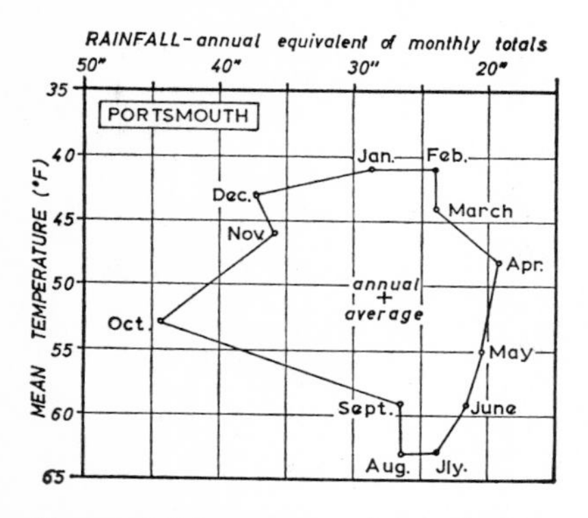

clinographic curve A graphed cu. illustrating the slope of an area as it varies with altitude, in practice plotting the average gradient in degrees between pairs of successive contours. The *f* indicates a method of finding the mean angle of slope between 2 successive contours, by drawing concentric circles equal in area to that enclosed by each contour. The c. c. is drawn using contour intervals as vertical components, and inserting each section of average slope between each pair of contours with a protractor. [*f*, *opposite*]

clinometer An instrument for measuring a vertical angle. The *Indian c.* consists of a base with a spirit-level, a peep-sight at one end, and a vertical leaf with a thin central slit at the other. Degrees above and below the horizontal (0°) are marked on the leaf. There are other more complex kinds of c.; e.g. the ABNEY LEVEL. Another type, with a graduated arc and a pendulum, is used for measuring the dip of strata.

clint A low flat-topped ridge between furrows or fissures (GRIKES) worn along the lines of the joints in the surface of a bare plateau formed of Carboniferous Limestone, down which rain-water (containing carbon dioxide) has percolated, thus enlarging them by CARBONATION-SOLUTION. E.g. above Malham Cove, Yorkshire, where the c.'s and grikes are aligned mainly N.W. to S.E. [*f* GRIKE]

close A specialized meaning applied to various types of enclosed space; e.g. around a cathedral, with lawns, and houses occupied by cathedral dignitaries, a choir-school, etc.

closed system An aspect of GENERAL SYSTEMS THEORY, whereby an assemblage of phenomena exists isolated from the rest by a boundary through which neither energy nor material can pass; ct. OPEN S. The s. is characterized by the destruction of any heterogeneity within it, and hence by a trend towards max. ENTROPY. For instance, in an isolated tank containing gas higher in temperature at one end, the temperature differential will gradually decay, with the ultimate establishment of a homogeneous condition. It is important to realize that some authorities claim that in geographical studies it is difficult or impossible to envisage any s. that is wholly c.

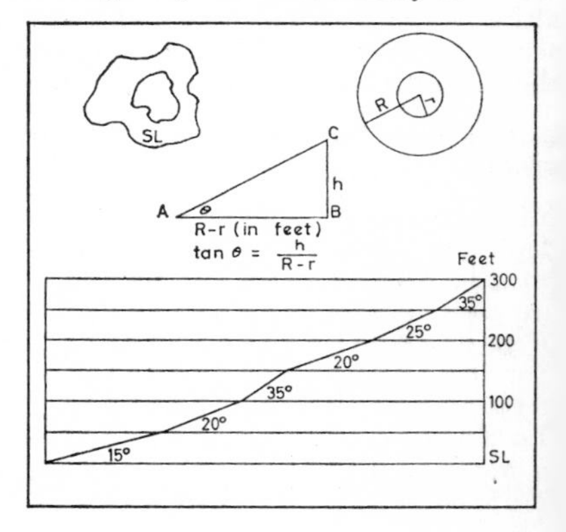

cloud A mass of tiny visible particles, usually of water (0·02–0·06 mm. (0·0008–0·0024 ins. in diameter), sometimes of ice, which form by condensation on nuclei such as dust and smoke particles, salt, pollen and negative ions. They float in masses at various heights above sea-level, ranging from near the ground (FOG or MIST) to over 12,000 m. (40,000 ft.). They are classified: (i) *by height*: low c.'s up to 2400 m. (8000 ft.), medium c.'s at 2400–6000 m. (8–20,000 ft.), high c.'s at 6000–12,000 m. (20–40,000 ft.); (ii) *by form*: feathery or fibrous (CIRRUS), globular or heaped (CUMULUS) and sheet or layer (STRATUS). Other types (see individual names) are distinguished by combinations of the 3 form names, by adding the suffix *alto* to indicate height, and *nimbus* to signify falling rain, thus forming 10 main genera. Other variations are lens-shaped LENTICULARIS, turret-shaped CASTELLATUS, breast-shaped MAMMATUS, tattered ragged clouds (FRACTO-), and many more. The amount and nature of the c. cover is recorded, expressed in terms of the proportion of the sky covered (in 10ths or 8ths) and indicated on a map by a proportionally shaded disc. Lines of equal cloudiness (*isonephs*) can be interpolated if adequate statistics are available. Definite c. sequences can be observed in connection with the passage of pressure-systems. The cloudiest parts of the world are in mid-latitudes on the W. coasts of continents, and near the equator. The least cloudy are the hot deserts, continental interiors, and the Mediterranean lands in summer.

cloudburst A fanciful name for a sudden, concentrated downpour of rain. In U.S.A. this is sometimes defined as a rate of fall of 100 mm. (3·94 ins.) per hour; see INTENSITY OF RAINFALL. The area affected is usually quite small, since the c. is caused by intense local convectional rising, usually accompanied by a thunderstorm.

clough, cleugh A steep-sided valley in N. England, esp. Derbyshire.

cluse (Fr.) A steep-sided valley cutting transversely across a limestone ridge, esp. in the Jura Mts. and in the Fore-Alps of Savoy. Its existence involves the development of ANTECEDENT or SUPERIMPOSED drainage; a river, existing before an anticline was upraised, maintained its course by erosion as uplift progressed. The drainage as a result reveals an alternation between the longitudinal valleys (*vaux*, sing. *val*) and the transverse *cluses*, with sudden changes of direction, elbow-bends, and frequent river-CAPTURES. Note the course of a river such as the Doubs. [*f*]

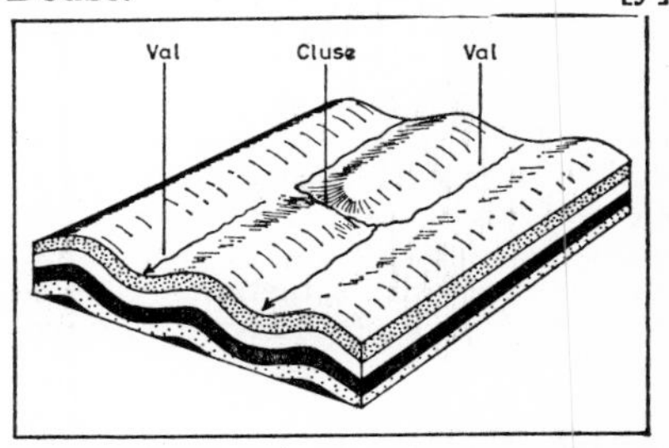

clustering A concentration of distribution, spatially as on a DOT MAP, or in the graphical presentation of a given series of values or events, as by a flattening in a LORENZ CURVE. See CONCENTRATION.

coal Applied to a wide range of combustible deposits derived from accumulated layers of vegetation, classified according to their content of carbon and volatile matter, together with their colour, hardness and age. The vegetation decomposes under water in the absence of air (ANAEROBIC conditions), with associated pressure and temperature increase resulting from its burial under overlying sediments and from earth-movements. The series comprises: PEAT, BROWN-COAL, LIGNITE (these are not usually included in overall production figures) CANNEL COAL, BITUMINOUS COAL, STEAM-COAL, and ANTHRACITE. See also CLARAIN, DURAIN, FUSAIN and VITRAIN structures. C.p. (excl. peat, brown-coal, lignite) = U.S.A., China, U.S.S.R., U.K., W. Germany, Poland, India, Japan, France.

coaling station A harbour where coal-burning ships could refuel; e.g. Gibraltar, Aden, Colombo, Singapore. Now they are largely replaced by oil-fuelling depots.

Coal Measures The series of coal seams and intervening strata of shale, clay, sandstone and ironstone, in the Upper CARBONIFEROUS formation (approx. equivalent to the Pennsylvanian in U.S.A.). Although c. m.'s (l.c.) also occur in the Lower Carboniferous of Scotland, these are not included in the C. M. series, which refer to the particular stratigraphical division. At the end of the Carboniferous period occurred the HERCYNIAN orogeny, which folded, faulted and in places crushed the C. M. Subsequent denudation removed them from more elevated areas, so that they survive in individual basins flanking, sometimes among, uplands, sometimes covered by new deposits (a CONCEALED COALFIELD).

coast A gen. and undefined term for the zone of contact between land and sea; the strip of land bordering the sea, appreciably wider than the SHORE. [*f* COASTLINE]

coastal plain (i) Used in gen. terms of any area of lowland bordering the sea or ocean, sloping very gently seaward. It may result from the accumulation of material: alluvium (*deltaic plain*), sand, mud and peat, with or without a relative fall of sea-level, together with the contribution of dyking and draining by man; e.g. the N. Sea coast of Europe from Denmark to near Calais; the Adriatic coast of N. Italy. (ii) In U.S.A. the term c. p. is restricted to an area of the CONTINENTAL SHELF which has recently emerged as a result of a relative fall of sea-level; e.g. S.E. U.S.A., extending inland for 160–480 km. (100–300 mi.) along the Atlantic and Gulf of Mexico for a distance of 3000 km. The outcropping sedimentary rocks (of Cretaceous and Tertiary age) reveal in plan a series of belts parallel to the coast, hence the term *belted c. p.*

coastline Often used as a gen. term (the edge of the land as viewed from the sea), but specif. it signifies either: (i) the line reached by the highest storm waves; (ii) the high-water mark of medium tides; or (iii) on a steep coast, the base of the cliffs. The Ordnance Survey uses (ii), and on the One-inch Series stipples the area between the mean high-water and low-water marks. [*f*]

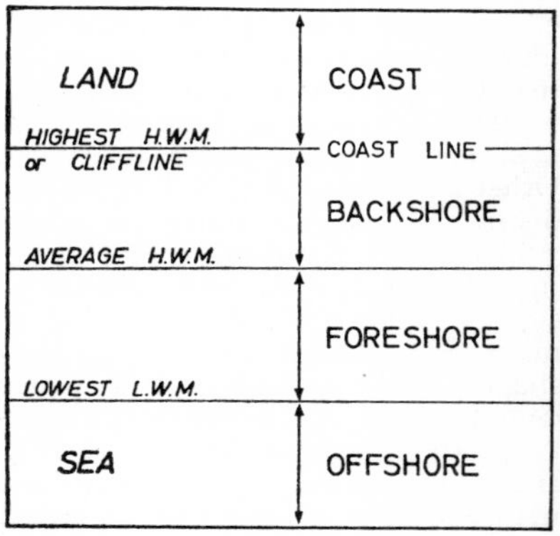

cobalt A metal obtained from its sulphide and arsenide ore, usually in association with copper and silver ores. Formerly it was chiefly used as a blue pigment, now as an ingredient in steel-alloys for high-speed cutting tools and jet-engine parts, and is often alloyed with chromium, tungsten and molybdenum. C.p. = Congo (Kinshasa), Morocco, Canada, Zambia, Finland.

cobble A water-worn stone, larger than a pebble, smaller than a boulder; usually 60–200 mm. (2·4–8 ins.) in diameter. In U.S.A. it is defined by the WENTWORTH SCALE as between 64–256 mm. (2·5–10·0 ins.).

coefficient of variability The degree of v. of statistics about a mean value. For example, the mean annual rainfall figure for a station represents the mean of actual annual totals though often these are widely dispersed on either side of the mean. Various statistical formulae are used e.g. (i) $c.\ of\ v.\ (\%) = \frac{\sigma}{M} \times 100$, where σ = standard deviation and M is the mean value; or (ii)

$$c.\ of\ v.\ (\%) = \frac{Interquartile\ Range}{Median} \times 100$$

coesite A dense form of silica first synthesized in the laboratory under very high pressure conditions by L. Coes. It was found naturally in the rocks beneath Meteor Crater, Arizona, obviously the result of impact metamorphism when a meteorite struck the surface of the earth at a very high velocity.

coffee A tree (the two main species are *Coffea arabica* and *C. robusta*) grown in tropical uplands, which yields clusters of dark red berries, each containing 2 'beans'. After drying and roasting, these are ground to produce the ingredient of a beverage. C.p. = Brazil, Colombia, Angola, Mexico, Uganda, Ethiopia, Ivory Coast, Salvador, Indonesia, Guatemala.

coire (Gaelic) See CIRQUE.

coke (i) *Metallurgical c.*, hard enough to withstand handling and the burden of the load in a blast-furnace without crushing, made in coke-ovens, with coke-oven gas as a by-product, using coals with 65–80% carbon. (ii) *Gas c.*, a softer more friable c., produced as a by-product of the gas industry in retorts, mainly from coals with a high volatile and lower carbon content.

coking coal A type of coal of which the residue after heating in the absence of air is coherent.

col (Fr.) (i) A conspicuous depression or notch in a ridge or range, providing a pass from one side to the other, commonly resulting from the development of back-to-back CIRQUES. Used widely in the French Alps; e.g. C. du Midi, C. du Bonhomme, C. de la Seigne. It is sometimes called a *pass* or *saddle*. A c. may develop in a CUESTA from the beheading of a back-slope valley by scarp-retreat, leaving a c. in the cuesta at the head of the latter; e.g. Cocking Gap (alt. 100 m.) in the W. of the S. Downs near Midhurst, and several distinct c.'s in the S. Downs near South Harting. (ii) ('*atmospheric*'): Used analogously with ref. to weather maps where the ridges represent high pressure, the valleys low pressure. In a c. the weather is extremely variable and presents a difficult problem to the forecaster. [*f*]

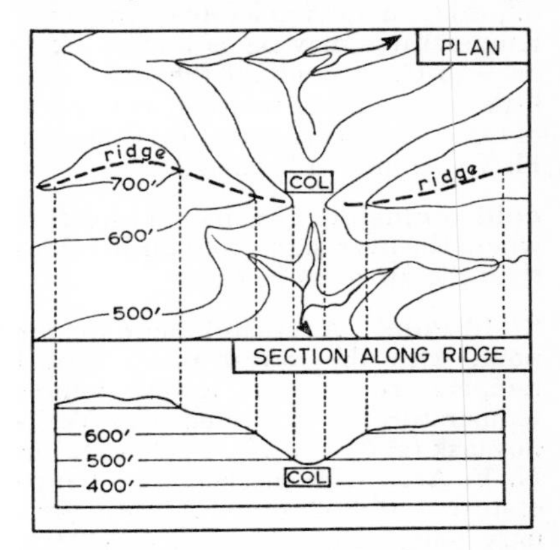

colatitude The complement of the latitude = 90° − latitude.

cold desert (i) A gen. term for TUNDRA and POLAR regions, where low temperatures limit or prohibit vegetation (W. Köppen's *ET* and *EF* types). (ii) A name sometimes used for areas in continental interiors, poleward of 50°N., and further S. in central Asia where plateaus are shut off by high mountains from maritime influences; these are defined by A. A. Miller as having one or more months each with a mean temperature below 6°C. (43°F.) (W. Köppen's *BWk* type); e.g. Luktchun in the Tarim Basin, precipitation 193 mm. (7·6 ins.), mean January temperature −11°C. (12°F.), mean July temperature 32°C. (90°F.).

cold front The boundary-zone between a mass of warm air and an advancing, undercutting wedge of cold air which forces the warm air upward; the rear of the warm sector of a DEPRESSION (ct. WARM FRONT). There is an appreciable drop in temperature, extensive

CUMULONIMBUS and FRACTO-clouds develop, rain falls in heavy showers (sometimes accompanied by thunder), and the wind freshens from a N. or N.W. direction (in the N. hemisphere). An extreme form of a c. f. is a LINE-SQUALL. See ANAFRONT, KATAFRONT. [*f* DEPRESSION]

'cold glacier' In ct. to a 'WARM G.', this is a moving ice-mass in which temperatures may be as low as −20° or −30°C. throughout the year, with little or no surface melting, as in parts of Greenland and Antarctica. Called by H. W. Ahlman a '*polar g.*'.

cold occlusion In an OCCLUSION [*f*] where the overtaking cold air is colder than the air-mass in front.

'Cold Pole' A fanciful name for the point with the lowest mean annual temperature or the lowest winter temperature; it is located near Verkhoyansk (at 67° 33′ N., 133° 24′ E.) in N. E. Asia, with mean annual temperature −16·3°C. (2·7°F.), mean January temperature −50°C. (−58°F.), mean min. January temperature −64°C. (−83°F.), lowest recorded temperature −70°C. (−94°F.), mean annual range 65·5°C. (118·6°F.). Another c. p. must occur somewhere in Antarctica.

cold wall A DISCONTINUITY LAYER between water of markedly contrasting temperature; e.g. in the N. Atlantic Ocean between the Labrador Current and the Gulf Stream; in the Pacific Ocean between the Okhotsk Current and the Kuroshio.

cold-water desert Where a hot desert extends to the W. coast of a continent, its climate is affected by equatorward-flowing cold currents and by upwelling cold water ('*cold-water coast*') (W. Köppen's *BWk* climatic type). These c.-w. d.'s are restricted to narrow strips along the coasts of N.W. Africa, S.W. Africa, N. Chile, N.W. Australia. Cool air moves from over the sea on to the land, producing summer temperatures low for the latitude; e.g. mean July temps. for Walvis Bay, 19°C. (66°F.), Iquique, 22°C. (71°F.). The seasonal range is therefore reduced; e.g. Walvis Bay, 5°C. (9°F.). Fogs may form over the sea and roll inland for a few km.; heavy dews are also common.

cold wave A sudden inrush of cold Polar air behind a DEPRESSION, causing a marked fall of temperature. In U.S.A., the term implies a specific fall below a definite figure in a certain time, depending on season and location.

collective farming A type of agricultural organization started in the U.S.S.R. and now also practised in other E. European countries, Communist China, N. Viet-Nam, N. Korea. The land is compulsorily amalgamated into large c. farms (*kolkhoz*, plur. *-y*), run by a manager, and worked by labourers as directed. A proportion of the produce is delivered to the Government, and the workers receive shares of the proceeds from the sale of the remainder as wages, according to the work done; ct. SOVKHOZ.

collectivism Political and economic systems based on central planning by the State and (to a greater or less degree) compulsory co-operation by its citizens.

colloid A substance in a state of extremely fine subdivision, with particles from 10^{-5} to 10^{-7} cm. in diameter, of both mineral and organic material. It plays a vital yet highly complex role in soil chemistry. Its function is: (i) *physical*, in that it imparts an adhesive quality to certain constituents, notably clay particles (ii) *chemical*, in that it can attract and hold ions of dissolved substances, esp. of bases such as calcium. These properties are the result of the electrical forces of different molecules combined along interfaces.

colluvium A heterogeneous mixture of loose, incoherent rock fragments, scree and mud, which has moved down to the base of a slope under gravity, the result of MASS-WASTING. In U.S.A. it is also called *slope-wash.*

colony (colonial, colonization) (i) Orig. a body of settlers and the territory they occupied away from their native land, usually initially undeveloped and either thinly populated or uninhabited. Many c.'s were founded for strategic or economic motives, some were conquered from other c. powers. In a political sense, the overseas land and people are still subject to the mother-country, though usually with some measure of self-government. In most British c.'s, the Sovereign is represented by a Governor, an executive council and a legislative council, including chief officers, members nominated by the Governor, and some elected members. The c. is subject to Acts passed by the British Parliament in London. Many British c.'s have become DOMINIONS, some have become independent within the Commonwealth (e.g. Ghana), others outside the Commonwealth (e.g. Burma). (ii) Biologically, a group of closely associated similar organisms; e.g. a c. of corals. A botanist uses colonization to indicate the gradual spread of a plant species into an area.

columnar structure The result of the cooling of igneous rocks, when nternal contraction has set up a series of regular JOINTS vertical to the cooling surface, thus producing columns; e.g. the hexagonal columns of basalt in the Giant's Causeway (N. reland) and Fingal's Cave (Staffa); he pentagonal columns of phonolite n the Devil's Tower (Wyoming), nd the Devil's Postpile in the E. ierra Nevada (California).

ombe, coombe (i) A CIRQUE in umberland; e.g. Birkness Combe, ear Buttermere. (ii) A short valley in . England, usually with a steep head, ommonly found in chalk country; .g. Pebblecombe in the N. Downs, E. f Dorking. (iii) A short steep valley nning down to the sea; e.g. Wollaombe and Combe Martin, in N. evon. (iv) A high-lying longitudinal depression along the crest of an ANTICLINE in the folded Jura Mts., cut by an actively eroding stream, with infacing cliffs (*crêts*) of limestone; e.g. C. Berthod, in which rises a headstream of the R. Bienne.

combe rock See COOMBE DEPOSIT.

comb-ridge A sharp-edged serrated ridge, near-horizontal, formed when two CIRQUES have developed back to back; an ARÊTE in a strict sense.

'comfort zone' The range of temperature and RELATIVE HUMIDITY which is physiologically most comfortable to human beings. In England this is around 15°C. (60°F.) and a relative humidity of 60%. As temperature rises, the relative humidity should be lower for comfort.

commensalism Used in urban geography to describe one form of interdependence in which each unit in an area is a competitor with all other units, because they all carry out similar work, or sell the same things. Specialized areas of e.g. jewellers or doctors within towns or cities demonstrate this kind of interdependence. The benefits derived from their proximity consist of mutual trade promotion, mutual improvement of the environment, advertisement (often indirectly by association) and propaganda. E.g. Hatton Garden (diamonds), Harley Street (doctors), Mincing Lane (tea), Fleet St. (newspapers), Carnaby St. (modern clothing), all in London.

commercial core Any nucleation of commercial activity within a specif. area.

commercial geography The study of the production, distribution and consumption of products. It is often regarded as syn. with ECONOMIC G., but the latter really has much wider implications.

comminution (i) The reduction of rock material to progressively smaller particles by agencies of weathering and erosion, or by earth movements.

(ii) The breaking of stones by mechanical means; e.g. crushing for road-metal, ballast, aggregate.

common, common land An area of land, often unenclosed or 'waste', over which members of a community may have certain rights, such as access, pasture and turf-cutting. In the New Forest (Hampshire), the Commoners (each of whom occupies not less than 1 acre (0·04047 ha.) of land to which C. Rights are attached) have various privileges: the C. of Pasture, Mast, Turbary, Marl and Fuelwood; the first of these is now the most important, and applies to about 180 sq. km. (44,500 acres) over which ponies, horses, cattle and donkeys (not goats, and sheep only in a few expressly defined areas) can graze, and pigs from 25 Sept. to 22 Nov. can PANNAGE on the mast (with breeding sows all the year). Many c.'s are owned by boroughs and parishes as open spaces; e.g. Southampton C.

community (i) A group of plants growing in a particular area, usually of distinctive character, and requiring certain physical conditions which satisfy them. The c. can be of different scales in the plant hierarchy: (*a*) plant formation (e.g. Temperate Deciduous Forest); (*b*) plant association (e.g. oak forest); (*c*) plant society, a local c. with special conditions and species. (ii) Used as a gen. reference to a group of people living in proximity.

commuter One who travels to and from a town from his residence some distance away, applied partic. to people making their daily journey to work, and orig. used of those who held a form of season-ticket for travel. The term is derived from U.S.A., where a person could hold a 'commutation' (i.e. season) ticket. Also vb. 'to *commute*', participle '*commuting*'.

compaction One of the processes of LITHIFICATION by which unconsolidated materials are consolidated, either by the weight of subsequent overlying deposits or by the pressure of earth-movements. This chiefly affects fine-grained deposits, such as silt and clay; these are compacted into mudstone and shale.

compage A REGION defined with ref. to all the features of the environment functionally associated with Man's occupance of the earth; a whole complex structure revealed by the compaction of the constituent parts. A term revived by D. Whittlesey.

compass An instrument used to find direction, consisting of a free-swinging magnetized needle which points to the N. and S. Magnetic Poles along the local line of magnetic force (see MAGNETIC DECLINATION), mounted on a card graduated in degrees, with the cardinal and ordinal points marked. More elaborate models have a peep-sight and prism (PRISMATIC C.). See also GYRO-C.

competence, of a stream A measure of the ability of a stream to move particles of a certain size, indicated by the weight of the largest fragment that it can transport. As stream velocity increases, the max. particle weight increases, though not in direct ratio. The SIXTH POWER LAW, postulated by W. Hopkins in 1842, suggests that the weight of the largest fragment that can be carried increases with the sixth power of the stream velocity. Thus for each stream velocity there is a corresponding max. weight of particle that can be carried. Likewise, for a given particle size there is a critical water velocity which must be attained before that particle can be picked up. However, once in motion the particle may be transported by a much lower velocity current. The difference between 'pick up' and 'carry' velocities is partic. marked with very small grain sizes; thus it is difficult for slow-moving water to erode a mud bank though mud particles can be carried by an extremely low velocity current. Another concept postulates that the diameter of a particle which a stream can carry varies with the square of the velocity. Load-carrying ability may also be measured by CAPACITY, which is the total load. A large slowly flowing stream may have a high capacity, but a low c.; its load consists of a large quantity of fine material in suspension.

stream velocity		*diameter of particle*		
(*km.p.h.*)	(*m.p.h.*)	(*mm.*)	(*ins.*)	
0·4	0·25	0·5	0·02	coarse sand
0·8	0·50	2·0	0·08	coarse grit
1·6	1·00	6·4	0·25	small stones

competent bed A rock-stratum that is sufficiently strong to bend during folding movements, rather than to be distorted by plastic flow and deformation. Ct. INCOMPETENT BED.

complex climatology An analysis of the climate of a place in respect of the frequency of weather types experienced, these being defined in terms of the various climatic ELEMENTS.

composite profile A p. constructed to represent the surface of any area of relief, as viewed in the horizontal plane of the summit-levels from an infinite distance, and so including only the highest points of a series of parallel p.'s. Ct. PROJECTED and SUPERIMPOSED P.'s. [*f* PROFILE]

composite volcanic cone A cone built up over a long period of time as the result of a number of ERUPTIONS, consisting of layers of ash, cinder and lava fed from the main pipe, which culminates in a crater; e.g. Stromboli and Etna (Italy), Fujiyama (Japan), Mt. Hood (Oregon), and indeed most of the world's highest c.'s. It is often known in U.S.A. as a *strato-volcano*.

comprehensive development The development or redevelopment of some specif. area as a phased operation within a c. plan for the area as a whole.

compressed profile A device by which a series of COMPOSITE P.'s is c. together, and when viewed at right-angles only those features not obscured by higher ones in the foreground are visible.

compressional movement, of the earth's crust A form of strain developing in a horizontal plane when rocks are exposed to stress. It involves a contraction of the surface rocks, resulting in: (i) FOLDS; (ii) REVERSED FAULTS; (iii) THRUST-FAULTS. It is believed that some RIFT-VALLEYS are caused by c. m.'s.

computergraphics Computer facilities used to process data to provide spatial distributions and tonal values in symbolic form. This is especially valuable in urban geography, where the spatial relationships are highly complex and the number of variables to be mapped is considerable. Note the Institute of C. at Harvard, U.S.A. See also SYMAP.

concealed coalfield An area of workable coal in deposits which are covered by newer rocks. The exposed portion of a c. was usually worked first; later mining spread to the c. c., as in the E. part of the Yorkshire Coalfield, and the N. part of the Ruhr. Some c.'s are wholly concealed; e.g. the Kempen C. of Belgium.

concentration In an ecological sense, the tendency of people and institutions to cluster in towns and cities. Because the term relates to the attraction of people or institutions into an area, it essentially deals with mass. The opposite situation, the breakdown of a centre, is referred to as *de-c*.

concordance of summit-levels See ACCORDANT SUMMIT-LEVELS (and *f*).

concordant Parallel to the general lines of the structure or 'grain' of the country, or to the general strata; e.g. a drainage pattern which has developed in a systematic relationship with, and consequent upon, the structure. See also C. COAST, C. INTRUSION. Ct. DISCORDANT.

concordant coast A coastline parallel to the general trend-lines of the relief; sometimes called longitudinal or Pacific type. Such a coast tends to be straight and regular, unless a considerable relative rise of sea-level occurs, when the outer ranges may become long lines of islands and the parallel valleys will form SOUNDS; e.g. the coast of British Columbia; the E.

Adriatic coast of Yugoslavia; Cork Harbour in S. Ireland. [*f*]

concordant intrusion An I. of igneous material that lies parallel to the stratification of the rocks into which it was intruded; e.g. a SILL. Ct. DYKE, which is a DISCORDANT I.

concretion A more or less rounded mass or nodule of hard material within a bed of different rock. It was probably formed by the localized concentration of a cementing material (calcite, dolomite, ferrous oxide, silica) during the consolidation of the bed; e.g. DOGGERS in various shales: Coal Measure, Corallian, Kimmeridge Clay, Lias. When the nodules are calcareous they can be used for cement, while ironstone nodules form valuable ores. See also GEODE.

condensation The physical process by which vapour passes into the liquid or solid form. It occurs either when air is cooled to its DEW-POINT or when air becomes saturated by evaporation into it. Further cooling will cause the excess vapour in the air to be condensed on nuclei as minute water droplets or (if the dew-point is below 0°C.) into the solid form of HOAR-FROST. See also DEW, MIST, FOG, CLOUD, RAINFALL and SNOW.

condensation trail (or **vapour trail**) A line or stream of white cloud-like particles formed behind an aircraft flying in cold, clear, humid air. It results from the c. of water-vapour derived from combustion of fuel through the aircraft's exhaust, and as a result of the reduction of pressure behind the wing-tips. Abbr. *contrail.*

conditional instability The state of a column of air with a lapse-rate less than the DRY ADIABATIC LAPSE-RATE, but greater than the SATURATED ADIABATIC LAPSE-RATE; the i. is conditional upon air becoming saturated by vertical uplift. See LATENT INSTABILITY.

condominium The joint government of a territory by 2 other states; the Sudan was so governed by Britain and Egypt until 1956; there is an Anglo-French c. in the New Hebrides.

conduction, thermal The process of direct heat transfer through matter from a point of high temperature to one of low, in ct. to CONVECTION and RADIATION.

cone (i) A volcanic peak with a broad base tapering to a summit; see ASH C., CINDER C., COMPOSITE C. (ii) See ALLUVIAL C.

cone of exhaustion A local lowering of the WATER-TABLE around a well as the result of pumping out water more rapidly than it can percolate laterally through the AQUIFER. In U.S.A. a *c. of depression.* [*f*]

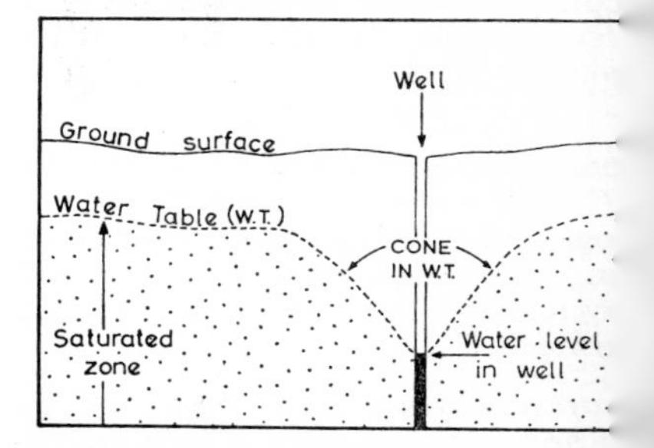

cone sheet A funnel-shaped zone of DYKES or fissures surrounding, in more or less arcuate form, a circular or dome-shaped igneous intrusion which has exerted pressure and so caused fractures within the country-rock. These are inclined inwards towards the top of the intrusion. E.g. in Mull, Skye (Black Cuillins) and Arran. Ct. RING-DYKE.

confluence (i) The point at which a tributary (or *confluent*) joins the main stream. (ii) The body of water so produced, hence a combined flood. (iii) By analogy, the term is used gen. of junctions of routeways, and city street intersections.

conformable Where one stratum lies parallel upon another in correct geological sequence, without the occurrence of any break or interruption by denudation or earth-movements. Noun: *conformity*. Ct. UNCONFORMITY.

conformal projection A class of p. (also known as *orthomorphic*) in which shape is maintained over a small area, and at any point the scale is the same in every direction, and angles around every point are correctly represented; e.g. MERCATOR, LAMBERT'S CONFORMAL, STEREOGRAPHIC P.'s.

congelifluction The flow of earth (see SOLIFLUCTION) under PERIGLACIAL conditions affecting permanently frozen sub-soil. A term introduced by K. Bryan in 1946, along with several others. Each has a precise scientific meaning, which perhaps compensates for their ugliness.

congelifraction Frost-splitting.

congeliturbation Frost action, involving frost-heaving and churning of the ground, and differential mass-movements such as SOLIFLUCTION, which leads to the disturbance of the soil and sub-soil. Various types of PATTERNED GROUND, such as *stone stripes* and *stone polygons*, are produced.

conglomerate A rock composed of rounded, waterworn pebbles, 'cemented' in a matrix of calcium carbonate, silica or iron oxide; ct. BRECCIA, FAULT-BRECCIA, AGGLOMERATE. It is popularly known as 'puddingstone'. The emphasis is on the rounded constituents, by comparison with the angularity of breccia, but there is no precise division.

conic(al) projection One of a group of p.'s in which part of the globe is projected upon a tangent cone, which in effect is opened up and laid out flat. Most c. p.'s have concentric circular parallels, some have straight, others curved, meridians. The cone touches the globe along one or more *standard parallels* (*s.p.*).

Radii of Standard Parallels
(earth's radius $R=1$)
$r=R.\cot$ *latitude*

latitude	
0°	∞
10°	5·671
20°	2·747
30°	1·732
40°	1·192
50°	0·839
60°	0·577
70°	0·364
80°	0·176
90°	0·000

For the ordinary (or *Simple*) c. p., the selected standard parallel is drawn to scale with the above radius, and divided truly, i.e.

$$1° \text{ of long.} = \frac{2\pi R.\cos \text{ lat.}}{360}.$$

Draw the central meridian, and divide this truly (i.e. $1° = \frac{2\pi R}{360}$), and draw other parallels concentric to the standard parallel through these divisions. Scale is true along all meridians and on the standard parallel, elsewhere it is too large. See C. WITH TWO S. P.'s, ALBERS' (C. EQUAL AREA WITH TWO S. P.'s), LAMBERT'S CONFORMAL WITH TWO S. P.'s, BONNE, POLYCONIC. [*f*]

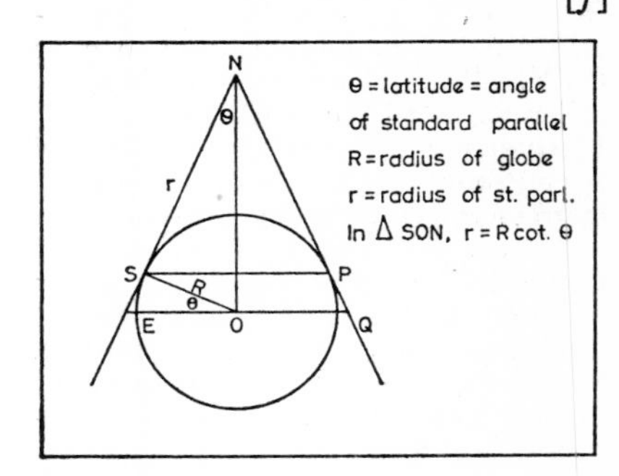

Conic(al) Projection with Two Standard Parallels A CONIC(AL) P. with 2 standard parallels, spaced their true distance apart. All parallels are concentric circles, meridians are radiating straight lines. Scale is true on each s. p., less between them, and progressively larger beyond them, so that the overall scale error, as cf. the C. P., is reduced. This is sometimes (incorrectly) called the SECANT C. P. The s. p.'s are chosen to fit a particular country or continent; e.g. for Europe, between 40°N. and 60°N.

coniferous forest A type of f. in which the trees are mostly evergreen (though not larch), cone-bearing (belonging to the order *Coniferales*), with needle-leaves, shallow root-systems, and soft-wood timber. They can grow under widely differing conditions of climate (from the tropics to the subarctic), of relief (from coastal plains and swamps to high steep mountains), and of soil (from heavy clays to poor sands). Many varieties are tolerant of thin, acid soils and extreme winter cold. They are generally rapid growing, and so are used for AFFORESTATION. The chief types of c. f. are: (i) HYGROPHYTIC, found in W. and S.E. of N. America, S. Chile, W. Europe, parts of China and Japan, S.E. Australia, New Zealand; these are mostly tall trees, incl. Douglas fir, sequoia, red cedar, Sitka spruce, hemlock, white pine, yellow pine; (ii) MESOPHYTIC, found in high latitudes, known as TAIGA, and at high altitudes, with extensive stands of species of pine, spruce, fir, larch; (iii) XEROPHYTIC, found at high elevations in semi-arid areas, e.g. juniper, pinyon pine (S.W. U.S.A.). Conifers grow in many other parts of the world; e.g. in the Mediterranean area (Corsican pine and Aleppo pine), in California (Monterey cypress), on tropical coasts (casuarina).

conjunction The position of 2 heavenly bodies in the same straight line as seen from the earth (SYZYGY). The earth, moon and sun in c. result in the new moon, when the tide-producing forces are at their max., hence SPRING-TIDES. Ct. QUADRATURE, OPPOSITION.

connate water W. which has been retained in sedimentary rocks since the time of their formation (sometimes called *fossil w.*).

connexité (Fr.) The concept of 'inter-connection', as advocated by the French scholar J. Brunhes; i.e. the functional relationship between society and environment. In some respects, cognate with the Germ. ZUSAMMENHANG.

consequent stream A s. whose direction of flow is directly related to or is consequent upon the original slope of the land; ct. SUBSEQUENT, OBSEQUENT. The term was first used by J. W. Powell in 1875. It was intended to signify something more than that water on a slope will flow downhill, but was used in relation to initial fold-structures. The term is now sometimes used as a noun, 'a consequent' (the word river understood). [*f* SUBSEQUENT STREAM]

conservancy (i) A noun derived from the act of CONSERVATION, as in the Nature C. of Gt. Britain (established 1949), which controls numerous National Nature Reserves. *Note:* The Nature C. now comes under the Natural Environment Research Council. (ii) The regulation of rivers e.g. the Thames C.

conservation The preservation from destruction of natural resources (soil, vegetation, animals) by careful control and management, esp. for the benefit of posterity. This is not so much a 'holding-back' as the maintenance of a favourable balance in the use of the environment.

consociation A vegetation unit dominated by a single species; e.g. beechwood, dominated by the beech tree. Ct. ASSOCIATION.

constant slope Used by W. Penc[illegible] and A. Wood for that part of a [illegible] profile of accumulation which le[illegible] below the FREE FACE [*f*] above, and ct[illegible] with the WAXING and WANING S.'s.

constructive wave One of a series of gentle w.'s rolling in steadily on to a coast at about 6–8 a minute, which has a powerful push of the SWASH, and because of frictional retardation a less powerful BACKWASH. It therefore tends to move material (esp. shingle) up a beach, so building up ridges.

consumer goods Manufactured articles for use by a consumer, in ct. to CAPITAL G.'s. The implication is that such g.'s are 'consumed' and then replaced.

contact metamorphism See THERMAL M.

continent One of the earth's major constituent land-masses, composed of SIALIC rocks, rising from the ocean floor. Structurally it includes shallowly submerged adjacent areas (the CONTINENTAL SHELF) and neighbouring islands; in this sense, the c.'s occupy about 30% of the earth's surface. Actual dry land comprises about 29%.

Areas (U.N. figures)

	million sq. km.	*sq. mi.*
Africa	30·6	11·8
N. America	17·9	6·9
S. and Central America	24·3	9·4
Asia	45·6	17·6
Europe	9·8	3·8
Australia (with N. Z. and Oceania)	8·5	3·3
Antarctica	11·4	4·4

Europe without the U.S.S.R. has an area of 5·0 million sq. km. (1·9 million sq. mi.), Asia without the U.S.S.R. of [illegible]8 million sq. km. (10·9 million sq. mi.), the U.S.S.R. of 22 million sq. km. (8·6 million sq. mi.).
Note: 'The Continent' indicates Europe to people of Gt. Britain.

continental air-mass An a.-m. whose source-region is a high-pressure area over a c. interior; it is usually of low humidity. It is denoted by *c* in the terminology, and may be of either high latitude (Polar), hence *Pc*, or low latitude (Tropical), hence *Tc*. In U.S.A. the letters are reversed, cP, cT.

continental climate The climate of a c. interior, characterized by seasonal extremes of temperature, with low rainfall occurring chiefly in early summer. This is mainly the result of great distance from the sea, hence *continentality*.

continental divide A major d. which separates the drainage basins of a continent. E.g. in N. America a 'T'-shaped d. separates streams flowing W. to the Pacific, E. and S. to the Atlantic, and N. to the Arctic. *Note:* Triple D. Peak, in Glacier National Park, Montana, is at the junction of the 'T'.

continental drift The hypothesis that c. masses have changed their relative positions, possibly the result of the fragmentation and moving apart of an orig. larger mass. The concept was put forward by A. Snider (1858), developed by F. B. Taylor (1908) and esp. by A. Wegener (1915). The earliest theories were based on the apparent similarity of coastlines along each side of the Atlantic Ocean, esp. the 'fitting' of S. America into Africa. It was suggested that GONDWANALAND and Laurasia were masses which broke up, and portions moved apart. The main problem was the nature of the energy required; later, convection currents created by the accumulation of radioactive heat were postulated. There has been a recent revival of interest in the hypothesis, based on the evidence of PALAEOMAGNETISM, and on the concept of an expanding (as distinct from a contracting) earth.

continental ice-sheet An i.-s. of c. dimensions: e.g. in Antarctica; the Quaternary i.-s.'s which covered the N. parts of Europe and N. America.

continental island An i. which stands close to, and is structurally related to, a continent, rising from the C. SHELF, formed by a relative rise of sea-level; e.g. the British Isles, Newfoundland, Ceylon. The contrast is with *oceanic i.'s*, rising from the floors of ocean deeps.

continental platform A continent and its surrounding shelf as far as the edge of the C. SLOPE; i.e. the true structural continent, of SIALIC material, in ct. to ocean basins.

continental sea A partially enclosed s., lying on or within a continent in the structural sense, linked with the open ocean; e.g. Baltic S., North S., Hudson Bay, Yellow S.

continental shelf The gently sloping (1° or less) margins of a continent, submerged beneath the sea, extending from the coast to a point where the seaward slope increases markedly. This outer edge has a depth variously ascribed between 120 m. (65 fthms.) and 370 m. (200 fthms.); an average of 130 m. (430 ft.) has been suggested. It is well developed off W. Europe [320 km. (200 mi.) W. of Land's End], 240 km. (150 mi.) off Florida, 1200 km. (750 mi.) off the Arctic coast of Siberia, 560 km. (350 mi.) off the coast of Argentina. Around some continents it is much narrower, or almost absent, esp. along coasts where fold mountains run parallel and close to the ocean (E. Pacific Ocean). Most c. s.'s are merely portions of the structural continent inundated by a slight relative rise of sea-level, though some parts may be the result of (i): marine planation (e.g. the Strandflat off N. Norway); (ii) glacial erosion during a period of low sea-level; (iii) the building-up of an offshore terrace or delta by river deposition; and even (iv) deposition by ice-sheets (to which the Grand Banks of Newfoundland are due in part). The definition of a c. s. has acquired a critical significance in connection with the international law of territorial waters, esp. with regard to fishing rights and the working of petroleum and gas resources.
[*f opposite*]

continental slope The marked slope from the edge of the C. SHELF to the deep-sea or abyssal plain, from about 180–3600 m. (100–2000 fthms.), in places much further [e.g. to 9000 m. (5000 fthms.) off the Philippines]. The slope is usually between 2° and 5°.
[*f* ABYSSAL]

contorted drift BOULDER-CLAY which exhibits foldings, twistings and irregularities, probably due to pressure from an ice-sheet. The contortions are usually drawn out in the direction of ice-movement; cf. PUSH-MORAINE. The proper name (with capital letters) is used to denote the upper of the two TILLS seen in the cliffs near Cromer, Norfolk, ascribed to the North Sea Glaciation (equivalent to the 2nd glacial period).

contour (-line) A line on a map connecting all points the same distance above (or below) a specific datum, loosely termed 'sea-level' (see ORDNANCE DATUM). A distinction is sometimes drawn between c.'s based on an instrumental survey, and *form-lines* sketched in on maps from gen. observations and from a few located spot-heights. It is also called an *isohypse*.

contour-interval The vertical distance between 2 successive c.'s on a map. This is chosen according to the amount of vertical height involved and the scale of the map.

contour-ploughing Ploughing along a slope rather than up and down it, to check runoff of rain which might wash away the soil; one of the measures to combat SOIL EROSION.

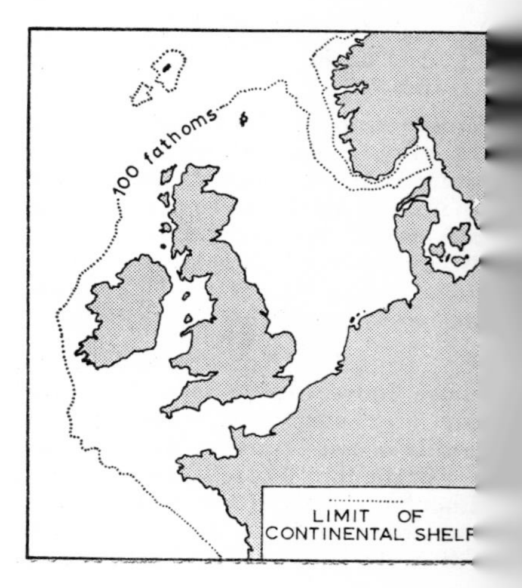

contrail See CONDENSATION TRAIL.

conurbation A term, orig. coined by P. Geddes (1915), for an extensive urban area, usually resulting from the coalescence of several orig. separate expanding towns or 'urban nuclei'. The term has also been more strictly defined as a continuously built-up area in which there is no apparent gap between several towns which have coalesced; this places emphasis on 'bricks-and-mortar' continuity. Some writers consider c. in a functional sense, with varying definitions. 8 c.'s are officially designated in Gt. Britain for Census purposes (population, 1961):

	(thousands)
Greater London	8172
S.E. Lancashire	2442
W. Yorkshire	1717
W. Midlands	2377
Merseyside	1392
Tyneside	856
Central Clydeside	1802
S. E. Wales	1908

Some writers greatly extend the use of the term, and postulate *minor* c.'s with as few as 50,000 inhabitants; others produce sub-divisions such as *uninuclear* and *polynuclear* c.'s.

convection (i) The mass movement of the constituent particles within a liquid or gas as a result of different temperatures and therefore different densities within the medium. The movement involves both the medium itself (*c. currents*) and the actual heat; the term is used for each. In a meteorological context, it involves vertical heat transference within the atmosphere (ct. ADVECTION, or horizontal movement). A *c. cell* is an updraught of heated air (syn. with a THERMAL), with a compensatory downward movement of cooler, denser air. (ii) *Forced c.* is due to wind turbulence over uneven terrain.

convection rain Rainfall resulting when moist air, having been warmed by CONDUCTION from a heated land surface, expands, rises, and is adiabatically cooled to the DEW-POINT. CUMULUS clouds develop into towering CUMULONIMBUS clouds with an immense vertical range, from which heavy r. or hail may fall, accompanied by thunder. C. r. occurs commonly during the afternoon near the equator, the result of constant high temperatures and high humidity. With increasing distance from the equator, the c. r. becomes associated more markedly with summer heating. In maritime temperate climates, most c. r. is associated with unstable Polar air-masses.

conventional projection A type of MAP P. constructed according to a mathematical formula, not a PERSPECTIVE P. The formula is selected to preserve some partic. property. See also SPECIF. P.'s.

conventional sign A standard s. used on a map to indicate a partic. feature, wherever the scale of the map is such that this feature cannot be drawn to scale. It may be a letter or a symbol. C.s.'s include boundary lines, which may or may not follow landscape features. A map carries a LEGEND showing the characteristic c.s.'s used.

convergence (i) In *oceans*, sharply defined lines separating converging masses of water, often of differing temperature and salinity; e.g. in the S. Ocean about 50°S., where cold dense Antarctic water meets warmer and more saline water spreading S. (ct. DIVERGENCE, DISCONTINUITY). (ii) In *climatology*, a type of air-flow such that in a given area at a given altitude inflow is greater than outflow, so that air tends to accumulate. If density remains constant, such horizontal c. must be accompanied by vertical motion; thus surface c. is usually accompanied by an ascending air current. C. may be produced either by STREAMLINES approaching each other (*streamline c.*), or by a single air current being subject to a progressive reduction of velocity (*isotach c.*). Areas with streamline c. usually have isotach c. as well; e.g. the INTERTROPICAL CONVERGENCE ZONE. Pure *velocity c.* occurs in the S. Indian

Ocean in July. (iii) *C. of species* (biological): the increase in degree of similarity between different species which are developing in such a way that their life forms become progressively more alike. (iv) The gradual decrease in thickness of a geological formation, so that the upper and lower horizons converge. [*f*]

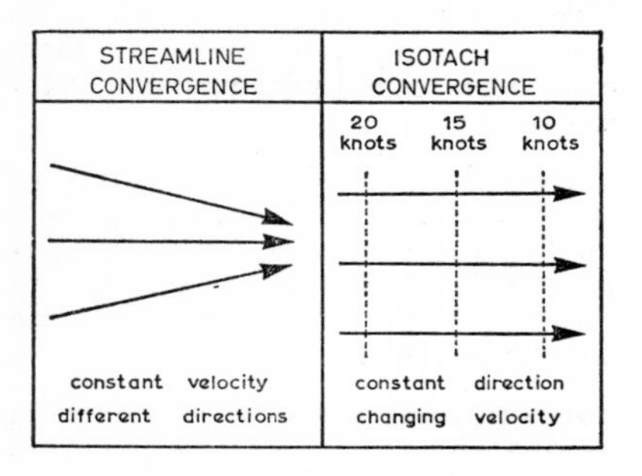

coombe See COMBE.

Coombe Deposit A mass of unstratified rubble from any rock type accumulating in a valley bottom and masking the lower slopes; the product of a past phase of solifluction. *C. Rock* is specifically the c. deposit of a Chalk landscape, being a compacted and hardened chalk mud (solifluction or melt-water) with added sand and flints. One of the main areas of occurrence is on the Sussex coastal plain. See also HEAD.

co-operation, co-operative system, co-operative An association of farmers, sometimes voluntary (Denmark, Netherlands), sometimes compulsory (E. European countries), to act as purchasing and distributing agents of seed, feeding-stuffs and fertilizers, and as grading and selling agents for the produce; e.g. creameries, bacon factories and egg-packing stations in Denmark and Ireland, fruit-handling c.'s in California, wine c.'s in S. France. The system allows farmers to operate their own holdings as individuals, yet has the advantages of bulk-purchasing, grading and standardization, and of large-scale contract marketing.

cop A small rounded hill in N. and central England; e.g. Mow C., 340 m. (1100 ft.) above sea-level, near Burslem, Staffs.

copper Almost certainly the earliest metal known to Man, since it occurred commonly on the surface as a free (*native*) metal, which could be used for ornaments and vessels, later alloyed with tin to make bronze (the 'Bronze Age'), and with zinc to make brass. It occurs in nature as a metal and in a group of over 300 different minerals (sulphides, carbonates, oxides), notably c. pyrites (c. sulphate or chalcopyrite, $CuFeS_2$), which commonly decomposes to form blue (*azurite*) and green (*malachite*) carbonates ($Cu_2Co_3 - OH_2$). The c. minerals are easily weathered, and the c. may be carried down to accumulate in a *zone of enrichment*, where it is mined in huge open-cast quarries. C. is widely used in the electrical industry and in alloys C.p. = U.S.A., U.S.S.R., Chile, Zambia, Canada, Congo (Kinshasa), Peru S. Africa.

coppice A small wood, notably o oak or hazel, periodically cut bac hard to the ground to produce stem for hurdles and fencing.

coprolite (i) The fossilized or petri fied excrement of creatures, esp reptiles (saurians), mainly of Jurassi and Cretaceous age. It consists c phosphatic nodules, and forms valuable source of fertilizer. (ii) Use more gen. for nodules or masses c calcium phosphate in sedimentar rocks, as in the Lower Greensand nea Cambridge and near Flamboroug Head.

copse A small cluster of trees an undergrowth.

coral A lime-secreting marine polyp mainly living in colonies in inte tropical seas. C.'s can only grow i clear, well-oxygenated water, wi plentiful supplies of microscopic life food; they cannot live in fresh or sil laden water, nor if its temperatu falls below 20°C. (68°F.). They ca not live at depths much exceedi

45–55 m. (25–30 fthms.), though are occasionally found as deep as 70 m. (40 fthms.), they cannot exist for long out of water, and are rarely found above low-tide level. They are confined to seas within 30°N. and 30°S., though exceptionally they grow further away (e.g. the Bermudas in the path of the warm water of the Gulf Stream), and are absent from the W. coasts of continents because of cool currents and up-welling cold deep water. The term is also used of the hard calcareous substance (*c. rock*), which is the accumulation of their skeletons. Throughout geological time, c.'s have been reef-builders; e.g. Wenlock Edge, Shropshire, which is in part a Silurian c. reef; the Corallian reefs of Jurassic age along the coast of Dorset near Osmington.

coral mud An accumulation of very fine fragments of c. found in the bathyal zone of the CONTINENTAL SLOPE around c. reefs.

coral reef A reef composed of c. limestone, the accumulated skeletons of c. polyp colonies. The three main types of r. are: (i) FRINGING R.; (ii) BARRIER R. [*f*]; (iii) ATOLL [*f*].

coral sand Comminuted fragments of c., ground small by the waves, forming white s. around reefs.

cordillera (Sp.) A series of mountain ranges, broadly parallel or closely *en échelon*, belonging to a single OROGENIC belt. Applied broadly to the whole 'Mountain West' of N. America. The C.'s de los Andes extend from Venezuela to Cape Horn, and are rarely less than 320 km. (200 mi.) wide. Also used in S. America for small individual ranges; e.g. C. de Corabaya, C. de Clonche, C. Negra.

core (i) The central mass of the earth, of about 3476 km. (2160 mi.) radius, bounded by the GUTENBERG DISCONTINUITY. It probably consists of a metallic mass of nickel-iron (NIFE), with a density of about 12·0. Recent work indicates the presence of an inner core, of 1380–1450 km. (860–900 mi.) radius, with a density of about 17·0 at the centre. See EARTHQUAKE [*f*]. (ii) Used of towns to signify the central functional area or business district (C.B.D.), hence *c. area*. (iii) Used of states to indicate the nucleus from which the state has grown (e.g. the Ile de France) or on which it depends; and of a region which has a focal area binding it together.

core sample A sample of soil, rock or ice, obtained by driving a hollow tube into the medium concerned, and withdrawing the sample intact. Such coring-tubes are even used on very long wires to obtain a c. s. of the material underlying the ocean floor.

Coriolis Force The effect of the apparent deflecting f. produced by the rotation of the earth upon a body moving on its surface, which is deflected to the right in the N. hemisphere, to the left in the S. (see FERREL'S LAW). The C.F. is proportional to $2vw \sin \phi$, where v = velocity of the object, w = angular velocity of the earth's rotation, and ϕ is the latitude. It was called after the French mathematician, G. G. de Coriolis, who discussed its effects in 1835, and the concept was developed by W. Ferrel in 1855.

corn Gen. grain, the seed of various cereal plants; specif. in N. America MAIZE, hence the C. Belt across the Midwestern states.

cornbrash A thin stratum of impure calcium carbonate in the Middle and Upper Jurassic, which weathers to form a stony soil. The Wiltshire dialect name for this soil was 'brash', and as it produced good crops of cereals the name was adopted for the formation by Wm. Smith (*c.* 1813).

cornice An overhanging edge of compact snow on the lee side of a steep mountain ridge, developed by eddying. Occasionally each side of a ridge may have a c.; e.g. the Obergabelhorn in the Pennine Alps, Switzerland.

corniche (Fr.) (i) Derived orig. from an architectural term; applied specif. to the road between Nice and

Mentone on the steep hill-side overlooking the Mediterranean Sea. It is now used for other similar coastal roads. (ii) Used in a physical sense, syn. with CLIFF.

corona A series of luminous concentric rings around the sun or moon, ranging from blue (inner) via green and yellow to red (outer), the result of diffraction of light by water-drops. Ct. HALO. Its angular diameter is much less than that of a halo. *Note:* The *solar c.* around the circumference of the sun is visible at a total eclipse.

corral (from Sp. *corro*) An enclosure, orig. of pioneers' covered waggons in Indian territory in N. America, for defence and for protecting stock. Later a wired enclosure for animals in U.S.A. and Latin America. Cf. KRAAL.

corrasion Mechanical erosion, or frictional wearing down, of a rock-surface by material moved under gravity or transported by running water, ice, wind and waves; c. is the process, ABRASION the result on the mass attacked. The term was first used by J. W. Powell in 1875.

corridor (i) A strip of territory of one state interrupting the territory of another to give access to some important port or to the coast; e.g. the pre-1939 Polish C. (ii) The term has been extended to cover an 'air c.', allowing airway access; e.g. three c.'s from W. Germany to W. Berlin, and for any prescribed international air-route over a country. [*f*]

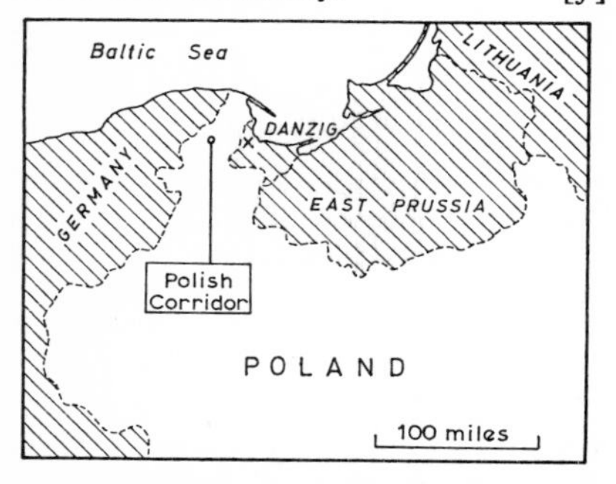

corrie (Scottish) See CIRQUE.

corrosion The wearing away of rocks by chemical processes, comprising: (i) SOLUTION (e.g. of common salt); (ii) CARBONATION (e.g. of limestone); (iii) HYDROLYSIS; (iv) OXIDATION; and (v) HYDRATION. The results are: (*a*) the conversion of original minerals into secondary weaker minerals usually more readily removable or soluble; and (*b*) the removal of 'cements' in sedimentary rocks, so that the consolidation or adhesion of the particles is weakened, and the rock tends to crumble.

corundum A mineral consisting of aluminium oxide (Al_2O_3), which in its finest form occurs as various gems, notably ruby and sapphire. It is No. 9 on the MOHS' SCALE of hardness, second only to diamond. It is used widely as an abrasive (emery), and for bearings on machinery. Most of the natural c. is mined in the N.E. Transvaal. Artificial c. is now made from fused bauxite.

co-seismal line A l. connecting points on the earth's surface at which an earthquake wave has arrived simultaneously. Ct. *isoseismal l.*, which is an indication of the amount of earthquake intensity.

cosmic App. to phenomena or features which occur or are situated beyond the earth's atmosphere; e.g. *c. dust*, *c. particles*.

cosmogony A theory or investigation into the origin of the solar system.

cosmography A word used extensively in the past to indicate the description and mapping of the universe, including the earth; a common title of many ancient 'physical geographies'.

cosmology The scientific investigation of the laws of the universe as an ordered entity by astronomers, physicists and mathematicians; e.g. 'the new c.' of F. Hoyle.

cosmopolitan Common to all the world, free from national prejudices, attachments, sympathies or limitations, with a world-wide outlook.

costa (Sp.) A stretch of coast, e.g. C. Brava; later applied to other resort areas along the coast of Spain; e.g. C. del Sol, Blanca, Dorada.

cost-benefit analysis A system of objectively comparing alternative proposals by quantifying in financial terms the total of costs and benefits that will accrue. These will include not only commercial costs, but also such factors as wasted time, nervous strain, traffic hazards and social benefits, for which *value judgements* must be made.

côte (Fr.) (i) An ESCARPMENT or steep-edge of a hill in France; e.g. C. d'Or. (ii) A section of coast; e.g. C. d'Azur.

coteau (Fr.) A name given by the early French explorers in N. America to a sharp ridge of hills or a prominent escarpment; e.g. C. des Prairies, Missouri C. Such names are still used.

co-tidal line A line on a tidal chart joining points at which high water occurs simultaneously; it radiates from an AMPHIDROMIC POINT, where the water remains at approx. the same level, the height of the tidal rise increasing outwards to the extremities of the c.-t. l. [*f* AMPHIDROMIC SYSTEM]

cottar, cottier A rural dweller in Scotland and Ireland, who, though not a CROFTER or farmer, rents a small plot of land (commonly less than half an acre) on an annual lease.

cotton An annual shrub, growing best in sub-tropical regions; it bears flowers, these leave bolls (seed-pods) which burst revealing a ball of fibres. The fibres vary in length from 64 mm. (*long staple*), producing fine c., to 24 mm. (*short staple*), yielding coarser grades of yarn. C.p. = U.S.A., U.S.S.R., China, India, Mexico, Brazil, Egypt.

cottonseed The seeds left after the fibre has been removed from the boll of the cotton plant. When crushed, they yield a valuable oil for making margarine and cooking-fat. C.p. = U.S.S.R., U.S.A., China, India, Brazil, Mexico.

coulée (Fr.) (i) A congealed lava flow, of basalt, rhyolite or obsidian, esp. in U.S.A. (no accent); e.g. in the San Francisco Mts., Arizona. (ii) An overflow channel which in the past carried melt-water from an ice-sheet, now dry; e.g. Grand C. in Washington State, the Ice-age channel of the Columbia R. (iii) A solifluction c., consisting of material moved by PERIGLACIAL action, having the appearance of a 'tongue' of debris.

couloir (Fr.) A steep, narrow gully on a mountain-side, esp. in the French Alps; e.g. the Whymper C. on the Aiguille Verte, above Chamonix.

counterdrift Deliberate planning to enable people living on the margins of large GROWTH AREAS to take part in the benefits resulting from such growth.

countertrades An obsolete word. See ANTI-TRADES.

country rock (i) A mass of r. which has been traversed and penetrated by later INTRUSIONS of igneous rock. (ii) In U.S.A., the gen. bedrock, cfd. unconsolidated material.

county (i) Territorial divisions of Gt. Britain and Ireland, many of which are of considerable antiquity derived from the pre-Norman shires. Now they are administrative units of the first order, though not including numerous large towns which are independent for local government purposes (c. BOROUGHS). Some cities have c. status; e.g. London, the city and c. of Bristol. Some larger c.'s have been subdivided for administrative purposes; e.g. N., W. and E. Ridings of Yorkshire, the 3 parts of Lincolnshire (Holland, Kesteven, Lindsey), E. and W. Sussex, E. and W. Suffolk. The total number of administrative c.'s in England (including the Isle of Ely, Isle of Wight, Soke of Peterborough and city of York) is 50, in Wales 13, in Scotland 33, in N. Ireland 6 (including c. of Londonderry). There are 26 c.'s in the Irish Republic. (ii) In U.S.A. the political and administrative unit in scale below a state (except in Alaska, S. Carolina and Louisiana); these total 3,051, with 5 more in Rhode Island which have no functions.

county borough A large town which has been given administrative status equivalent to that of a c., of which it may form a geographical but not an administrative part. There are 79 in England (1968) (of which 43 are also cities), 4 in Wales (2 cities), 2 in N. Ireland.

cove (i) A small rounded bay, usually with a narrow entrance; e.g. Lulworth C., Dorset. (ii) A steep-sided rounded hollow among the mountains of the English Lake District; e.g. West C., Green C. and Hind C. on Pillar Mtn., Ennerdale. (iii) Any steeply walled, semicircular opening, esp. at the head of a valley; e.g. Malham C., Yorkshire.

cover-crop A quick-maturing crop grown between main crops as a protective mat over the soil, so as to reduce the danger of SOIL EROSION.

'cradle-area' A region where an early civilization emerged and evolved; e.g. the Tigris-Euphrates valley.

crag (i) A steep, rugged rock projecting from a mountain side, esp. in the English Lake District; e.g. Dow C. (Coniston), Gimmer C. (Langdale). (ii) The name for compacted shelly sands in E. Anglia, hence the proper name of certain geological formations: the Coralline C. of Pliocene age, the Red, Norwich and Weybourne C.'s of Pleistocene age.

crag-and-tail A mass of rock ('the crag'), lying in the path of an oncoming ice-sheet, so protecting the softer rocks in its lee from erosion, and leaving a gently sloping 'tail'; e.g. the hard basalt plug of Edinburgh Castle Rock (the c.), with a t. of Carboniferous Limestone lying on Old Red Sandstone, sloping E., along which now extends the Royal Mile. The town of Clackmannan in Scotland has the ruins of a tower on a 58 m. (190 ft.) high c. with the main street extending to the E. down the t. The t. may also comprise TILL preserved in the area of stagnation in the lee of the protective mass; e.g. Arthur's Seat, Edinburgh. See STOSS. [*f*]

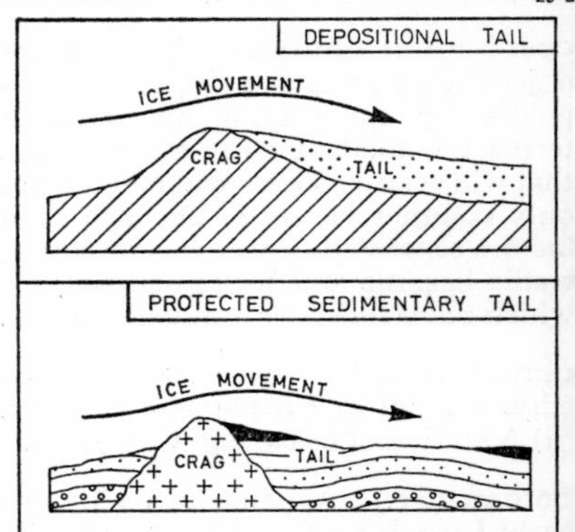

crater (i) The rounded funnel-shaped hollow at the summit of a volcano; cf. CALDERA. (ii) A depression caused by the impact of a meteorite; e.g. Meteor C., Arizona, U.S.A.

crater-lake A l. which has accumulated in the crater of a volcano; e.g. L. Toba in N. Sumatra; Öskjuvatn in Iceland; C. L., Oregon, 600 m. (2000 ft.) deep, the water surface at 1877 m. (6160 ft.) above sea-level, with a perimeter of about 48 km. (30 mi.). [*f*]

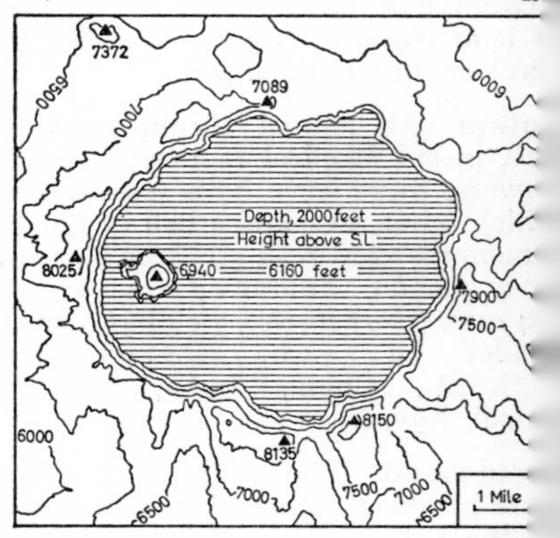

craton A stable and gen. immobile part of the earth's crust, usually on a large scale; syn. with SHIELD. Hence *cratonic*, *cratogenic*.

creek (i) A narrow tidal inlet, espec. on a low-lying coast among mud-banks; e.g. Ashlett C., Southampton Water. (ii) A small stream in U.S.A. (iii) An intermittent stream in Australia; e.g. Copper C., which (when there is water) drains into L. Eyre.

creep The slow gradual viscous movement downhill of soil and REGOLITH, lubricated by rain-water, under the influence of gravity. This may be observed in the landscape as an accumulation of material against stone walls on the up-slope side, leaving bent posts and trees, and in rock outcrops a distinct line of distortion. Basically the result of gravity, c. may be caused by numerous minor agencies working together. See CAMBER.

creole A person born in the W. Indies, but not of aboriginal descent; e.g. of French, Spanish or African negro origin. Initially no concept of colour was implied, but the word has become associated particularly with people of mixed blood.

crepuscular rays (i) Clearly defined r.'s of sunshine which break through chinks in a heavy layer of STRATOCUMULUS clouds towards the earth's surface. This is a transferred term, since it lit. means 'twilight'. (ii) Dark and light bands, seen just after the sun has passed below the horizon, and diverging upwards from the sun's position. The light bands are areas still illuminated by sunshine, the dark bands indicate where hills or clouds cut off the sun's r.'s.

crescentic dune See BARKHAN.

crêt (Fr.) An in-facing ESCARPMENT in the French Jura, forming a wall of a COMBE worn along the crest of an anticline by river erosion, culminating in a near-vertical limestone cliff (*la corniche calcaire*), below which a gradual slope, covered with scree and in part wooded, descends to the floor of the combe. The highest point of the Jura is a summit on such an escarpment, the C. de la Neige (1723 m., 5653 ft.).

Cretaceous The 3rd of the geological periods of the Mesozoic era, and the system of rocks laid down during that time, dated *c.* 135–70 million years ago. The constituent series in England are as follows:

Chalk: Upper, Middle, Lower	Pure white limestone
Upper Greensand	Sandstone
Gault	Bluish clay
Lower Greensand	Sandstone
Weald Clay	Thick clay
Hastings Sands	Sands and clays

The boundary between the Lower and Upper C. is variously interpreted; some geologists place this at the base of the Lower Chalk, as being the horizon of the most marked lithological change, others include the Gault and the Upper Greensand in the Upper C.

crevasse (Fr.) A deep crack or fissure in a glacier, trending either transversely across it where the slope increases, or longitudinally down it where the glacier spreads out because of the widening of its valley; in either case, differential movement within the ice causes tension and shearing. Intersecting c.'s on a steep slope will produce an ICEFALL; pinnacles of ice isolated by c.'s are SÉRACS.

crevasse filling A straight ridge of stratified sand and gravel, formed by the f. of a c. in a stagnant ice-sheet which later melted. This is very similar in form to an ESKER (*sensu lato*), but is not winding or branching.

crib (Welsh) A high summit ridge in the mountains; e.g. C. Goch on the Snowdon 'horse-shoe' of peaks.

critical mass Used in economic planning to indicate the degree of development that must be present before appreciable economic growth can occur.

critical path analysis An attempt to establish a pattern of a sequence of events and the time required for its completion, involving various steps, stages and factors. E.g. in the transport between the suburbs and the CENTRAL BUSINESS DISTRICT; in geomorphology the sequence of glaciation and deglaciation in a particular area. C.p.a. can be used in planning any particular project, in making a feasibility study, etc.

critical temperature A t. of specif. importance for vegetation; e.g. freezing-point (0°C., 32°F.), since many plants are vulnerable to frost, esp. when in blossom. Another is 6°C., 43°F., since for most plants active growth cannot take place below this point.

croft A small farm-holding in Scotland, hence *crofter*, the man who rents (or owns) and cultivates a c., and *crofting*, this system of farming of a barely subsistence character.

cromlech The Celtic form of a megalithic burial chamber (DOLMEN), with a flat roof-stone resting on vertical stones. From Welsh *crom* (crooked), *llech* (flat stone).

cross-cut A horizontal passage in a mine, driven at right-angles to the gen. trend of a vein of metal ore or a seam of coal. Ct. DRIFT, which follows it.

cross-cutting relationships, Law of An igneous rock is younger in age than any other rock across which it cuts; this is one method of establishing the relative age of rocks.

cross-faulting Two intersecting series of FAULTS.

cross-profile, of a valley A profile drawn transversely across a river valley, roughly at right-angles to the stream on its floor. The form of the c.-p. is determined by: (i) the nature of the rocks; (ii) the erosive power of the stream; (iii) the effects of weathering and rainwash on the valley sides; (iv) the stage which the processes in (ii) and (iii) have attained in their effects on (i). [f]

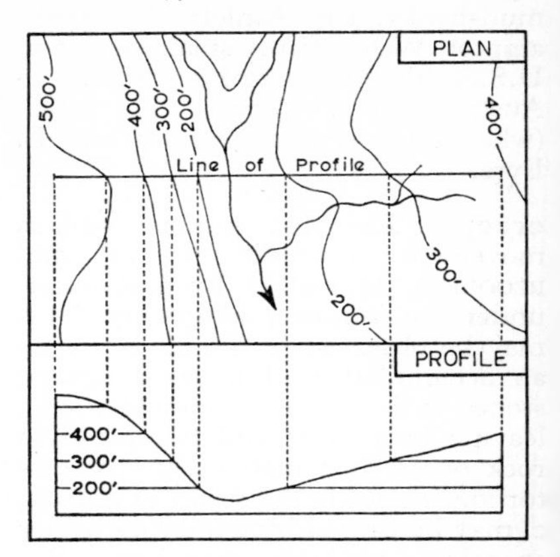

cross-wind A w. blowing more or less at right-angles to the course of a moving object or person; this is esp. important to aircraft.

crude capacity The capacity of a street or district to accommodate traffic, both moving and stationary without taking into account the maintenance of prescribed environmenta standards.

crumb-structure, of soil The accumulation of fine soil-particles int crumbs, thus coarsening the textur of a clay, and helping to make it mor workable. This is assisted by the floc culating effect of lime.

crush-breccia See FAULT-B.

crust, of the earth The upp granitic (SIAL) and intermediate bas (crustal SIMA) layers of the earth, abo 16–48 km. (10–30 mi.) in thicknes lying above the MOHOROVIČIĆ discon tinuity, and so forming the outermo 'shell' (or LITHOSPHERE). The usage this term dates from the time wh geologists thought of the earth as a h liquid body, with an outer cool, so 'skin'. It is now realized that this is t simple a picture. The analysis of t rocks comprising the c. by weight (i) *by elements:* oxygen 47%; silico 28%; aluminium, 8%; iron, 5 calcium, 3·5%; sodium, 2·5%; potassium, 2·5%; magnesium, 2·2%; titanium, 0·5%; hydrogen, 0·2%; carbon, 0·2%; phosphorus, 0·1%; sulphur, 0·1%; (ii) *by oxides:* silica, 59·1%; alumina, 15·2%; iron, 6·8%; lime, 5·1%; soda, 3·7%; magnesia, 3·5%; potash, 3·1%; water, 1·3%; titania, 1·0%.

[*f* ISOSTASY]

cryergic App. to physical phenomena resulting from frost action.

cryolaccolith See HYDROLACCOLITH.

cryopedology The study of structures in the ground resulting from intensive frost action, including: (i) the processes which have formed them; (ii) their occurrence; and (iii) the civil engineering problems produced in overcoming the difficulties which such structures present. The term and others of the same character were coined by Kirk Bryan in 1946.

cryoplanation Land reduction by intensive frost action, including SOLIFLUCTION and the work of rivers in transporting material produced by such action.

ryoplankton Microscopic organsms, both vegetable and animal, iving under conditions of permanent now and ice. See CRYOVEGETATION.

ryoturbation Frost action which isturbs and modifies the superficial ayers in the PERIGLACIAL zone. This erm has been gen. replaced by CONELITURBATION, which has a more recise scientific meaning.

ryovegetation Plant communities eveloping in permanent snow and ice, ence *cryophyte*; e.g. the purplishrown 'bloom' on the glaciers of S. laska, denoting the presence of algae.

yptovolcano An area of volcanic tivity, where the explosive energy of ses has shattered the basement rocks great depths, forming numerous PES as a result of FLUIDIZATION, and has ejected large quantities of shattered material (PYROCLASTS). E.g. in the Swabian Jura (W. Germany), esp. the Ries Basin, where activity is still manifested as hot springs around its margins.

crystal A solid aggregate of molecules with an orderly atomic arrangement, which may be bounded by symmetrically arranged plane surfaces; e.g. quartz is a 6-sided prism with pyramid terminations, a diamond is 8-sided, pyrite is cubic.

crystalline rock A r. with constituent minerals of crystalline form, developed either through cooling from a molten state (i.e. most igneous rocks, esp. plutonic ones which cool slowly at depth), or as the result of metamorphism (limestone recrystallized to form marble, shales and slates to form schist).

cuesta (Sp.) A ridge with a steep SCARP-SLOPE (or *escarpment*) and a gentler BACK-SLOPE. The term is now used particularly in connection with an asymmetrical ridge resulting from the differential denudation of gently inclined strata; a more resistant bed (limestone, chalk) may stand out as a low ridge separated by intervening vales worn in clay. In this case, the gentle back-slope of the ridge is practically parallel to the dip of the strata. This term is now more commonly used than *scarp* or *escarpment*, which should refer to only the steep slope. It is more succinct than 'scarped ridge'. [*f*]

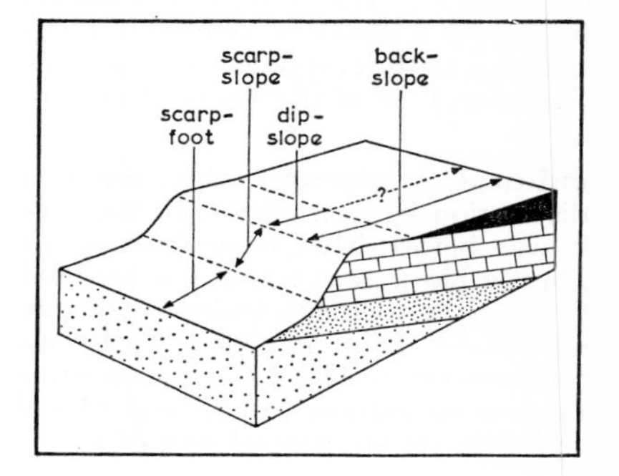

culmination (tectonic) When the strike of a NAPPE (i.e. a direction at right-angles to the folding movements) is traced, it seems to undulate, with higher c.'s alternating with DEPRESSIONS. The c. thus carried the overlying rocks of the nappes upwards, so that they have been more exposed to denudation and therefore largely removed, usually, as a result, exposing crystalline basement rocks in the Alps; e.g. Aiguilles Rouges—Mont Blanc, the Aar—Gotthard massifs. [*f*]

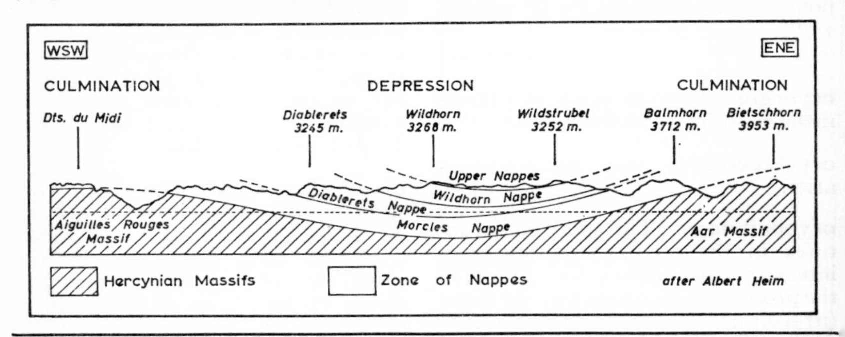

Culm Measures (i) A formation, chiefly of shales, sandstones and thin layers of impure anthracite, found mainly in Devon and E. Cornwall. (ii) The name C. is used for fine and dusty coal in S. Wales.

cultigen A plant which has been deliberately bred by man, and does not occur as a wild form in nature, e.g. cabbage.

cultural diffusion The areal spread of c. characteristics from the point of innovation or inception; see DIFFUSION (ii).

cultural geography The definition of this term has perhaps aroused more emotion and controversy than most. Some would accept everything in g. other than PHYSICAL G., others would limit it to a study of economic technologies, of the c. appraisal of resources (*c. ecology*), or of the study of culture areas.

cultural landscape Sometimes a distinction is drawn between the l. as it was before Man came (*natural* or *physical*) and that which has resulted from his settlement (*cultural*). Some authorities maintain that any distinction between them is undesirable, since the l. is an indivisible unit and Man's activities are an integral part of it.

cumulative frequency curve A f. c. in which values are added successively to each other and then converted into percentages. The 50th percentile on such a curve is the MEDIAN.

cumulo-dome A dome-shaped, apparently craterless, volcano, built up of successive flows of viscous acid lava; e.g. Grand Sarcoui in Auvergne, France. Syn. with MAMÉLON.

cumulonimbus (*Cb*) A CUMULU cloud which develops to an immens vertical height 10–11 km. (6–7 mi.) often with the upper part spread ou like an anvil; it is usually associate with thunderstorms and torrential rai or hail. From the side the cloud i dazzling white, from below its bas may be almost black. Its upper portio is popularly called a 'thunderhead'.

cumulose deposit A superfici mantle of organic d.'s, such as peat ar other swamp materials.

cumulus (*Cu*) A convection clou which grows vertically from a flat ba (corresponding to the condensati level) into a large white globular domed summit, sometimes of im mense height. 'Fair weather c.' di away in the evening as convecti currents lessen, others may devel into CUMULONIMBUS.

cupola (i) An irregular mass of rock protruding from the 'roof' of a BATHOLITH. (ii) A small isolated mass of intrusive rock lying near, though separate from, a batholith.

current (i) The distinct and defined movement of water in the channel of a river. (ii) The vertical motion of air in an air-mass (CONVECTION c.). (iii) A permanent or seasonal movement of surface sea-water (DRIFT); e.g. N. Atlantic Drift, Florida C., Benguela C. (iv) A minor ocean or sea c., esp. flowing through a strait, the result of differences in temperature and salinity at each end; e.g. a surface c. flows E. through the Straits of Gibraltar to make good the evaporation loss in the Mediterranean Sea and the W. deep movement of saline water. (v) A *tidal c.* through a restricted channel.

current bedding Thin strata or LAMINAE inclined at varying oblique angles to the general stratification, esp. in sandstone, resulting from the changing currents, both of water and wind, which were responsible for the deposition of constituent sand-grains. At intervals, a change of direction of wind or water may truncate the existing beds, and new layers at a different angle may be added later. Syn. with cross-bedding and false-bedding. [*f*]

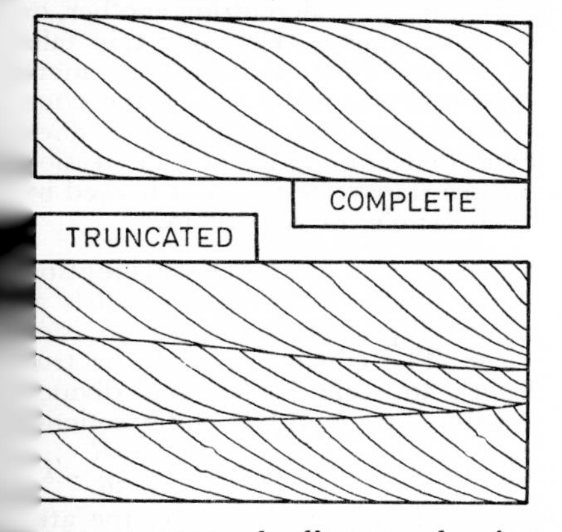

current rose A diagram showing for any point in the sea or ocean the % frequency of c. flow in each direction by means of proportional radiating rays.

curve-parallels (i) Superimposed graphs (in series) of related climatic phenomena; e.g. rainfall with relative humidity, vapour pressure or evaporation. (ii) Graphs used to demonstrate similarities of climatic trends over a period of time at different stations, with a horizontal time scale and a vertical scale in degrees of temperature or ins. of rainfall.

cusec A unit of measure of the discharge of a river; i.e. the number of cu. ft. per second passing a particular reach. 1 cusec = 538,000 gallons per 24 hours. The number of cusecs is obtained by multiplying the rate of flow in ft. per second by the cross-section in sq. ft. of the river at the point of observation.

cuspate delta A d. formed where a river reaches a straight coastline along which wave-action is vigorous, so that the material is spread out uniformly on either side of the river mouth; e.g. R. Tiber, Italy.

cuspate foreland A broadly triangular f. of shingle or sand projecting into the sea, caused by the convergence to an apex of separate SPITS or beach-ridges broadly at right-angles, formed by 2 sets of powerful constructive waves; e.g. Dungeness; Cape Kennedy (Canaveral) in Florida; and the Darss in E. Germany. [*f*]

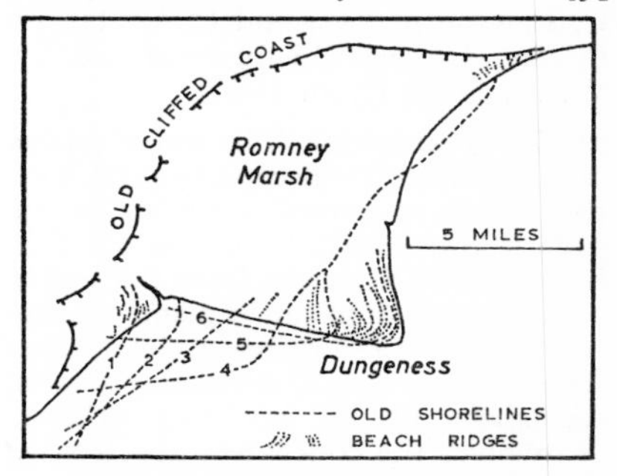

customs union A grouping between two or more states, whereby they create a common tariff area; e.g. the Belgo-Luxembourg Economic Union, succeeded by Benelux; the European Economic Community; the European Free Trade Area.

cusum chart A diagram in which cumulative sum techniques are used to analyse a series of slope-angles produced at regular distances along a SLOPE-PROFILE LINE.

cutoff A channel cut by a river as a short-circuit across the neck of an acute MEANDER, leaving an OXBOW. [*f* OXBOW]

cuvette (Fr.) A large-scale basin in which deposition of sediment has taken or is taking place; e.g. the Anglo-French C. (S.E. England and N.E. France).

CVP Index An i. involving the relationship between climate, vegetation and productivity, with specif. ref. to timber growth. Factors involved include the mean temperature of the warmest month, the mean annual temperature range, precipitation, EVAPOTRANSPIRATION and length of growing season.

cwm (Welsh) See CIRQUE.

cycle While this would appear strictly to mean a complete round of events or circumstances, its use is well established in geographical literature for a succession of stages or events which occur repeatedly in the same order. It is used partic. of the C. OF EROSION or (on a smaller scale) the C. OF SEDIMENTATION. It was used by A. C. Lawson in 1894 and C. W. Hayes in 1899, but was chiefly developed by W. M. Davis (from 1889).

cycle of erosion The modification of the physical landscape as a result of the action of natural agencies in an orderly progressive sequence; the full hypothetical c. ranges from the uplift of the land into an upland to an ultimate low, almost featureless plain. The concept was first developed and formalized by W. M. Davis (1889). It involves the concept of age: *youth, maturity, old age, rejuvenation.*

cycle of sedimentation The deposition of material in a basin during one complete phase of a marine TRANSGRESSION; i.e. dry land, shallow water, deep water, shallow water, dry land, with their respective associated deposits.

cyclogenesis The atmospheric process by which an intense tropical storm develops at a local heat-source (or 'warm-core') over the ocean; a tremendous vortical disturbance is set up, indicated by a whorl of towering CUMULUS clouds.

cyclone A small tropical low pressure system, with a diameter of 80–400 km. (50–250 mi.) occurring in the Arabian Sea and the Bay of Bengal between about 6°N. and 20°N. The BAROMETRIC GRADIENT is very steep, falling to 965 mb. (28·5 ins.) or lower, sometimes by about 40 mb. in a few hours. The centre (or 'eye') of the storm is a small area about 20 km. (12 mi.) across, but round it whirl winds of tremendous force (120–280 + km.p.h., 75–175 + m.p.h.). Torrential rain, associated with thunderstorms, occurs. It forms a great menace to shipping. At one time the term c. was used to denote any small unit-area of low barometric pressure, but in middle latitudes it should be avoided in favour of DEPRESSION, 'low' or 'disturbance', although the term CYCLONIC RAIN is still used.

cyclonic (frontal) rain Precipitation along the FRONTS and in the warm sector of a DEPRESSION in middle and high latitudes, caused by one air-mass overriding or undercutting another, or by CONVERGENCE. Drizzling rain falls in a broad belt, followed by more concentrated squally rain as the COLD FRONT passes. This rainfall is intensified by the effect of relief, as a depression crosses transversely a coast backed by uplands; e.g. in W. Britain.

Cylindrical Equal Area Projection A p. with meridians and parallels as straight lines, perpendicular to each other. The equator is made its true length to scale ($2\pi R$), and divided equally for meridian intersections; e.g. 10° meridian interval $= \frac{2\pi R}{36}$. To make this an equal area p., the area between any 2 parallels is made the same as on the globe to scale. Distance of each parallel from the Equator $= R \sin \theta$, where $\theta =$ latitude

The intervals between the parallels decrease from the equator, so that the increasing E. to W. exaggeration produced by the parallel meridians is balanced by compression in a S. to N. direction by increasing the closeness of the parallels. This p. is not much used, except in its oblique form.

cylindrical projection A type of p. in which the globe is regarded as projected upon a cylinder at a tangent to or intersecting its surface. In effect this is opened up and laid flat. The meridians and parallels are straight lines intersecting at right-angles; e.g. MERCATOR, GALL, CYLINDRICAL EQUAL AREA, EQUIRECTANGULAR (or *Simple Cylindrical* or *Plate Carrée*) P.

cymotrichous The quality of having 'wavy' hair; hence in some classifications of mankind (e.g. by A. C. Haddon) one main category is of the *Cymotrichi*. Ct. LEIOTRICHOUS, ULOTRICHOUS.

dacite A fine-grained EXTRUSIVE igneous rock, of the same composition as quartz-diorite. Such a volcanic peak as Mt. Lassen (Cascade Mtns., U.S.A.) is largely formed of it.

dairy-farming The rearing of cattle for milk, hence the production of .iquid milk (for sale), butter, cheese, ınd condensed, evaporated or dried nilk. Pigs and poultry are often ıssociated as side-lines. It is very widely distributed (e.g. around most arge towns, serving their big local narkets), but the economy of some ountries (Denmark, New Zealand) lepends on it to a large extent.

lale A broad open valley, as in the 'ennines (e.g. Swaledale, Wensleyale) and the English Lake District :.g. Borrowdale, Langdale).

'almatian coast A coast-line where ıe trend of the relief is broadly arallel to the c., and a rise of sea-level as affected its margins. The outer ınges have become lines of islands, ıe parallel valleys long inlets; e.g. the type-region of the Yugoslavian coast of the Adriatic Sea (hence 'Dalmatian'). [*f* CONCORDANT COAST]

dam A barrier built across the course of a river, to control and impound the flow of water. It may simply be an earth embankment (the largest earth-filled d. is the Mangla in Pakistan, part of the Indus Basin Project), or be constructed of rock with an impermeable cover or core, or made of solid masonry. It may retain water either through its weight (*gravity d.*), or because it is structurally braced (*arch d.*). D.'s have also been built of other materials, e.g. steel or timber. D.'s may serve a single or a number of purposes: (i) storing water for irrigation or human consumption; (ii) the creation of a head of water and a permanent supply for hydro-electricity production; (iii) maintaining supplies of water to canals or waterways; and (iv) flood control. E.g. *Gravity d.*: Aswan d. (R. Nile); Wilson d. (Tennessee R.); Grand Coulee (Columbia R.). *Arch d.:* Kariba (Zambezi R.); Chastang (Dordogne R.). See also BARRAGE.

Daniglacial stage One of the stages of the retreat of the Quaternary ice-sheets in N.W. Europe, *c.* 18–15,000 B.C. The ice left Denmark during this stage, though Norway and Sweden were still covered.

Darcy's law An equation expressing the rate of flow of ground water through an AQUIFER, put forward in 1856 by a Frenchman, H. Darcy. $V = P\frac{h}{l}$, where V = velocity, h = HEAD, l = length of distance of flow between 2 points, and P = coefficient of PERMEABILITY of the particular aquifer.

dasymetric technique A method of drawing a map to show density of population, departing from large administrative units with mean figures, and using 'reasoned guesses' to produce realistic categories, for which densities can be estimated.

data bank A storage system (or 'library') within a computer, which enables mappable d. to be stored either as lines (strings of co-ordinates on magnetic tape) or as points (pairs of co-ordinates either on punched cards or tape). The d. b. is an integral part of the process of AUTOMATED CARTOGRAPHY.

datum A fact, point or level which is known or taken for granted as the basis for reasoning, deduction or measurement. Thus datum-level is the zero from which land altitudes and sea depths are determined; see ORDNANCE DATUM. The scientific analysis of data has become a vitally important part of QUANTIFICATION in geography incl. d. collection, storage, retrieval and processing. Pl. *data*.

day-degree See DEGREE-DAY.

Daylight Saving Time See B.S.T. (British Summer Time).

dead ground An area invisible to an observer because of the form of the intervening land. [*f*]

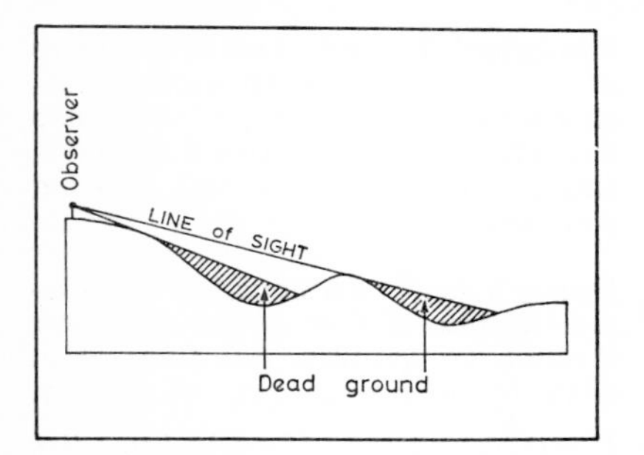

dead ice Stagnant i., usually covered with earthy material and boulders, at the margin of a stationary glacier or ice-sheet. This slowly melts, depositing the material as GROUND-MORAINE.

dead reckoning Navigation of a ship by carefully calculating from its speed and direction, with no external assistance such as radio or astronomical observations.

deadweight tonnage The weight in tons of the cargo, stores and fuel carried by a merchant vessel when down to her Plimsoll line, an indication of the cargo carrying and earning capacity of a ship. This measurement is gen. used for tankers. Ct. GROSS REGISTER, NET REGISTER, DISPLACEMENT T.

death-rate, crude A significant vital statistic in the changing total of population, expressed as the average number of deaths per thousand inhabitants. This is not a very refined measure of mortality, because the age-structure is not taken into account. An 'ageing' population, the result of better hygiene and medical developments, will ultimately cause the c. d.-r. to increase.

debris (Fr.) A superficial accumulation of loose material, disintegrated rocks, sand and clay that have been moved from their orig. site and re-deposited. Now used in English without an accent.

deciduous (forest) The property of annual autumnal leaf-shedding ('the fall') of broad-leaved trees and a few varieties of conifer in middle and high latitudes, as opposed to evergreens. In a Tropical Rain-forest leaf-shed may occur at any time, in a Monsoon Forest during the hot season.

decken structure (in Germ., *Decken-struktur*) A series of RECUMBENT FOLDS or OVERTHRUST masses, lying above each other.

declination (i) Of a *heavenly body* the angular measurement along a GREAT CIRCLE through both Celestial Poles and the body, measured from the Celestial Equator to the body; see CELESTIAL SPHERE. D. is therefore analogous to latitude in a terrestrial sense. The d. of the sun varies from $23\frac{1}{2}$°N. (about 21 June) to $23\frac{1}{2}$°S (about 22 Dec.); this can be found in the *Nautical Almanac*. The complement of the d. (90°−d.) is known as the *Polar Distance*. The d. of stars varies very little during the year. (ii) *Magnetic* the angular distance (either E. or W between true N. and magnetic N.; see AGONIC [*f*]. [*f* RIGHT ASCENSION]

décollement (Fr.) A type of earth-folding of a superficial nature, so that the overlying strata move easily over a basement surface of low friction. E.g. in the Jura Mtns., where the basement beds are of Tertiary ROCK-SALT and ANHYDRITE.

decomposition In geology, the break-down of the minerals in rocks, and hence of the rocks themselves, by chemical weathering (CORROSION).

deep, ocean An elongated trough or trench below the deep-sea plain, exceeding 5500 m. (3000 fthms.) in depth, usually near island arcs or coasts closely bordered by fold mountain ranges, hence sometimes called a *fore-deep*. Most occur in the Pacific; e.g. the Mariana Trench off Guam (11,033 m., 36,198 ft., the deepest known sounding), the Emden D. off the Philippines (10,794 m., 35,412 ft.), the Ramapo D. (10,554 m., 34,626 ft.), the Mansyu D. in the Mariana Trench (9866 m., 32,370 ft.). The d.'s are commonly asymmetrical, with a steeper slope near the land than the open ocean side. [*f* ABYSSAL ZONE]

deepening In a meteorological context, used to express a decrease in the atmospheric pressure at the centre of a low pressure system; e.g. 'a d. disturbance', involving a worsening of the weather, an increase of wind-speed, and probably an increase in precipitation. The opposite of *filling up*).

deep focus The SEISMIC FOCUS of an earthquake occurring at a depth greater than about 300 km.

deep-sea plain An undulating p. lying at depths between 3600 m.–5500 m. (2–3000 fthms.), occupying about 2/3rds of the entire ocean floor. Its surface is gen. covered with PELAGIC OOZES. From it rise extensive submarine plateaus, curving ridges, SEAMOUNTS, GUYOTS and volcanic islands.

deferred junction, of rivers A tributary, unable to join a main river because of LEVEES along the latter's course, may flow parallel for a considerable distance before effecting its confluence. It usually occurs on a flood-plain; e.g. the R. Yazoo enters the Mississippi Bottomlands and flows parallel to the main river for 280 km. (175 mi.) before it can effect a j. [*f*]

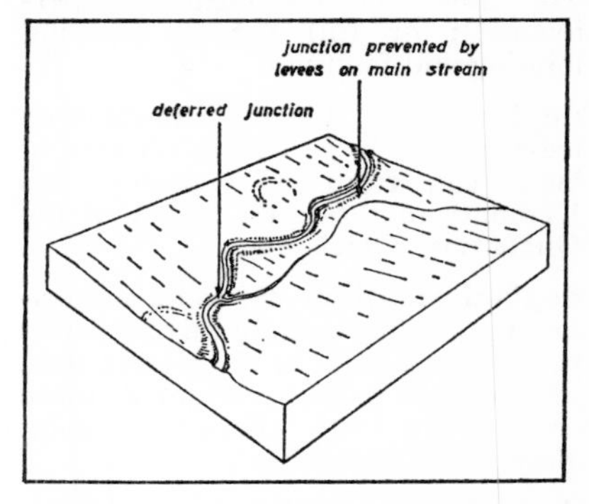

defile A narrow steep-sided pass through the mountains.

deflation The removal of dry unconsolidated material, esp. dust and sand, from the surface of the earth by wind. The finest material is borne high in the air and carried for many miles; coarser material is swept away in sand-storms; still heavier material moves along the surface in a series of swirling hops. D. can take place in any arid or semi-arid area, esp. where the protective mat of vegetation has been removed (e.g. the 'Dust Bowl' of S.W. U.S.A.), or along a sand-dune coast (e.g. the Landes of S.W. France). See BLOW-OUT. [*f*]

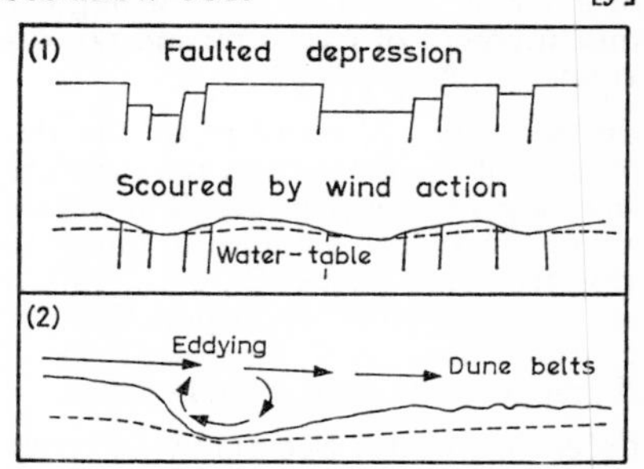

deforestation The complete felling and clearance of a forest.

deformation Any change in the shape, volume or structure of a mass of rock by folding, faulting, compression, shearing or solid flow, all the result of STRESS. The d. may be: (i) *elastic* (the rock will return to its original form); (ii) *plastic* (the rock will 'flow' and the change is permanent); or (iii) *rupture* (causing jointing and faulting).

deglaciation The withdrawal of an ice-sheet *in the past* from an area of land; e.g. from N. Germany and Denmark in the latter stages of the Quaternary Glaciation.

deglacierization By some authorities, the gradual withdrawal of a glacier from a land-mass *at the present time.* The term is partic. relevant to those areas of the earth where the present glaciers are shrinking.

degradation (i) The gen. lowering of the land-surface by physical processes (esp. by rivers), and the removal of material to be deposited elsewhere. (ii) The term has also been used in the more limited sense of the vertical erosion by rivers in order to maintain a GRADED state (ct. AGGRADATION). See also DENUDATION.

degree (i) A unit of temperature on any of the thermometric scales (FAHRENHEIT, CENTIGRADE or CELSIUS, RÉAUMUR and ABSOLUTE or Kelvin). For conversion table from Fahrenheit to Centigrade, see the entry under the former. (ii) A unit of angular measurement equal to 1/360th of a circle. (iii) The unit of angular measurement of LATITUDE and LONGITUDE.

degree-day A measure of the departure of the mean daily temperature from a given temperature. The values above or below significant CRITICAL TEMPERATURES (e.g. 0°C., 6°C., 18°C.) are added together for the period under consideration, and expressed as a single ACCUMULATED TEMPERATURE.

dejection cone See ALLUVIAL CONE.

dell A wooded hollow, used in a literary sense with no particular connotation. In Wisconsin, U.S.A., 'the D.'s' are valleys in the 'Driftless Area', eroded by the Wisconsin R.

delta A tract of alluvium formed at the mouth of a river where the deposition of some of its load exceeds its rate of removal, crossed by the divergent channels (DISTRIBUTARIES) of the river. It was orig. named after the Nile D., whose outline broadly resembles the Gk. letter Δ, though not all d.'s are of that shape (ct. BIRD'S FOOT D.). Not only is a river current checked where it enters the sea, causing deposition of sediment, but fine particles of clay coagulate and settle rapidly when they mix with salt water. The deposition of TOPSET, FORESET and BOTTOMSET [*f*] BEDS follows a regular sequence. The chief types of d. are: (i) ARCUATE [*f*]; (ii) BIRD'S FOOT (or birdfoot) [*f*]; (iii) CUSPATE. See also LACUSTRINE D. [*f*]

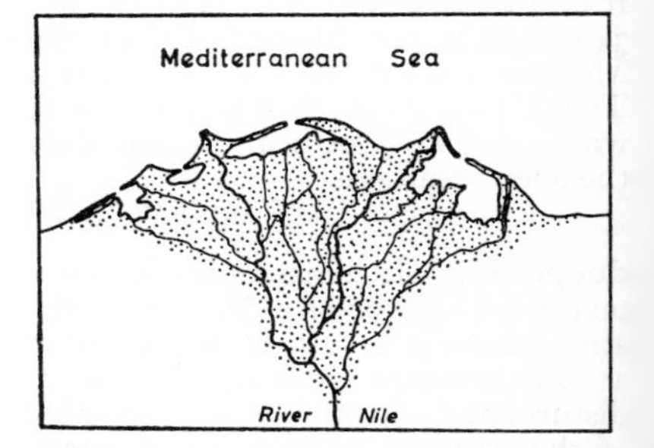

Demangeon's Coefficient of Dispersion Devised for the study of settlement patterns in France, using data from the Fr. Census, and employed to construct a map in the Fr National Atlas. It is obtained from the formula $C = \frac{E \times N}{T}$, where E = population of each *écart* (i.e. the population of a commune minus that of its chief place), N = number of *écarts* − 1, T = total population. This formula is possible because the Fr Census gives 2 returns of population for each commune: one for the chief place and one for the *écart*.

demersal fish A species, such as cod, haddock, whiting, halibut, plaice and sole, which lives near the bottom of shallow seas, esp. on banks; caught mainly in trawls.

demesne The part of a medieval manor retained by the lord for his own occupation or that of his servants, in contrast to that occupied by the villagers.

demographic coefficient An index designed to give a measure of future population pressure in any region; e.g. $C = dR$, where C is the d. c., d is the density in people per sq. mi., and R is the NET REPRODUCTION RATE. The same index for two generations distant is $C = dR^2$. Another d. c. index is $C = dt$, where t represents the rate of natural increase per 1000 inhabitants. Index values can be plotted for each administrative unit, and a choropleth map drawn.

demography The study and description of population; hence *U.N. Demographic Year Book.*

démoiselle (Fr.) An earth-pillar capped by a boulder which has protected the underlying material from weathering. These are partic. common in TILL and in volcanic BRECCIA; e.g. in the Chamonix valley, French Alps.

dendritic drainage A tree-like pattern of converging tributaries upon a main river (from Gk. *dendron*, a tree), usually where the rocks do not vary over its basin, or where there is no structural control; i.e. a type of INSEQUENT D. [*f opposite*]

dendrochronology The working-out of the climatic chronology of the past, using the evidence of tree-rings, since each reflects in its width the temperature and precipitation of the year in which it was formed. E.g. in S.W. U.S.A., where a master-index has been constructed from the Douglas fir; this is of great value to archaeologists dating pre-historic Indian cultures.

dendrogram A diagram used in the classification of data, to bring out associations and relationships in diagrammatic form. The result is a pattern of branching detail, hence the alt. name '*linkage tree*'. See LINKAGE ANALYSIS.

dene, den, dean (i) A steep-sided wooded valley; e.g. Lockeridge D., well-known for its SARSENS, and the Dean near Aldbourne, both in Wiltshire. (ii) An area of sand-hills near the coast; e.g. the Denes near Yarmouth.

density The concentration of matter, expressed as mass per unit volume, in gm. per cu. cm. The unit of d. employed is that of water; at 0°C., 1 c.c. of water weighs 0·999878 gm. which is near enough to 1·0. D. is an *absolute* quantity, while SPECIFIC GRAVITY (numerically the same) is a *relative* quantity. Warm water is less dense; e.g. d. at 15°C. = 0·999154 gm. per c.c. D. of ice at 0°C. = 0·91752; of charcoal = 0·34; of granite, 2·70; of lead, 11·36; of platinum, 21·53. D. of SIAL = 2·65–2·70; of SIMA = 3·0–3·3; of olivine, 3·27–3·37; of the earth as a whole = 5·527; of the core of the earth = about 12·0; of air at 0°C. and 766 mm. barometric pressure = 0·001293. The d. of sea-water depends on both salinity and temperature. For a particular salinity, d. varies inversely with the temperature. The mean d. of surface water for the whole ocean = 1·0252; from 3600 m. (2000 fthms.) down it remains constant at 1·0280.

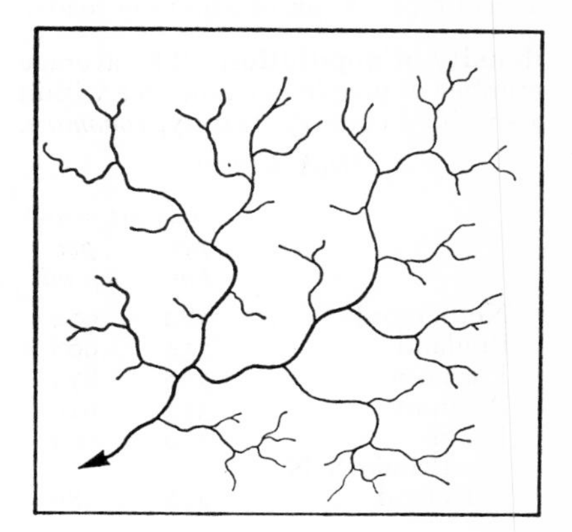

At 15·5°C., fresh water has a d. of 1·000; water with salinity of 30‰ = 1·0220; of 40‰, 1·0300 gm. per c.c. Muddy water has a greater d. than clear. The d. of a gas is proportional to its molecular weight. If the d. of air is taken as 1·0, that of hydrogen = 0·0693, and for all other gases d. = 0·0693 × molecular weight. *Conversion:* 1 lb. per cu. ft. = 16·019 kg. per sq. m.; 1 kg. per cu. m. = 0·062 lb. per cu. ft.; 1 gm. per cu. m. = 0·036 lb. per cu. in.

density current An ocean c. caused by differences in d. of the water (the result of varying salinity and temperature); e.g. salty dense water from the Mediterranean Sea flows W. along the bottom of the Straits of Gibraltar, while less saline, less dense water flows E. on the surface. It is estimated that the entire water in the Mediterranean is thereby changed every 75 years. Again, dense cold water from the Arctic Ocean sinks and flows S. under the less dense warm water of the Atlantic Ocean.

density, in town planning *Residential d.* is expressed as: (i) persons per acre; (ii) houses per acre; (iii) habitable rooms per acre; and (iv) persons per habitable room. The d. of a *commercial* or *business* district is expressed as either: (a) *plot ratio* = total area of floor space within buildings on plot/total ground area of site; or (b) *floor space index* = total area of floor space/total area of plot + ½ area of adjoining roads.

density of population The average number of people per unit area within a specified country, county, *commune.*

High density

	(recent year) per km.	per sq. mi.
Netherlands	382	972
England	354	908
Formosa	338	875
Belgium	313	810
Japan	270	703
Gt. Britain & N. Ireland	226	586

Low density

	per km.	*per sq. mi.*
Sweden	18	46
Norway	11	30
New Zealand	10	26
Saudi Arabia	3	7
Bolivia	4	9
Canada	2	6
Iceland	2	5
Australia	2	4

Very much higher and lower figures are often characteristic of smaller units of area; e.g. (very high) Singapore, Hong Kong, Gibraltar; (very low) Tristan da Cunha, Spitsbergen.

denudation The operation of all natural agencies by which parts of the earth's surface undergo destruction, wastage and loss through WEATHERING, MASS-MOVEMENT, EROSION and TRANSPORT. (Sometimes d. is used syn. with erosion, but the latter excludes weathering.) The term orig. meant the 'laying bare' of rocks by removal of the material covering them. Now it is commonly used as syn. with DEGRADATION, though W. M. Davis distinguished between d. as the *early* very active processes in the cycle of landform development, and degradation as the *later* more leisurely processes. Others regard denudation as the actual *processes,* degradation as the collective *result* on the landscape.

denudation chronology The study in geomorphology of the sequence of events leading to the formation of the present landscape.

dependency A territory and its population subject to rule by another country; this term, or better 'dependent territory', is now commonly used instead of 'colony'; e.g. Falkland Islands D.'s.

deposition The laying-down of material which has been transported by running water, wind, ice, the tides and currents in the sea; it is the complement of DENUDATION in the whole sum of processes of physiographic change. D. includes not only the mechanical laying-down of material, but also chemical precipitation, the formation of a crust or layer by evaporation, and the growth, accumulation and decay of living organisms (e.g. coral, humus, peat).

depressed area An a. which suffers economic distress as the result of massive unemployment, commonly the result of the decline of a major basic industry on which the area has long depended. Hence also '*Distressed A.*'. Both terms were unfortunate and were obviously disliked by the inhabitants, and in Britain they were changed officially first to '*special a.*' and then to the more promising DEVELOPMENT A.

depression (i) (*physiographic*) Any hollow in the earth's surface; more specif. a low-lying area wholly surrounded by higher ground, with no outlet for surface drainage; i.e. a basin of INTERNAL (or inland) DRAINAGE; e.g. the Lop Nor d. in central Asia. (ii) (*tectonic*) In a d., NAPPE structures are carried downwards and are therefore largely preserved in the present mountain ranges; see CULMINATION [*f*]. (iii) (*atmospheric*) A low-pressure system in mid and high latitudes, formerly called a CYCLONE, now often a 'low' or a 'disturbance'. This term is used more commonly in Gt. Britain than elsewhere. [*f*]

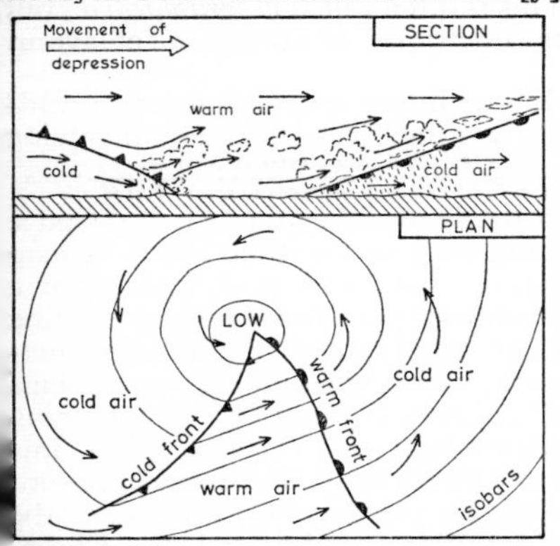

depth hoar The surface area of a snowfield is cooled at night, because of RADIATION, while the lower layers of snow retain relatively warmer air within the pores. This air rises through the snow layers, and becomes cool and supersaturated; its moisture is deposited directly on to snow crystals which grow as prisms and cups, with weak points of contact. The result is the creation of a layer of fragile ice, which may form a 'bridge' across a CREVASSE, dangerous for a mountaineer to attempt to cross.

deranged drainage In drift-covered areas, the irregular deposition of material may produce a disordered drainage pattern, with a confused intermingling of isles, marshes and streams; e.g. in Finland; the Canadian Shield.

desalinization, desalination The removal of salt from soil or water. (i) After reclamation of areas formerly covered by the sea, the marine clays are salt impregnated; this is removed by repeated deep ploughing, exposure to weathering, the continual pumping away of rain-water, the application of gypsum (which makes a highly soluble compound, calcium chloride), and the cultivation of salt-tolerant and absorbing plants; e.g. in the sea-polders of the Netherlands. (ii) In an area of constant irrigation in a hot climate where evaporation is rapid, the saline content of the soil steadily increases; this is avoided (e.g. in Imperial Valley, California) by flushing fresh water at intervals through miles of pipe-drains. (iii) The production of fresh from salt water; e.g. in Kuwait.

desert An area of land with scanty rainfall and therefore with little vegetation and (under natural conditions) limited human use. The main categories of d. are: (i) the hot, trade wind d.'s, areas of high atmospheric pressure, with airnfall less than 250 mm. (10 ins.), high summer temperature (W. Köppen's *BWh* group); e.g. the Sahara; (ii) coastal d.'s, on the W. margins of continents, in latitudes 15°–30°, with cold offshore currents, and summer temperatures (about 18°C., 65°F.) low for latitude (*BWk*); e.g. Atacama, Kalahari; (iii) mid-latitude d.'s of continental interiors, with high summer and low winter temperatures (*BWk*, *BSk*); e.g. the Gobi D.; (iv) the ice and snow d.'s of Polar lands (*EF*); e.g. Greenland, Antarctica. See also ARID.

desert pavement An area of REG, or pebbly d., where the abrasive action of the wind carrying a load of quartz-sand has ground and polished the upper surfaces of the pebbles, leaving a closely packed layer; sometimes this is called a *d. mosaic*. The pebbles are commonly cemented together by the precipitation of salts drawn to the surface in solution by capillarity, and left by evaporation.

desert varnish The hard, glazed surface of a rock in a d., the result not of 'sand-blast' scouring, but of a film of iron oxide, sometimes with manganese oxide, deposited by evaporation from the strong solutions brought to the surface by capillarity.

desiccation (i) Gen., drying up. Partic. this implies the progressive increase in aridity of an area, usually the long-term result of a climatic change towards a decreasing precipitation, e.g. the d. of central Asia, but sometimes the result of deforestation, over-cropping, or a failure to continue irrigation (e.g. due to war or pestilence). (ii) The loss of moisture from pore-spaces in a soil or a sediment.

desiccation breccia A mass of angular material, the result of the drying-out and cracking into irregular sun-baked polygons of an area of formerly wet clay or mud; these broken slabs were later washed away by flood waters, redeposited and compacted.

designation The process laid down in the 1947 and 1962 Town and Country Planning Acts, under which sites required for compulsory acquisition were defined in a DEVELOPMENT PLAN. The term is now obsolete, but is still frequently and erroneously used to describe all the proposals of the development plan.

desilication The removal of silica (SiO_2) from the soil in areas of heavy rainfall; thus the soils of tropical rain-forests are gen. deficient in silica, as well as in bases.

desire line In transportation studies, a straight l. drawn on a map joining 2 points between which there is a desire or reason to travel, though not necessarily the actual route to be followed. On a d. l. diagram, one l. is drawn for each separate movement, contiguous to the next, so that the overall thickness of the combined l.'s gives an indication of the total number of desired movements.

destructive wave One of a series of storm-w.'s, following each other in rapid succession, with an almost vertical plunge of water on to the beach as each wave breaks, and with a more powerful BACKWASH than SWASH. Thus they 'comb down' the beach, and move material seaward.

determinism The philosophical doctrine that man is largely conditioned by his environment, which therefore determines his pattern of life; ct. POSSIBILISM. Strictly, the concept is that of *environmental d.* Usually a partic. set of environmental factors, less than the total environment, is used to explain man's conduct, esp. the factors of the physical environment. This is often called *geographical d.*, but as a doctrine it is now gen. out of favour.

detour index In transportation studies, a numerical statement involving

$$\frac{\text{the shortest distance between 2 points}}{\text{the most direct route between the same 2 points}} \times 100$$

in terms of either the actual distance (km., mi.), or the time taken. A map may be produced of d. indices for a whole series of pairs of nodes, resulting in an informative transportation network, which will emphasize problems of physical obstacles, intervening attractions, etc. Technological improvements may result in significant changes in the network; e.g. the Mont Blanc road-tunnel, the Chesapeake Bay Bridge-Tunnel, the Verrazano Narrows Bridge (New York Harbor), a possible English Channel Tunnel or Humber Bridge.

detritus Fragments of disintegrated rock-material that have been moved from their orig. site; DEBRIS is now more commonly used.

development Under the Town and Country Planning Act (1968), d. refers to the use of land for building, re-building, mining and engineering operations in gen. A planning authority exercises statutory control over such d. in accordance with the D. PLAN in force.

Development Area A specified region to which, because of large-scale unemployment in a basic industry, Government encouragement for new industrial activity has been directed: Merseyside, W. Cumberland, N.E. Coast, Clyde Valley, N.E. Scotland, S. Wales, N. Ireland. Orig. created under the Special Areas Development and Improvement Act (1934), various alterations of nomenclature have been made as a result of later legislation and changes in the economic position. Thus under the Local Employment Act (1960), *D. Districts* were scheduled, applying to local administrative units in which unemployment was over 5% of the insured population. In 1965, when regional planning councils were established, the individual D.A.'s lost their identity; thus W. Cumberland became part of a larger D. A. under the Northern Economic Planning Council (comprising Cumberland, Westmorland, Northumberland, Durham). *Special Districts* are within the D.A.'s (e.g. Workington, Whitehaven and Millom), which qualify for extra grants to establish new industries.

development control The process whereby a local planning authority carries out its statutory duty to control d. in accordance with the policies laid down in its D. PLAN by granting or withholding consent to applications for planning permission.

Development Plan A statutory p. which a local planning authority (county or county borough) is required to prepare under the Town and Country Planning Acts (1947, 1962, 1968), indicating the manner in which land in the authority's area should be used and the steps by which d. should be carried out. It comprises a Written Statement, amplified at present by 1 in. to 1 mi. county maps and 6 in. to 1 mi. town maps, shortly to be superseded by STRUCTURE PLANS and LOCAL PLANS. The p. has to show proposals for about 20 years ahead and must be revised at intervals of not more than 5 years.

deviation (i) Used by P. Haggett (1965) with reference to AREAL DIFFERENTIATION as the traditional and fundamental approach to geography, from which the 3 d.'s are 'the landscape school', 'the locational school', and 'the ecological school'. (ii) (*Magnetic*) See DECLINATION (ii).

Devonian The 4th of the geological periods of the Palaeozoic era, dating from *c.* 400–350 million years ago, and the system of rocks laid down during that time. In the British Isles these rocks occur in Cornwall; Devon (whence the name); S. and central Wales; the S. Uplands, the Midland Valley and N.E. of Scotland; the Orkneys; and S.W. Ireland. In Devon and Cornwall the system consists of sandstones, grits, slates and limestones, for the most part laid down in the sea. Elsewhere it comprises red and brown sandstones, conglomerates, marls and limestone, probably laid down in lakes; these rocks are known as Old Red Sandstone. The Caledonian orogeny, which began in the Silurian, continued well into the D. It was a period of vulcanicity, with intrusions of granite in the Grampians, the S. Uplands (Criffel), the Cheviots, and the English Lake District (Shap, Eskdale).

dew The condensation of water-droplets on the surfaces of plants and other ground objects, the result of cooling by nocturnal radiation to a temperature below the D.-POINT of the layer of air immediately resting on the earth's surface. D. forms partic. under conditions of a clear, still atmosphere; it requires high humidity at the surface, and favourable radiating surfaces surrounded by air; e.g. blades of grass. The d. of spring and early summer is mainly derived from water-vapour in the lowest layers of the atmosphere, while in autumn it comes from the ground itself, both directly and transpired by plants.

dew-mound Used in the drier parts of Israel, Libya and other desert lands, esp. in the Middle East. A mound of earth is covered with flat stones on which d. condenses, trickling between them into the earth, which is thus kept moist. Olives, oranges and other fruit-trees grow out of the mounds, and are supplied with water.

dew-point A critical temperature at which air, being cooled, becomes saturated with water-vapour (i.e. the RELATIVE HUMIDITY = 100) and below which condensation of the excess vapour causes the formation of minute drops of water, provided that some nuclei for condensation are present. D.-p. can be determined using a HYGROMETER, with tables, or by a special d.-p. hygrometer, using a polished metal surface which can be cooled until a film of moisture appears at a temperature which is recorded.

dew-pond A shallow artificial hollow containing water, usually lined with puddled clay and straw, found commonly in the chalklands of S. England. Most of the water is derived from rainfall, some by condensation of sea-mists from the Channel, but only a small amount actually from dew.

diabase The American term for dolerite in gen., but used specif. in Britain for dolerite which has been much altered by the decomposition of the felspars and the mafic minerals.

diaclinal Applied by J. W. Powell in 1875 to streams and valleys which have a direction at right-angles to the strike of the rocks, crossing the axis of a folded structure.

diagenesis Syn. with LITHIFICATION.

diagonal scale A scale in which the unit-lengths shown can be subdivided with precision. In the [*f*] below, the upper arrow indicates a length of 22 mi., 4 furlongs, the lower of 37 mi., 7 furlongs. [*f*]

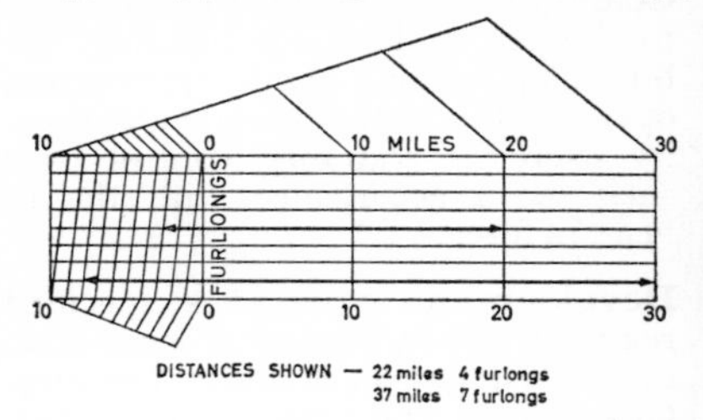

diamond A crystalline form of carbon, formed under conditions of extreme heat and pressure, the hardest known mineral, with the highest number (10) in the MOHS' (HARDNESS) SCALE. It is usually found embedded in 'pipes' of igneous rock, or may be washed out and redeposited in 'placers'. Small, poor quality and artificial stones are used widely as industrial abrasives. Gemstones have an artificially maintained value; they are measured in points and CARATS. The largest d. ever discovered was the Cullinan, weighing 3032 carats (*c.* $1\frac{3}{4}$ lb.). Ct. carat as used for gold. C.p. = Congo (Kinshasa), S. Africa, Ghana, S. W. Africa, Angola, Sierra Leone, Tanzania.

diapir The upward piercing of ar anticlinal fold, forming cracks anc fissures, by either (i) igneous material (ii) masses of salt forced up througl cracks from a SALT-DOME (e.g. S.W France); or (iii) mud (hence MUD VOLCANO).

diastrophism, diastrophic Th forces which have disturbed or de formed the earth's crust, includin folding, faulting, uplift and depres sion, though not vulcanicity. Th forces are classified as: (i) EPEIRC GENIC; (ii) OROGENIC. Ct. TECTONIC.

diatomaceous earth A friab deposit consisting largely of aquat vegetable organisms (*diatoms*) whic secrete an external casing of silica; on death these accumulate on the floors of lakes and the sea. They are so minute that 40–70 million form 1 cu. in. It is found in Kentmere in Westmorland, in N.E. Skye, Japan, and in vast beds (over 300 m. thick) of Miocene age in California. The hydrous earth is dried to form *diatomite*, of chalk-like appearance; its chemical inertness, porosity, low thermal conductivity and lightness make it a good filling medium, insulator and filter material.

diatom ooze A siliceous o. consisting of skeletons of microscopic plants (*diatoms*) flourishing in the cold waters of oceans, deposited on the ABYSSAL ZONE of their floors. It is found in a continuous band round the S. Ocean in latitudes 50°S. to 60°S., and in the N. Pacific Ocean.

die-back A diseased condition of plants, often applied gen. to the dying-off of large tracts of similar species at the same time, sometimes for no apparent reason; e.g. *Spartina Townsendii* on the marshes around the coast of S. England, partic. in the Lymington and Beaulieu estuaries, where both '*pan d.-b.*' and '*channel d.-b.*' can be seen. The phenomenon is difficult to explain and may result simply from some slight environmental change.

differential denudation The results of weathering and erosion on an area where the rocks display very varied resistance; different rock-types are worn down to differing degrees, sometimes resulting in striking relief-forms. [*f opposite*]

diffluence, difluence The lateral branching of a glacier, so that part flows away from the main ice-stream. This usually results from some down-valley blocking, either by a narrowing of the valley profile or at the junction of a tributary glacier. The ice in the main glacier will build up, and when thick enough it may flow over a col in the valley side into a neighbouring valley. This ice-flow may cut down the col so much that in post-glacial times the river occupying the valley may take a new course through the overflow-channel. E.g. in Scotland, between the Cairngorms and Grampians, the upper Feshie flows N.W. through a channel across a pre-glacial watershed, the result of past d. In the Karakoram Mtns. the N. Rimu glacier sends off a lobe of ice N.E., its melt-water flowing to the Yarkand R., and on into the Lop Nor depression; the other lobe joins the main Rimu glacier, hence to the Shayok R. and the Indus.

diffraction In meteorology, the process of radiation spreading akin to scattering.

diffusion (i) Used in meteorology to describe the apparently random mixing of air bodies, either by *molecular d.* (which is a slow process of mixing of little relative importance), or by *eddy d.* (the result of turbulent motion). (The term is also used in respect of liquids and light.) (ii) Used to refer to the spread or propagation of some feature from a given point or points over a progressively wider area: plants, animals, peoples, ideas, cultures, languages, techniques, religions, etc. Many attempts have been made to produce random d. patterns and models, involving problems of distance, the nature and location of physical, economic and other obstacles, etc., in the light of changing technological availabilities.

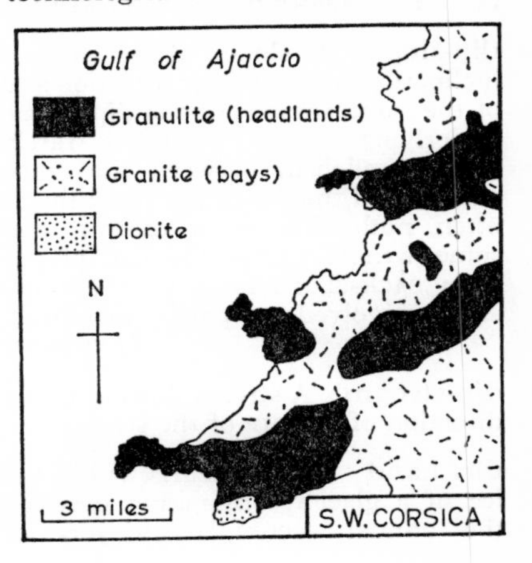

dike See DYKE. The form dike is usual in U.S.A.

diluvium, diluvial An almost obsolete term, orig. applied to materials thought to be deposited by the Genesis Flood, then to glacial drift. It is still sometimes used to denote the older Quaternary deposits, in ct. to the younger ALLUVIUM.

Dines pressure tube anemometer A self-recording instrument measuring the direction and velocity of the wind. It consists of a prominently placed vane, with a shaft going down to a recording unit; the front end of the vane has an opening, kept facing into the wind, communicating changes of pressure (and thus velocity) to the recorder.

diorite A coarse-grained intrusive igneous rock, with much plagioclase felspar, the rest various ferro-magnesian silicates; it is intermediate in composition between acid and basic rocks. It occurs in N. Wales, in the Malvern Hills, and near Nuneaton (Warwickshire) where it is quarried.

dip The maximum angle of slope of a stratum of sedimentary rock at a given point. The angle of inclination is given in degrees from the horizontal, not the surface slope of the land, and direction is expressed as a compass bearing. The direction of d. is at right-angles to the STRIKE. Ct. TRUE D. and APPARENT D. [*f* STRIKE]

dip-slope A slope whose surface inclination is in the same direction and of the same amount as the dip of the underlying strata. There is rarely exact parallelism between the land-surface and the dip of the rock; this is often noticeable on a CUESTA, and some writers prefer the use of the term BACK-SLOPE to d.-s. where the angles involved are partic. divergent. [*f* CUESTA]

dip-stream A stream flowing in the direction of the DIP of the strata.

dirt band A b. of debris within the ice of a glacier, originating between the annual accumulation-layers of fresh FIRN. Its exposure at the surface of the glacier may form one type of OGIVE.

disappearing stream A surface s. which vanishes underground, usually down a SWALLOW-HOLE when it passes on to limestone or other particularly pervious rock; the water works its way down through joints, enlarging them by both carbonation-solution and erosion, towards the base of the limestone, where it emerges as a RESURGENCE; e.g. the R. Jonte in the Grands Causses, Central Massif of France, which disappears underground near Meyrueis and reappears at Les Douzes. [*f* SWALLOW-HOLE]

discharge, of a river The quantity of water passing down a stream, depending on its volume and velocity, expressed in CUSECS, or in U.S.A. in *c.f.s.* (the number of cu. ft. per second passing a specific section), or in cu. m. per second. River d. is measured at a gauging-station, using a current-meter and a gauge for ascertaining the depth of water. See RATING CURVE.

disconformity A type of UNCONFORMITY involving a non-sequence of beds but not a contrast in their dip and strike. Sedimentary rocks were subject to prolonged denudation which removed certain strata, and then material was deposited on them to form new strata parallel to the underlying ones, without any folding or other earth-movements involving angular change either of the under- or over-lying rocks. A d. may be ver difficult to recognize, usually requirin the evidence of distinctive fossils. E.g the section revealed in the walls of th Grand Canyon, S.W. U.S.A., where (i) the Redwall Limestone of Devonia age was deposited directly on the Mua Limestone of Cambrian age, rep resenting an interruption of depositio of about 150 million years; and (ii) th 'Great Disconformity', where Cam brian rocks rest on the eroded surfac of the Pre-Cambrian basement rock [*f* UNCONFORMIT

discontinuity (i) A plane or thi layer of separation in the ocea between masses of water of contrasting temperature and salinity; e.g. in the N.W. Atlantic between the Labrador Current and the Gulf Stream Drift. (ii) A FRONT or frontal zone between AIR-MASSES of contrasted temperature and humidity; e.g. the Pacific and Atlantic Polar Fronts, Mediterranean Front, Arctic Front. (iii) A sudden change of character in the structure of the earth at great depths; the two main d.'s are the MOHOROVIČIĆ and the GUTENBERG. (iv) In recent geomorphological studies, a marked change of slope is referred to as a d., most commonly the result of an outcrop of a resistant band of rock, or to the interaction of different slope processes above and below the d., or to different erosional histories of various parts of the slope PROFILE. [*f* ISOSTASY]

discordance (i) A DISCORDANT INTRUSION. (ii) An ANGULAR UNCONFORMITY; i.e., where the BEDDING-PLANES of adjacent strata are not parallel.

discordant (adj.) Cutting across the general lines of the structure or 'grain' of the country. A drainage pattern which has not developed in a systematic relationship with, and is not consequent upon, the present structure. See also D. COAST, D. DRAINAGE, D. INTRUSION. Ct. CONCORDANT.

discordant coast A coastline trending transversely across the 'grain' of ridges and valleys (a transverse or 'Atlantic' coastline); e.g. S.W. Ireland, N.W. Spain. [*f* RIA]

discordant drainage A river which is discordant with the present geological structures over which it flows, the result of SUPERIMPOSITION, ANTECEDENCE, CAPTURE or GLACIAL DIVERSION. [*f* ACCORDANT DRAINAGE]

discordant intrusion An intrusion of igneous rock that cuts across the general stratification of the rocks into which it is intruded; e.g. a DYKE. Ct. SILL, which is an accordant i.

discordant junction Where a tributary stream abruptly joins a main river flowing at a markedly different level, as from a HANGING VALLEY. Ct. ACCORDANT JUNCTION.

dislocation Syn. with FAULT, where the strata are displaced relative to each other.

dismembered drainage A complex d. system that has been broken up by the submergence of its lower parts by the sea, thus creating a series of streams draining individually to the sea; e.g. the R.'s Frome, Stour, Avon, Test, Itchen, Hamble, Meon, which may once have been tributaries of the so-named 'Solent R.', now reach the English Channel, Solent and Southampton Water independently.

dispersal The movement of industry and population from large congested centres into declining country areas. In order to overcome the inherent attraction of large towns, d. will normally need positive inducement, and much of the work has been done by the establishment of New Towns, town expansion schemes, the work of the Location of Offices Bureau, and the encouragement of the Board of Trade.

dispersed city A situation in which a group of c.'s exist at an approx. similar functional level. Although the THRESHOLD POPULATION of the whole area is large enough to support a higher order place, this is absent. Instead, the higher functions are scattered amongst these lower order places. Hence the group of c.'s functions as a single unit, and are said to form a d. c. The term must not be confused with URBAN SPRAWL or with D. SETTLEMENT.

dispersed settlement A pattern of rural s., with isolated farms or cottages not grouped in villages or hamlets; ct. NUCLEATED S. E.g. in the provinces of W. and E. Flanders in Belgium.

dispersion diagram A d. used to indicate the distribution of any quantity for any unit of time over some period. In a rainfall d. d. a vertical column is used, with a scale of ins. or mm., in which one dot is placed to indicate each year's or month's total precipitation for a specif. place. From it the median and quartile values can be established.

displacement tonnage Used of warships in terms of the weight of water displaced; i.e. in effect, the weight of a vessel when fully laden. Actually this figure varies according to the density of the water (i.e. the degree of salinity), and a standard figure is used.

dissection In geomorphology, the cutting-up of a land-surface by eroding streams, specif. of an uplifted PENEPLAIN.

distributary An individual channel formed by the splitting of a river, as in a delta, which does not rejoin the main stream, but reaches the sea independently; e.g. the Grande Rhône and Petite Rhône; the Rosetta and Damietta branches of the Nile.

disturbance In meteorology, a DEPRESSION or 'low' of no particular intensity.

disturbance line An intense linear pressure system which travels W. across W. Africa, esp. in spring and autumn, which brings heavy falls of rain.

diurnal range The difference between the max. and min. value within a period of 24 hours, esp. of temperature; e.g. there is a high d. r. in a hot desert, where the clear skies allow great INSOLATION during the day and rapid terrestrial RADIATION by night, with a fall of 14–17°C. (25–30°F.) in the 2 hours succeeding sunset. Total d. r.'s of 33–40°C. (60–70°F.) are commonly recorded. Azizia in Tripoli has the highest ever recorded d. r., between 52°C. and −3°C. (126°F. and 26°F.).

diurnal tide In some areas of the oceans (e.g. in the Gulf of Mexico, around the Philippine I., off the coast of Alaska, and off parts of the coast of China), only one high and one low t. occur in each 24 hours. In these areas, because of the shape of the water-body, the *diurnal* component of the t.-producing forces is dominant. [*f* TIDE]

divagation The lateral shifting of a river's course as a result of extensive deposition in its bed, esp. in conjunction with the development of MEANDERS.

divergence (i) In the oceans, a line or zone from which the surface water moves away under the influence of wind-drift, thus causing the upwelling of deep water. (ii) In climatology, a type of air-flow such that in a certain area at a given altitude the outflow is greater than the inflow, so that the mass of air contained tends to decrease. If the density remains constant, such horizontal d. must be accompanied by some compensating inflow from a descending air current. Like CONVERGENCE, d. takes two forms: *streamline d.* and *isotach d.* D. is considered as a positive quantity, convergence as a negative form of d. An example of d. is in the area of descending and outblowing air in the HORSE LATITUDES. (iii) See GLACIAL D. (iv) In biology, the *d. of species*: the evolution of orig. similar species in such a way that their life forms become progressively less similar. [*f*]

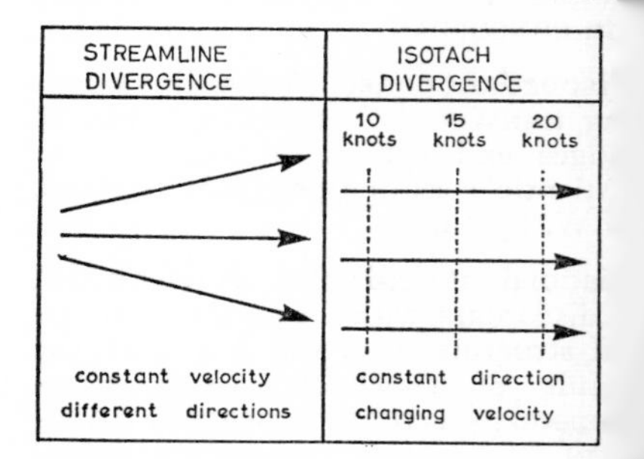

divide A ridge or area of hig[h] ground between river basins; th[e] highest line of an INTERFLUVE. Syn[.] with the English use of WATERSHE[D]. It is used specif. in N. America a[s] the *Continental Divide*, a line o[f] separation between Pacific an[d] Atlantic drainage.

divided circle A diagrammatic device whereby a c. is divided into sectors, each of which is proportional in size to the value it represents. The several categories can be shaded and labelled. [*f*]

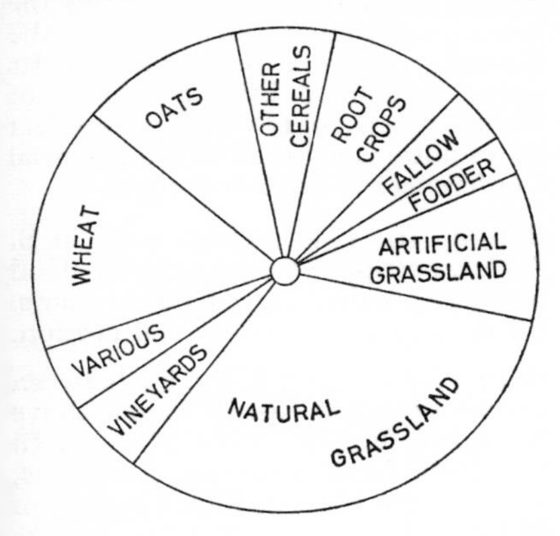

doab A low-lying, alluvium-covered area between two converging rivers, specif. in the Indo-Gangetic plain of India and Pakistan.

Doctor The HARMATTAN wind in West Africa.

dod, dodd A rounded hill or summit in N.W. England and S. Scotland; e.g. Dodd on the S. slopes of Skiddaw, Starling Dodd on the N. side of Ennerdale, Great Dod in the N. part of the Helvellyn range.

'dog-days' A popular name, esp. in U.S.A., for the hottest time of summer, associated with the star Sirius ('the Dog Star').

dogger A large nodule or CONCRETION found in sedimentary rocks under various conditions; e.g. ironstone nodules in the Jurassic rocks of Yorkshire; siliceous d.'s in the Coralian beds of Dorset, and in the Wenlock Shales of Silurian age in Shropshire.

doldrums A belt of light indeterminate winds in equatorial latitudes, notably over the E. part of the oceans, though displaced a few degrees of latitude with the seasons, lagging behind the overhead sun. The belt is characterized by high temperatures and humidity, and forms part of the INTER-TROPICAL CONVERGENCE ZONE.

dolerite A basic igneous rock (with 52–45% silica), of HYPABYSSAL occurrence in minor intrusions, such as SILLS and DYKES. It is usually dark coloured and fine or medium textured, and sometimes solidifies in columns; e.g. the Gt. Whin Sill of N. England, where the rock is known as 'whinstone' by quarrymen. It is called *diabase* in U.S.A.

dolichocephalic Long-headed. This is defined by anthropologists as a skull with a CEPHALIC INDEX of 75 or less. Hence *dolichocephaly*. Sometimes abbrev. to *dokeph*.

doline (Fr.), alt. **dolina** (It., Slavonic) A shallow funnel- or saucer-shaped depression found in limestone country (KARST). It commonly leads down into a vertical shaft. Recent international usage defines it as large enough to contain cultivable soil on its floor. A *collapse-d.* is a depression on the surface of the ground caused by subsurface solution-enlargement, with subsequent collapse.

dolmen A prehistoric erection consisting of two or more vertical stones set in the ground, with a horizontal stone resting on them; it served as a Neolithic burial chamber.

dolomite (i) A yellowish or brownish mineral consisting of calcium magnesium carbonate, $CaMg(CO_3)_2$, with up to about 44% of $MgCO_3$. It was named after an 18th-century French geologist, D. G. Dolomieu. Its origin is varied; in some cases it is precipitated directly from sea-water, possibly under warm shallow conditions; in others 'dolomitization' of limestone has taken place, i.e. the post-depositional replacement has occurred of some of the calcium by magnesium. D. rock is sometimes called *magnesian limestone*, occurring in the Permian; it forms cliffs along the Durham coast, and a distinctive escarpment, over 180 m.

high, overlooking the Durham coalfield to the W. (ii) The D.'s, the proper name of ranges of pinnacled mountains in N.E. Italy.

dome (i) Any d.-shaped land-form; e.g. the English Lake District. (ii) A broad, short ANTICLINE or PERICLINE; e.g. the Kingsclere Pericline in N. Hampshire; the Weald. (iii) An underground d.-shaped structure containing salt, with oil and gas; in places great plugs of salt were forced upwards through sedimentary strata, with a cap of limestone, gypsum and anhydrite; e.g. the SALT-D.'s of Texas and Louisiana. (iv) An *oil-d.* in gently flexed sedimentary strata; e.g. Rock Springs D. and Teapot D. in Wyoming; the Dominguez Hills near Los Angeles, California. (v) An igneous intrusion of d. shape, including a LACCOLITH (Navajo Mtn., Utah) and a BATHOLITH (the Black Hills of Dakota). (vi) A rounded granite peak; e.g. the d.'s of Yosemite (California); Stone Mtn. on the Piedmont upland, Georgia; the *bâlons* in the Vosges, France. (vii) A rounded snow-peak; e.g. Dôme de Gouter (Mont Blanc), Dôme de Neige (Dauphiné), in the French Alps. (viii) An acid volcano; e.g. Puy de Dôme, Auvergne, France. *Note:* A d. may be either structural or physiographic, or both. Thus the Weald is a dome structure, not a physiographic dome, while the Black Hills of S. Dakota are both.

dominant The characteristic species in a partic. plant COMMUNITY, contributing most to the general 'plant landscape', and influencing which other plants will also grow there. In an oak-wood, the d. is the oak; in a heathland, it is usually ling (*Calluna*). The d. tends to be the largest or strongest plant in the community.

dominant wave The largest w.'s along a section of coast; i.e. those capable of the most powerful work.

dominant wind The most *significant* w. at a given place, not necessarily the one which blows most frequently (i.e. PREVAILING). Thus a shingle-beach may be aligned at right-angles to the d. w., not necessarily to the prevailing w.

dominion A self-governing independent state, formerly a colony of the United Kingdom, but now quite independent, though recognizing the loose association of which the Sovereign of the U.K. is the head. Its status is defined by the Statute of Westminster, 1931, which gave effect to resolutions adopted at the Imperial Conferences of 1926 and 1930.

donga (i) A steep-sided gully in S. Africa, occupied temporarily by flood waters, so called by its Bantu name. (ii) A gully produced by soil erosion.

dore An opening in a ridge between masses of rocks; e.g. D. Head, above Mosedale, English Lake District. In the Gaelic form, *dorus;* e.g. An Dorus, a gap in the Black Cuillin ridge, I. of Skye.

dormant volcano A v. which has not erupted in historic times, but is not thought to be extinct; e.g. most of the v.'s in the Andes and the Cascade Mtns.

dormitory town Residential t.'s within reach of large t.'s or cities where live the people who travel in to the latter to work. See COMMUTER. E.g. Hayward's Heath, Reigate and many others around London.

dot map The representation of quantities or values by d.'s of uniform size, each having a specif. value. The d.'s are inserted either evenly and uniformly within the boundaries of an administrative unit for which statistical information is available, or more precisely when detailed information is available concerning the exact location of the values (though a certain amount of subjective judgement is required). The success of the m. depends largely on the value given to the d.; it should not be so low that there will be difficulty in inserting all the d.'s where quantities are great, nor so high that there will be units with few d.'s. If several types of distribution are to be shown, d.'s of different colours may be used. See also MILLE M., PERCENTAGE D. M.

double tide A tidal régime in which the t. rises to a max., falls slightly and then rises to a 2nd max.; e.g. in Southampton Water, along the Solent and as far as Bournemouth, and on the French side of the Channel near Honfleur. Alt., a d. low t. may occur, as at Portland, Dorset (the GULDER). The d. t. is attributed to the effects of: (*a*) the configuration of the coast; (*b*) shallow water, causing the deformation of a PROGRESSIVE WAVE and the superimposition of a quarter-diurnal t. upon a semi-diurnal t. If the harmonic curves of a semi-diurnal and quarter-diurnal t. are combined, they will produce a d. high t. if the initial phase difference between the constituents is near 180°. If the initial phase difference is near zero, a d. low t. results. [*f* TIDE]

doup A rounded hollow in N. England, esp. in the English Lake District; e.g. Great D. beside the Pillar Rock, Ennerdale.

down (land) (i) A gently undulating upland, usually of chalk; e.g. N., S., Dorset, Hampshire D.'s. (ii) Temperate grasslands in Australia (e.g. the Darling D.'s) and New Zealand. (iii) The proper name for part of the N. Sea near the Goodwin Sands ('The Downs').

downthrow The vertical change of level of the strata in a fault; the strata are lowered on the d. side. [*f* FAULT]

downtown An American term for the central area, esp. the main business and shopping area, of a town or city. See CENTRAL BUSINESS DISTRICT.

down-warping A gentle downward deformation of the crust, without any distinct folding or faulting. This can be caused by the weight of a continental ice-sheet; e.g. in the Gulf of Bothnia, where the 'recoil' resulting from the relief of weight by ice-melting has averaged 11·18 mm. per year for the last 150 years. Similar d.-w. and recoil have taken place in the Great Lakes area of U.S.A. and Canada. D.-w. may also be caused by widespread sedimentation; e.g. beneath the Mississippi delta. Several extensive shallow lakes are the result of d.-w. or 'sagging'; e.g. L. Victoria (E. Africa); L. Eyre (Australia); L. Chad (W. Africa). D.-w. contributed to the creation of GEOSYNCLINES.

drag fold (i) A minor f. formed subsidiary to a main f., or along the sides of a FAULT where the vertical displacement has resulted in flexures and puckers in the rocks. (ii) Used in geology with specif. ref. to COMPETENT and INCOMPETENT strata, the d. f. being produced in the latter by the movement of the adjacent more rigid competent beds.

drain An artificial channel used for carrying off excess water from an area; e.g. in the English Fen District (the Hundred Foot D., also called the New Bedford R., N.W. of Ely); in the Somerset Levels (the North D., near Wedmore; the King's Sedgemoor D.).

drainage (i) The act of removing water from a previously marshy area; e.g. the d. of the Fens. (ii) The discharge of water from any area through a system of natural streams; hence *d. basin* or area, a unit-area drained by a single river system; *d. pattern, system* or *network*, the actual arrangement of the main river and its tributaries; *d. density*, the ratio of the total length of all streams in a single system to the area drained by that system.

Dreikanter (Germ.) A stone scoured on 3 sides or facets by the sandblast effect of the wind (i.e. a VENTIFACT) in a desert. Ct. EINKANTER.

drift (i) All materials (boulders, sand, clay, gravel) derived from glacial erosion, and deposited either directly from the ice or by melt-water; i.e. including both glacial and fluvioglacial deposits. (ii) Used in a wider sense by the British Geological Survey to distinguish all superficial deposits from the solid bed-rock. Maps are published in 'drift' and 'solid' editions. (iii) A horizontal passage in a mine, following a vein of metal ore or a seam of coal; ct. CROSS-CUT, which intersects it. (iv) The slow movement of surface ocean water under the influence of prevailing winds; e.g. the N. Atlantic D., moving at about 5 nautical mi. (about 9 km.) per day. (v) Any surface movement of loose material by the wind, with accumulations; e.g. *snow-d.*, *sand-d.*

drift-ice Loose, detached pieces of floating ice, separated by open water, moving with the wind or a current, through which a ship can readily penetrate.

drizzle Fine continuous rainfall, in which raindrops are very small, usually defined as less than 0·5 mm. in diameter. It is esp. associated with a WARM FRONT.

drought A continuous period of dry weather, specif. defined as ABSOLUTE DROUGHT, PARTIAL DROUGHT or DRY SPELL.

drove An ancient track with the right of free access for animals; hence a 'drover', one who took cattle and sheep long distances along such ways to market. *D.-roads* in the form of ancient trackways can be traced in many parts of Britain; e.g. the 'green lanes' of the Pennines with stone walls on each side, along which sheep were moved.

drowned valley A v. filled with water by a positive change of sea-level, the result of either subsidence of the land or rising of the sea. See FIORD, RIA, DALMATIAN COAST.

drumlin (Irish) A smooth elongated hummock of boulder-clay, with a long axis parallel to the direction of the moving ice-sheet responsible for its deposition; it varies in size from a small mound to a hillock 2 km. long and 90 m. high. They commonly occur in swarms *en échelon*, sometimes called 'basket of eggs' relief. The ice deposited each mass of clay, possibly from part of the ice-base which was locally more loaded with material, because friction between the clay and the underlying floor was greater than that between the clay and the over-lying ice. The shape was then stream-lined by subsequent ice-movement. In gen., where the ice moved most rapidly and exerted most pressure, the d. was most elongated. E.g. in N. England (the Aire Gap, the coastal plain of Morecambe Bay, the Solway Plain); the Midland Valley of Scotland; N. Ireland. See also ROCK-DRUMLIN. [*f*]

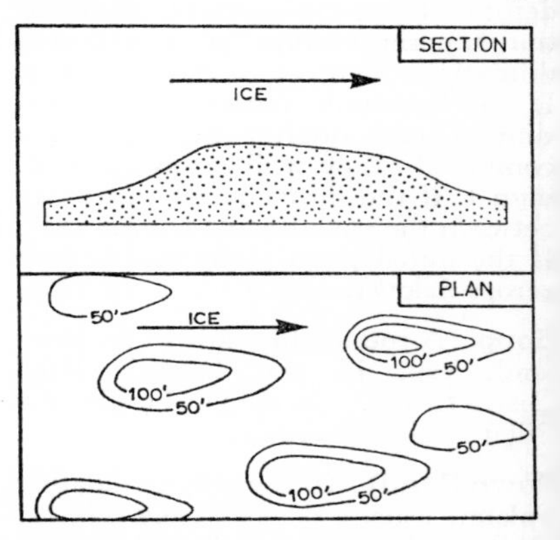

dry adiabatic lapse-rate A L.-R. of temperature with height, occurring when a 'parcel' of unsaturated air rises through the atmosphere in equilibrium, so expanding and cooling dynamically. The d. a. l.-r. is 1°C. in 100 m. (5·4°F. in 1000 ft.) of ascent, and is a physical constant. See ADIABATIC.

dry-bulb thermometer An ordinary mercury t., used in conjunction with a WET-BULB T. to obtain RELATIVE HUMIDITY.

dry delta See ALLUVIAL CONE, ALLUVIAL FAN.

dry-farming Agriculture in a semi-arid area, without the help of irrigation, by conserving soil moisture through mulching, the maintenance of a fine tilth (so preventing the baking and cracking of the surface), and the utilization of two years' rain for one crop.

dry-gap See WIND-GAP.

dry-point settlement A settlement on a higher area of dry land among low-lying marshes; e.g. Ely in E. Anglia; Veurne in W. Belgium.

dry spell Any period of drought; sometimes defined specif. as 15 consecutive days, none of which has more than 1·0 mm. (0·04 in.) of precipitation (British definition). In U.S.A. there is no measurable precipitation for 14 days.

dry valley A v., notably in chalk and limestone, which contains no permanent stream, though one may break out following heavy rainfall (see BOURNE). In chalk country, the pattern of the d. v.'s is like that of a normal river system. Many of them dissect chalk CUESTAS, have steep walls and abrupt heads, and sometimes follow zig-zag courses; e.g. Rake Bottom, near Butser Hill, Hants. There are various theories of their origin: (i) a gradual lowering of the WATER-TABLE following a general reduction of precipitation; (ii) the cutting back of an escarpment, with a resultant lowering of the SPRING-LINE; (iii) surface erosion under PERIGLACIAL conditions, when the chalk was frozen and therefore impermeable; (iv) powerful SPRING-SAPPING, cutting back along lines of weakness such as jointing, which helps to account for the frequent angularity of the course and the steep valley-head. D. v.'s in limestone may also be due to: (i) a former surface stream disappearing down a joint; (ii) the collapse of an underground cavern; (iii) a lowered water-table.

dumpy level An instrument for LEVELLING in survey work, consisting of a short telescope fixed to a base, with an attached spirit-level.

dune A low ridge or hillock of drifted sand, mainly moved by the wind, occurring: (i) in deserts; e.g. the ERG of the Sahara and the KOUM of Turkestan; and (ii) along low-lying coasts, above high-tide level; e.g. the Culbin Sands along the Moray Firth; the Formby d.'s in Lancashire; the Landes in S.W. France; along the coast of N.W. Europe from the E. Baltic to the Belgo-French frontier. Factors in d. formation include: (*a*) the load of sand; (*b*) the strength and direction of the wind; (*c*) the nature of the surface over which sand is moved (deep sand or bare rock); (*d*) the presence of an obstacle as a nucleus; (*e*) the presence of vegetation, wild or deliberately planted (acacia, eucalypts in deserts, marram grass and pines on coast); (*f*) the presence of ground-water reaching the surface. See also BARKHAN, SEIF-D. Elaborate classifications of d.'s include HEAD-D., TAIL-D., ADVANCED-D., LATERAL D., WAKE-D., STAR-D., SWORD-D. [*f*]

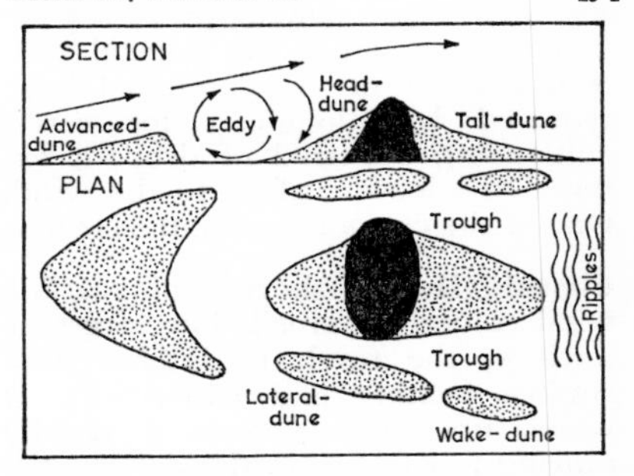

dunite An ultrabasic rock consisting mainly of the mineral *olivine* (magnesium iron silicate, Mg_2SiO_4 with Fe_2SiO_4).

durain structure, in coal In banded coal, layers of dull grey to brown material, hard and tough in quality.

duricrust A concentration in the upper part of the soil of various materials (aluminous, calcareous, siliceous, ferruginous, magnesian), drawn up in strong solutions by capillarity; these form a hard surface layer. It is usually found in semi-arid areas with a brief rainy season and a long hot dry season.

dusk Gen. the time of day between full light and darkness. More specif., the period between TWILIGHT (civil) and darkness.

dust Small (less than 0·6 mm. in diameter) particles of comminuted matter, fine enough to be carried in suspension by the wind. *Cosmic d.* is of meteoric origin, *volcanic d.* is the product of an eruption, *hygroscopic d.* (i.e. with a marked affinity for water) acts as nuclei for the condensation of water-vapour. An average 1 cu. mi. of air contains 4000 tons of d. Hence D.-STORM.

dust bowl A semi-arid area, from which surface soil is being or has been removed by the wind, esp. where vegetation has been destroyed by injudicious cultivation or over-grazing. Orig. this was a proper name for parts of S.W. and W. U.S.A., esp. in Kansas; it is now used for any area where similar processes are in operation. Once the original protective cover of grassland had been removed by ploughing for cereal cultivation, a period of drought exposed the loose surface to the wind. One of the categories of SOIL EROSION.

dust-devil A short-lived swirling wind round a small low pressure nucleus, the result of intense local surface heating and convection. It whips up dust to form a rapidly moving pillar; e.g. in the Sahara, the Kalahari, central and W. Australia, parts of the U.S. Mid-West.

dust-storm A storm in a semi-arid area, in which winds carry dense clouds of dust, sometimes to a great height, often obscuring visibility. It is often of whirlwind form, or may comprise a 'wall' of d. up to 3000 m. high. Termed *haboob* in the Sudan. In a d.-s., a cu. mi. of air contains about 16,000 tons of d. Ct. SANDSTORM.

dust-well A small hollow on the surface of a glacier, formed by a patch of d. absorbing the sun's rays, causing a rise of temperature, and so melting the surrounding ice. The d. gradually sinks down into the ice.

Dwyka Tillite A widespread t. found in S. Africa, in places resting on an ice-worn rock-pavement, and containing striated boulders and ERRATICS. It is of Upper Carboniferous age, and is regarded as clear evidence of a Carboniferous-Permian glaciation. Similar t.'s have been found in Orissa in India (the *Talchir T.*), Australia, Argentina and Brazil. It is thought that there were at least 3 distinct advances of the Carboniferous-Permian glaciation.

dyke (dike) (i) A mass of intrusive rock which cuts discordantly (i.e. transgressively) across the bedding-planes of the country rock. Sometimes d.'s occur in parallel 'swarms', as in N.W. Scotland, Mull and Arran. When affected by denudation, a d. will either stand up as a ridge (where the d.-rock is more resistant than its surroundings), or be worn away to form a ditchlike depression. (ii) A drainage ditch or watercourse; e.g. the Friar D., Vale of Pickering, Yorkshire. (iii) An artificial embankment to protect low-lying land from flooding; e.g. around parts of the English Fen District, the Netherlands [*f* POLDER] W. Denmark. (iv) An embankment in the flood-plain of a river parallel to its course; e.g. the Mississippi, the lower Rhine. (v) A man-made defensive earthwork; e.g. Offa's D., built in the 8th century from the estuary of the Dee to that of the Wye. [*f*

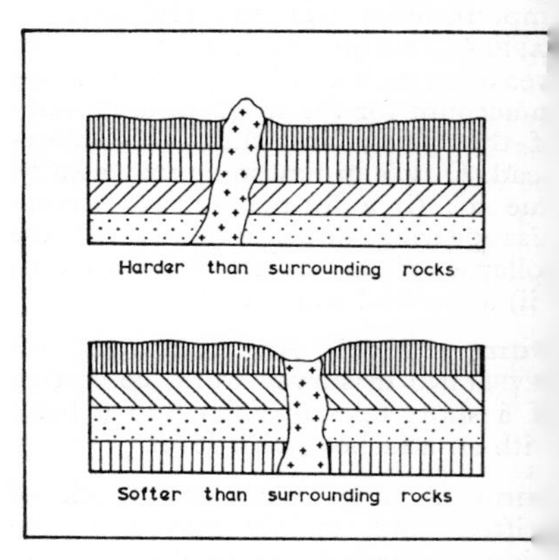

dyke-spring A spring thrown o[ut] along the line of a DYKE formed of [an] impermeable rock such as basalt [or] dolerite, which penetrates a perm[e]able sedimentary rock or a pervio[us] igneous one; e.g. in the Black Cuillins, (I. of Skye) made of well-jointed GABBRO, with several high-level springs along the lines of dykes. [f]

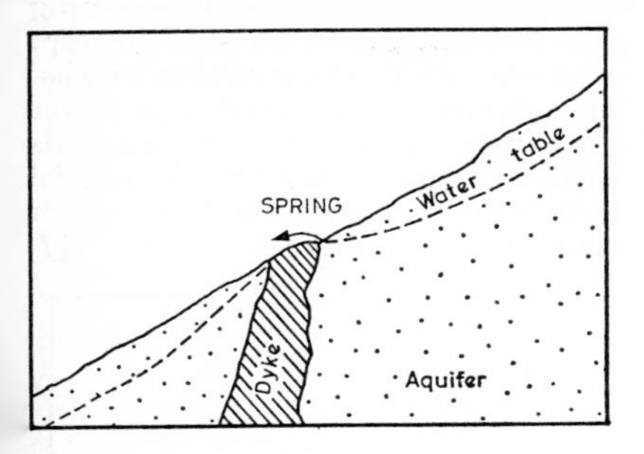

dyke swarm A collection of d.'s, often parallel to each other, near and around a PLUTONIC intrusion; e.g. in the I.'s of Mull and Arran, in W. Scotland.

Dymaxion Projection A mathe-natically derived p. formulated and leveloped by R. Buckminster Fuller, he name obtained from a contraction f *dynamic*, *maximum* and *ion*. It claims o have numerous advantages, the hief being the provision of a world nap of negligible dimensional dis-ortion. The p. is based on an all REAT CIRCLE grid, so producing a asic pattern of equal-sided great ircle triangles, with matching edges n which scale is true. All spherical reat circle co-ordinates are trans-rred to a plane surface by means of e mathematically complex Fuller ransformation; this adjustment is rried out by interior contraction of e data, rather than by the exterior retching used by most other p.'s. he triangles can be reassembled in a ntinuous 'flat skin' or 'world mosaic' any selected way, so as to bring cus to bear on any of the dynamic terrelationships of the world's sur-ce. The radii of the p. are always rpendicular to the transforming rfaces, which facilitates the accurate nsference of astronomical data to e plane surface. This means that the is highly suitable for assembling a mprehensive mosaic of aerial photo-aphs, and also for automatically plotting and guiding missiles and aircraft on a 'world-around' scale. One application of this p. (the *D. Airocean World Fuller Projective-Transformation*) was published in 1954.

dynamic climatology C. approached from the viewpoint of its relationship to the general circulation of the atmosphere, energy processes, and atmospheric dynamics and thermodynamics. Ct. SYNOPTIC C.

dynamic equilibrium The concept that conditions of balance (e.) can develop in the physical landscape; e.g. between the rate of rock weathering and the rate of removal of the weathered products on a slope. The existence of such an e. may be indicated by slope angles within a small region of uniform relief, rock-type, climate and vegetation. The e. is not maintained forever, but may be disturbed, e.g. by a climatic change affecting the rates of weathering and removal. However, it is believed that in time the e. will be restored by an adjustment of the slope angle to meet the requirements of the new situation. Thus the e. is d. (changing). 'Acceptance of d. e. theory implies the abandonment of landscape analysis in terms of structure, process and stage . . . , and its replacement by explanation of landforms simply as the outcome of an interaction between structure and process' (R. J. Small). Cf. STAGE.

dynamic lapse-rate A r. of temperature change of ADIABATIC character, applying to rising 'parcels of air'. Ct. STATIC LAPSE-RATE.

dynamic metamorphism The alteration of rocks by pressure, associated with earth-movements, usually on a large scale (hence *regional* m.). This can produce both complex mineralogical changes, and a reshaping and rearrangement of the rock fabric; e.g. shale is turned into slate, coarse-grained sandstone into quartzite, and crystalline rocks into gneiss. Many rocks (esp. schist) show a 'wavy-grain'

(FOLIATED) structure, others (slate) a LAMINAR structure.

dynamic rejuvenation The r. of a river caused by a change of BASE-LEVEL, due to a relative rise or fall of sea-level, or from an alteration of the land surface through TECTONIC causes (folding, faulting, tilting, uplift). Ct. STATIC REJUVENATION.

dyne An absolute unit of force, producing an acceleration of 1 cm. per second per second when acting upon a mass of 1 gm.; i.e. *1 gm. cm. per sec.*^{2}. At sea-level and at 45°N. and 45°S., a mass of 1 gm. experiences a gravitational force of 980·616 d.'s. On the S.I., 10^5d.'s = 1 NEWTON.

eagre See BORE.

earth (i) The planet inhabited by Man, with its orbit around the sun between those of Venus and Mars. Its shape is an oblate spheroid (i.e. a flattened sphere), with its equatorial diameter 12,755 km., (7926 mi.), polar diameter 12,714 km. (7900 mi.), with an oblateness of 1/293 0·33%. The International ELLIPSOID OF REFERENCE (1924) has an ellipticity of 1/297. The Airy spheroid (1870), used by the British Ordnance Survey, has 1/299. The Clarke spheroid (1866), used for N. America, has 1/295. The Clarke spheroid (1880), used by the British Admiralty, has 1/293·5. Equatorial circumference, 40,076 km. (24,902 mi.), polar circumference, 40,008 km. (24,860 mi.). Area = $510{\cdot}1 \times 10^6$ sq. km. ($197{\cdot}0 \times 10^6$ sq. mi.). Mean density = 5·517; mass = $5{\cdot}882 \times 10^{21}$ tons. Volume, $259{\cdot}9 \times 10^9$ cu. mi. (ii) The solid material on the surface of this planet, in ct. to water. (iii) Loose, disintegrated surface material, in ct. to solid rock, sometimes even applied to soil; e.g. 'the good earth'. (iv) A proper name for certain amorphous fine-grained materials; e.g. Fuller's E.

earth-flow A type of MASS-MOVEMENT of surface material down a slope under the influence of gravity, with lubrication from rain-water. A mass of saturated material pulls away from the slope as a SLIP, leaving a low arcuate cliff, and moves down as a plastic mass, forming bulging lobes or tongues, and lower down becoming a MUD-FLOW. Shallow e.-f.'s are esp. common over impermeable rocks. Even when turf-covered, the lower slope may bulge out, with the sods wrinkling and cracking. A special form of e.-f. is SOLIFLUCTION, under PERIGLACIAL conditions. [*f*]

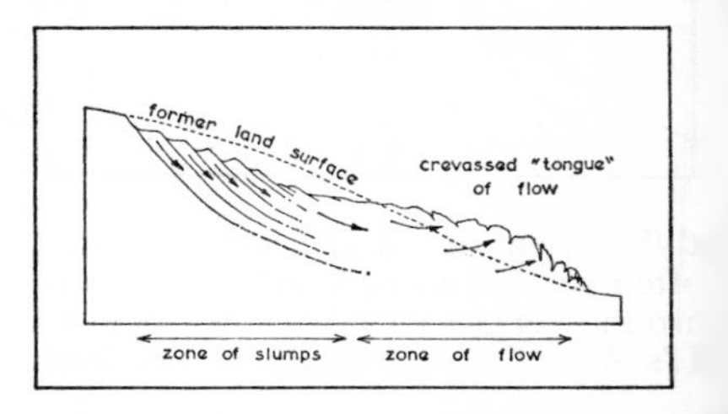

earth-movement A movemen caused by the internal forces (compressive, tensional, uplifting, depressing, folding, faulting), on both a majo and minor scale, and both rapic (earthquakes) and slow, which affec the CRUST of the earth; these are knowı collectively as DIASTROPHISM.

earth pillar A mass of soft earth material, capped and protected by rock, the result of rapid subaeri denudation, esp. in a semi-arid BAD LAND area (see DÉMOISELLE). It i common in boulder-clay, where th mixture of fine material and boulde provides a suitable medium for th development of such phenomena; e. in the Chamonix valley, French Alp near Bolzano, Italian Tyrol.

earthquake A rapid and detectab tremor, movement and adjustment and within the rocks in the earth crust, causing the propagation of series of elastic shock-waves outwar in all directions. These are classifi as: (i) *primary* (longitudinal, push, '*P*' waves); (ii) *secondary* (transvers SHAKE or '*S*' waves); (iii) *surface* (' waves). Many subdivisions are di tinguished. The most severe e.'s a associated with distinct fault-lin e.g. the San Francisco e. (1906) along the San Andreas Fault. [*f*]

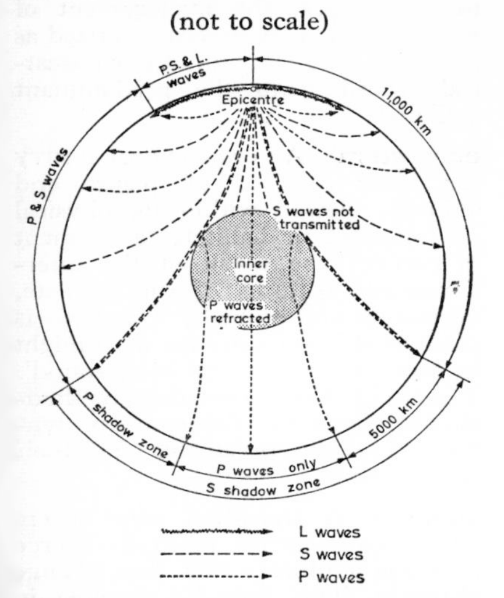

earth-slide A type of mass-move- ment where a large block of e. moves downhill *en masse*, but retains its character without breaking up. It is similar in character to a snow-slide.

easting The distance E. from the origin of a GRID, thus providing the first half of a standard grid-reference.

ebb channel The route by which a tidal current flows seawards after high tide. This may differ from the flood t., esp. among complex offshore banks, thus causing progressive changes in the form of these banks. E.g. the effects of the interplay of ebb and flood tidal currents on the Goodwin Sands, which tend to rotate anticlock- wise, the N. end moving W., the S. end moving E.

ebb tide The recession or outward movement of water, after high t., as the level falls towards low t.; the reverse of FLOOD T. In estuaries, harbours and other inlets, the ebbing of the t. may cause a powerful outward-flowing tidal stream or current, followed by a period of slack, and then by the rise of the next flood t. Where the range of the t. is large, as in shallow marginal seas and estuaries (e.g. 13 m. at Avon- mouth, 21 m. in the Bay of Fundy), the amount of water transferred is large, and the ebb t. is powerful.

echo-sounder A surveying instru- ment used to determine depths of oceans, whereby sonic or ultra-sonic vibrations are transmitted down through the water to the ocean floor, to return in the form of an echo which is electrically recorded as an *echogram*. For very accurate work the ship is anchored, but for normal purposes a continuous profile is obtained while the ship is under way. It is also used, carried mounted on a sledge, for measuring the thickness of ice, as in Greenland and Antarctica, and for measuring the depth of different den- sities of soil or rock.

Eckert Projections A series of 6 p.'s, used for world maps, broadly similar in appearance to MOLLWEIDE, with each pole represented by a line half the length of the equator, rather than by a point, the parallels straight lines. In the Fourth E.P. the meridians are ellipses passing through the equator, which is divided truly, and the parallels are spaced so as to make the p. equal area, as follows:

Distance of parallels from Equator

0°	0·0
10°	0·155
20°	0·308
30°	0·454
40°	0·592
50°	0·718
60°	0·827
70°	0·915
80°	0·976
90°	1·0

The shape of the continents is quite good, much better than on Mollweide. In the Sixth E. P., the meridians are *sine*-curves.

eclipse *Of the moon*, caused by the interposition of the earth when it is in line between the sun and the moon; *of the sun*, caused by the interposition of the moon when it is in line between the sun and the earth. The e. may be *total* or *partial*.

ecliptic The apparent path of the sun through the sky during a complete revolution in a year (about 365¼ days); the plane of this path (the plane of the e.) is tilted at an angle (which varies slightly) of about 66½° to the earth's axis; i.e. an obliquity of about 23½° from the plane of the earth's equator. It has been calculated that this obliquity varies in about 40,000 years between 22° and 24·5°. The ecliptic is divided into 12 sections, each denoted by a Sign of the ZODIAC. [*f* AXIS, OF EARTH]

eclogite A gabbroid rock, composed chiefly of pyroxene and garnet, which has been subject to intense DYNAMIC METAMORPHISM. It has an attractive green and red colouring. It is found among the Lewisian rocks in N.W. Scotland.

ecoclimate The study of climate in relation to plant and animal life. Cf. ECOSYSTEM.

ecology The science of the mutual relationship of organisms to their environment. Hence *ecological region*. *Human* or *cultural e.* stresses the relationship of 'people and place' (as defined by geographers), of 'people and people' (as defined by sociologists). Without any qualification, the term e. is botanical in concept. Note the *Journal of Animal Ecology*, published in the U.K.

economic geography That aspect of g. dealing with patterns of economic distribution, and with the factors and processes affecting the areal differentiation of these patterns on the earth's surface. Very often the term means little more than g. with an emphasis on the exploitation of resources, production and trade. Sometimes it is regarded as syn. with *commercial g.*

economic planning The p. of the distribution of e. resources on a wide scale, in relation to population and national finance, with the main emphasis on quantitative values. Ct. PHYSICAL PLANNING.

economy The administration of the resources of a place or community, espec. the management of wealth. An e. may be characterized as primitive, or as agricultural or industrial according to the predominant activities.

ecosystem An organic COMMUNITY (or *biotic complex*) of plants and animals, viewed within its physical environment or habitat; 'a segment of nature', the result of the interdependent features of soil, climate, vegetation and animals. The term is often used in ECOLOGY for what might be termed the 'physical background'. The study of an e. provides a methodological basis for the complex synthesis between organisms and their environment.

ecotone A transition zone where plant COMMUNITIES naturally merge into one another rather than change abruptly. This may be an area in which the two communities actually compete, thereby giving an impression of gradual mergence.

ecotope In one use, syn. with ECOSYSTEM; in another, the area actually occupied by an ecosystem.

ecumene (œcumene) Orig. the habitable world as known to the ancient Greeks; now it usually refers to the inhabited or the most densely settled parts of the earth's surface (as ct. *non-ecumene* [or ANECUMENE], the uninhabited or only temporarily inhabited areas).

edaphic App. to soil features which affect the growth of plants and other organisms. These include variations in texture, acidity and alkalinity, the presence of minerals, and the water content. It is also used gen. as relating to, or app. to, the soil. From Gk. *edaphos*, basis or floor.

eddy A movement of a fluid substance, partic. air and water, within a larger one. Thus winds are thought of as having partic. characteristics such as force and direction, these being the net result of the movement of e.'s. The scale of e.'s may vary considerably, from DEPRESSIONS and ANTICYCLONES within the gen. circulation of the atmosphere to current e.'s in a small air-stream, channelled by the earth's rotation into a predominantly E.–W. direction.

edge Used in place-names for arêtes, ridges, mountain crests and plateau margins; e.g. Swirral E. on Helvellyn; Cross Fell E. in the N. Pennines; the E., the N. margin of Kinder Scout near Edale; Wenlock E., Shropshire.

edge-line, of relief The depiction of a sudden sharp change or break of slope on a relief map by a heavy line; known in Germany as *Kantographie*, after *Kanten* (edges), and in U.S.A. as *kantography*.

effectiveness of precipitation, effective p. (i) The actual total p., ninus the max. possible evaporation. A high evaporation rate obviously reduces the value of the p. for agriculture and water supply. E.g. in N. Ceylon, the mean annual rainfall is 270 mm. (50 ins.), which would be heavy in middle latitudes, yet the rea is known as the 'Dry Zone', with nuch semi-arid scrub-jungle, beause of the high evaporation rate. (ii) n hydrology, that portion of the p. vhich flows away in a stream channel. ii) In irrigation, that portion of the p. hich passes into the soil and is available for cultivated plants.

fficiency, of a stream An inication of the load-carrying ability of stream, the joint result of its OMPETENCE and CAPACITY. G. K. ilbert suggested (1914) a formula, here e = capacity in grams per second/ scharge in CUSECS × percentage slope channel bed.

fluent (i) A stream issuing from a ke. (ii) The outflow from a sewage orks. (iii) The outfall from the land ter irrigation. (iv) The waste proicts, such as a liquid or foam, from a ctory (*industrial e.*). The problem of llution of rivers by industrial e. or treated sewage is very serious in rts of Gt. Britain.

re See BORE.

Einkanter (Germ.) A stone on the surface of the desert upon which the scouring effect of wind-blown sand has cut a single facet; this implies a steady, unchanging wind-direction. Ct. DREIKANTER.

ekistics (Gk. *ekos*, habitat) Introduced in 1959 in a town-planning context, implying the science of human settlements, their evolution and development.

elastic rebound When rocks are subject to stress, the pressure or strain accumulates until they reach breaking point, when they tear apart, to snap back into a position of little or no strain. This is the cause of FAULTS and EARTHQUAKES. It has been estimated that strain accumulated along the San Andreas Fault for a century before the sudden release of e. energy resulted in the San Francisco earthquake of 1906.

E-layer The HEAVISIDE-KENNELLY LAYER.

Elbe glaciation The 1st glacial period in the N. European Plain, by those authorities who support a 4-fold glaciation in that area and ascribe certain morainic remnants near the Elbe valley to it; i.e. they equate the E.g. with the Günz in the sequence of Alpine glaciations. Others deny that the Scandinavian ice-sheet advanced so far S., and believe that the 1st glaciation to affect the N. European Plain was the Elster; i.e. equivalent to the 2nd or Mindel Alpine glaciation.

elbow of capture See CAPTURE, RIVER.

electricity Power generated in dynamos, in which energy is derived from: (i) hydraulic turbines in the case of hydro-e.; and (ii) steam-turbines in the case of thermal e. For the latter, heat is obtained from: (*a*) coal, esp. sub-standard varieties, inc. lignite and brown-coal (W. and E. Germany); (*b*) peat, as in Ireland; (*c*) oil (e.g. Marchwood, near Southampton); (*d*) nuclear power (e.g. Calder Hall, W. Cumberland); (*e*) geothermic power from subterranean heat (e.g. Lardarello in Italy). Tidal-power has been harnessed at a station on the Rance estuary near St. Malo, France, nad its use in the Bay of Fundy, N. America, is under discussion. *Note:* U.S.A. produces one-third of the world's total electricity, followed by U.S.S.R., Japan, U.K., W. Germany, Canada, France, Italy.

electrometeor Used by the U.K. Meteorological Office for a visible or audible manifestation of atmospheric electricity; e.g. LIGHTNING, AURORA, ST. ELMO'S FIRE.

element (i) A combination of protons, neutrons and electrons; each is identified by the number of protons (atomic number) in its nucleus. 103 e.'s are known, 92 occurring in nature, 11 made in the laboratory. They range from hydrogen (atomic number 1) to lawrencium (103). See also ISOTOPE. (ii) Each of the physical constituents which make up the sum total of climate: temperature, pressure, wind, humidity and precipitation; hence 'the elements'.

elevation (i) The height or altitude above some particular level; e.g. above sea-level, or specif. above some DATUM. (ii) The vertical angle between the horizontal and some higher point, as in surveying or astronomical observations.

ellipsoid of reference The figure of the EARTH expressed as the ratio between the difference between the equatorial and polar radii, and the equatorial radius. If equatorial radius $=a$, and polar radius$=b$, then ellipticity$=\frac{a-b}{a}$. If the International E. of R. (1924) is used, ellipticity$=$ 1/297. For the alternative values, see EARTH.

Elster glaciation A main stage in the g. of the N. European Plain, correlated with the 2nd Alpine (Mindel) and American (Kansan) glacial periods. The moraines of the E. g. are deeply weathered, and form part of the Older Drift in Europe.

eluviation The removal of material, esp. bases, in solution (*chemical e.* or LEACHING), or of the finer fractions in suspension (*mechanical e.*), from the upper to the middle and lower layers of the soil profile by downward percolation of water, or horizontally through the same layer. The upper horizon from which material has been eluviated is the *eluvial* (or *A*) *horizon*; ct. ILLUVIATION. *Note*: In U.S.A. chemical solution is known as *leaching*, and e. is confined to the mechanical removal of fine particles or colloids. [*f* SOIL PROFILE]

elvan A vein of intrusive igneous rock, usually of quartz-porphyry, found in Cornwall and Devon, penetrating the granitic rocks.

embayment (i) An open rounded bay in the coastline. (ii) In a structural sense, a large-scale 'sag' near the border of a continental mass, where a considerable thickness of sediment may be deposited. (iii) Also in a structural sense, an area of sedimentary rocks projecting into a mass of crystalline rocks; cf. the Germ. use of '*Bucht*'; e.g. Kölnische Bucht.

emergence A rise of the level of th land relative to the sea, so that area formerly under water become dr land; specif. emerged coasts, char acterized by RAISED BEACHES, and som coastal plains; e.g. S.E. U.S.A.

emigrant A person who leaves hi native country to settle in anothe hence *emigration*.

emplacement The process of crea tion or formation in position of igneous rock masses, either by dis placement, replacement or META MORPHISM of the country rock; e. the e. of the granitic rocks of th Ben Nevis massif during early Devon ian times, the result of the down faulting of a central core among old metamorphic rocks, and its infilling b successive upwellings of MAGMA.

empolder(ing) The act of creati a POLDER.

enclave (Fr.) (i) An outlying po tion of a State entirely surrounded territory of another State, used in reference to this surrounding area; e.g. Baarle-Hertog, Belgian territory lying within the Netherlands. Ct. EXCLAVE [*f*]. It may be used also of other administrative units; e.g. with reference to the outlying N. part of Worcestershire around Dudley, an e. within Staffordshire. (ii) A small concentration of an ethnic group surrounded by others; e.g. Italians in Trieste.

enclosure The process by which common fields, meadows and pastures were enclosed into small fields by fences or hedges. Ct. INCLOSURE, the term for the legal measures by which this was carried out.

endemic Specif. an infectious disease which is a native of, or is normally confined and restricted to, a partic. limited area and liable to occur there at any time; e.g. cholera in parts of E. India, yellow fever in W. Africa, dengue in Burma. At any time an outbreak may flare up and spread, resulting in an EPIDEMIC.

endogenetic (endogenic, endogenous) The 'internal' forces of contraction, expansion, uplift, depression, distortion, disruption and outpouring which have contributed to the present landforms. Ct. EXOGENETIC.

Endrumpf (Germ.) The end-product, the final land-surface of denudation, as envisaged by W. Penck. Though Penck actually equated it with the PENEPLAIN of W. M. Davis, it may be likened rather to L. C. King's PEDIPLAIN.

nglacial Contained within the mass of a glacier or ice-sheet (debris, boulder, moraine, stream). Ct. SUBGLACIAL.

glacial river A melt-water stream flowing through a tunnel within a glacier or ice-sheet.

gulfment The inward collapse of a volcanic cone, the result of molten lava beneath it being drawn off under the surface of the earth or through a fissure in the flanks of a volcano. This is one way in which a CALDERA may be formed. E.g. Crater L., Oregon, formed from a peak (now denoted by the name Mt. Mazama) which was partially destroyed by an eruption and accompanying e. about 6500 years ago. It was once believed that this peak 'blew its top' by a PAROXYSMAL ERUPTION, but it is now thought that it collapsed inwards; of an estimated 17 cu. mi. of material which disappeared from the probable cone, only 2 cu. mi. can be accounted for as debris spread around; the rest must have collapsed within the underlying magma chamber.

entrenched meander See INTRENCHED M.

entrepôt A centre to which goods in transit are brought for temporary storage and re-exportation. It is also used as an adj., as in *entrepôt port*; e.g. Singapore, Rotterdam, Antwerp.

entropy Used in GENERAL SYSTEMS THEORY to describe the amount of free energy in a system. It is a definition with a negative quality, because maximum e. relates to the minimum amount of free energy in a system, whereas systems possessing a great deal of energy have minimum e. A decreasing amount of free energy in a system (i.e. the trend towards max. e.) signifies the progressive destruction of the heterogeneity of the system, the levelling of differences that formerly existed. Thus W. M. Davis's cyclical concept of landscape development relied on PENEPLANATION, the progressive levelling of differences within the system, and thus involved a trend towards max. e.

environment The whole sum of the surrounding external conditions within which an organism, a community or an object exists. The term is often used in a limited way in geography; e.g. the *natural e.*, meaning either the non-cultural and non-social e., or the landscape before Man came. The *geographical e.* means the factors of the e. whose relationships are considered in terms of spatial location. The *physical e.* includes all phenomena apart from Man and the things he creates, while the *non-human e.* includes everything not in a social system, whether made by Man or not. These are all slightly different, and unqualified use of the term can be misleading.

environmental capacity The c. of an area to accommodate motor vehicles while maintaining an acceptable standard of environment, that is, the gen. comfort, convenience and aesthetic quality of the area.

environmentalism A philosophical doctrine which stresses the influence of the environment on Man's pattern of life (see DETERMINISM). The term is frequently employed without a clear definition, and many writers use it not in the sense of total environment, but of a partial, physical environment. Thus e. is often thought of in terms of the effect of the physical environment only.

environmental lapse-rate The actual rate of decrease of temperature with increasing altitude at a given moment, with a mean of about 0·6°C. per 100 m. (1°F. for 300 ft. or 3·56°F. for 1000 ft.). See PROCESS L.-R.

environmental perception The way in which an individual or culture group regards its ENVIRONMENT. Each group tends to regard its 'home environment' as normal, and sees its resource possibilities in terms of its own cultural traditions.

Eocene The 1st (lowest) of the geological periods of the Cainozoic or Tertiary era, lasting from about 70 to 40 million years ago, and the system of rocks laid down during that time. In England it is represented in the London and Hampshire Basins, its deposits chiefly comprising marine, deltaic and estuarine sands, clays and loams.

Hampshire Basin	*London Basin*
6 Barton Beds	—
5 Bracklesham Beds	Middle and Upper Bagshot Sands
4 Lower Bagshot Sands	Lower Bagshot Sands
3 London Clay	London Clay
2 Reading Beds	Woolwich and Reading Beds
1 Absent	Thanet Sand

Note: In U.S.A., the E. is regarded as an epoch within the Tertiary (i.e. in a lower grade in the geological hierarchy than in Britain), preceded by the Palaeocene (70–60 million years ago) and succeeded by the Oligocene, a policy followed by A. Holmes in his time-scale of 1959. Some American authorities, however, dispense with the Palaeocene, and regard the E. as the oldest Tertiary epoch and series of rocks.

Eogene Syn., esp. in U.S.A., with PALAEOGENE.

eohypse A 'restored contour' of a former land-surface (e.g. a dissected plateau), inserted by plotting the surviving portions of the land and by the EXTRAPOLATION of the origina contours.

eolith A flint found in E. Angli (specif. within the CRAG formations o the Tertiary), which is claimed b some authorities to have been chippe and shaped by very early Man; i.e. i is the earliest known artifact. Other believe that an e. is simply a naturall formed flint, though possibly it wa used by early Man without being i any way fashioned or modified by hin

Eozoic (i) Applied to the earliest er corresponding to the Pre-Cambria group of rocks; from Gk., 'dawn (life'. This is in the sequence E Palaeozoic, Mesozoic and Cainozoi (ii) Some authorities narrow i application to the earliest era of Pr Cambrian time, followed by tl Archaeozoic and Proterozoic.

epeiric sea A relatively shallow on the continental shelf, the result an EPEIROGENIC movement; syn. wi *epicontinental;* e.g. the North S., Iri S., Baltic S., Hudson Bay.

epeirogenic, epeirogenetic (fr Gk. *epeiros*, a continent) Large-sc continent-building forces, hence applied to all *en masse* vertical or radial crustal movements, both of uplift and depression. Strata are not thereby folded or crumpled, though they may be slightly tilted or warped. The contrast is with OROGENIC. E.g. the e. sinking of the N. part of N. America, hence the formation of the Arctic islands and Hudson Bay.

epicentre A point on the surface of the earth vertically above the SEISMIC FOCUS (or ORIGIN) of an earthquake shock. [*f* EARTHQUAKE]

epicontinental sea See EPEIRIC S.

epidemic A sudden outbreak of a large number of cases of an infectious disease, which in due course dies out; e.g. the influenza epidemic of 1917–19. Ct. ENDEMIC, PANDEMIC.

epigene Geological processes taking place at the surface of the earth, and the rocks formed there. Hence *epigenic, epigenesis*.

epigenetic drainage See SUPERIMPOSED DRAINAGE.

epigenetic mineral An ore m. formed later than the rock which contains it.

epiphyte A plant growing on another, not parasitic, but using it for support; e.g. lichens, mosses, orchids.

epoch The 3rd category of the sub-divisions of geological time, in the hierarchy *era, period, e., age.* The rocks deposited during an e. are called a *series;* e.g. Coal Measures, Bunter, Lias Clay. 42 e.'s are distinguished in Gt. Britain.

equal-area projection A type of MAP P. in which areas are accurately shown, though at the expense of shape and direction, resulting in considerable distortion; e.g. MOLLWEIDE, ALBERS', BONNE, AZIMUTHAL EQUAL AREA, CYLINDRICAL EQUAL AREA.

equation of Time The difference between MEAN SOLAR T. and APPARENT (or *local*) T. (i.e. as shown on a sundial) at noon. On about 24 Dec., 15 April, 14 June and 1 Sept. these t.'s coincide. The max. positive value (sun 'fast') is 14 min. 25 sec. on 11 Feb., and the max. negative value (sun 'slow') is 16 min. 22 sec. on 2 Nov. The value for every day of the year is given in the *Nautical Almanac*; see ANALEMMA [*f*]. The variation in the interval between successive meridian passages of the sun occurs because: (i) the earth's orbit round the sun is not circular; (ii) the earth's velocity in its orbit is not constant; (iii) the ECLIPTIC is inclined to the equator.

equator The parallel of latitude 0°, midway between the Poles in a plane at right-angles to the earth's axis, length 40,076 km. (24,901·92 mi.); a GREAT CIRCLE.

equatorial climate A climatic type occurring between about 10°N. and 10°S. near sea-level, with constant high temperatures and humidity, and with about 12 hours day and night. There is little seasonal change, since the climate is dominated throughout the year by a near overhead midday sun, converging tropical maritime air-masses associated with atmospheric disturbances, and heavy CONVECTIONAL rainfall. Under the W. Köppen system, this is type *Af*, with *Am* where monsoonal influences are experienced (as in parts of Indonesia and the coast of W. Africa).

equatorial current A surface movement of oceanic water in e. latitudes: (i) towards the S.W. or W. in the N. hemisphere (the N. E. C.); (ii) towards the N.W. or W. in the S. hemisphere (the S. E. C.); and (iii) an E.-moving counter-current between the two in the DOLDRUMS. The N. and S. E. C.'s flow at about 28 km. (15 nautical mi.) per day.

equatorial rain-forest The luxuriant forest found within about 7° latitude N. and S. of the equator, with constant rainfall, high temperatures, and no seasonal periodicity. It is found mainly in the Amazon Basin, Congo Basin, and parts of Indonesia (esp. Borneo and New Guinea). It is characterized by a profusion and variety of species, and a distinctive 'layered' arrangement, with a 'canopy' or upper layer of foliage. As there is virtually no seasonal climatic change, the general aspect of a monotonous uniform greenness varies little. The most valuable trees, though scattered, are hardwoods; e.g. rosewood, mahogany, greenheart, ebony, ironwood.

equicorrelative An isopleth indicating correlations of sets of data; e.g. the degree of relationship between sets of rainfall statistics can be expressed in the form of *correlation coefficients* between pairs of stations, the values plotted, and isopleths interpolated.

equifinality The state of a system of interrelated objects which start from a variety of initial conditions, and suffer a series of changes so that they end in the same result. Applied partic. to landforms: e.g. a PENEPLAIN, a DRY-VALLEY, an amphitheatre-like hollow (which may be the result of a CIRQUE-GLACIER, ROTATIONAL MASS-MOVEMENT, SPRING SAPPING, etc.). GRANITE may be formed from a variety of parent materials.

equilibrium A state of balance, when various forces have created a state or form which will not be altered with the passage of time unless the controlling factors change; otherwise, if a temporary tendency towards change is introduced, it is countered by an opposing force which tends to restore the system to its former state; e.g. isostatic e., the shore-profile of e., the river long-profile of e., stable and unstable e. of a column of air in the atmosphere. See DYNAMIC E.

equinoctial The Celestial Equator.

equinoctial gale A period of strong winds which may be experienced about each EQUINOX. A period of storm in mid-September does frequently occur in Britain. G. Manley found in the records for 52 years he examined (1889–1940) such stormy periods in 31 years, with a frequency peak about 20 Sept.; on 16–17 Sept., 1935, winds up to 158 km.p.h. (98 m.p.h.) were experienced in Cornwall. The last week in March is also notably stormy. No physical cause for this has been discovered, but it is a well-established sailors' tradition, possibly related to the frequency of Caribbean hurricanes in September.

equinox One of the 2 points of intersection of the sun's apparent path during the year with the plane of the terrestrial equator; hence the time when the sun is directly overhead at noon at the equator. This occurs at about 21 March (the *spring* or *vernal e.*) and 22 Sept. (the *autumnal e.*). At this time, day and night are approx. equal throughout the world, since the circle of illumination by the sun of half the globe passes through the Poles, hence each parallel is half in light, half in darkness.

equipluve A line on a map joining places with the same PLUVIOMETRIC COEFFICIENT.

Equirectangular Projection A very simple p. with a network of horizontal parallels and vertical meridians. A standard parallel near the centre of the area to be shown is divided truly $= \frac{2\pi R . \cos lat.}{meridian\ interval}$. On a world map the equator ($2\pi R$) can be used as the standard parallel. Meridians are drawn vertically through the divisions of the standard parallel, themselves divided true to scale, and the other parallels are drawn as horizontal lines. If the equator is the standard parallel, the graticule will consist of squares; an other standard parallel will produc rectangles with the N. to S. dimensio the longer. The p. is neither equ area nor conformal. It is used fo large-scale city and estate maps fo geographic reference. The U.S. A Force have used since 1951 a wor map on the E. P., divided into 1 squares, starting at an origin at 18 longitude and 90°S. latitude in t S.W. corner (the GEOREF system).

equivalence The quality of a PR JECTION in which the product of t scale ratios *a* and *b*, at right-angles to each other, is everywhere the same; i.e. areas of any size are represented in correct proportion to each other.

equivariable An isopleth interpolated to join places with equal COEFFICIENTS OF VARIABILITY: e.g. with a similar deviation from their average climatic conditions, usually expressed as a percentage.

era A major division of geological time: Palaeozoic, Mesozoic, Cainozoic and Quaternary. Pre-Cambrian times are sometimes: (i) described under a single era, Eozoic; (ii) divided into 3 e.'s: Eozoic (oldest), Archaeozoic and Proterozoic. An e. in time corresponds to a *group* of rocks.

erg (Arabic) The sandy desert of the Sahara, with areas of dunes, sand-sheets and undulating 'sand-seas'.

ergograph Coined by A. Geddes to describe a graph which shows the amount and nature of seasonal human activities. It may take the form of a circle, with the circumference divided into 30° sections to represent each month, with concentric graph-lines to show the proportions of time spent in each activity (a square-root scale from the centre of the circle to the perimeter = 24 hours). An ordinary line graph can also be used. (*f* compiled by W. R. Mead.)

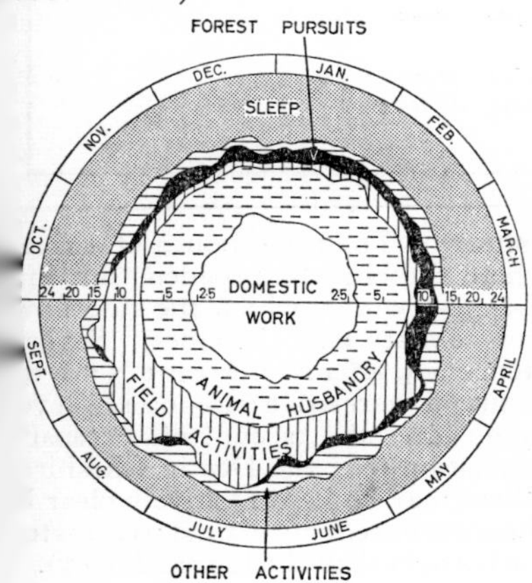

erosion Processes of earth-sculpture by agents (running water, ice, wind, waves) that involve the transport of material, not including static WEATHERING nor MASS-MOVEMENT through gravity. It includes both *physical e.* (CORRASION) and *chemical e.* (SOLUTION, CARBONATION). It is commonly (but quite incorrectly) used as syn. with DENUDATION.

erosion surface Commonly used in the sense of a level s. formed by e., rather than by deposition. The term can be misleading, since most of the earth's land-s. is in some sense an e. s., whereas the term is usually intended to refer only to near-horizontal portions. In this context the term PLANATION surface is preferable. E. *platform* is sometimes used for a s. of limited extent.

erratic, erratic block A fragment of rock carried by a glacier or ice-sheet and deposited some distance from the outcrop from which it was derived (see also PERCHED B.); e.g. boulders of Shap granite found in the Scarborough area, Ribble valley and near Wolverhampton; riebeckite-microgranite from Ailsa Craig (in the Firth of Clyde) found in Merseyside, Anglesey, the I. of Man, S. Wales, S.E. Ireland; dark Silurian boulders lying on Carboniferous Limestone in the Pennines (the Norber Stone); boulders of red jasper, found in Boone County, Kentucky, at a distance of nearly 1000 km. (600 mi.) from the nearest solid rock of this type, N. of L. Superior. The Madison Boulder in New Hampshire, U.S.A., which was moved 3·2 km. by the ice-sheet, weighs 4662 tons.

eruption The process by which solid, liquid or gaseous materials are extruded or emitted on to the earth's surface as a result of volcanic activity. The e. may be: (i) *central*, from a single vent or a group of closely related vents; (ii) *linear* or FISSURE [*f*], when lava wells up along a line of crustal weakness. The materials erupted include gaseous compounds of sulphur, hydrogen and carbon dioxide; steam and water-vapour; SCORIA,

PUMICE, CINDERS, DUST, ASH; acid and basic LAVA (hence *eruptive* rocks). The e. may range from quiet outflows and outwelling, to explosive and paroxysmal activity. An e. may be small-scale, *periodic* or *intermittent*; see GEYSER.

escarpment (i) The steep side of a CUESTA, equivalent (but preferable) to SCARP; more loosely the whole cuesta. (ii) A commonly used term applied to any steep slope interrupting the general continuity of the landscape; e.g. a line of sea-cliffs; a 'step' or edge where a stratum of hard limestone, or an igneous SILL, appears on the surface.

---

esker (sometimes **eskar** in U.S.A.) Used in a wide sense for all glacial and fluvioglacial sands and gravels which occur in elongated ridges and mounds. It was derived from the Irish word *eiscir*. The general term e. includes: (i) *ås*, pl. *åsar* (Swedish), anglicized to *os*; (ii) KAME-TERRACE; (iii) KAME-MORAINE. See also BEADED ESKER. In the diagram, the type of e. shown is an *os*. [*f*]

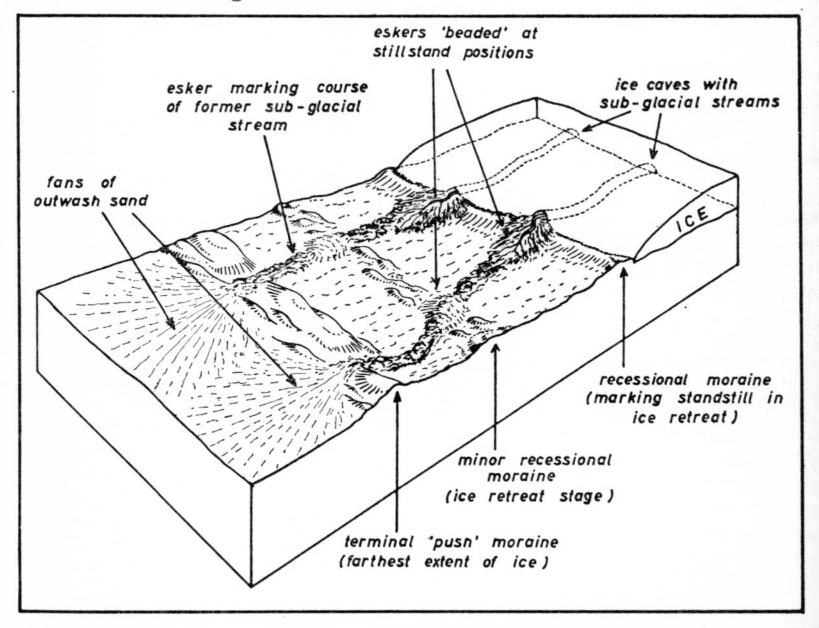

---

espinal (S. Am.) An area of thorny scrub.

estancia (Sp.) A large farm in the Argentine, used for cattle-rearing on an extensive scale. These are now being broken up and enclosed; there is more intensive rearing of animals, using alfalfa rather than natural grass, and much feeding with corn and oil-cake.

estuary The tidal mouth of a rive where the channel broadens out in V-shape, within which the tide flov and ebbs; e.g. Thames, Sever Hence adj. *estuarine*. Most e.'s are tl result of a relative rise of sea-leve Some are the scene of extensi deposition (e.g. the Dee in Cheshire others may be kept relatively clear the scouring effects of tidal streams (e the bottleneck of the lower Mersey).

étang (Fr.) A shallow lake, esp. one lying among sand-dunes or beach-gravels; e.g. the E. de Berre, N.W. of Marseilles; the E. de Biscatosse along the Landes coast of France.

etesian winds, etesians Strong N. to N.W. w.'s blowing at intervals during summer in the E. Mediterranean Sea, the result of a steep pressure gradient towards the thermal low pressure over the Sahara. They are at their strongest (20–30, sometimes 45 knots) in the late afternoon when convection is at its max., but they weaken and die away towards evening. They frequently cause rough seas, and over land may bring clouds of dust, sometimes with fog in coastal areas.

ethnography The description of the distribution of the races and cultures of mankind.

ethnology The scientific study of the races and cultures of mankind.

eugeosyncline A GEOSYNCLINE associated with VULCANICITY during its infilling by sedimentation. Ct. MIO-GEOSYNCLINE.

eustasy, eustatism, *adj.* **eustatic** A world-wide change of sea-level, indicating an actual fall or rise of the sea (e.g. by abstraction of water to form ce-sheets or the return of water after heir melting), but not by a local novement of the land itself. At their nax., the Quaternary ice-sheets must have lowered sea-level by 90 m., and heir remnants still contain enough vater to raise it a further 30 m.; this s known as *glacial-e.* The term was reated by E. Suess (1888). He also ncluded movements of the ocean-oors, and the filling in of their basins y sedimentation, thus causing changes f capacity.

vaporation The physical process f molecular transfer by which a quid is changed into a gas. In imatology, the rate of e. of water is function of: (i) VAPOUR-PRESSURE;) air temperature; (iii) wind; (iv) e nature of the surface; ct. bare soil d rock, where the rate of e. is high, with a fine protective surface tilth or plant-cover, where the rate is low. High e. rates occur in hot deserts; e.g. Atbara (Sudan) has a potential e. rate of 6250 mm. (246 ins.), Helwan (Egypt) of 2390 mm. (94 in.). Rates are lowest in equatorial climates (500–750 mm., 20–30 ins.) because of the high humidity, and in cool mid-climates (London, 330 mm., 13 ins.) because of the relatively low temperatures.

evaporimeter A device for measuring the amount and rate of evaporation. In the *Piche e.*, distilled water in a tube is allowed to evaporate from a piece of porous paper, and the loss in a certain time is measured on a graduated scale along the tube. Another method is to expose water in a large shallow tank, and measure the falling level at intervals. Syn. with *atmometer* in U.S.A.

evaporite A SEDIMENTARY rock composed of minerals precipitated from a solution, and dried out by evaporation; e.g. GYPSUM, ROCK-SALT.

evapotranspiration The loss of moisture from the terrain by direct evaporation plus transpiration from vegetation; e.g. a single maize plant in U.S.A. evapotranspired 54 gallons between 5 May and 8 Sept.; an acre of maize lost 324,000 gallons in that time, equivalent to 11 ins. of rainfall (W. W. Robbins, T. E. Weier, C. R. Stocking, 1957). At Slaidburn, Yorks., the Fylde Water Board found that in a spruce plantation, with 984 mm. (38·75 ins.) rainfall, only 255 mm. (10·75 ins.) penetrated the soil; i.e. there was a loss through e. of 1 million gallons a day on 1500 acres. It is necessary to distinguish between: (i) *potential e.*, a theoretical max. loss assuming a continuous supply of water to the surface and the vegetation (e.g. by irrigation); (ii) *actual e.*, the observed or true amount of e., which will lessen as soil-moisture diminishes or if no rainfall or irrigation water is received. Actual e. will equal potential e. in areas of high rainfall and low evaporation. If evaporation exceeds precipitation, the area is subject to desiccation.

everglade(s) A marshy area, with tall grass, canes and some trees, specif. (proper name) in Florida and along the Gulf coast of S. U.S.A.

eversion Erosion by eddies in the rock-bed of a stream, forming POT-HOLES.

exaration Glacial erosion carried out by ice alone, i.e. PLUCKING, not by ice laden with débris. Ct. ABRASION.

exceptionalism (i) Coined by F. K. Schaeffer (1953) to denote a methodological approach (with which he did not agree) to geography, by which it was regarded as a pure science, but with a uniqueness through its principles of AREAL DIFFERENTIATION which distinguished it from all other sciences (i.e. made it exceptional). R. Hartshorne commented on Schaeffer's paper: 'This . . . leads the reader to follow the theme of an apparent major issue, e., which proves to be non-existent.' (ii) More recently P. Haggett (1965) refers to a modern approach to e. by which MODELS can be created, from which exceptions and deviations can be computed quantitatively.

exclave An outlying portion of a State, entirely surrounded by the territory of another State, used in ref. to the State of which it forms a political part (ct. ENCLAVE). E.g. Llivia is an e. of Spain, a Spanish enclave within French territory. [*f*]

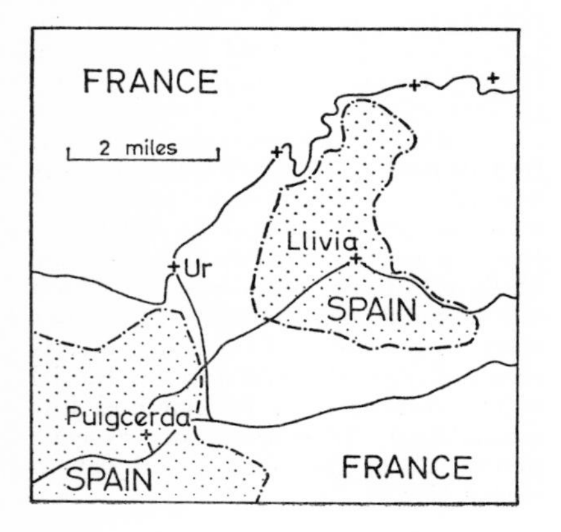

exfoliation The heating of a rock-surface by the sun and its rapid cooling at night produce strains which cause a 'shell', 'scale' or concentric sheet to pull away and split or peel off, a process known as e. This may be partic. effective on the face of a crag which receives and loses the sun's heat rapidly, esp. where the rock is composed of different minerals, with varying coefficients of expansion, resulting in complex strains. The process of e. probably requires the presence of moisture to be effective; some authorities now insist that e. involves chemical change. The term *spalling* is used in U.S.A. E. is limited to a superficial 'scaling'; cf. GRANULAR DISINTEGRATION.

exhumation In geomorphology, the uncovering by denudation of a former surface or feature which had been buried by later deposition, hence 'exhumed relief', 'e. surface', 'e. landscape'; sometimes the result is a 'fossil erosion surface'; e.g. the e. of Charnwood Forest (Leicestershire), formerly buried under the Triassic rocks; of parts of the chalk dip-slopes of the London Basin; of parts of the Central Massif of France.

exogenetic (exogenic) The external forces of denudation (weathering, mass-wasting, erosion, transport, deposition), which combine with internal (ENDOGENETIC) forces to produce land-forms.

exosphere A zone in the IONOSPHERE above about 500–750 km (300–470 mi.), characterized by the presence of neutral atomic oxygen, ionized oxygen and hydrogen atoms.

exotic A plant or animal which has been introduced from some other area where it flourished naturally; e.g. rhododendron into England from the Himalayas; citrus fruit into the Mediterranean lands from China; the rabbit into Australia.

exotic river A r. deriving most of its volume from headstreams in another region, as from snow-melt or heavy rainfall on distant mountains e.g. the Colorado R. in S.W. U.S.A.; the Blue and White Niles; the Tigris and Euphrates.

expanded-foot glacier A g. which issues beyond the mouth of a valley to expand into a broad lobe of ice on an adjacent plain; e.g. Skeidarajökull on the edge of Vatnajökull, in Iceland. An e.-f. g. is a small version of a PIEDMONT G.

expanded town A t. where large-scale public development is undertaken under the Town Development Act (1952), primarily for the purpose of relieving congestion or over-population elsewhere. Development is financed by the local authority concerned, rather than by the Government, as is the case in development under the New Towns Act (see NEW TOWN). E.g. Swindon, Haverhill, King's Lynn, Thetford, Basingstoke.

exports Items shipped out of a country for sale abroad as part of its trade; these are *visible e.'s. Invisible e.'s* comprise payments for services, transport, insurance, loan interest, revenue from foreign investments and tourism.

exposure (i) The position of any place relative to sunshine, prevailing winds, oceanic influences; cf. ASPECT. (ii) In a geological context, a place where solid rocks are naturally or artificially exposed to view. *Note:* A rock covered with soil derived from its own disintegration would be mapped as an OUTCROP, not as an e.

extensive agriculture Farming in large units, usually highly mechanized, employing little labour, with relatively low yields per acre, but with a large total yield and a high yield per man; e.g. the wheatlands of Canada and U.S.A.

extinct volcano A v. formed in long past geological time, in an area where there is now no sign of activity. The orig. form may have been largely destroyed by denudation; e.g. Arthur's Seat and Castle Rock, Edinburgh; Mont Dore in the Central Massif of France.

extractive industries Mining and quarrying; some authorities also include forestry.

extrapolation Extending the values of a series of variables on either side of some known values. In a cartographic or diagrammatic sense, the term implies the reconstruction of past or future patterns or trends by extending and developing present patterns or trends as mapped or graphed. E.g. the extension and projection of a graph of population trends; the reconstruction of former contours (EOHYPSES) from surviving fragments of a dissected land-surface; the reconstruction of the former profile of a river valley before rejuvenation by e. from the KNICKPOINT. It can also be based upon finding a mathematical expression to fit a segment of river profile, and then e. of the curve downstream on that basis.

extraterritorial App. to certain agreed rights and immunities enjoyed by people living within another State, exempting them from internal jurisdiction; e.g. European residents in the TREATY-PORTS of China between 1843 and 1943.

extreme climate A c. with a considerable difference or *range* between the temperatures of the warmest and coolest months. This occurs mainly within continental interiors, far from moderating marine influences; e.g. Cold Continental C. in central Asia (Verkhoyansk, Jan. mean, —51°C. (—59°F.), July mean 15·5°C. (60°F.), with an absolute minimum of —69·5°C. (—93°F.); this is W. Köppen's *Dwd* type.

extremes, of temperature The highest and lowest shade t.'s recorded during a day (*diurnal e.*), month, year or any period of available records at a meteorological station. It may be used of the mean highest and mean lowest t. over a period of time. E.g. Ajaccio (Corsica), in Dec., extreme max. = 20°C. (68°F.), extreme min. = −2°C. (28°F.). *Note*: the mean monthly t. for Ajaccio in Dec., the figure usually quoted in climatic tables, is 9°C.

(48°F.). The world's extreme max. t. = 57·8°C. (136°F.) (Azizia, Tripoli), the extreme min. = −88·3°C. (−127°F.) (in Vostok in Antarctica in 1960). Gt. Britain's extreme max. = 38·1°C. (100·5°F.) (Tonbridge, Kent), extreme min. = −27·2°C. (−17°F.) (Braemar, Scotland).

extrusion The emission of solid, liquid or gaseous materials on to the earth's surface during an eruption; hence EXTRUSIVE ROCK.

extrusion flow, of ice Movement caused by the great thickness and resultant pressure of ice within an ice-sheet. This causes an *en masse* subsidence within the ice, which towards the margins becomes a more horizontal and outward movement, probably facilitated by PLASTIC DEFORMATION in the basal layers. So in the Pleistocene an ice-sheet gradually crept S. over extensive lowlands (N.W. Europe, N. America), despite the gentle gradient.

extrusive rock An igneous r. formed by the pouring out of lava on the surface, where it solidified; syn. with volcanic r., as opposed to PLUTONIC or INTRUSIVE. The cooling is rapid, resulting in small-crystalled or even glassy r.'s; e.g. basalt, obsidian, rhyolite, andesite.

exudation basin A depression in the surface of the ice at the head of a glacier where it leaves an ice-sheet; e.g. in Greenland.

'eye' Used of a HURRICANE or other tropical storm to describe the central area, a small region 20–60 km. (12–40 mi.) across, of calms or light variable winds, surrounded by the storm-area. The atmospheric pressure at the centre may be as low as 965 mb.

eyot See AIT.

fabric Used to indicate the physical composition or makeup of some compound; e.g. TILL f., an assemblage of clay, stones and sand, which may be statistically analysed in various ways, incl. the orientation and dip of the stones within its matrix.

facet (i) Orig. one face of a stone (esp. when worn, as by the wind); ct. DREIKANTER, EINKANTER. (ii) Any eroded plane surface which interrupts a general slope; e.g. *a faceted spur.* (iii) Used specif. as a polished surface of a gemstone.

facies (i) Appearance, nature or character, applied esp. to descriptions of the composite character of a rock. (ii) A certain classificatory implication, distinguishing one stratigraphic body from another, with ref. both to lithological character and fossil content. (iii) Applied to rocks of a certain age where lateral changes of lithology imply changing conditions of formation, with accompanying differences in the fossil content.

factor Used in 2 senses: (i) a gen. cause or control contributing to a partic. result, effect or condition; e.g. in climate, f.'s include latitude, altitude, distribution of land and sea, ocean currents, presence of lakes, influence of relief barriers. (ii) In the analysis of situations involving numerous simultaneous variables, the term implies one category or 'family' of variables; *f. analysis* is an important aspect of geographical research, enabling data to be mathematically analysed, by means of a computer, on a very large scale.

Fahrenheit A graduated scale o[f] temperature, introduced about 172[0] by Gabriel Fahrenheit, on which [0] represents the melting-point of ice in [a] mixture of sal-ammoniac and wate[r], 32° represents the melting-point of ic[e] in water, and 96° was taken as th[e] temperature of a healthy perso[n] (subsequently found to be 98·4°). H[e] did not use the boiling-point of wate[r] but found by experiment that on h[is] scale it equalled 212°. Fahrenheit [is] remembered also for his improveme[nt] of the barometer and thermometer.

Conversion: °F. = $\frac{9}{5}$(C.° + 32°)
°C. = $\frac{5}{9}$(F.° − 32°)

°F	°C	°F	°C
100	37·7	46·4	8
98·6	37	45	7·2
96·8	36	44·6	7
95	35	42·8	6
93·2	34	41	5
91·4	33	40	4·4
90	32·2	39·2	4
89·6	32	37·4	3
87·8	31	35·6	2
86	30	35	1·6
85	29·4	33·8	1
84·2	29	33	0·5
82·4	28	32	0
80·6	27	31	−0·5
80	26·6	30·2	−1
78·8	26	30	−1·1
77	25	28·4	−2
75·2	24	26·6	−3
75	23·8	25	−3·8
73·4	23	24·8	−4
71·6	22	23	−5
70	21·1	21·2	−6
69·8	21	20	−6·6
68	20	19·4	−7
66·2	19	17·6	−8
65	18·3	15·8	−9
64·4	18	15	−9·4
62·6	17	14	−10
60·8	16	12·2	−11
60	15·5	10·4	−12
59	15	10	−12·2
57·2	14	8·6	−13
55·4	13	6·8	−14
55	12·7	5	−15
53·6	12	3·2	−16
51·8	11	1·4	−17
50	10	1	−17·2
48·2	9	0	−17·7

Vote: −40° is common to both scales.

airway The main navigable chan-el of a river or estuary, usually uoyed and lighted, leading up to a ort or harbour.

ll (i) Used in U.S.A. for autumn. i) See WATERFALL.

ll-line A line or narrow belt join-g the points where a series of near-rallel rivers descends by falls or pids from a plateau edge on to a wland; e.g. the F.-l. in S.E. U.S.A., tween the ancient rocks of the Pied-ont Plateau and the newer rocks of e Atlantic coastal plain. [*f opposite*]

fallow Arable land which is allowed to rest uncropped for one or more seasons.

false-bedding See CURRENT B.

false cirrus A type of thick, greyish CIRRUS cloud, often associated with the upper part of a thunder-cloud, and gen. a sign of bad weather.

false drumlin A mass of rock over which a thin cover of DRIFT has been laid by an ice-sheet, thus giving the superficial appearance of a true D. [*f* DRUMLIN]

false origin A point of o. in a GRID system, from which the position of any place is uniquely defined in terms of its co-ordinates. The o. is transferred from the intersection of the projection axes to the 'false' position to avoid negative quantities. The f. o. of the British National Grid lies just S.W. of the Scilly I's., transferred 400 km. W. and 100 km. N. from the intersection of the central meridian 2°W. and standard parallel 49°N. The f. o. of American military maps on the U.T.M. GRID for each of the 60 zones is a point 500 km. W. of the central meridian on the equator for the N. hemisphere, 10,000 km. S. of this point for the S. hemisphere.

fan See ALLUVIAL F.

fan-folding An ANTICLINORIUM with the axial planes of the folds converging towards a centre.

fanglomerate Material deposited in an ALLUVIAL FAN, consisting of heterogeneous fragments of all sizes, which have been compacted and/or cemented into solid rock.

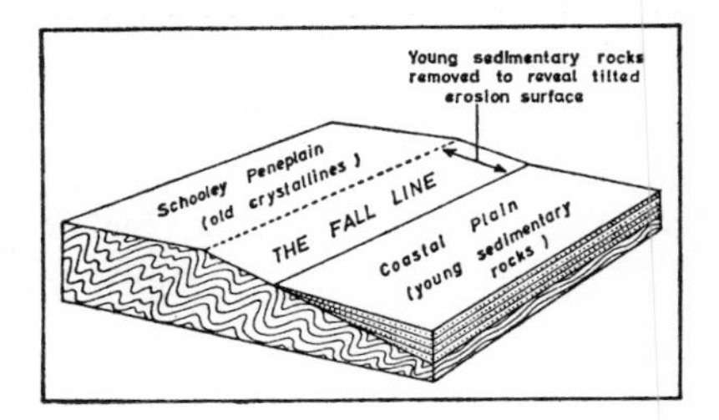

fast ice I. which covers the frozen surface of the sea, but remains f. to the actual coast.

fathom A nautical measurement of depth = 6 ft. = 1·829 m.; 100 fthms. = 1 cable; 10 cables = 1 nautical mile.

fault A surface of fracture or rupture of strata, involving permanent dislocation and displacement within the earth's crust, as a result of the accumulation of strain; see ELASTIC REBOUND. Hence *faulting*. See also NORMAL, REVERSED, TEAR and THRUST F., HADE, HEAVE, THROW. [*f*]

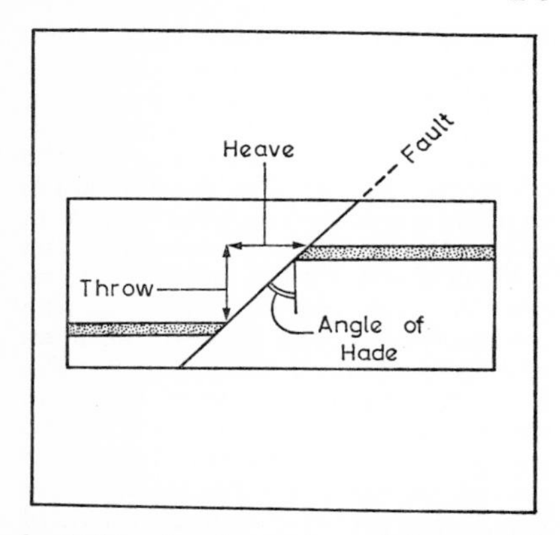

fault-block A section of the earth's crust sharply defined by faults; it may stand up prominently and loftily (BLOCK MOUNTAIN, HORST), be tilted (TILT-BLOCK), or be denuded to the level of the surrounding country (faulted inlier). Ct. GRABEN.

fault-breccia The rock occurring in the SHATTER-BELT of a fault, consisting of crushed angular fragments. [*f* HANGING WALL]

fault-line scarp Where faulting brings rocks of varying resistance into close juxtaposition, differential denudation may wear away the less resistant rock on one side of the fault, thus forming a cliff or scarp along its line. Ct. FAULT-SCARP. The resultant higher ground may be on either the upthrow or downthrow side. E.g. a f.-l.s. has developed along the line of the Mid-Craven F. in Yorkshire. To the N. is a plateau of Carboniferous Limestone, which has been down-faulted to the S., so that less resistant Bowland Shales have been brought into juxtaposition along the line of the f. Denudation has caused the line of the f. to stand out as a f.-l. s., in places 300 ft. high, including Malham Cove and a line of 'scars'. The E. face of the Sierra Nevada, California, is a major f.-l. s., as is the E. face of the Grand Tetons, Wyoming. As denudation progresses, the f.-l. s. may develop into an OBSEQUENT F.-L. S. (facing the opposite direction), then into a RESEQUENT F.-L. S. (facing the original direction). [*f*]

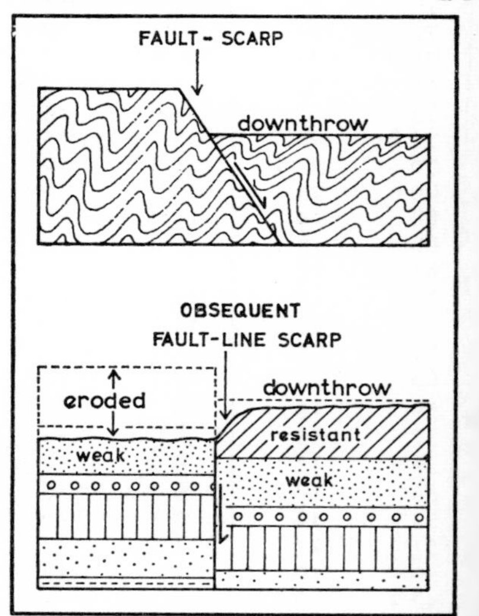

fault-scarp A visible steep edge or slope of a recent fault, due to earth-movements. It is an initial land-form present only in the earliest stages of denudation succeeding the earth-movement which caused it; e.g. two f.-s.'s were produced in Montana and Wyoming by the Madison earthquake of 1959, forming long 'steps' across country about 4–6 m. high. Most f.-s.'s are soon obliterated by weathering and erosion, or they later develop into FAULT-LINE SCARPS. [*f*]

fault-spring A SPRING thrown out along a fault, where a permeable bed such as sandstone, is brought up against an impermeable rock, such as shale.

fauna The animal life of any region, geological formation or period.

feather-edge The fine edge of a bed of rock, usually sedimentary (though it includes also igneous intrusions), where it thins or PINCHES-OUT and disappears. It can be seen on the surface of a DIP-SLOPE, as in the N. Downs, where an Eocene f.-e. occurs, from beneath which the Chalk emerges; and at the base of the Coal Measures in N. Staffordshire ('F.-e. Coal').

fell (Norwegian) An open hill- or mountain-side in N. England, esp. the English Lake District, hence *f.-pasture*, and '*the f.*' as the upper part of a sheep-farm. It is also used for culminating summits; e.g. Scafell, Bowfell (English Lake District); and Crossfell (N. Pennines).

Felsenmeer (Germ.) A continuous spread of large angular boulders, hence *block-field*, *block-spread*, *boulder-field*, formed *in situ* by acute frost action on well jointed rocks, covering the surface of a high plateau or flat-topped mountain. Lit. a 'rock-sea'; e.g. the summits of the Scafell Pike range in the English Lake District, and of Glyder Fach and Glyder Fawr in N. Wales.

felspar, feldspar A complicated group of minerals consisting mainly of silicates of aluminium, with those of potassium, sodium or calcium. *Alkali f.* includes orthoclase (mainly potassium), soda-orthoclase (containing some sodium), and anorthoclase (mainly sodium). *Soda-lime f.* (with silicates of sodium and calcium) is known as plagioclase-f., a series varying in composition from albite (sodium) to anorthite (calcium); the opalescent variety of albite is the moonstone. The plagioclase-f.'s form important constituents of such sialic rocks as diorite and andesite, and occur also in a calcic form in such basic rocks as gabbro and basalt.

felucca (Arabic) A small boat used in the E. Mediterranean Sea, the Red Sea and on the Nile, propelled by a lateen sail, oars or both.

fen A water-logged area, with pond-weeds, reeds, rushes and alders, in which peat is accumulating, but in which the ground-water is alkaline (usually with calcium carbonate) or neutral; e.g. the F. District of Lincolnshire and Norfolk; the Lancashire Mosses; the Somerset Levels; the '*laagveen*' of the Netherlands. When drained, it affords good humus- and mineral-rich soil, of excellent texture, dark in colour. Ct. BOG, MARSH.

fenland An area of FEN. A proper name for the area in E. England around the Wash.

Fenster (Germ.), **fenêtre** (Fr.), *lit.* a window An opening worn through the upper strata by denudation, where rocks have been overfolded, thus exposing the underlying younger rocks in its floor. The term should not be used simply for the exposure of older underlying rocks by denudation, as along the crest of an anticline; see INLIER. E.g. the Fenêtre de Theux in the Ardennes, Belgium, is a small massif consisting mainly of Devonian rocks, outlined by perimeter faults. It is a piece of a NAPPE, revealed by the removal of overthrust Cambrian-Silurian rocks by denudation. In the E. Alps, denudation has cut through the Austride nappes, exposing the underlying nappes in 3 places: in the Hohe Tauern, the Semmering and Lower Engadine districts, the classic 'windows of the Alps'. [*f*]

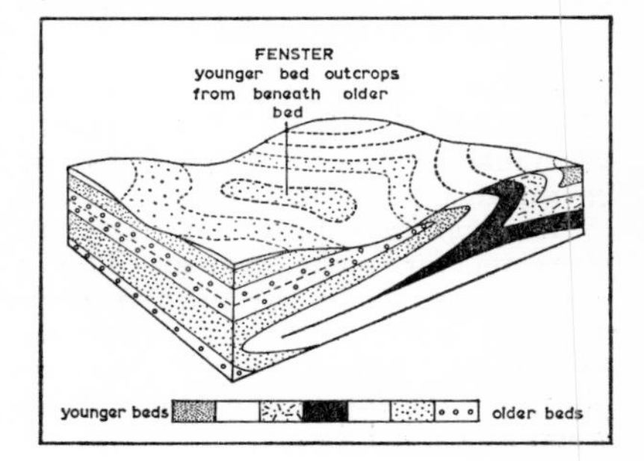

Ferrel's Law As the result of the earth's rotation, a body moving on its surface is subject to an apparent deflection to the right in the N. hemisphere, to the left in the S. (the CORIOLIS FORCE). This apparent force is zero at the equator, but increases progressively N. and S. Its effects are shown most specif. in air and water. The law was postulated by W. Ferrel, an American scientist (1856).

ferric In chemical terms, a trivalent (or tervalent) iron compound. The two most economically valuable are *haematite* (red f. oxide, Fe_2O_3) and *limonite* (hydrated f. oxide, $2Fe_2O_3.3H_2O$).

ferromagnesian mineral A m. characterized by an abundance of iron and magnesium, of which the chief are the AMPHIBOLE, OLIVINE and PYROXENE groups. Gen. these m.'s are dark in colour. Syn. with *mafic*.

ferrous App. to iron compounds that are not saturated with oxygen; in chemical terms, a divalent (or bivalent) compound of iron. They are usually formed by the reduction of ferric compounds; e.g. *siderite* (f. carbonate, $FeCO_3$), the ores found in the Jurassic oolites of England.

ferruginous Rocks containing some iron, often reddish in colour.

fertility ratio In population studies, the number of young children in the population related to the number of women of child-bearing age. A valuable indication of future trends is thus gained.

$$f.\,r. = \frac{no.\ of\ children\ under\ 5\ years}{no.\ of\ women\ aged\ 15\text{–}50} \times 1000$$

fertilizer A substance of animal (dung, dried blood, bone meal), vegetable (compost, spent hops) or chemical ('artificials') origin, added to the soil to supply the necessary elements for plant growth. The 3 main requirements are *nitrogen* (ammonium salts, nitrates); *phosphorus* (bonemeal, basic slag, superphosphates); *potassium* (sulphate of potash, wood ashes, kainite).

fetch The distance of open water over which a wind-blown sea-wave has travelled, or over which a wind blows. This helps to determine the height and energy of a wave, and hence its erosive and depositive effects on the coast. Many coastal features of deposition seem to be orientated to face the direction of max. f.; e.g. a CUSPATE FORELAND [*f*], such as Dungeness. Wave attack is emphasized and concentrated on headlands exposed to a long f.

fiard (fjard) (Swedish, *fjärd*) A coastal indentation resulting from a rise of sea-level on the margins of a glaciated rocky lowland or PENEPLAIN, with lower shores and broader profiles than a FIORD, usually with a threshold, and deeper than a RIA, and with numerous fringing islands; e.g. along the coast of S. E. Sweden. This English usage has taken on a more specialized meaning not apparent in Sweden, where the term means simply large open areas of water surrounded by islands, as in this map of an area near Stockholm. [*f*]

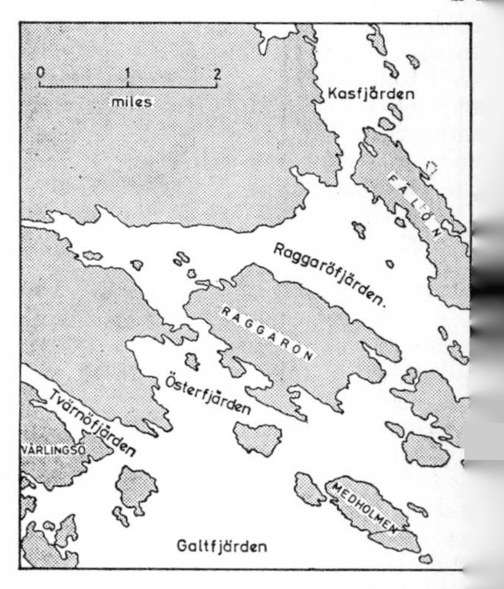

ffridd (Welsh) An enclosed field adjacent to farm-buildings, beyond which lies rough hill-grazing.

filling An increase in the atmospheric pressure near the centre of low pressure system (DEPRESSION), hence 'filling-up', or dying away, as opposed to '*deepening*'.

finger lake A long, narrow l. occupying a U-shaped glacial trough; e.g. in the English Lake District; N. Wales; the Highlands of Scotland (*lochs*); and N. Italy (Maggiore, Como, the upper part of Garda). Most of these are the result partly of the glacial erosion of the containing valley in solid rock, partly of the deposition of a morainic dam across the lower end of the valley. The Finger Lakes of U.S.A. occur in New York State, near Syracuse; they now drain N. to L. Ontario, though their preglacial valleys drained S. to the Susquehanna R. [*f*]

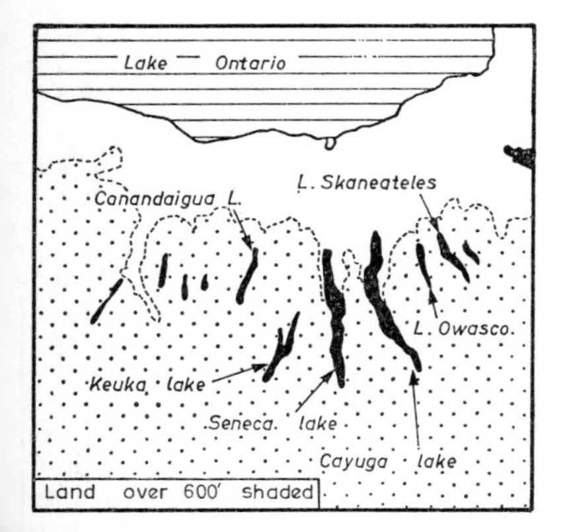

Finiglacial stage The 3rd stage in the retreat of the continental ice-sheet from Scandinavia, from about 8000–6500 B.C. During this stage, the ice left the area which is now Finland. The land was gradually uplifted by ISOSTATIC recovery, and the Baltic Sea was reduced to the land-locked L. Ancylus.

fiord, fjord (Norwegian) A long narrow arm of the sea, the result of the submergence of a deep glacial valley, with steep walls (the angle of which continues beneath the water), deep water virtually to its head, numerous rectilinear branches, and a bar or threshold near its mouth (usually of solid rock, with sometimes a cover of glacial debris). The f.'s of Norway are deeply cut into the high plateau of Scandinavia. During the Ice Age, a glacier eroded along a line of least resistance, such as a preglacial river valley, a major fault-line, or a line of weak unresistant rocks (e.g. the Hardanger F., which lies along a syncline of schist between two masses of hard crystalline rock). The Sogne F. is 160 km. (100 mi.) long and mostly less than 5 km. (3 mi.) in width, with walls sloping at 28°–40° from the 1500 m. (5000 ft.) plateau-surface to over 900 m. (3000 ft.) below sea-level. Other f.'s are found in W. Scotland (the sea-lochs), Greenland, Labrador, British Columbia, Alaska, S. Chile, and the W. of S. I. of New Zealand. [*f*]

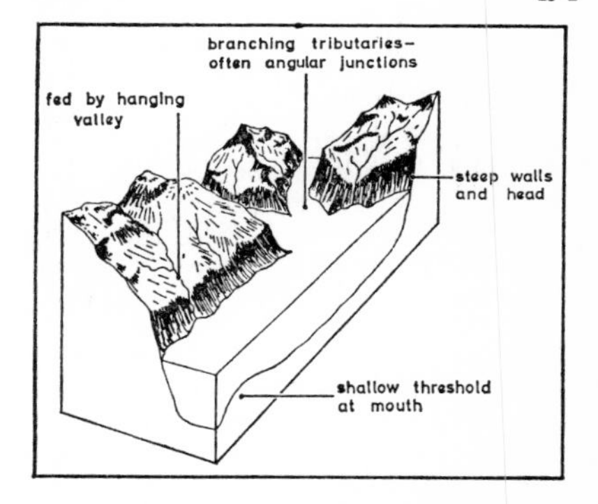

fire-clay A clay which is capable of withstanding great heat, and is used for making refractory bricks for the linings of furnaces. It occurs commonly as an underclay beneath seams of coal in the Lower Coal Measures, and may be the bed in which the roots of the 'coal forest' grew. See also GANISTER.

fire-damp An explosive mixture of air and methane (CH_4), which may occur in coal-mines.

firn (*Germ.*) Syn. with *névé*, but many glaciologists prefer f. Others define f. as the compacted snow, *névé* as the area of f. Lit. 'last year's' or 'old' snow. Where snow is able to accumulate in a basin or hollow, both from direct snowfall and from avalanches down the surrounding slopes, it is compressed by the weight of the addition of successive layers, and is gradually changed into a more compact form. Air is retained between the particles, forming a mass of whitish granular ice. During surface melting on summer days, water percolates into the f. and then re-freezes at night to form a more compact mass. SUBLIMATION assists, whereby molecules of water vapour escape from the snow-flakes and re-attach themselves, so that granules of crystals become progressively more tightly packed. Its density is greater than 0·4, less than 0·82. (Some American glaciologists give a precise density of 0·55.) Some degree of stratification can be seen, representing each year's contribution of snow. From f. other terms are derived: *firnification*, the creation of f.; the *f.-line* or *f.-limit*, the highest level attained by the snow-line during the year or alt. the edge of the snow-cover at the end of summer; the *f. equilibrium line*, the level at which ABLATION exactly balances ALIMENTATION; the *f.-field*, the actual mass of accumulated compact snow; the *f.-basin*, in which it accumulates.

firth A Scottish word for some types of water-area, notably: (i) the lower part of an estuary (e.g. the F. of Forth, Solway F.); and (ii) a strait (e.g. the Pentland F.).

fissile The quality of a rock being easily split, where the BEDDING-PLANES, JOINT-PLANES, LAMINAE or CLEAVAGE are well developed and defined; e.g. slate (a property of which a quarryman takes advantage), shale, flags, schist and oolitic limestones (as in the Cotswolds, where they are used for roofing, and are locally called 'slates', e.g. the Stonesfield Slate, a sandy limestone of the Great Oolite Series). F. bedding has laminae less than 2 mm. thick.

fissure eruption A linear volcanic e., in which lava, gen. basic and of low viscosity, wells up to the surface along a line of crustal weakness, usually without any explosive activity. The gen. result is an extensive basalt plateau. In the Laki (Iceland) f. e. of 1783, an outpouring of lava occurred along a f. 30 km. (20 mi.) long, and also a string of ASH CONES (or 'eruptive conelets') was built up along its line, as in the diagram. [*f*]

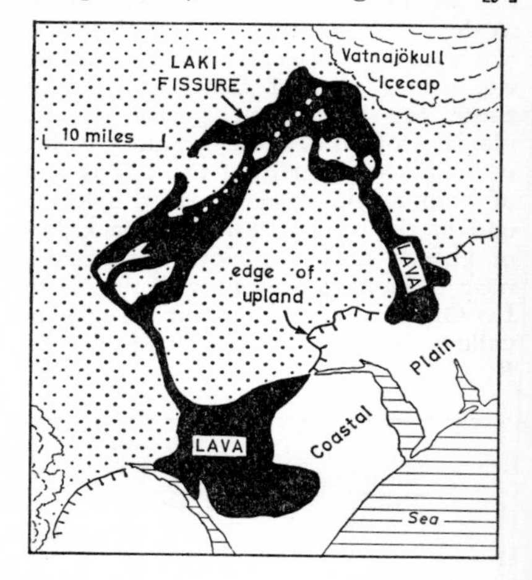

fjard See FIARD.

fjeld, fjell (Norwegian), **fjäll** (Swedish) Cognate to *fell* in N. England, and broadly similar in character, though the Scandinavian usage implies a plateau surface, often rocky, bleak and monotonous, above the tree-line, covered with snow in winter. Poor pasture occurs on lower parts, and lichens and mosses on the higher, with occasional patches of dwarf birch, willow, alder, juniper and berried plants of a TUNDRA aspect. It is used by Lapps in the N., and for summer grazing of cattle and goats on the lower areas further S. E.g. the Dovre F. lying at 900–1200 m. (3–4000 ft.).

fjord (Norwegian) See FIORD.

flagstone, flag (derived from Old Norse, *flaga*, a slab) A sandstone or sandy limestone which splits along the BEDDING-PLANES, and is hard enough to be used for paving and building. This is esp. the case when flakes of mica have been deposited on the bedding-planes. E.g. the Old Red Sandstone (though grey in colour) of

Caithness (N. Scotland); the Carboniferous sandstones of Yorkshire; the Silurian flags of S. Cumberland and Westmorland.

Flandrian Transgression A general rise of sea-level from *c.* 8000 B.C., when sea-level was approx. 55 m. below the present. The rise covered the basin of the N. Sea lowlands, and formed the N. Sea and the English Channel. The Straits of Dover were breached *c.* 5000 B.C. By *c.* 3000 B.C., sea-level stood about 6 m. lower than at present.

flash (i) A sudden rise of water in a stream. (ii) A small lake in a hollow caused by subsidence due to underground mineral working, esp. salt; e.g. Bottom F., Top F. near Winsford, in Cheshire.

flash-flood A sudden but short-lived torrent in a usually dry valley, notably in a semi-arid area after a rare, brief, but intense rainstorm, carrying an immense load of solid matter, the product of desert weathering. The stream turns into a mud-flow and soon comes to rest. The flow of water may be torrential for a short time, and there are records of people being swept away and drowned. E.g. in Arizona, Utah, S. California. It may also be caused by the collapse of a dam; e.g. Fréjus in France (1959); or of an ice- or log-jam, as commonly in N. Canada.

flat (i) A low-lying area of marsh or swamp in a river valley; e.g. Altcar F.'s, Lancs. (ii) A mud-bank exposed at low tide (*mud-f.*). (iii) In geomorphology, any nearly level area within a region of marked relief. (iv) In mining, a horizontal mass of ore deposited along a BEDDING-PLANE, as in the case of the galena and calamine (*smithsonite*) deposits of the N. Pennines in Carboniferous Limestone. The f.'s are usually linked by vertical fissures (RAKE). (v) A bank of alluvium or gravel in the bed of a stream, containing alluvial tin or gold.

flatiron A triangular mass of rock outcropping on the end of a spur where the strata are inclined steeply. The term is used esp. in U.S.A.; e.g. the Flat-Irons, consisting of great slabs of coarse sandstone and conglomerate, the Fountain formation of Pennsylvanian (Upper Carboniferous) age, outcropping along the Front Range of the Rockies, and overlooking the town of Boulder, Colorado. The smooth steep E. faces are actually the tilted bedding planes. [*f*]

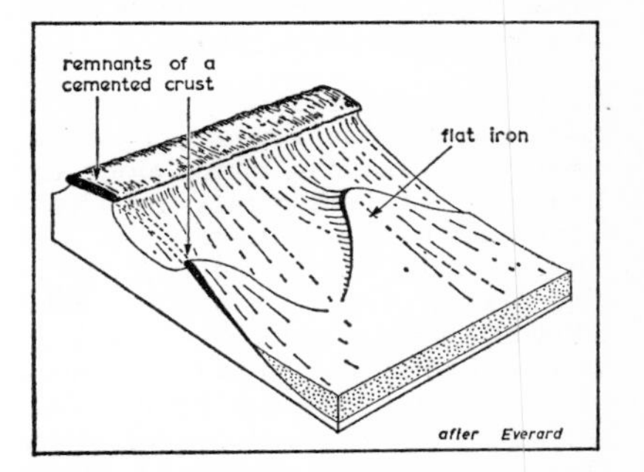

flatland ratio map A m. constructed by gridding a contour m., calculating the percentage of each square occupied by land below any selected critical slope, plotting these values, and inserting isopleths.

flax An annual plant of the genus *Linum*, esp. *L. usitatissimum*, which grows to about 1 m. high, with lance-shaped leaves and small blue flowers; the former yield the linen fibres, the latter develop seeds, from which LINSEED oil is obtained. It is grown on rich soils in cool temperate latitudes, notably in the U.S.S.R. (about $\frac{3}{4}$ of the world total), Poland, E. Germany, France, N. Ireland and Belgium. About 700,000 tons of flax fibre are produced each year. The stalks are pulled, combed to remove the seeds, retted (i.e. rotted) to get rid of the resinous matter by soaking the leaves in soft water (e.g. in a river such as the Lys, now more usually in tanks) for some weeks, dried, the woody core removed by 'scutching' (i.e. beating), and the fibres separated. These, which are often 0·6 m. long, are spun into thread and woven to form linen cloth.

fleet (i) A small lagoon behind the coastline, usually of salt or brackish water, separated from the open sea by a long sand- or shingle-bank parallel to the coast; e.g. the Fleet behind Chesil Bank, Dorset. (ii) A small creek or inlet, the name preserved in the case of Fleet St., London, which was called after the Fleet R. Also Northfleet, Kent.

flexure The gentle bending of strata as a result of tensional stress.

flint A black or dark grey mass or nodule of a dense, fine-grained silica, usually occurring in bands along the bedding- or joint-planes in chalk. They appear prominently in the Upper Chalk, but are rarely found in the Lower Chalk. Probably small siliceous sponge-spicules, originally scattered throughout the Chalk, were carried away in solution by percolating water, and were redeposited where some silica already existed; i.e. at the bedding-planes, so that large amounts of silica were concentrated there. Thus the f. appears to grow as in the manner of a concretion. The silica in f. is very pure. A nodule is hard and tough, but under impact it breaks with a conchoidal fracture, providing a sharp edge. F. was the chief raw material of tools and weapons of the Stone Ages. Cf. EOLITH.

flocculation, *vb.* to **flocculate** (i) The process by which soils and soil-colloids coagulate or aggregate into 'crumbs', thus coarsening the texture of a fine clay soil; this is stimulated by liming. (ii) The deposition of clay particles in salt water.

floe A thin, detached, floating horizontal sheet of sea-ice.

flood The inundation of any area which is not normally covered with water, through a temporary rise in level of a river, a lake or the sea. (i) A *river* floods when its channel is inadequate to accommodate the discharge from its CATCHMENT; i.e. it rises from BANKFULL to f.-stage, so spreading over its f.-plain. This may occur after exceptional rainfall or snow-melt; e.g. on the R. Ohio, Jan., 1963. On some rivers, flooding is part of the seasonal régime; e.g. the R. Nile, on which irrigation depends. If a river flows above its f.-plain as a result of sedimentation in its bed, so that it is contained between natural or artificial embankments (LEVEES), the danger of widespread flooding if the levees are breached is very great; e.g. the lower Mississippi; the R. Po (esp. in the winter of 1951–2 and the autumn of 1966); the Hwang-Ho ('China's Sorrow'). Concentrated rainfall may produce exceptional local f.'s; e.g. in the valleys of the E. and W. Lyn R., N. Devon, following heavy rain on Exmoor, 18 Aug., 1952; discharge of water here exceeded 18,000 CUSECS for a few hours. (ii) A *lake* may be enlarged by exceptional rainfall, so flooding the adjacent low-lying land; e.g. Derwentwater and Bassenthwaite (English Lake District) were joined in Aug., 1938, flooding the inter-lacustrine land; L. Eyre (Australia) expanded to 8000 sq. km. (3000 sq. mi.) of water in 1950–1. (iii) The *sea* may flood low-lying coastlands if a high tide coincides with a STORM-SURGE, esp. if dikes (dykes) collapse; e.g. 31 Jan.–1 Feb., 1953, when the coastlands of the Netherlands and S.E. England were inundated. (iv) The *collapse of a dam* may cause disastrous f.'s; e.g. Dolgarrog, Wales (1925); Fréjus in France (1959); near Los Angeles (Dec., 1963). A natural dam across the Gros Ventre R., Wyoming, formed by a land-slide in 1925, collapsed in 1926; the resultant f. destroyed the town of Kelly.

flood control The prevention a far as is possible of inundation c land by flooding, including: (i) th reduction of concentrated runoff b maintaining a vegetation cover; e.g the Tennessee Valley, U.S.A.; (ii) th creation of storage basins to hol water temporarily; e.g. in Iraq; (ii the creation of barrages to hold u water until it can be passed dowr stream; e.g. the Ohio R.; (iv) th dredging and straightening of channels to facilitate outfall, and the creation of more direct outfalls to the sea; e.g. the R. Ouse in E. Anglia.

flood-plain The floor of a valley over which a river may spread in time of flood, depositing ALLUVIUM. The f.-p. is present in the lower valleys of 'old' rivers, usually bounded by low BLUFFS well back from the channel; it is an area of gentle or even imperceptible gradients, across which wanders the alluvium-laden river, beyond whose channel lie marshes, OXBOWS and stagnant creeks, with man-made embankments set well back. The alluvium may be deposited as only a thin veneer, or in a great thickness; in the lower Nile valley no boring has ever reached the rock-floor. The f.-p's of the Nile, Tigris–Euphrates, Indus, Ganges are intensively cultivated.
[*f* MEANDER-BELT]

flood-stage The state of river flow beyond the BANKFULL stage; i.e. the water overflows beyond the banks.

flood tide (i) The rising t. across mud-flats, sand-flats and the beach between the low-water and high-water marks on a gently shelving coast; this may be very rapid; e.g. in the Solway Firth, Morecambe Bay and the Dee estuary. (ii) The inflowing tidal stream up an estuary or other inlet, usually flowing strongly for about 3–4 hours until a period of slack occurs at high t.; ships usually 'come up with the t.' There are anomalies; e.g. in Southampton Water the normal f. t. is interrupted by a period of slack-water occurring 1½ hours after low water and lasting for 1½ hours, known as the '*young flood stand*'. The f. t. in an estuary into which a river flows with a powerful current may cause a BORE.

floor space index A comparative measurement of the quantity of accommodation in buildings on different sites, derived from the formula

$$\frac{\textit{total floor space of building}}{\textit{area of site, incl. half the area of adjoining road.}}$$

flora (i) The plant-life of any region, formation or period. (ii) A work listing the plant-life by species, or describing the plant-life of a partic. area.

flow The movement of a fluid. (i) The f. of water or lava, under the influence of gravity. (ii) The movement of air (wind), broadly from areas of high to low pressure, modified by the effects of the CORIOLIS FORCE and friction with the earth's surface. (iii) The *plastic f.* of solid rocks under the influence of stress, without any fracture or rupture. (iv) The movement of glacier-ice; see GLACIER FLOW.

flow diagram, chart A d. in which a sequence of interlinked topics, events or items is presented to show the development or evolution of some theme, objective or product; e.g. the production of a map, from raw field-data, through the drawing and reproducing stages. This enables adherence to specification, cost control, quality control, assessment of progress, etc., to be rigorously maintained at all stages, thus effecting economies in time and cost.

flow-line map A m. showing movement of freight, passengers, tonnage of shipping. A line indicates the trend of the routeway concerned, while the quantitative indication of traffic is given by the thickness of the line drawn to scale, using such values as TONS, TON-KILOMETRES, or ton-kilometres per kilometre.

fluidization Derived from an industrial process, and used by D. Reynolds (1954) in connection with igneous activity when hot gases pass through fine-grained material, causing it to flow, and resulting in the creation and enlargement of cracks in rocks by both physical and chemical means. This is probably one way in which PIPES are formed, as circular enlargements of cracks or faults.

flume (i) An artificial stream channel constructed for industrial purposes: to provide water for power, to float logs and for water supply. (ii) Used in parts of U.S.A. for a narrow ravine or gorge.

fluorspar, fluorite Calcium fluoride (CaF_2), a glassy translucent mineral, usually in the form of cubic crystals, occurring in clusters or aggregates, commonly as a GANGUE-filling or 'veinstone' in a vein of a metallic mineral, notably of lead and zinc. It has a wide range of colour, from white to deep purple; a massive dark blue variety, mined at Castleton in Derbyshire, is known as 'Blue John'. It occurs commonly in the limestone of England, esp. Derbyshire, and in Illinois and Kentucky in U.S.A. It is widely worked for ornamental purposes, and is also used as a flux in the basic open-hearth steel-making process, for making hydrofluoric acid, and in the ceramics industry for 'vitriolite'. It has given its name to the property of *fluorescence* (i.e. the power to absorb light of one wavelength and emit light of another longer wavelength, notably when exposed to ultra-violet light).

flush (i) A sudden rush of water down a stream. (ii) An area of minerally enriched soil; see FLUSHED SOIL.

flushed soil The s.'s on lower hill-slopes, areas of accumulation which are base-rich because of the downhill movement to them of the products of leaching from higher up. F.s.'s are classified as: (i) a *damp f.*, found around spring-sources and rivulets, where water with mineral salts in solution enriches a small area, often indicated by a patch of bright green vegetation; (ii) a *dry f.*, consisting of freshly weathered rock particles, commonly at the foot of a scree, stone-slide, gully or crag, where new rock is constantly exposed to weathering, thus yielding a continuous supply of bases.

fluvial, fluviatile App. to a river. There is no real difference between them, but the practice is to use fluvial in respect of river flow and erosive activity, and fluviatile for the results of river action (*f. deposits*), and for river life (*f. fauna* and *flora*).

fluvioglacial App. to the effects of melt-water streams issuing from a glacier-snout or an ice-sheet margin. In U.S.A. the term *glaciofluvial* is preferred, since logically the *glac-* precedes the *fluv-*. (i) *F. erosion*: (*a*) the cutting of a channel or trough by melt-water along the edge of an ice-sheet (e.g. URSTROMTAL); (*b*) the cutting of an overflow channel (SPILLWAY) across a pre-glacial watershed. (ii) *F. deposition*: the laying-down of an OUTWASH plain of gravels, sands and clays, in that order from the edge of the glacier; the laying down of fine materials in a PROGLACIAL LAKE, forming lacustrine sediments (e.g. 'L. Agassiz' in N. America); the deposition of VARVES.

Flysch Coarse sandstones, calcareous shales, conglomerate, marls and clays, occurring on the borders of the Alpine ranges. These are of early Tertiary (possibly also late Cretaceous) age, and are composed of materials eroded from the rising fold-ranges before the fold mountain-building max. in Miocene times. The ct. is with the MOLASSE, composed of materials worn away and redeposited *after* the orogenic max.

focus See SEISMIC FOCUS.

Foehn See FÖHN.

fog Obscurity of the ground-layers of the atmosphere, the result of the condensation of water-droplets, together with particles of smoke and dust held in suspension. Under the International Meteorological Code, the term is defined as a visibility of less than 1 km. In Gt. Britain a thick f. has a visibility of less than 200 m (220 yds.). See also ADVECTION F., ARCTIC SMOKE, FRONTAL F., RADIATION F., SMOG, STEAM F. Ct. MIST, HAZE.

fog-bow A type of rainbow formed with the sun behind the observer, and with an area of f. in front. The droplets are so small that the break-up of light by refraction and reflection into the colours of the spectrum is prevented, and the bow is thus colourless, giving a white effect.

fog-drip A form of precipitation of water droplets from a bank of wet f., esp. along a COLD WATER DESERT coast, where the humidity is high; e.g. California, N. Chile and Peru, S.W. Africa. The amount of moisture may be enough to nurture some vegetation; cf. DEW-MOUND.

foggara An underground channel which leads water from an AQUIFER near the foot of a mountain to a neighbouring lowland for irrigation, esp. in the Sahara Desert. Cf. QANAT in Persia and KAREZ in Baluchistan.

Föhn, Foehn (Germ.) A warm, very dry wind, descending esp. the N. slopes of the Alps. It blows when a depression lies N. of the Alps. Moist air is drawn from over the Mediterranean Sea, and rises over the mountains, cooling at the SATURATED ADIABATIC LAPSE-RATE. Air-mass turbulence on the crest and leeward side causes eddying and descent of the air. On descent it warms at the DRY ADIABATIC LAPSE-RATE. Temperatures can rise by 10°C. or more in a few hours; a rise of 17°C. (30°F.) in 3 minutes has been recorded. In spring snow is rapidly melted, so clearing the pastures. Sporadic periods of f. early in the year can do much harm, causing avalanches (as in early 1951 in E. Switzerland and Austria), and premature budding of trees and plants. The term *f.-effect* is used of any similar wind; e.g. SAMUN (Persia), NOR'WESTER (New Zealand), BERG (S. Africa), SANTA ANA (California), CHINOOK (E. Rockies of N. America), ONDA (Argentine).

hrde (Germ.) (*plur.* **Föhrden**) long straight-sided inlet in a oulder-clay coastline, the result of ubmergence by a rise of sea-level of narrow valley along a low-lying coast. 'hese valleys were probably eroded by vers flowing in tunnels under the e-sheet of Quaternary times. They e confusingly known as *fjords* in E. enmark. E.g. Flensburger F., Kieler . [*f opposite*]

ld, folding The bending or umpling of strata as a result of mpressive forces in the earth's crust, usually along well-marked zones which indicate lines of weakness. The f. ranges from a gentle flexure (sometimes referred to as an *inflexion*), through simple upfolds (ANTICLINES) and downfolds (SYNCLINES), to ASYMMETRICAL F.'s, OVERFOLDS, RECUMBENT F.'s and NAPPES. See also ANTICLINORIUM, SYNCLINORIUM, PERICLINE, OROGENY. [*f* ANTICLINE]

foliation (from Lat., *folia*, leaves) (i) A wavy laminated or banded fabric in such rocks as schist and gneiss, the result of METAMORPHISM re-crystallizing and segregating different minerals into parallel layers. (ii) A wavy structure in bands of ice in the deeper parts of a glacier.

foot A measure of length = 12 ins., 1/3 yd., 0·30479 m.; 1 cu. ft. = 1728 cu. ins., 28316·1 cc., 6·23 gallons of water; 1 ft.-pound weight = a unit of energy, the work done in lifting 1 lb. mass through 1 ft. vertically; 1 sq. ft. = 144 sq. ins. = 0·093 sq. m.

foothills A transitional line of hills, lying between and more or less parallel to a main range of mountains and a plain, the result either of the intermediate zone of uplift or of active denudation; e.g. the f.'s of the Rockies, Himalayas.

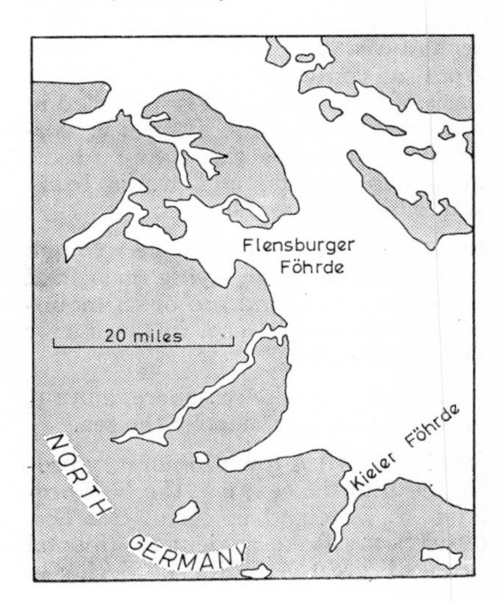

foot wall (i) The lower side of a FAULT; i.e. beneath the fault. (ii) A term used by miners to denote the solid rock beneath a vein, lode or other ore-body. [*f* HANGING WALL]

force (from Old Norse *fors*) A name for a waterfall in the English Lake District and the Pennines; e.g. Scale F., Buttermere; High F., Teesdale.

ford The shallow part of a river which can easily be crossed; hence a common place-name for a riverside settlement; e.g. Oxford. Note the stages of development (f., then bridge) implied by Fordingbridge (Hampshire).

forecast Adopted since 1860 for the weather anticipated for the future, to avoid such terms as 'prophecy'. In 1963 the Meteorological Office in Gt. Britain introduced 'long range f.'s' for a month ahead, following a long period of 'daily f.'s'. A f. is in fact a probabilistic statement issued with some degree of confidence on the basis of the available evidence. In Europe the f. tends to be a definite prediction; in U.S.A. the probability is stated; e.g. a 50% probability of rain during the next 24 hours. (ii) Used increasingly in the social sciences in connection with future needs, planning and decision making, using statistical and other decision-making techniques. An *explorative f.* starts from a contemporary and definite basis of knowledge and is projected in anticipation into the future; a *normative f.* assesses the future needs, aims and targets, and works backwards to the present.

foredeep A deep elongated trough in the ocean floor, lying near and parallel to an island arc or to mountainous land areas, esp. in the Pacific Ocean. See DEEP.

foredune On the sea-shore, a DUNE, or a line of dunes, nearest the sea.

foreland (i) A low promontory projecting into the sea; e.g. the N. Foreland, S. Foreland in Kent. See also CUSPATE F. (ii) An ancient continental mass of great stability bordering a GEOSYNCLINE [*f*]. Orogenic forces were directed towards the f. and folded ranges were squeezed on to it during mountain-building periods; e.g. the Hercynian F. of Europe, during the Alpine orogeny. Earlier authorities thought that the motive force came from one side of a geosyncline only (the *backland* or *hinterland*), directed on to the other side (the f.). This concept has been largely replaced by the idea of two f.'s moving together, as in the case of the Himalayas. Even so, in the Alps it is clear that the force was mainly directed to the N., and the Hercynian continent acted as a f. (iii) *Glacial f.*: an area of lowland, lying beyond mountain ranges which during the Quaternary glaciation nourished glaciers; these moved down and out as PIEDMONT GLACIERS on to the f.; e.g. the Bavarian F., the Swiss Plateau, the N. part of the Plain of Lombardy in Italy, the Lannemezan on the French side of the Pyrenees, the Carpathian F. in S. Poland.

foreset beds Inclined b. built outward and forward in a DELTA, each one above and in front of the previous one. [*f* BOTTOMSET BEDS]

foreshore The area extending from the lowest spring tide low-water line to the average high-water line. [*f* COASTLINE]

forest (i) A continuous and extensive tract of woodland, usually of commercial value. (ii) In Gt. Britain an area of former f., now largely cleared for agriculture and settlement though perhaps with some small surviving portions of woodland, but the district retains the f. name; e.g. F. of Bere (Hampshire), Ashdown F. (Kent), Sherwood F. (Nottinghamshire). (iii) A royal hunting ground, outside the common law and subject to f. law, e.g. the New F. (iv) A waste or uncultivated area of heath or moorland used for hunting and stalking (as in Scotland); e.g. Mamore F. (Inverness-shire), F. of Atholl (Perthshire), F. of Mar (Aberdeenshire).

forked lightning A type of cloud-to-ground electrical discharge, with distinctive down-pointing and branching flashes.

formal region A unit area with a certain uniformity of characteristics, in contrast to a FUNCTIONAL R. characterized as a sphere of activity. Thus an area of mountains may constitute a f. r., though it may also be part of several functional ones.

formation The 4th category of the geological division of the hierarchy of rocks, corresponding to the time interval of an age. By some geologists, a f. is called a *stage*. It consists of a stratum, or a series of strata, with some distinct and well-defined lithological properties. It can be gen., e.g. a coal-seam; or specif., with its own name, e.g. the Nothe Grit, Nothe Clay, Bencliff Grit and Osmington Oolite are f.'s within the Corallian Series within the Jurassic system within the Mesozoic group.

form-line A contour which has been drawn in by eye, usually with the help of spot-heights, but not instrumentally surveyed.

form-ratio A simple relationship, defined by G. K. Gilbert (1914), between the depth and width of a river, expressed as a fraction (1/100) or ratio (1 : 100).

Fortin barometer A special pattern of mercury b., though the KEW pattern is gen. used. The main eature of the F. b. is that the level of he mercury in the cistern is brought o zero on the scale by a screw adjustnent before a reading is taken.

osse (i) From Lat. *fossa*, a ditch or rench around an ancient earthwork, r forming a line of defence. It surives in place-names; e.g. the F. Way. ii) In French usage, a linear deep or ubmarine trench; e.g. F. de Cap Breton in the Bay of Biscay [*f* SUBIARINE CANYON]. (iii) In U.S.A., a epression between the side of a alley-glacier and the containing-wall f a valley, coined by W. D. Thornury (1954), along the line of which a AME TERRACE [*f*] may develop. (iv) In N. England a waterfall, usually spelt 'foss'; e.g. Janet's Foss, Malham, Yorkshire; probably a corruption of FORCE.

fossil The hard part of an organism, or its exact replacement by mineral matter, or its impression, preserved in sedimentary rocks. Some f.'s may have pyritized shells; e.g. ammonites in the Jurassic. F.'s include fauna, flora, foot-marks of animals, and impressions of soft-bodied animals; e.g. worms. The science of PALAEONTOLOGY is the study of f.'s, since 'they stamp the sediments with a characteristic mark' (F. Hodson).

fossil erosion surface A s. which has been worn down, buried by subsequent deposition, and later 'exhumed' by the removal of the 'cover' of newer deposits by denudation; e.g. the sub-Eocene s. in S.E. England. Occasionally small patches of the 'cover' may survive on the s.

fossil water See CONNATE W.

Foucault's pendulum A p. consisting of a heavy metal ball suspended from a wire, set swinging in a certain direction. Its path appears to move gradually round to the right, and in due course it arrives back at its original direction. This is the result of the earth's rotation. A p. located at the Pole would show one complete rotation in 24 hours (i.e. 15° in 1 hour), in latitude 50° it takes about 31 hours, at the equator there would be no turning at all. The amount of turning indicated by the p. varies with the *sine* of the latitude; i.e. number of degrees of turning per hour $= 15 \times \sin \theta$, where θ is the latitude. It was originally devised by L. Foucault (1851).

fractional crystallization The separation of the constituents of MAGMA as it cools into successive crystallized minerals. Gen. the order is the inverse of the melting point, one of increasing acidity: first the accessory minerals, then ferromagnesian silicates, felspars and finally quartz. The whole process is very complex, differing with the chemical nature of the magma and whether solidifying at depth or on the surface. Some magmas solidify in 2 distinct phases.

fractional distillation The process by which petroleum is split into its 'fractions' (groups of hydrocarbons) by heating, then passing in a gaseous form into a fractionating tower, where it rises. The various fractions liquefy at different temperatures, the highest ones (remaining as gases) passing out at the top.

fracto- cloud, fractus A prefix attached to other c.-names (as *fractonimbus*, *fractostratus*, *fractocumulus*) to indicate tattered, shredded, ragged c.'s, usually an indication of high winds and stormy conditions in the upper atmosphere, which may possibly affect the weather near sea-level.

fracture A clean-cut break in a rock stratum which has been subjected to strain resulting from either tension or compression. Cf. FAULT.

frazil Needle-like ice-crystals which develop as a spongy mass in supercooled water which is in motion, so that sheet-ice cannot be formed.

fragmental rock A r. made from recognizable fragments of others, compacted or cemented together; syn. with CLASTIC R.; e.g. conglomerate, sandstone, clay. See also PYROCLAST.

free face A rock wall too steep for weathered material to rest upon; this material falls to the bottom, where it forms a SCREE. The f. f. is one of the slope elements distinguished by W. Penck and A. Wood as part of the hill-side slope profile. See also CONSTANT SLOPE, WAXING SLOPE, WANING SLOPE. [*f opposite*]

free port A p. where goods can be unloaded and then loaded for re-export without payment of customs duty; e.g. Singapore, Hong Kong, part of Copenhagen.

freestone A fine-grained, even-textured rock that can be sawn freely in any direction into blocks, occurring in thick beds; ct. the thin beds of FLAGSTONES. The category includes several Jurassic limestones (Bath, Ham, Portland Stones); and various sandstones (e.g. New Red in the Penrith valley and in the neighbourhood of Liverpool, where they are being used for the new Anglican cathedral; Fell Sandstone in the Lower Carboniferous of Northumberland). Some types, esp. limestone, can be cut easily, but harden on exposure; e.g. the limestones of Malta.

freeway A MOTORWAY in U.S.A. Other terms with various definitions are *expressway*, *thruway*, *limited-access highway*, *parkway*, *interstate highway*.

freeze-thaw An important and very active form of weathering under PERIGLACIAL conditions, in the mountains, and in temperate latitudes in winter, wherever the temperature fluctuates above and below 0°C. The action of frost is to break up the rock, while melt-water removes the fragments. Results include the production of hollows by NIVATION, the general break-up of the surface, the sorting of coarse and fine material, the formation of PATTERNED GROUND, and the movement of material by SOLIFLUCTION.

freezing-point The temperature at which a liquid changes to the solid state. Water freezes at 0°C. (32°F.). Ct. mercury at −39°C., nitrogen at −210°C.

frequency curve A graph showing the number of occurrences of values (e.g. of temperature, population density, summit-levels); usually the horizontal axis indicates the range in size

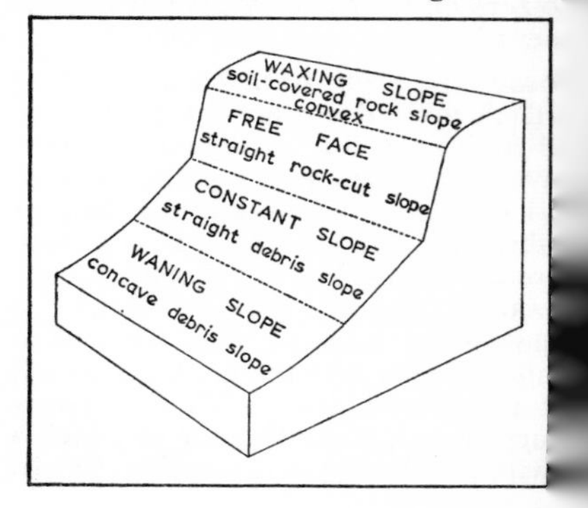

of the variable, the vertical axis either the actual or percentage frequency for each value of the variable. The values may be grouped into classes, so forming a HISTOGRAM.

fresh breeze Force 5 in the BEAUFORT WIND SCALE.

freshet A sudden surge of flood-water down a small stream, the result of a heavy rainstorm in the upper basin, or of snow-melt following a rapid rise of temperature.

fret A sea-mist that rolls up suddenly over a shore, but does not extend far inland.

friable (i) App. to a rock which is easily crumbled; e.g. a soft, poorly cemented sandstone. (ii) Also applied to loamy soils, with a well-marked crumb-structure.

friagem A cold strong wind experienced on the CAMPOS of Brazil and in E. Bolivia, the result of an anticyclone developing, with air movement from the S. It may cause a distinct cold spell during the dry season in winter.

frigid climate One of the 3 climatic :ypes or zones postulated by classical ınd later scholars, the others being *'orrid* and *temperate*. Currently the erm is applied in a gen. sense to an ırea with a snow cover for much of the 'ear and with a permanently frozen ubsoil (PERMAFROST), or to the polar-.rctic group of climates (W. Köppen's *ET* and *EF* types).

ringing reef An uneven platform f CORAL, fringing and attached to the oast, with a shallow narrow lagoon etween it and the mainland, or with o lagoon at all, and with its seaward dge sloping quite steeply into deep 'ater. [*f opposite*]

'ont The boundary surface, interıce or transition zone which separates vo AIR-MASSES of markedly different mperature and humidity. This may :cur: (i) on a large scale, the f.'s :tween major air-masses; e.g. *Pacific olar F., Atlantic Polar F., Mediter-nean F., Arctic F.*; (ii) on a smaller ale in a local DEPRESSION: WARM F., OLD F., OCCLUDED F. The term was originally used by V. and J. Bjerknes in 1918. [*f* DEPRESSION]

frontal fog Short-lived f., really a thick, fine drizzle, sometimes associated with the passage of the WARM FRONT of a DEPRESSION. The warm rain falls into the cold underlying air near the ground, saturating the air to form fog.

frontal rain See CYCLONIC R.

frontier (i) A long narrow zone which 'fronts' or faces a neighbouring country; otherwise called the BORDER or MARCH. It should be distinguished from BOUNDARY, which is the actual line of definition between two countries, though as a *political f.* it is used by many as syn. with boundary. (ii) In U.S.A. it indicates a thinly populated or little developed zone on the margins of more settled lands. In American history, the officials of the Census Bureau use a definition of a zone with 2–6 people per sq. mi. (8–23 per sq. km.) representing a stage in the W.-ward expansion of settlement and population.

frontogenesis The physical processes in the atmosphere by which a FRONT is developed or intensified, part of the formation of an individual DEPRESSION. Its appreciation demands the study of conditions in the upper atmosphere, partic. the development of CELLS, and of the JET STREAMS which may separate them.

frontolysis The dissipation of a frontal zone or FRONT.

frost (i) The condition of the atmosphere when it is at or below the freezing-point of water. Usually des-

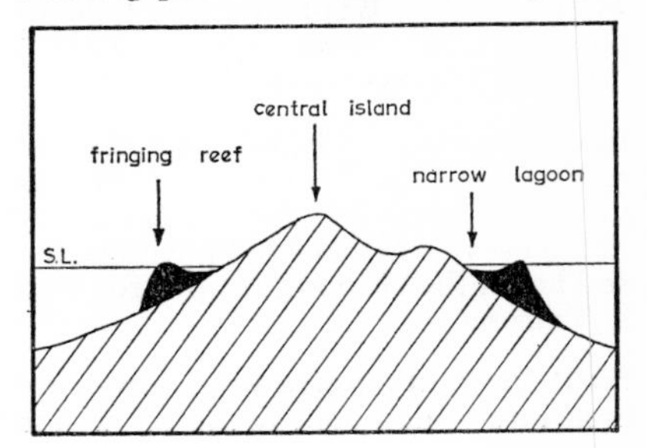

criptive adjectives are applied to the degree of f.: *light, heavy, hard, sharp, killing, black.* (ii) An agent of weathering which freezes water contained in cracks and fissures in rocks and the soil. As the volume of water when frozen increases by about 10%, it exerts great pressure and the rock tends to shatter. This is esp. potent in high mountains. (iii) Minute ice-crystals formed by the freezing of dew, fog or water vapour, sometimes called *white f.* See GLAZED F., HOAR-F., RIME. Many self-evident words are used with f. as a prefix: *f. shattering, f. cracking, f. splitting, f. stirring, f. weathering, f. wedging, air-f.,* GROUND-F.

frostbite The freezing of parts of the body, espec. the extremities, the result of absence of protection (e.g. loss of a glove), restriction of circulation by tight boots, etc. This may lead in severe cases to loss of toes or fingers through gangrene. Ct. *frost-burn*, the result of touching metal with bare fingers at below-freezing-point temperatures, causing a sudden loss of heat from the flesh, and therefore the effect of burning.

frost boundary A b. indicating the occurrence and incidence of f. It may indicate the limit of areas: (i) which have never experienced f.; (ii) with mean minimum temperatures above 0°C.; (iii) with a growing season with a specific number of f.-free days; e.g. cotton needs 200; (iv) with a lowest mean monthly temperature above 0°C.; (v) with no month with a mean temperature above 0°C.; i.e. the boundary of a polar climate of perpetual f. (W. Köppen's *EF* type of climate). All these are important limits of various types of activity.

frost heaving The lifting of the soil surface by f. action within it; this is gen. the result of the formation of 'ice-lenses' which expand within the soil layer, thus forcing it upwards into a *f. mound.*

frost 'pocket', hollow A low-lying area into which cold dense air drains by gravity, as radiation of heat into space on a clear night causes rapid cooling of the air on the upper hill-slopes. The 'p.' may thus contain air below freezing-point, while the temperature on the higher slopes is still above; see INVERSION OF TEMPERATURE. Fruit-growers seek to avoid such sites, since f. may cause serious losses should it occur during blossom time, and the chance of this happening is much increased in a f. 'p.'

fuelling-station A port regularly used by shipping to refuel, formerly with coal (hence coaling-station), now oil; e.g. Aden, Curaçao.

full A ridge of sand or shingle, roughly parallel or at a slight angle to the shore. It is formed by constructive waves just in front of the line of their break-point, aligned at right-angles to their direction of approach. The f.'s are separated by long shallow depressions known variously as SWALES, *lows, slashes, furrows* or *runnels.* During a storm, with strong onshore winds and destructive waves, the f.'s are usually combed down and destroyed. They are found on the Lancashire coast, esp. at Blackpool and Formby (sand); on the Dorset coast and at Dungeness (shingle).

fumarole (It.) A small hole or vent in the surface, from which issue steam, hydrochloric acid, sulphur dioxide and ammonium chloride, usually in the form of powerful jets. Ct. SOLFATARA. E.g. the Valley of 10,000 Smokes, Alaska; Bumpass Hell on the flanks of Mount Lassen in N.E. California.

functional region A r. distinguished by its unity of organization or the interdependence of its parts. Usually the unifying force is one of movement between the parts, as in a river basin, a marketing r., or a city HINTERLAND. These r.'s depend not on the prevalence of certain features to be found throughout the area, but on how they are organized. Ct. r.'s representing approx. uniformity of characteristics which are called FORMAL.

fundamental complex An obsolescent term for the great 'shield' areas of Pre-Cambrian rocks in the earth's crust, largely replaced by BASAL C.

funnel cloud A whirling cone of dark grey c. which projects downwards from the low-lying base of a CUMULONIMBUS C., gradually elongating until it touches the surface of the sea; it may become part of a WATERSPOUT. It is often associated with a TORNADO.

furlong From Anglo-Saxon *furh*, a furrow, and *lang*, long, hence the length of a ploughed furrow in a common field. As measure of length, 1 f. = 10 chains, 40 rods, 220 yds., 1/8th of a statute mile, 201·16 m.

fusain In banded coal, layers of 'dirty' material consisting mainly of flakes of charcoal-like substance, which is extremely friable and dusty, with a high ash content.

gabbro A dark-coloured plutonic rock, basic in composition (i.e. with about 45–55% silica), with calcic plagioclase, forming a sharp-textured crystalline mass. The Black Cuillins of Skye and the mtns. of Rhum are of g.; there are also several large masses in N.E. Scotland (Insch Mass, 80 sq. km. (70 sq. mi.) in extent, Huntly Mass), and in N. Guernsey, Channel I.'s.

gale (i) In gen. parlance, a strong wind. Specif. on the BEAUFORT SCALE: (a) *moderate* g. (Force 7), mean 5 m.p.h.; (b) *fresh* g. (Force 8), 42 m.p.h.; (c) *strong* g. (Force 9), 50 m.p.h.; (d) *whole* g. (Force 10), 59 m.p.h. Winds of Force 11 and 12 are popularly referred to as g.'s, but in the Beaufort Scale are called 'storm' and 'hurricane'. (ii) A royalty formerly paid in the Forest of Dean for a plot of land, giving the right to mine thereon for coal or iron-ore, or to quarry for stone.

galena The chief ore of LEAD, lead sulphide (*PbS*), soft and heavy, with a dull lustre, commonly occurring in crystals of cubic form, and found usually in Carboniferous Limestone.

gallery forest A dense tangle of trees fringing the banks of a river in what is otherwise open SAVANNA country. The vegetation meets overhead, giving a tunnel-like appearance, hence the Sp. term *galeria*, used in S. America, from which 'gallery' is corrupted.

gallon A measure of volume. 1 g. of water at 15°C. weighs 10 lb., occupying 277·3 cu. ins. or 0·16 cu. ft.; 1 g. = 4545·96 cu. cm., 160 liquid oz., 8 pints; 1 g. (British or Imperial) = 1·2009 g.'s (U.S.A.) = 4·55 litres.

Gall's Stereographic Projection A type of CYLINDRICAL P. in which the cylinder intersects the globe at 45°N. and S.; along these parallels the scale is correct, between them it is too small, and poleward of them it is too large. The parallels are parallel straight lines. Their separation is obtained: (i) graphically by projecting stereographically (i.e. from the opposite side of the equatorial diameter) on to the edge of the cylinder, intersecting at 45°N. and S.; (ii) trigonometrically from the formula $1{\cdot}7071\ R.\ \tan \frac{1}{2}\theta$, where θ is the latitude, which gives the distance from the equator. The meridians are vertical lines, drawn through the equator, which is divided truly; e.g. 10° meridian interval

$$= \frac{2\pi R. \cos 45^\circ}{36}$$

The p. is neither equal area nor orthomorphic, but it exaggerates shape and area in high latitudes less than does the MERCATOR P. [*f*]

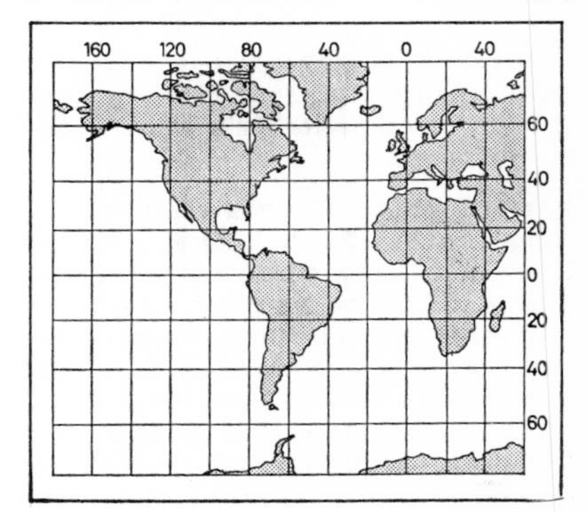

gangue The worthless mineral matter, known as *veinstone*, in a vein, commonly quartz, within which lie the metallic ores.

ganister, gannister A rock with a high silica content, found as an underclay in the Lower Coal Measures, and used for making heat-resistant refractory bricks for furnace-linings.

gap A break in a ridge (see WATER-GAP, WIND-GAP). In the Appalachians in U.S.A., which are crossed by rivers flowing to the Atlantic, such g.'s are frequent; e.g. the Cumberland G. in Kentucky, a defile 1600 ft. deep, first followed by the Wilderness Road, a trail blazed by Daniel Boone in 1775.

gap town A t. situated in, or near, or commanding, a gap in a ridge; e.g. Guildford, Dorking. G. t.'s frequently owe their origin not only to their ability to command routes through the gap along the valley floors, but also to their commanding position with respect to routes wishing to cross the 'g.' from high ground on either side of a marshy valley-plain. Ancient routes often run transverse to the g., not through it; e.g. Lincoln in the gap cut by the R. Witham, with the N. to S. Ermine St. along the crest of the oolitic limestone ridge, while through the town passes the Fosse Way (A46) and the much later railway.

gara (*pl.* gour) (Arabic) A mushroom-shaped rock found in desert lands, resulting from undercutting by wind abrasion, caused by heavy sand particles carried near the ground. This is partic. well marked if in a bedded mass of rock a soft stratum near ground-level underlies a much more resistant one.

garden city A town carefully designed and built to maintain something of an open rural character, with a relatively low housing density of only about 10 per acre, open spaces and trees, facilities for recreation and social life, and with carefully planned industrial development, usually of a 'light' character; e.g. Welwyn G. C. (Herts.); G. C. (Long I., U.S.A.); the *cités ouvrières* of the Kempen coalfield, Belgium. Hence also 'g.-suburb'; e.g. Hampstead (London), Wavertree (Liverpool), implying a spacious, well laid-out housing-estate with trees and lawns.

garigue, garrigue (Fr.) A stunted evergreen xerophytic scrub, found on limestone in drier areas of Mediterranean climate, incl. stunted evergreen oak, thorny aromatic shrubs, prostrate prickly plants and tuberous perennials, separated by bare rock. This Provençal term has also been applied to the uncultivated land with calcareous soil on which such scrub grows, as in S. France, Corsica, Sardinia and Malta.

garth A close, croft, garden or enclosed space near a farm (from Icelandic). Hence a common suffix in N. England; e.g. Gatesgarth in Buttermere, English Lake District.

gas, natural Gaseous hydrocarbons in the form of ethane and methane, found in the earth's crust, frequently associated with petroleum. Afte purification it can be used for domesti and industrial heating, but it is also raw material of the petrochemical in dustry, and by means of polymerizatio a wide range of substances can be made C.p. = U.S.A., U.S.S.R., Canada Roumania, Mexico, Italy, Nether lands.

gat (i) A strait or channel betwee offshore islands or shoals, notabl between the Frisian I.'s (where the are known as *zeegaten*), through whic flow powerful tidal streams; e. Texel Zeegat. (ii) An opening in th cliffs leading inland from the sea. I Kentish place-names, this usually b comes 'gate'; e.g. Margate, Ramsgate.

gate (i) A valley through a low ran of hills (Anglo-Saxon *geat*, an opening hence Reigate, Rogate. It is sometim modified to *yate*; e.g. Markyate, whe A5 enters the Chilterns. (ii) An ope ing, wider than a gap, between hi areas; e.g. the Midland G. between t S. end of the Pennines and the Wreki the G. of Carcassonne (*f* below). (i

A restricted section in a river valley; e.g. the Iron G. on the R. Danube. (iv) An entrance to a bay or harbour between promontories; e.g. the Golden G., San Francisco. [*f*]

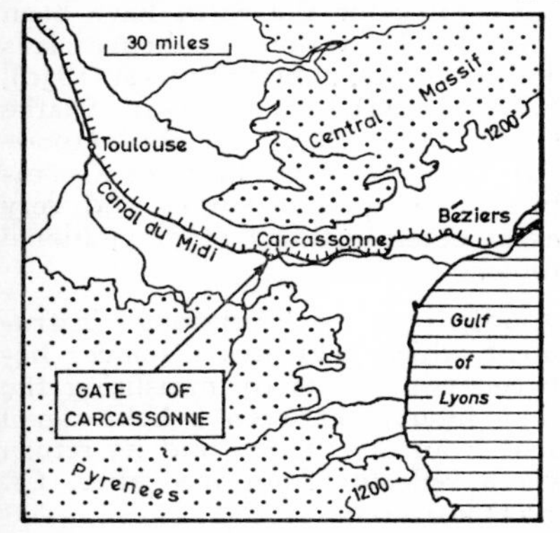

gauge, of railway The distance apart of the inner edges of r. lines: *standard* (4 ft. 8½ ins.), *broad* (5 ft. or more), *narrow* (1 m., 3 ft. 6 ins. and smaller).

Gauss Conformal Projection The TRANSVERSE MERCATOR P.

Gaussian Curve A mathematically derived frequency distribution which possesses perfect symmetry about the central value.

gazetteer An alphabetical list of places, with ref. to their positions, either by LATITUDE and LONGITUDE as in an atlas, by a REFERENCE NET or by GRID REFERENCE; e.g. *G. of Great Britain*, published by the Ordnance Survey Office, with grid references.

geanticline, geoanticline An upfold on an earth scale. GEOSYNCLINES do not have g.'s as complementary features, and the term is often used for upfolds developing within a geosyncline, as the latter is laterally compressed.

geest (Germ.) An area of coarse sand and gravel, mostly under sparse HEATH vegetation, occurring mainly in N. Germany (where the word occurs frequently in regional names) and in adjacent parts of the Netherlands, Denmark and Poland. The sands and gravels are of fluvioglacial origin, laid down as part of the 'Older Drift' cover by streams issuing from the continental ice-sheets, and later cut into individual areas, partly by fluvioglacial channels, partly by the modern river valleys; e.g. the Lüneburg Heath between the Weser and Elbe valleys, the G. of Oldenburg and Hanover between the Ems and Weser valleys.

gendarme (Fr.) A sharp rock pinnacle projecting from a ridge, orig. in the French Alps, now used widely by mountaineers; e.g. the Grand G. on the Grépon (Mont Blanc massif).

general systems theory A concept developed by L. von Bertalanffy (1951) as a framework for modern science, orig. with partic. ref. to the biological sciences. It represents an effort to generalize ideas, processes and functional relationships in any partic. s. of interrelated objects or ideas, to work with analogies from one discipline to another, to produce a scientific doctrine of 'wholeness'. Inevitably the approach stresses the relationship between form and process, as well as the multivariate character of phenomena, and is being increasingly utilized by geographers. A certain critical reaction to g. s. t. can be discerned; e.g. 'g. s. t. seems to be an irrelevant distraction' (M. Chisholm). See CLOSED S., OPEN SYSTEM.

generative city A c. that stimulates the economic or cultural development of the region in which it is situated.

generic Phenomena which are closely linked and similar in type. Thus the term 'Mediterranean climate' is a g. concept summarizing certain climatic features, which may be used to describe climates elsewhere which are broadly similar and belong to the same type.

generic region A r. distinguished by certain criteria of a given type, found as a repeating pattern; e.g. a chalk cuesta, a horst, an area of savanna, and an area of Mediterranean climate. Ct. a SPECIFIC R., which is unique, precisely located, and usually has a proper name.

genetic description, of landforms An analysis of landforms on the basis of their origin, in terms of the processes which are sculpturing the landform, and of the stage which these processes have reached; the systematic study of landforms.

geo, occasionally **gio** (from Norse *gya*, a creek) A long narrow steep-sided inlet, running inland from the edge of a cliff. It has been worn by marine erosion at the base of the cliff along a line of weakness, such as a major joint-plane (as in Old Red Sandstone) or a minor fault-plane. The cave is thus driven inland, and its roof is cut away along the same plane by waves surging into it, together with the compressional effect on the air within. Ultimately the roof collapses. E.g. in the Orkneys, Caithness, S. Skye and Soay, all cut in Old Red Sandstone; the Huntsman's Leap, Pembrokeshire, in Carboniferous Limestone.

geochronology The dating of past events in the earth's history, as indicated in the record of the rocks; the 'science of earth time'. It includes *relative* chronology, in order of occurrence or formation, and *absolute* chronology, involving dating in years, using the measurement of RADIOACTIVE DECAY (uranium/lead, rubidium/strontium, potassium/argon), and RADIOCARBON DATING. Partic. attention is paid to the precise dating of events of the Quaternary by geologists, archaeologists and botanists. The precise measurement of geological time is sometimes called *geochronometry*.

geode A hollow, near-spherical nodule, between 2·5 cm. and 28 cm. or even more in diameter, commonly found in limestone. Its interior is commonly lined with crystals; occasionally a loose piece of material inside may make an intriguing rattling sound.

geodesy The branch of mathematics dealing with the shape and size of the earth, or with substantial portions of it. From this can be obtained data which will enable the exact fixing of control points for TRIANGULATION and LEVELLING of a high degree of accuracy, the basis of any major topographical survey; hence *geodetic*, the application of g. to surveying. Standard figures for the earth have been produced by various mathematicians and surveyors, notably Everest (1830), Bessel (1841), Airy (1849), Clarke (1858, 1866, 1880), Hayford (1909–10). Geodetic work involves a combination of precise survey and very exact determinations of gravitational force.

geodimeter A surveying instrument which estimates distances between two points by measuring the time interval between a light signal sent from one station and its return via a reflector from the other. Cf. TELLUROMETER.

geodynamics The study of deformation forces or processes within the earth.

geographical mile Syn. with NAUTICAL MILE.

geographical momentum, inertia The tendency for activities to be maintained or increased when the original reasons for their introduction have changed, have been modified, or have disappeared. Once an industry is established, it may maintain itself because of its associations and traditions; e.g. steel manufacture in Sheffield, St. Etienne; the tobacco industry in Bristol. See INDUSTRIAL MOMENTUM

geography The concept and scope of g. have undergone considerable change, and it is highly unlikely that any definition would satisfy everyone. Most are agreed that it comprises the study of the earth's surface in its areal differentiation as the home of Man; how much it is a 'science of distributions', physical and human, of areal and spatial relationships, how much Man in his spatial setting is the crux, and to what extent the study of the region is the core of the subject, are all matters for debate. The geographer seeks to describe the diverse features of the earth's surface to explain if possible how these features have come to be what they are, and to discuss how they influence the distribution of Man with his multifarious activities. G., standing as it does transitionally yet centrally between the natural sciences, the social studies and the humanities, is thus in its concept and content an integrated whole. The field of g. has been viewed in different ways; while the traditional position is that of AREAL DIFFERENTIATION, it is possible to regard it in terms of LANDSCAPE, natural and cultural (e.g. by Carl Sauer), in ecological terms (see ECOLOGY) (e.g. by the French human geographers), and in locational terms (esp. the development of LOCATIONAL THEORY). See also EXCEPTIONALISM, SET THEORY.

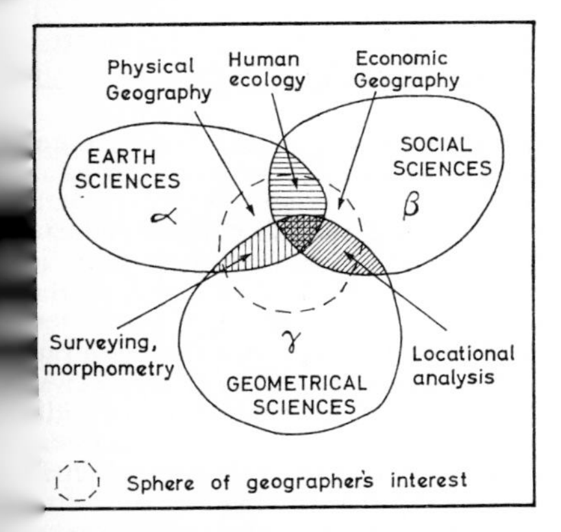

eohistory A term coined by F. 'randel to cover his study of the Iediterranean basin in the second alf of the 16th century, using a series cross-sections. It is really syn. with ne concept of HISTORICAL GEO-RAPHY.

oid The terrestrial spheroid, an arth-shaped body', regarded either: as a mean sea-level surface con-nued through the continents; or (ii) as undulating surface determined in avitational terms (*geopotential*), ther higher than the surface of the heroid under the continents, rather wer under the oceans.

geological time The chronology of the earth's history, organized in a hierarchy of time intervals, as follows (in descending scale): *era, period, epoch, age, moment*, corresponding in terms of the rocks to *group, system, series, stage, zone*. See GEOCHRONOLOGY.

geology The history, composition, structure and processes of the earth; earth science. Its accepted sub-divisions are: (i) *Stratigraphy*; (ii) *Palaeontology*; (iii) *Mineralogy*; (iv) *Petrology*; (v) *Physical G.*; (vi) *Structural G.*; (vii) *Geophysics*; (viii) *Engineering G.*; (ix) *Economic G.* An alternative sub-division is (*a*) *physical g.*; (*b*) *historical g.* Physical g. is itself sometimes divided into *structural g.* and *dynamic g. Geomorphology* is regarded as a branch of g. in U.S.A., but more usually as a branch of geography in Gt. Britain.

geomagnetism The earth's magnetic field and the study thereof.

geometronics Used in U.S.A. for aspects of cartographical work involving *geo* (the earth) and *metron* (measurement), utilizing electronic and other modern techniques.

geomorphology The scientific interpretation of the origin and development of the landforms of the earth; the modern development of *physiography*.

geopacifics Introduced by T. Griffith Taylor in 1947, partly as a reply to the misuse of GEOPOLITICS, to denote the study of geography in the cause of peace, freedom and humanity.

geophysics The scientific study of the physics of the earth's crust and its interior, involving consideration of earthquake waves, both natural and those made by deliberate explosions (SEISMOLOGY), magnetism, gravitational fields and electrical conductivity, using precise quantitative methods. Applied g. is mainly concerned with the techniques involved in the discovery and location of structures associated with economic minerals (petroleum, iron ore, radio-active ores) by means of geophysical surveys. A recent tendency, esp. in U.S.A., is to widen the term to include the physics of the earth's environment also, incl. meteorology, astrophysics, etc.; this does not meet universal acceptance.

geopolitics Derived from the Germ. term *Geopolitik*, which began with the study of geographical factors in political systems, and developed in Germany into concepts of racial superiority, such as *Lebensraum*. It is a convenient term in English for political geography, esp. emphasizing the geographical relations of states.

Georef system A world reference s. for the location of points on the earth's surface, used by the U.S. Air Force for the direction of long-range aircraft and missiles, and for other strategic purposes, introduced in 1951. The world-map is covered with a rectangular graticule, with 15° meridian intervals, each band lettered A to Z (omitting I and O), starting at 180° and working E., and with 15° parallel intervals, lettered A to M (omitting I), starting from the S. Pole. The origin of the map reference is the bottom left corner, i.e. the intersection of 180° and the line representing the S. Pole. Thus 288 quadrangles are located by two letters; Gt. Britain is in sq. KM. Each quadrangle is sub-divided into degrees, then minutes, then 100ths of minutes (not seconds).

geoscope A data storage and display device, conceived by R. Buckminster Fuller, affording a comprehensive inventory of the world's raw and organized resources, together with the history and trend patterns of people's movements and needs. One type of g., made of light metal trussing on which basic geographical data are marked, is linked to a computer storing all available data. The 1st g. was made at Cornell in 1952, consisting of a 20 ft. globe. Others were installed at the universities of Minnesota (1954–6), Princeton (1960) and Colorado (1964). The Princeton g. utilized the DYMAXION PROJECTION.

geostrophic flow Used in meteorology for the concept of a wind blowing parallel to the isobars as a result of the forces exerted by the pressure gradient in one direction and by the CORIOLIS deflection in the opposite direction. Winds approaching pure g. f. are found in the upper atmosphere, but nearer to the surface of the earth friction causes them always to blow at an oblique angle to the isobars towards low pressure. The term was coined by Sir Napier Shaw in 1915. See FERREL'S LAW. [*f*]

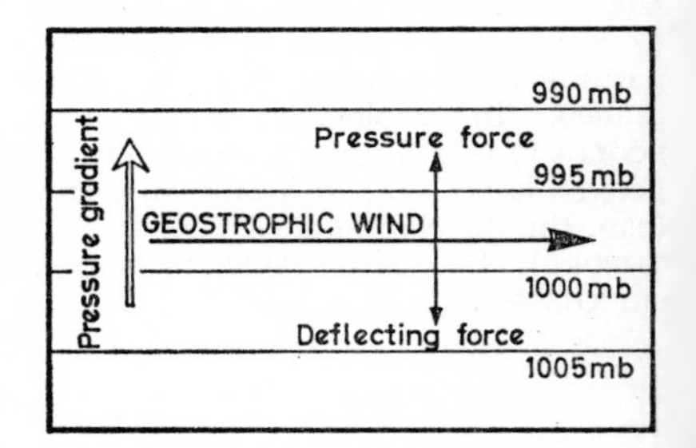

geosyncline A linear depression o a downfold in the crust on an eart scale, a name first used by J. D. Dan in 1873. A g. probably developed as slow continuous down-warping o subsidence, the floor of which gradu ally sank over a long period o geological time at much the same rat as thicknesses of sediment, worn awa from the land-masses on either sid were deposited within it. Some main tain that this deposition was an actu cause of the sinking of its flo through its own weight, others that was merely a consequence of th existence of a convenient repository. [*f* MEDIAN MAS

geothermal gradient The increa in temperature in the rocks of th earth's crust with increasing dept Various figures are quoted for thi evidence indicates that the g. g. is n constant, but steepens with increasi depth. Estimates vary from 1°C. 28 m. (1°F. in 51·1 ft.) to 1°C. in 40 (1°F. in 72·9 ft.); an average for t SIAL layer could be 1°C. per 28·6 (1°F. in 52·1 ft.). The temperature 986°C. (1806·8°F.) has been given the base of the SIALIC rocks.

Gestalt concept (Germ.) Used in synoptic climatology to denote a complex of climatic elements occurring in a well-known and recognizable pattern; e.g. a COLD FRONT and its weather pattern.

geyser (derived from the Icelandic word, *geysir*, lit. a 'gusher' or 'roarer') An intermittent fountain of hot water ejected from a hole in the earth's crust with considerable force, accompanied by steam. Some g.'s erupt at regular intervals, other more irregularly and spasmodically. Superheating in the long, bending pipe of a g. causes the build-up of water pressure at a temperature above 100°C., until suddenly part is converted into superheated steam, and the water in the upper part of the pipe is violently emitted. Cooler water flows into the pipe, and the heat increase begins again. A cone of *sinter*, deposited from the hot water, accumulates around the vent. E.g. in Iceland, Yellowstone National Park (Wyoming, U.S.A.), N. Island of New Zealand. 'Old Faithful', in Yellowstone, has an average eruption interval of about 65 mins., ranging between as little as 33 mins. and as much as 95 mins.; it throws 10–20,000 gallons of near-boiling water, accompanied by steam, 37–55 m. (120–180 ft.) into the air, lasting from 2 to 5 mins. Between 1870 and the end of 1968, Old Faithful erupted approx. 790,000 times. [*f*]

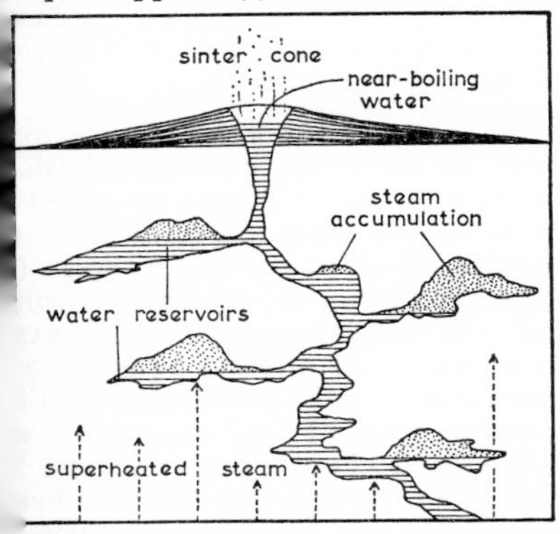

geyserite A siliceous deposit (*sinter*) formed around thermal springs and geysers; e.g. a low cone has been built up around the vent of Old Faithful, Yellowstone National Park. Ct. TRAVERTINE, which is calcareous.

ghat (Hindi) A mountain pass in India; e.g. the Thalghat and the Bhorghat behind Bombay. Probably the term was transferred by Europeans to the mountains through which the passes led, hence W. Ghats and, by analogy, E. Ghats.

ghetto The Jewish quarter of a town; often also used of any racial group concentration in a town.

gibli, ghibli (Arabic) An extreme form of SIROCCO in Tunisia; it is sometimes so intense that it feels like the hot blast from an opened oven-door.

gill, ghyll (Norse) (the latter spelling is obsolescent) (i) A swift-flowing mountain torrent in the English Lake District (e.g. Piers G., Lingmell G., both in Wasdale), and in Yorkshire (e.g. Tennant G., in Craven). (ii) A fanciful name for a wooded ravine.

Gipfelflur (Germ.) The surface of uniformity of summit levels, independent of structure and rock type, found in the Alps (*'die G. der Alpen'*) and in the Rockies, but not thought to imply the remnant of a PENEPLAIN. This term was coined by A. Penck and used by his son W. Penck.

glacial App. to a glacier, ice-sheet or ice-age, its effects and results; hence g. epoch, advance, cycle, drift, erosion, deposition, lake. See under specif. terms.

glacial control theory The t., postulated by R. A. Daly, explaining the occurrence of CORAL at depths below its normal habitat by a fall in sea-level occasioned by the withdrawal of water contained in the ice-caps of Quaternary times. This meant: (i) that the water in tropical latitudes must have been so much cooler that all existing living coral was destroyed; and (ii) that pre-glacial reefs and other islands were planed down by marine erosion to the sea-level of the time. These platforms thus provided bases for the upward growth of coral, as the temperature of the sea again increased and the declining ice-sheets returned their melt-water to the oceans, thus causing a gradual rise of sea-level with which the growing coral could keep pace.

glacial divergence The interruption of a drainage pattern by the advance of a glacier or ice-sheet; e.g. the R. Derwent (Yorkshire) once flowed E. to reach the N. Sea, but when the N. Sea ice-sheet created a barrier, the river was forced to turn S. into the Vale of Pickering, a course which it has maintained in post-g. times. [*f* OVERFLOW CHANNEL]

glacial drainage channel A stream-cut c. associated with glaciation and melt-water. Formerly it was suggested that such c.'s were mainly OVERFLOW C.'s from glacially dammed lakes, but more recent research has indicated that streams flowing under and within glaciers may give rise to c.'s on the glacier bed which are revealed on ice retreat.

glacial lake A small sheet of water ponded in an angle between a valley-wall and the edge of a glacier; e.g. the Märjelen See on the margin of the Aletsch Glacier; the many l.'s around the edge of the Vatnajökull ice-sheet in Iceland (e.g. Vatnsdalur); Tulsequah L., Alaska. [*f*]

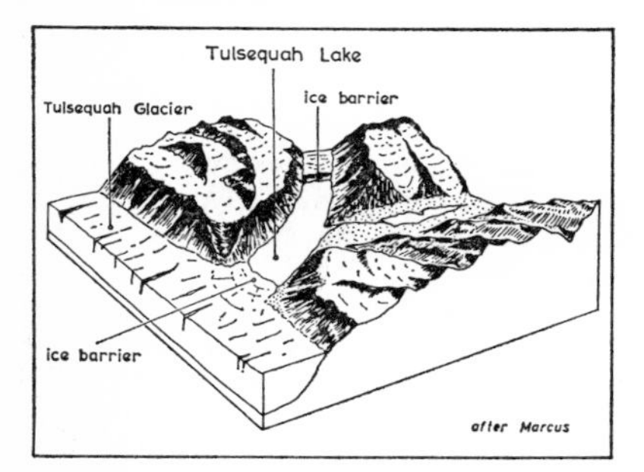

glaciation (i) The covering or occupation in the past of an area by an ice-sheet or glacier. (ii) The period of time during which this occurred; hence the Quaternary g. (iii) A gen. term for associated processes and results of all glacial activity.

glacier (Fr.) A mass of ice of limited width (ct. ICE-SHEET), moving outward from an area of accumulation. It is sometimes called a mountain-g., a VALLEY-G., or an alpine-g. It moves continuously from higher to lower ground, and is enclosed within distinct valley walls. G.'s vary in length from tiny tongues which barely protrude from a FIRN-basin, to the Petermann G. in Greenland (200 km., 125 mi.), and many in the Antarctic over 160 km. (100 mi.) long. The longest in Europe is the Aletsch (16 km., 10 mi.) in Switzerland. The world's longest is the Lambert G. (400 km., 250 mi.), discovered in 1957. See CIRQUE-G., EXPANDED-FOOT G., GLACIERET, HANGING G., PIEDMONT G., TIDAL G., WALL-SIDED G. Also there are many self-evident compound words and phrases. See also 'WARM GLACIER' and 'COLD GLACIER'.

glacier band A banded structure o some kind which may be seen in a g. either in plan on its surface, or ir section in a CREVASSE; it may be a dir b., or a b. of ice of contrasting colou and texture. See OGIVE.

glacieret A small glacier, esp. i U.S.A.; e.g. in the Sierra Nevada c California and the Olympic Mtns. c N.W. Washington State.

glacier flow A not very satisfactor term to cover the complex physic forces involved in the outward move ment of ice from the FIRN-field c origin. This may be either: (EXTRUSION F.; or (ii) gravity f., involv ing: (*a*) REGELATION; (*b*) INTERGRAN ULAR TRANSLATION; (*c*) PLASTIC D FORMATION; and (*d*) LAMINAR F. Th average rate of movement of a Alpine g. is about 0·3 m. per day. I Greenland 30 m. per day has bee measured, and 18 m. per day common. The Black Rapids G. Alaska attained 76 m. per day for short time in 1937. The record measured movement for 1 year was 1710 m. (5610 ft.) for the Storström G. in Greenland. The sides of a g. move less rapidly than the centre, sometimes at only half the rate, as the result of friction.

glacierization Coined by T. Griffith Taylor and used both in Gt. Britain and U.S.A. to indicate the gradual spread of glaciers or ice-sheets over an area during the present time; ct. DEGLACIERIZATION. The term GLACIATION refers to past time, together with: (i) all the processes involved; and (ii) the resultant features.

glacier milk A white turbid stream of melt-water, laden with pulverized rock ('rock-flour'), issuing from a glacier SNOUT.

glacier mill (from the Fr. *moulin*) See MOULIN.

'glacier mouse' A small rounded stone, almost covered with moss, found and named by J. Eythorsson on certain Icelandic glaciers. The stones either lie on superficial morainic material, or may have rolled off on to the adjacent ice.

glacier table A block of stone on a g., protecting an underlying stalk of ice from melting by the sun's rays, thus being left on a pedestal.

glacio-eustatism A change of sea-level resulting from the advance or shrinkage of ice-sheets, hence the withdrawal from or return to of water in the oceans. See EUSTASY.

glaciofluvial, alt. **glacifluvial** More commonly, and perhaps more logically, used in U.S.A. as syn. with FLUVIOGLACIAL.

glaciology The scientific study of ice, its form, nature, distribution, action and results.

glacis (i) A gently sloping bank, esp. on a mountain-side. (ii) An extension of political control by a state across a mountain divide; e.g. Ticino in Switzerland.

glauconite A greenish mineral, a hydrous silicate of iron and potassium (from Gk. *glaukos*, bluish-green), commonly found in rocks of marine origin, to which it may give a greenish appearance; e.g. in the Upper Greensand and the Chalk Marl (where the g. grains seem to be casts of foraminifera), and in other sandstones and arkoses. At present g. seems to be forming in submarine banks, and therefore sandstones containing it are probably of marine origin.

glazed frost, glaze A layer or coating of clear ice formed: (i) when rain falls on to a surface (e.g. a road) which has a temperature below freezing ('*black ice*'); (ii) when supercooled droplets freeze on impact with telegraph wires, branches, the leading edges of aircraft wings; (iii) by renewed f. after a partial thaw. G. f. is known in U.S.A. simply as *glaze*.

glebe The area of land vested in a clergyman as part of his living or benefice.

glei (gley) soil A s. HORIZON where intermittent waterlogging, the result of poor or impeded drainage, reduces oxidation or causes deoxidation of ferric compounds, so that the resulting ferrous compounds have a bluish-grey mottled appearance, with a sticky, clayey, compact and usually structureless texture. Hence the ugly word *gleiization*, used esp. in U.S.A. for the development of g. characters. It occurs commonly in bog-s.'s and meadow-s.'s (*humic-g.*).

glen A long steep-sided flat-bottomed valley, esp. in Scotland, narrower than a STRATH; e.g. G. More, G. Brittle, G. Roy. It is also commonly used in a compound place-name; e.g. Glencoe, Gleneagles, Glenfinnan.

glint-line A marked edge at the margins of much denuded rocks; used specif. for the boundary between an ancient SHIELD and younger rocks; e.g. in Scandinavia and along the edge of the Canadian Shield. It is derived from the Norwegian *glint*, a boundary. [*f*]

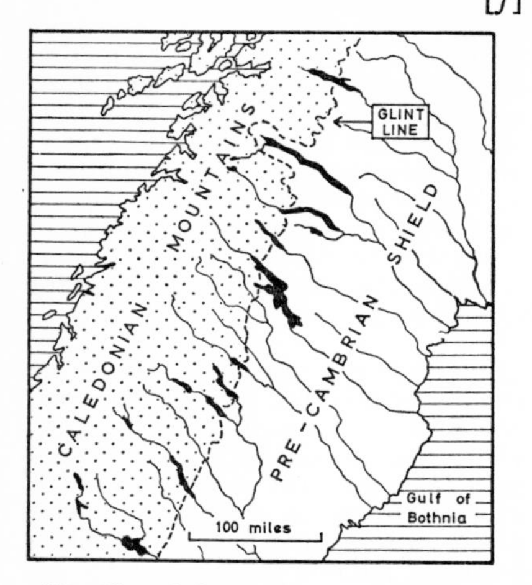

glint-line lake A l. formed along a g.-l. In Scandinavia a series of these long, narrow l.'s is strung out along the valley of each stream flowing more or less parallel to the Baltic Sea. The steep edge of the Baltic ice-sheet at one stage of the Quaternary Glaciation lay to the E., forming a dam, ponding water between this and the main watershed to the W. Some melt-water escaped W., eroding deep overflow channels. When the ice withdrew, the rivers resumed their pre-glacial courses to the E. and the overdeepened valleys remained l.-filled. Morainic blocking at their E. ends contributed. E.g. Store Lule L., 185 km. (115 mi.) long. [*f* GLINT-LINE]

globe (i) A commonly used name for the earth, as 'sailing round the g.' (ii) A spherical model on which is shown the pattern of continents and oceans. The earliest g. to which reference is made was by Crates (*c.* 160 B.C.), which has not survived. The earliest existing g. was by Martin Behaim (1492), made in Nürnberg, which does not show the New World. A modern g. is made with a shell of several alternative media (metal, card, plastic, glass, rubber).

Globigerina ooze A calcareous deposit on the deep-sea floor, though not in the great ocean depths, consisting of the remains of various minute one-celled foraminifera, *Globigerina* being the most common. It is very widely spread in the N. and S. Atlantic, Indian and the S. (but not the N.) Pacific Oceans. Its rate of deposition is very slow, taking about 3250 years for a thickness of 25 mm.

Globular Projection A p., neither equal-area nor conformal, but with small distortion. It is commonly used in pairs in atlases, showing each hemisphere. Draw a circle of radius $r=\sqrt{2}\,.\,R$, for $\pi r^2=2\pi R^2$ (half the area of a sphere), where R=radius of the earth to scale; put in the equator and the central meridian for the hemisphere; divide these and the circumference equally; and draw parallels and meridians as arcs of circles.

glory See BROCKENSPECTRE.

gloup, gloap See BLOW-HOLE.

gneiss A coarse-grained crystallin rock, gen. of foliated texture, and o streaked, wavy or banded appearance It is formed by the DYNAMIC META-MORPHISM of granite and other igneou rocks ('*orthogneiss*'), or of sedimentar rocks ('*paragneiss*') that have bee penetrated by *magmatic intrusio* ('*injection g.*'). The term covers a wid range of gneissic rocks derived fro granite, syenite, diorite and gabbr Note '*augen-g.*', which contains por phyritic crystals surrounded by mic or hornblende in an elliptical patter giving the appearance of an eye.

Gnomonic Projection An AZI-MUTHAL or ZENITHAL P., which i constructed by projecting from th centre of the globe on to a tange plane. The Polar case is best know and easiest to construct [*f* below Lines are drawn from the centre of circle of radius R (=earth, to scale) to a tangent plane touching at the Pole. The distances of these intersections from the Pole are the radii of the respective parallels, which are concentric circles, their distances apart increasing outwards. Alternatively, the radius of each parallel = $R \cdot \cot \theta$, where θ is the latitude. Meridians are radiating straight lines from the centre. Scale greatly increases from the centre and it cannot be practicably used more than 45° from the Pole. In the equatorial case, all parallels other than the equator are hyperbolas. In the oblique case, parallels in higher latitudes than the parallel of tangency are ellipses, the parallel of tangency is a hyperbola, and the parallels between the parallel of tangency and the equator are also hyperbolas. In all 3 cases, GREAT CIRCLES are straight lines, and all straight lines are great circles. But a RHUMB-LINE is curved (except for directions from the Pole in the Polar case). A navigator lays out his route as a straight line on the G. P., then transfers it to a MERCATOR P. The U.S. Hydrographic Office publishes charts for the world's oceans on the G. P. They are also used for radio and seismic work (as waves travel virtually in great circle directions) and for star-maps. [*f*]

old A precious metal known since arliest times, mentioned in *Genesis*, nd regarded throughout history as a ymbol of wealth, as well as being used as ornaments because of its untarnishable quality. Orig. it was wholly derived from alluvial gravels (PLACER deposits), the g. originating in a VEIN or REEF from which it was removed naturally by weathering and river-action. Now 80% of the world total is mined from veins, found especially in the AUREOLE surrounding granite. G. is too soft for most purposes, and is hardened by the addition of other metals (silver, palladium, nickel). Pure g. = 24 carat. A sovereign is 22 carat, with 2 parts of copper. C.p. = S. Africa, U.S.S.R., Canada, U.S.A., Australia, Ghana.

gold standard A system of currency by which paper money is exchangeable for a fixed amount of gold. Until 1914 Europe and America were on the full g. s., which facilitated international trade because currency exchange rates were fixed and definite, and all debts could be settled in g. Most countries still have a g. reserve to back their currencies. It is no longer possible in many countries to exchange notes for g.; in U.K. it is illegal for a private citizen to possess g. coins or bar g., but must surrender it to the Bank of England. In March, 1968, after a tremendous run on g., the 'G. Pool' introduced a double value, $35 per ounce fine for all official transactions between governments and central banks, and a free rate for private transactions.

Gondwanaland A single land-mass, the S. part of the Pre-Cambrian PANGAEA, from which the S. continents are thought to have been formed in Palaeozoic times. Evidence from many fields (geology, PALAEOBOTANY, PALAEOMAGNETISM) is indicative of a former unit consisting of Africa, Madagascar and India and probably parts of S. America and Australasia. Their suggested break-up and movement is part of the theory of CONTINENTAL DRIFT.

Goode's Interrupted Homolosine Projection An equal-area p., based on the MOLLWEIDE and SINUSOIDAL

P.'s, using the Sinusoidal from the equator to 40°N. and 40°S., Mollweide in higher latitudes. The oceans are 'interrupted' to allow the continents to be recentred on several meridians, so as to attain good overall shape. It is used widely for maps of economic distributions. [*f*]

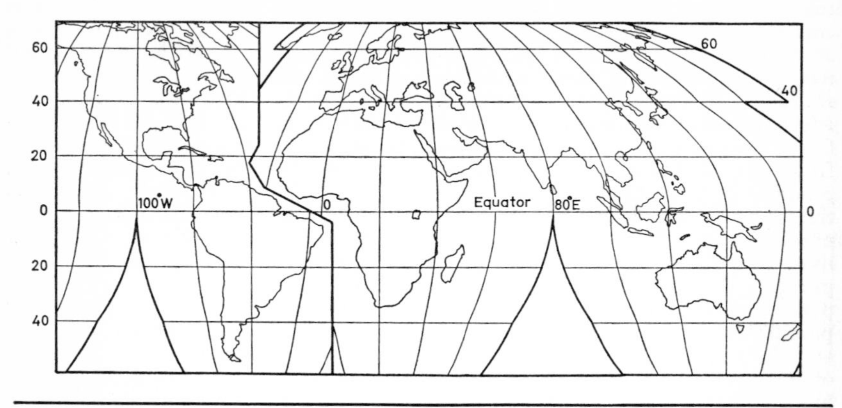

gore A triangular piece of paper or thin card on which is printed a section of a map of the world, bounded by meridians and tapering to the Poles. Twelve or 24 g.'s for each hemisphere can be glued on to a sphere to fit reasonably well, and so make a complete printed globe.

gorge A deep, steep-sided, rocky river valley; e.g. the Aar G. in Switzerland; the Rhine G. in W. Germany.

Gotiglacial One of the stages of the retreat of the Quaternary glaciation in Scandinavia, lasting from *c.* 15,000–8000 B.C. The ice-sheet left S. Sweden, and a rise of sea-level resulted in an enlarged Baltic Sea, known as the Yoldia Sea.

gouge (i) A fault-plane clay or other fine material, in ct. to FAULT-BRECCIA. (ii) The soft earthy material alongside the containing wall of a mineral vein, which facilitates mining of the vein itself, after the miner has 'gouged' it out.

Graben (Germ.) A narrow trough, let down between parallel faults, with throws in opposite directions. It is often regarded as syn. with RIFT-VALLEY, but strictly a g. is a structural feature, and is not necessarily a valley.

grad A measurement of latitude and longitude found on a number of old maps, esp. of France and Turkey, using 1g. = 1/400th of 360°, rather than a degree.

gradation (i) Syn. with DENUDATION, esp. in U.S.A. (see DEGRADATION). (ii) The process of bringing the land-surface to a uniform GRADE.

grade A concept of EQUILIBRIUM applied esp. to stream long-profiles and to ground-slopes. A graded stream is regarded as one whose flow is such that at a given point there is a balance between material eroded and material deposited. A change in volume, velocity and load, tending to increase either erosion or deposition will produce a change in the gradien of the long-profile in such a directio that the balance is ultimately restored In this theory g. can only be achieve in the later stages of the CYCLE O EROSION, and will be represented by flattened parabolic curve, concave upwards. It is now thought that g. ma be achieved equally well by change in factors other than the long-profile e.g. channel cross-section and rough ness, and the nature of the load. Thu a smooth profile is not essential for th production of g., and this state may b reached at any stage during a stream's development. It is also stressed that g. is a condition achieved only as the average state over a period of time; at any given time a graded stretch of river may not represent an exact balance between erosion and deposition. Similar arguments are applied to the production and removal of waste on ground-slopes. [*f* REJUVENATION]

grade separation The carrying of one highway over or under another; i.e. a *fly-over* or *fly-under* junction.

gradient (i) The steepness of a slope, expressed either as a proportion between its vertical interval (*VI*) (reduced to unity) and its HORIZONTAL EQUIVALENT (*HE*) (e.g. 1 in 20), or as an angular measurement from the horizontal. The latter can be computed by expressing the g. as a fraction, reducing it to a decimal, and then looking up the angle corresponding to to this computed tangent in a table. To give an approximate conversion (reasonably correct to about 7°) from the g. as a fraction to an angle, multiply by 60. Thus,

a slope of 1° = 1 in 60 (actually 57·14)
a slope of 2° = 1 in 30 (actually 28·65)
a slope of 3° = 1 in 20 (actually 19·08)

G. can also be expressed as a percentage; e.g. a 5% slope is 1 in 20 (about 3°). (ii) The degree of variation in various phenomena: temperature, atmospheric pressure (*barometric g.*), density, velocity. [*f* SLOPE-LENGTH]

gradient profile A p. along a road, railway or river, showing mileages, slopes of different character and degrees of slope, with added names *en route* and possibly geological outcrops.

gradient wind The w. resulting from the balance between the horizontal atmospheric pressure force, the CORIOLIS FORCE, and the centrifugal effect due to the curvature of the ISOBARS. This balance is modified by the effects of friction near the surface.

grain (i) The gen. trend of structure, relief and physical features of any area; e.g. the 'g.' of N. Scotland is N.E. to S.W. Ct. the transverse g. of a RIA coastline with the longitudinal g. of a DALMATIAN COAST. (ii) A small mineral particle, as sand-g. (iii) As an adjective, the coarseness or fineness of a rock, as 'fine-grained' clay, 'coarse-grained' sandstone. (iv) A measure of weight: (*a*) *avoirdupois*: 7000 g.'s. = 1 pound; 1 g. = 0·000143 lb. = 0·0648 gm.; 1 gm. = 15·4323 g.'s; (*b*) *apothecaries*' and *troy*: 480 g.'s = 1 ounce. (v) A general term for the seed of a cereal crop (wheat, barley, oats, rye, millet, rice), or even for the crop itself; g.-crop is sometimes used.

gramme (gram) 1 gm. *mass* = 15·432 grains; 1000 gm. = 1 kilogramme = 2·20462 lb.; 1 gm. = 1 cu. cm. of water at 4°C.; 1 oz. = 28·35 gm.

granite A PLUTONIC rock, coarse-grained, consisting mainly of quartz, orthoclase felspar and micas. It may be formed either by the slow cooling of a large intrusion of MAGMA (see BATHOLITH), or through *granitization* by the transformation of the pre-existing country rock by magmatic emanations, or by some other process of alteration not yet wholly understood. The chemical composition is markedly acid (65–75% silica). It occurs as large masses in uplands, exposed by denudation; e.g. Dartmoor, Bodmin Moor and Land's End, Shap, the Cheviots, the Cairngorms, the Wicklow Hills, the Mourne Mtns., the Mont Blanc massif, the mountains in Rocky Mtn. National Park, the Sierra Nevada. The term is also used in a wide and general sense for any coarse-crystalled, light-coloured igneous rock. Thus micas may be replaced by hornblende, hence a *hornblende-g.* When the dominant mafic minerals are micas, not hornblende, *biotite-g.* or *muscovite-g.* is formed. Granodiorite, granulite and GRANOPHYRE are all broadly included in g. *sensu lato.*

granophyre A very fine-grained granite or quartz-porphyry, with a micrographic structure (i.e. the quartz and felspar are closely inter-penetrated on a microscopic scale). G.

represents a stage in the GRANITIZATION of some other rock. E.g. the Ennerdale G., in the English Lake District.

granular disintegration The breaking down or crumbling of porous rocks into a g. mass, as a result of freezing following the absorption of water into the pore-spaces.

graphicacy The art of depicting spatial relationships in 2, sometimes 3, dimensions, by means of maps and diagrams (CARTOGRAPHY), skilfully chosen photographs, and graphs. '... Graphicacy spills over into literacy on the one hand and numeracy on the o her without being more than marginally absorbed in either....' (W. G. V. Balchin and A. Coleman).

graphite A soft, black opaque form of carbon, occurring in veins or lenticular masses, and found in Korea, U.S.S.R., Madagascar, Ceylon and Mexico. Mixed with fine clay, it is used for 'lead' pencils. Its other main uses are for making carbon crucibles, as a lubricant, in paint and in dry batteries.

grassland, natural A category of vegetation region, occurring in areas which experience a season of prolonged drought, but which have some rainfall (though gen. inadequate for tree growth) coinciding with the period of growth. The major types are: (i) *tropical* g. (SAVANNA); (ii) *mid-latitude* g. (STEPPE, PUSZTA, PRAIRIE, PAMPAS, VELD, DOWNS); (iii) *mountain* g., an altitudinal zone above the tree-line. In Britain g. occurs widely, replacing to a large extent what once must have been forest. It can be classified in various ways. *General character*: (i) *permanent* g., orig. sown by Man, but which has developed like a natural plant community, yet is subject to grazing; (ii) *short-ley* g., sown by Man, and remaining down for a few years within the arable rotation; (iii) *upland* g., such as down- and fell-grazing; all these may be improved by liming, fertilizing and drainage. *Type of grass*: (i) *turf* grasses; (ii) *meadow* grasses; (iii) *tussock* grasses. *Variety of grass*: (i) *neutral* g., the permanent grass of the lowlands in fields, with perennial ryegrass (*Lolium perenne*), meadow grass (*Poa pratensis*), timothy (*Phleum pratense*) and white clover (*Trifolium repens*); (ii) *basic* g., on chalk and limestone, with sheep's fescue (*Festuca ovina*) and red fescue (*Festuca rubra*); (iii) *acid* g., growing on base-deficient soils on siliceous rocks of N.W. Britain, with common bent (*Agrostis tenuis*) and sheep's fescue; (iv) *moor* g., on poorly drained peaty soils, with white bent (*Nardus stricta*) and wavy hair grass (*Deschampsia flexuosa*); (v) *grass heaths*, with sheep's fescue and wavy hair-grass.

graticule The net of parallels and meridians drawn on a specific MAP PROJECTION.

graupel A form of soft HAIL, or pellets formed of opaque ice-particles, usually falling in showers.

gravel An assemblage of water-worn stones, hence usually rounded, ranging in diameter from 2 to 50 mm. (0·08 in. to 2·0 ins.), of fine, medium and coarse grades. Some authorities limit g. to sizes of 2–10 mm. (0·4 in.), referring to the larger 10–50 mm. stones as pebbles. One definition in the U.S.A. uses a diameter of from 4·76 mm. to 76 mm. (0·167 in. to 3 ins.), others only up to 2·5 ins. There is in fact no real agreement. The Wentworth Scale of particle size in the U.S.A. excludes the term (a pebble is 4–64 mm.). A. N. Strahler equates g. with pebbles (0·167 in. to 2·5 ins.).

gravel train A VALLEY TRAIN composed mainly of g.

gravity, gravitation(al) Expressed as Newton's Law (1686): each body in the universe attracts every other body with a force directly proportional to the product of their masses and inversely proportional to the square of the distance between them, reckoned from their centres of mass along a line joining these centres. G. is used in compound words when downhill movement under its force is involved; e.g. *g. flow* (of a glacier) *g. movement* of material on a slope, esp. when lubricated by water (see MASS-MOVEMENT); *g. water* held in soil-pores which will drain away downward if and when free drainage conditions develop; *g. gliding* of strata which may produce OVERTHRUSTS, RECUMBENT FOLDS, NAPPES; *g. sliding* or downhill SHEARING; *g. wind* (see KATABATIC WIND).

gravity anomaly In geophysics, the difference between a computed and observed terrestrial g. value; a significant anomaly is evident where a mountain range produces less disturbance of gravity than was calculated. This must be compensated for by a deficiency of mass in depth; i.e. implying mountain 'roots' of less dense rocks; see ISOSTASY. An *excess* observed g. is *positive*, a *deficiency* is *negative*.

gravity collapse structure The c. and down-sliding of strata on either side of an ANTICLINE, partic. when some erosion has taken place, and where resistant strata are separated by INCOMPETENT BEDS. E.g. in the mountains of S.W. Iran.

gravity model In HUMAN GEOGRAPHY, a mathematical construction relating the movement between any 2 or more places to the rel. sizes of these places and the distances separating them.

gravity slope That portion of a hillside s. which is relatively steep; it is used esp. of a receding ESCARPMENT, usually steeper than 22°, by H. A. Meyerhoff, in considering W. Penck's concepts of S. RETREAT. Such s. commonly lies at the angle of rest of the material eroded from it, and may be equated with the CONSTANT S. [] of A. Wood.

great circle A c. on the earth's surface whose plane passes through its centre; the shortest distance between any 2 points on the surface is an arc of a g. c. An infinite number of g. c.'s can be drawn on the surface of a sphere, but only 1 g. c. can be drawn through any 2 points on a sphere unless they are ANTIPODAL, in which case an infinite number is possible. Every complete g. c. bisects any other at 2 antipodal points. On the GNOMONIC PROJECTION all g. c.'s appear as straight lines. *Note*: these properties are only approx. correct, since the earth is not a perfect sphere. [*f*]

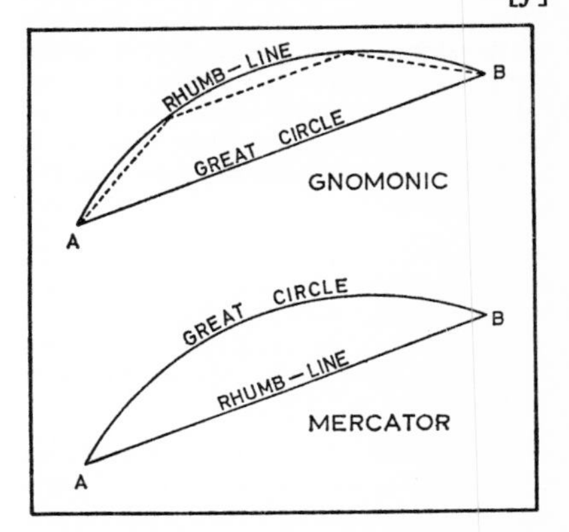

great circle route A long-distance air-route which follows approx. a g. c. e.g. London–Los Angeles, Amsterdam–Vancouver (both over Greenland), Amsterdam–Tokyo (over Greenland and Alaska), the so-called 'polar routes'. In practice, a g. c. r. is broken down into a series of RHUMB-LINES in order to avoid continuous navigational changes of direction.

Great Interglacial The longest of the i. periods in the Quaternary glaciation, the MINDEL-RISS, lasting for 190–240,000 years. [*f* MINDEL]

green belt A deliberately maintained zone of open country around a town, or separating two towns; e.g. the g. b. in the county of Hampshire, surrounding the city of Southampton, and separating the latter from Portsmouth. It is not merely an area not to be built on, with fields, woodlands and commons, and including golf-courses and playing-fields; it is also an area of open countryside, with farms and villages in which further building of houses and introduction of industry would not normally be allowed.

'greenhouse effect' A popular term for the important fact that short-wave RADIATION can pass easily through the atmosphere to the surface of the earth, while a proportion of the resultant heat is retained in the atmosphere, since the outgoing long-wave re-radiation cannot penetrate the atmosphere as easily, esp. when there is a cloud cover. Thus hard frosts in an English winter occur during clear nights, when outgoing radiation is at its maximum, but during a cloudy night frost is unlikely. Thus the atmosphere, partic. if a cloud-layer is present, acts like the glass in a g.

green lane A broad way, usually between dry-stone walls, grassed and not metalled, used for the movement of stock, notably across the Pennines in N. England; e.g. across Malham Moor.

Green Mud A fine-grained mud on the continental slope, at depths of 180–3300 m. (100–2000 fthms.), esp. off the W. coasts of Africa and N. and S. America, deriving its colour from glauconite, and including a considerable proportion of calcium carbonate.

green village A small v. clustering around an area of grass. A type in E. Germany is the 'long-green v.' (*Angerdorf*), where farm-houses are arranged in facing rows with a long narrow g. between.

Greenwich Mean Time M. SOLAR T. at Greenwich Observatory (zero meridian), used as standard t. for the British Isles and for parts of western Europe. Standard m. t.'s for almost all other countries and zones are exact hours and half-hours fast or slow on Greenwich. See ZONE STANDARD T.

Greenwich meridian The Standard or Prime m. which passes through the old Royal Observatory at Greenwich (a brass line is inset in the ground there); =0° longitude, from which meridional values for almost all the world are calculated. Longitude is expressed E. or W. of this to 180°. Other m.'s which have been employed in the past include Ferro (Canaries), used by Ptolemy, *c.* A.D. 150 (17° 14′ W. of Greenwich); the Panthéon Observatory in Paris; Philadelphia.

gregale A strong blustery N.E. wind, accompanied by showers, in the central Mediterranean Sea. It is sometimes called the *greco.* It is partic. frequent in the Tyrrhenian Sea, and around the coasts of Sicily and Malta; it caused St. Paul to be wrecked on the coast of the latter island. Occurring most commonly in winter, it is associated with the passage of a depression to the S.

'grey area' More correctly *intermediate area.* An a. within the U.K. where the low rate of economic growth and the high unemployment rates give rise to concern, but are not sufficient to justify their designation as DEVELOPMENT A.'s. G.a.'s were the subject of the Hunt Report (1969), which recommended that priority should still be given to the development a.'s, but that some relaxation of control was desirable in the case of g. a.'s, esp. the Northwest, Yorkshire and Humberside.

grey-brown podzol A type of soil transitional between the p.'s and the BROWN FOREST SOIL, more acid than the latter because of leaching under humid conditions, but with less pronounced leaching than the former, and with a greater organic content. It is found widely in Gt. Britain, W. Europe and N.E. U.S.A.; it carries good grassland and is used for mixed farming.

grey earth A group of soils (*serozem*), found in arid areas of mid-temperate latitudes, where the low rainfall results in little vegetation and a small organic content, but little leaching. Calcium carbonate is common, esp. in the upper layer, where it may form a lime-crust. The soils are highly deficient in nitrogen, and have low organic and inorganic colloidal content. They are found in the Great Basin of Utah, U.S.A.; S. Argentine; N. Libya; the area between the Aral and Caspian Seas; parts of Mongolia

N.W. Pakistan; S.W. and S. Australia. They are of little value for agriculture unless they are irrigated, adequately drained to flush off salt and so prevent further salt accumulation, and heavily fertilized with organic materials and nitrogenous artificials.

greywacke, graywacke, grauwacke (Germ.) A rather old-fashioned term applied to certain dark, strongly cemented, coarse sandstones or gritstones, containing large angular particles, hard and resistant. They are commonly of Lower Palaeozoic age.

greywether Syn. with SARSEN. Such blocks were orig. given this name because of their appearance to grazing sheep, as they occur scattered over grassy downlands; they are partic. common in Wiltshire.

grid (i) A network of squares covering a map-series, formed by lines drawn parallel and at right-angles to a central axis, and numbered E. and N. of an origin, from which the position of any point can be stated uniquely in terms of its easting and northing. Britain uses the NATIONAL G. Each American state has a g. (or more than one if its longitudinal extent exceeds 150 mi.), based on either the TRANSVERSE MERCATOR or the LAMBERT'S CONFORMAL projection, on which a g. of 1000-ft. squares is drawn, enabling lists of 'state-g. co-ordinates' to be compiled. This is used by the U.S. Land Office, hence is known as the U.S.L.O. g.; see also LAND SURVEY system. The U.S. military g. system, which is metric, is based on 60 zones, each of 6° longitude, drawn on a TRANSVERSE MERCATOR PROJECTION; hence known as the U.T.M. G. See also U.P.S. G., for polar regions within 80°N. and 80°S., drawn on a Polar STEREOGRAPHIC projection. (ii) A uniform pattern (usually of squares, equilateral triangles or hexagons) which covers a surface on which data have to be mapped, in order to carry out a spatial analysis of these data e.g. of slopes or altimetric frequencies), or to compute values at the node of each grid-unit for the interpolation of ISOPLETHS. [*f opposite*]

grike, gryke A deep groove crossing an area of limestone 'pavement', bounded by ridges (CLINT). It is the result of solution by CARBONATION through the action of acidulated rain water, concentrated along well-marked joints; e.g. in the N. Pennines, esp. near Malham in the Craven district of Yorkshire. [*f*]

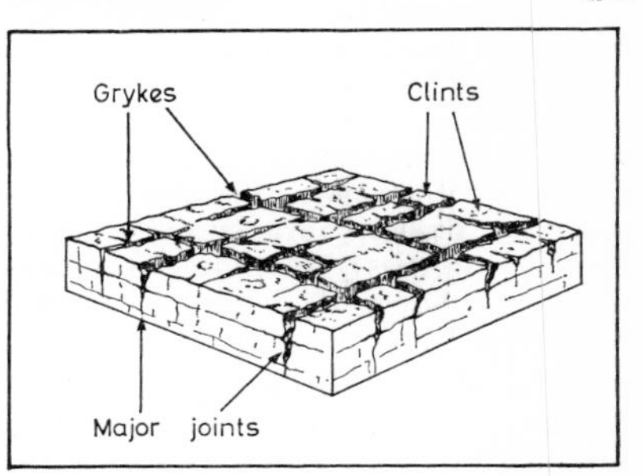

grit, gritstone A coarse sandstone, usually massive, with grains of uneven size, probably a compacted marine deltaic deposit. It is used as a proper name for Millstone Grit, the middle series in the Carboniferous system. The rock forms extensive moorland plateaus in the central Pennines and cappings on some mountains rising from the limestone plateaus; e.g. Ingleborough, Pen-y-Ghent and Whernside. The term is sometimes (incorrectly) used for certain limestones; e.g. Calcareous G., Pea G.

gross domestic product A measure of the value of the goods and services produced in a country before providing for depreciation or capital consumption.

gross national product An evaluation of the overall performance of a country's economy, indicated by the total value of all goods and services produced in that country, together with net receipts from services at home and abroad (commerce, insurance, finance, transport, professional activities), including also interest, profits and dividends earned abroad.

gross register tonnage (g. r. t.) The capacity of all the enclosed parts of a ship, including its superstructures, the unit of measurement being 1 g. r. t. = 100 cu. ft. This is the standard measurement for merchant ships.

gross reproduction rate A ratio obtained by relating the number of girl babies (i.e. potential mothers) to the number of women of child-bearing age:

$$g.r.r. = \frac{\text{no. of girl babies born}}{\text{no. of women aged 15–50}}$$

e.g. in France, 1935–6, on an average 1004 girl babies had been born to every 1000 women of child-bearing age, therefore g.r.r. = 1·004. This is used in population studies as an indication of future trends. But see NET R. R.

Grosswetterlage (Germ.) A large-scale circulation pattern of the atmosphere, within which the 'STEERING' over the region remains basically unchanged during a period of time.

grotto A picturesque or poetic term for a natural or artificial cave.

ground fog A low-lying RADIATION F., found partic. in hollows and depressions; by definition, it does not extend vertically to the base of any clouds.

ground-frost A temperature appreciably below freezing recorded on the grass or ground surface, though the air temperature may remain above freezing-point; this figure is sometimes defined as −1°C. (30·2°F.), and is likely to be harmful to tender vegetation.

ground ice (i) I. formed on or attached to the bed of a sheet of water. (ii) A mass of clear i. found in frozen ground, esp. under conditions of PERMAFROST. Cf. PIPKRAKE.

ground-moraine Debris carried at the base of a glacier or ice-sheet, deposited as a horizontal sheet of boulder-clay when this melts rapidly.

groundnut The seed of a small annual plant, which develops a few inches below the surface. When crushed, these nuts yield a high vegetable oil content (40% of their weight), which is used for margarine and cooking-fat; the residue forms a protein-rich cattle-cake, and is also used for making artificial fibres. C.p. = India, China, Nigeria, U.S.A., Brazil, Senegal.

ground swell Waves passing into shallow water, with a resultant increase in wave-height.

ground water The body of w. derived from percolation, contained in the soil, subsoil and underlying rocks above an impermeable layer. (*Note:* this excludes subterranean rivers.) Syn. with *phreatic w.*

group The rocks formed during the major time-division of an era; i.e. PRE-CAMBRIAN, PALAEOZOIC, MESOZOIC, CAINOZOIC, QUATERNARY.

growan The coarse-grained product of granitic decay, so called on Dartmoor by this local name.

growing season That part of the year with temperatures high enough to allow plant growth, gen. thought of in terms of cultivated crops. The g. s. is usually defined as the number of days between the last 'killing frost' of spring and the first one of autumn; e.g. 200 days are required for cotton. The g. s. has an important effect on agricultural systems and patterns; e.

whether spring or winter wheat is grown; barley can be grown further N. than wheat because of its short g. s.

growth area, point In planning, that a. or p. in an urban region where economic g. is started and continues with some form of stimulation.

groyne A timber, concrete or iron framework running out to sea for the purpose of arresting LONGSHORE DRIFT, and so maintaining material on the beach in some quantity. On the S. coast of England, g.'s cause an accumulation of material on their W. sides, since material moves towards the E. under the influence of prevailing W. and S.W. winds.

grus Partially decomposed granite.

G-scale A scale of geographical measurement proposed by P. Haggett, R. J. Chorley and D. R. Stoddart (*Nature*, 1965), based on the earth's surface area and derived by successive subdivisions of this area in terms of the power of 10. The G-value rises with decreasing size; thus the earth's surface has a G-value of 0 (in areal terms $1{\cdot}968 \times 10^8$ sq. mi.), the U.S.S.R. of 1·82, Yorkshire 4·51, Trafalgar Square approx. 10·0.

gryke See GRIKE.

guano The accumulation of bird-droppings, forming a valuable source of phosphatic fertilizers; e.g. along the coasts of Peru and Chile; along the coast of S.W. Africa; on Christmas I. in the Indian Ocean; on Navassa I. and Sombrero I. in the West Indies. The g. in Peru and Chile is granular, light-coloured and porous, that in the West Indies has been leached, and in places is hard and compact. On Christmas I. some of the coral reef limestones have been converted into calcium phosphate by percolation from the overlying g. beds.

gulch A deep rocky ravine in W. U.S.A.

gulder A double low tide, experienced near Portland, Dorset. See DOUBLE TIDE.

gulf A large inlet of the sea, usually more enclosed and more deeply indented than a bay; e.g. G. of Mexico, G. of Carpentaria.

gully A well-defined waterworn channel on a hill-side.

gully erosion The effects of a sudden rainstorm which has produced a localized concentrated run off, thus creating a deep gash in the land-surface, esp. when this consists of soft material; this is one important aspect of SOIL EROSION. Also *gullying*. This is seen well in the Badlands of S. Dakota, where numerous g.'s have been eroded in the Oligocene shales and limestones.

gumbo A type of sticky fine-grained clay soil in the midwestern states of the U.S.A., when saturated with water.

Günz The earliest of a series of 4 periods of fluvioglacial deposition, distinguished in the Bavarian Fore-land by A. Penck and E. Brückner (1909) as individual phases of distinct glacial advance during the Pleistocene. They correlated the 1st main glacial advance with a distinct gravel deposit, the Older Deckenschotter. The name is now applied gen. to the 1st major glacial period, though it is realized that the picture of glacials and inter-glacials is in fact much more complicated; see also MINDEL, RISS, WÜRM. The G. is equivalent in age to the NEBRASKAN in N. America, and probably to the ELBE of N. Europe.

[*f* MINDEL]

gust A temporary, short-lived increase in wind speed, followed by a lull. In U.S. terminology, the wind speed must attain 16 knots.

gustiness factor In meteorology, a measure of the intensity of gusts, obtained by dividing the range of wind speeds between gusts and the intervening calm or light wind periods, by the mean wind speed.

gut A narrow channel opening into the sea or a large estuary, esp. in E. U.S.A.

Gutenberg Discontinuity The d. at a depth of about 2900 km. (1800 mi.) from the surface of the earth, between the MANTLE and the CORE, called after the scientist (B. Gutenberg) who discovered it in 1914. At this depth the S-waves of earthquakes disappear, while the P-waves travel on at a reduced speed; i.e. it is likely that the G. discontinuity marks a change from a solid to a liquid medium, though of much greater density and under enormous pressure. Gutenberg also discovered the *G. Channel*; at a depth between 100 and 200 km. (60–120 mi.) below the upper surface of the mantle is a layer of less rigid and more plastic material, in which the speed of earthquake waves drops (P-WAVES from 8·2 to 7·85 km./sec., S-WAVES from 4·6 to 4·4). [*f* EARTHQUAKE]

guttation dew D. formed on vegetation, not by condensation from the air, but from water transpired from the plants themselves.

guyot A flat-topped mountain, rising from the floor of the Pacific Ocean to within 0·8 km. (0·5 mi.) of the surface; a few have also been discovered in the Atlantic Ocean. It differs from a SEAMOUNT, which has a pointed summit. There are estimated to be 10,000 g.'s and seamounts in the Pacific (H. W. Menard), some of which rise 3·2 km. (2 mi.) above its floor. They probably originated as volcanoes, their summits were worn down by marine planation, and were later covered with water either by a rise of sea-level, or by the subsidence of their foundations. Many g.'s are quite old, as shallow-water Cretaceous and Miocene material has been dredged from some summits. The name was given by H. H. Hess in 1946 after Arnold Guyot, a Swiss geographer who taught at Princeton, U.S.A.

gypsey A BOURNE.

gypsum Natural calcium sulphate ($CaSO_4.2H_2O$), one of the class of SEDIMENTARY rocks known as EVAPORITES. In a fully crystalline form it is *selenite*. *Alabaster* is a very fine-grained variety. Without its water of crystallisation it is *anhydrite*. G. occurs in the Permian and Triassic beds of N. England. It is used for making plaster of Paris, as a filler, as an ingredient in cement to retard its solidification, and for insulation purposes. G. is also spread on recently reclaimed marine clays (e.g. the Dutch polder lands) to assist in desalinization; this forms a highly soluble compound, sodium sulphate, which can be flushed away. ANHYDRITE is a raw material of the chemical industry, used to make ammonium sulphate and sulphuric acid.

gyre A closed circulatory system or 'cell', occurring in each of the major ocean basins, between about 20°N. to 30°N. and 20°S. to 30°S. Its movement is generated by: (i) the convection flow of warm surface water poleward; (ii) the deflective effect of the earth's rotation; (iii) the effects of prevailing winds. The N. Atlantic g. involves the Gulf Stream (N.), the N. Atlantic Drift (E. to N.E.), the Canaries Current (S. to S.W.), and the N. Equatorial Current (W.). The other ocean basins have similar g.'s, except for the landlocked Indian Ocean, where the triangular peninsula and the MONSOON change of wind direction cause a double g. moving in seasonally opposite directions; another double g. occurs in the S. Pacific.

gyro-compass A compass which does not make use of the earth's magnetism, consisting of a rotating wheel which has a constant direction of axis and plane of rotation, and hence can be used to find direction.

haar A cold mist experienced in spring and early summer, esp. in E Scotland.

habitat Used syn. with ENVIRONMENT, esp. in an ecological context, as an area in which the requirements of a specif. animal or plant are met.

hachure A line on a relief-map drawn down a slope, made thicker and closer together where the gradient is steepest. It enables minor but important details to be brought out which would be lost within the contour-interval on a contour-map, and can show striking relief in a dramatic manner. It lacks specif. information about altitude. [*f*]

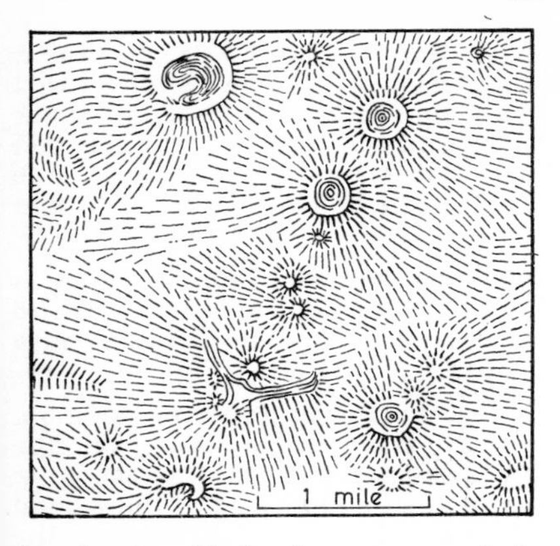

hacienda (Sp.) A term used in Spain, parts of Latin America, and the Philippines for a large agricultural estate, ranch or plantation.

hadal Sometimes applied to the deepest parts of the ocean below about 5500–6100 m. (18–20,000 ft.), i.e. below the ABYSSAL zone. Cf. DEEP, ocean.

hade (i) The angle of inclination of a FAULT-plane from the vertical. (ii) Used in mining for the angle of a VEIN or LODE, measured from the vertical. [*f* FAULT]

Hadley cell A feature of the mean atmospheric circulation, comprising a thermally driven circulation c. extending from the equator to about latitude 30° N. and S., with rising air at the equator, poleward high-altitude flow, descending flow at about 30° N. and S. and surface equatorward flow. The concept was put forward as long ago as 1686 by Edmund Halley, improved and modified in 1735 by G. Hadley in his explanation of the Trade Winds. [*f* ATMOSPHERIC CIRCULATION]

haematite, hematite An ore of IRON, Fe_2O_3, grey, black or reddish in colour. It has been worked for many years in W. Cumberland and Furness, the ore averaging a content of iron 40–62%, silica 5–15%, and a very low phosphorus content; here it occurs as veins, flats and vertical 'sops'; about 150 million tons have been mined to date. Sometimes it occurs in large kidney-shaped nodules, hence 'kidney ore'. The largest deposits of h. in the world are in the 'iron ranges' of the L. Superior district. In its cubic form, *martite*, h. is also the main ore at Krivoi Rog, Ukraine.

Haff (Germ.) A shallow fresh-water or brackish coastal lagoon, formed by the growth of a sand-spit (NEHRUNG) across the mouth of a river, sometimes enlarged by the submergence of adjacent lowlands by the sea. The classic examples are on the Baltic coast of E. Germany, Poland and U.S.S.R.; e.g. Kurisches H., Frisches H. [*f* NEHRUNG]

hag, hagg A steep-sided mass or bank of peat, found on a moorland where erosion is now active; e.g. on the gritstone Pennines, such as the Kinder Scout plateau.

ha-ha A boundary wall so constructed around a park or garden that it cannot be seen. A wide ditch is dug, or an embankment constructed, and the outward-facing side is made into a vertical wall.

hail, hailstone An ice-pellet, sometimes defined as having a diameter of 5 mm. or more, which falls from a CUMULONIMBUS cloud at the passage of a COLD FRONT, and in summer or in hot climates (India, S. U.S.A.) after exceptional heating at the earth's surface, convectional overturning, and rapidly ascending air-currents. It is clear that each h. forms around an 'embryo', a frozen droplet rather bigger than the unfrozen droplets, in a cloud. At first it is swept upwards by the updraught, until it attains a dimension which enables it to fall through the updraught, hence out of the cloud. This accounts for ordinary h.'s, but the growth of very large h.'s is still only partially understood. In section many h.'s show concentric layers of ice, alt. clear and 'milky'. The 'clear' layers are caused by the freezing of water under conditions of 'dry-growth', i.e. with low humidity in that part of the cloud, while the opaque layers are produced by 'wet-growth', i.e. in part of the cloud with high humidity so that small droplets freeze on impact. The largest h. recorded in Gt. Britain fell near Horsham on 5 Sept., 1958; in section it resembled a half grapefruit and weighed $6\frac{1}{2}$ ozs. One weighing 1·5 lb., with a circumference of 17 ins., was recorded in U.S.A. in 1928. One weighing $7\frac{1}{2}$ lb. has been reported in India, and one of 10 lb. in China. They can do much damage to orchards, crops and glass-houses, and in U.S.A. and India men and animals have been killed.

Haldenhang (Germ.) A term introduced by W. Penck to describe a section of a slope at the foot of a rock-face, often beneath a TALUS layer, and which is less steep than the slope above.

half-life That period of time in which half the atoms of a radioactive mineral disintegrate and change into the ISOTOPE of another element. The h.-l. of uranium^{238} is $4{\cdot}51 \times 10^9$ years, of potassium^{40} is $1{\cdot}19 \times 10^9$, of rubidium^{87} is $47{\cdot}0 \times 10^9$, of thorium^{232} is $13{\cdot}9 \times 10^9$, of carbon^{14} is only 5600 years.

halite Rock-salt, or sodium chloride, which forms thick layers in association with sedimentary rocks such as sandstone and shale.

halo A ring or rings of light, concentric to the sun or moon, when the sky is thinly cloud-veiled, the result of refraction of light by water-drops or ice-crystals; the ring may be white, or tinted red on the inner side, or range from red to blue (outer). Ct. CORONA.

halophyte A plant growing in salt-impregnated soil on the shore of an estuary or in a salt-marsh, or able to survive in the presence of salt-laden spray. The category includes marsh samphire (*Salicornia herbacea*), sea manna-grass (*Glyceria maritima*), sea aster (*Aster tripolium*), perennial rice-grass (*Spartina townsendii*). Sand-dune plants are not included in this category.

hamada (Arabic) A rock desert, consisting of a bare wind-scoured pavement, diversified by relict masses such as YARDANGS and ZEUGEN, and with little sand. H. plains cover large areas in the Sahara, W. Australia and the Gobi Desert in Mongolia.

hamlet A small group of houses in the English countryside, too small to be called a village, usually without a church.

Hammer Projection A variation, by E. Hammer in 1892, on the ZENITHAL EQUAL AREA (LAMBERT) P., made by doubling the horizontal distances along each parallel from the central meridian. This transforms the circular shape of the Lambert into an ellipse, similar in appearance to a MOLLWEIDE P., but with all parallels curved except the equator, which is a straight line. This p. is sometimes called Aitoff-Hammer, but it should be credited to Hammer alone; the AITOFF is quite distinct. [*f, page 171*]

hanger (i) A wood (usually beech) on the steep slopes of chalk country; e.g. Selborne H. (ii) The steep slope itself.

hanging glacier A short g. protruding from a basin or shelf high on a mountain-side, from which masses periodically break off as ice-avalanches. These may form one of the main objective dangers to the ascent of some peaks.

hanging valley A high-lying tribut-ary v., which leads with a marke[d] steepening of slope into a main v. This commonly occurs in a glaciate[d] area, where the main v. has bee[n] over-deepened by a glacier; e.g. th[e] side-v.'s of the upper Rhône; of th[e] Lauterbrunnen (both in Switzerland) and of the Yosemite valley (California). As a result, a stream flowing down [a] h. v. suddenly falls as a cascade int[o] the main valley; e.g. the Staubbac

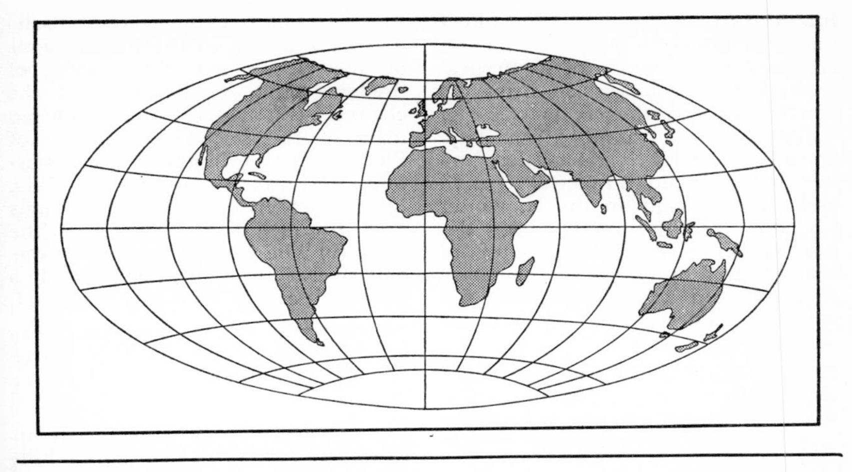

(Lauterbrunnen); Yosemite and Bridal Veil Falls (Yosemite). Other factors may cause minor valleys to join main ones in this discordant manner. River erosion in the main v. may result in its overdeepening. Sea-cliffs may contain h. v.'s cut off by marine erosion; e.g. along the Devonshire coast near Hartland, where Litter Water has a 23 m. (75 ft.) fall from the cliff to the shore. [*f* ALP]

hanging wall The upper face of a FAULT or VEIN. [*f*]

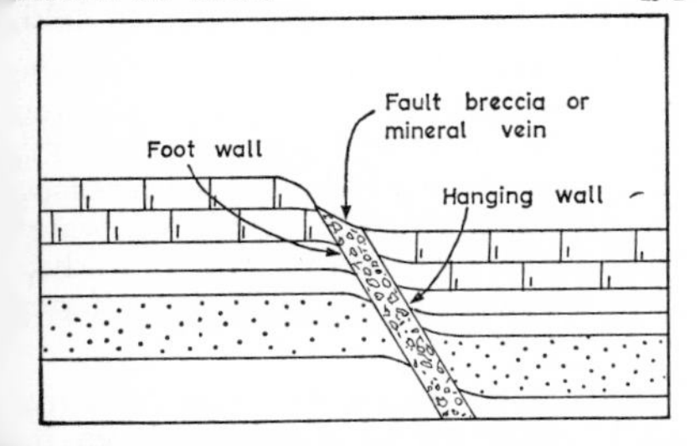

hardness scale The MOHS' s. of hardness of minerals, in which 10 selected minerals have been arranged in an ascending scale: 1. talc; 2. gypsum; 3. calcite; 4. fluorite; 5. apatite; 6. orthoclase; 7. quartz; 8. topaz; 9. corundum; 10. diamond. All others can be ascribed accordingly; e.g. 2·5 galena; 5·5 uraninite.

hard core, urban In u. geography, that area of the CENTRAL BUSINESS DISTRICT which represents the peak of functional intensity within the district.

hardpan A thin hard stratum within or beneath the surface soil, consisting of: (i) sand grains or gravel cemented by ferric salts deposited by percolating solutions (*ironpan*); (ii) compact redeposited humus compounds (*moorpan*); (iii) washed-down or synthesized clay (*claypan*); and (iv) a layer of calcium carbonate (*limepan*).

hard water W. which does not form a lather with soap, because of the presence in solution of calcium, magnesium and iron compounds. It may be: (i) *temporarily* h., containing soluble bicarbonates; (ii) *permanently* h., containing sulphates. Water flowing from limestone uplands is h., and cannot be used in the textile industry.

hardwoods Trees and timber of broad-leaved deciduous forests, affording a hard, compact wood. They are gen. much slower growing than softwoods, and the wood is more expensive. Temperate varieties include walnut, maple, cherry, poplar and oak; in Mediterranean climates are evergreen oaks, in Australia the eucalyptus; monsoon climates produce teak; tropical forests ebony, rosewood, greenheart, mahogany. Most h.'s are heavy, but not all; e.g. balsa.

harmattan A dry N.E. wind blowing from the Sahara towards the W. African coast; because of its dryness and relative coolness it brings a healthy contrast (hence its name of 'Doctor') to the prevailing humidity. Inland the wind is dust-laden and unpleasant. It blows much further S. and more strongly in the N. winter (to about 5°N.); dust has been deposited on ships in the Gulf of Guinea. In July (when the planetary pressure- and wind-belts move N.), it usually blows only as far as about 18°N.

hatchet-planimeter An instrument for measuring an area, as on a map; the point of a tracer-bar is carefully traced round the perimeter of that area.

haugh Esp. in Scotland and N. England, a piece of low-lying land, notably along the flood-plain of a river.

Hawaiian eruption A volcanic e. in which great quantities of basic lava flow quietly out, with no explosive activity, hardening to form a large, low-angled SHIELD VOLCANO, as in Hawaii.

Hawaiian 'high' One of the atmospheric high-pressure cells of the HORSE LATITUDES in the N. Pacific. It is much more markedly developed in summer than in winter.

haze Obscurity of the lower layers of the atmosphere as a result of the condensation of water droplets on dust, salt or smoke particles, with a visibility of more than 1 km. but less than 2 km. The term is also frequently used to describe a general obscurity which results from something other than moisture in the atmosphere; e.g. 'heat h.', 'dry h.'

head (i) A body of water at some altitude above an outfall, either natural (e.g. a high-lying lake) or held behind a high dam, used for hydro-electricity production. (ii) The actual difference of vertical height between the water surface (intake) and the point where it is used (discharge). (iii) The limit of ocean navigation: Bremen, Hamburg, New Orleans, Bordeaux; of river navigation: Basle. These terms depend on the dimensions of the vessels and the amount of dredging and regularization of the channels. (iv) (with cap.) A compacted mass of rubble, sand and clay of Pleistocene age, found in valley bottoms, on lower slopes and sea-cliffs, formed under periglacial conditions as a product of SOLIFLUCTION ('the H.'). (v) The leading wave of a BORE on a river. (vi) A cape, more usually a headland; e.g. St. Bees H., Beachy H.

head-dune A sand-dune which has accumulated in the 'dead-air' space on the windward side of some obstacle. [*f* DUNE]

head-dyke A dry stone wall in Scotland, separating the upper hill-pasture from the lower arable and meadowland.

heading A horizontal tunnel driven into an AQUIFER such as chalk to tap ground-water penetrating the fissures, in order to supply wells or reservoirs; e.g. a heading near Frinton 2 mi. long supplies water to Eastbourne; Brighton has 5 wells into which lead a total of 8 km. (5 mi.) of h.'s.

headland (i) A promontory with a steep cliff-face projecting into the sea. (ii) The unploughed section between ploughed land on a medieval open field.

headwall The back of a CIRQUE, rising steeply from its floor, separated from the FIRN-field by a RANDKLUFT. The steepness of the h. is increased by freeze-thaw, the result of water making its way down the randkluft into crevices in the rock-wall, hence *h. recession.*

headward erosion The cutting back upstream of a valley above its original source by rainwash, gullying and SPRING-SAPPING. Thus the source of the stream gradually recedes and ultimately may notch the ridge which forms the orig. watershed; this may ultimately lead to river capture.

headwater The upper part of a river system, used more commonly in the pl., denoting the upper basin and source-streams of a river.

heaf An unenclosed sheep-pasture on the hills of N. England, usually with its own flock born and bred there.

'Heartland' The geopolitical concept of the central part or 'core' of the Old World (the 'world-island' of Europe, Asia and Africa), introduced by H. J. Mackinder in 1904, and developed by him in 1919.

'Who rules East Europe commands the Heartland.
Who rules the Heartland commands the World-island.
Who rules the World-island commands the World.'
(*Democratic Ideals and Reality*)

heat balance The equilibrium condition which exists in average terms between the radiation received from the sun by the earth and the atmosphere, and that which is re-radiated or reflected (see ALBEDO). This is a concept of overall balance but varies with latitude and the seasons. Gen. lands between 35°N. and S. and the equator receive more radiation than they lose, while lands poleward of 35°N. and S. receive less than they lose. Heat is transferred from low to high latitudes by air-masses and ocean currents.

heath, heathland (i) Plants of the genus *Calluna* (ling or heath), though not botanically heather (*Erica*). (ii) The plant formation dominated by *Calluna*, which includes *Erica*, bilberry (*Vaccinium myrtillus*), dwarf gorse, broom, whin, wavy hair grass (*Deschampsia flexuosa*), and lichens such as *Cladonia*. (iii) The whole landscape (*heathland*) of uncultivated, gravelly or sandy soils, leached and acid in character, with a shrubby vegetation of the species listed in (ii). Silver birch and dwarf oak may occur sporadically. E.g. the Dorset h.'s; the Breckland of Norfolk–Suffolk; the centre of the Weald. The upland h.'s or 'moors' of N. England and Scotland can be included; e.g. the N. York Moors; the grouse-moors of the Scottish Highlands.

'heat-island' Used by T. J. Chandler (1962) for the area of central London above which the air temperature is slightly higher than surrounding areas, the result of the dissemination of h. from the concentration of buildings. Hence similarly in any large city.

heat-wave A popular term for a spell of several days of exceptionally hot weather, not specif. defined.

heave, of fault The forward displacement of strata in an inclined normal fault, expressed as the distance apart in feet of the ends of the disrupted strata. [*f* FAULT]

Heaviside-Kennelly Layer A section of the IONOSPHERE at about 100–120 km. (60–75 mi.), otherwise known as the *E-layer*. The ionosphere is characterized by high ion density, but variation of electron density allows it to be subdivided into separate l.'s. (There is a certain amount of doubt as to whether these zones of characteristic electron densities are in fact continuous l.'s or not.) Its practical value is that it reflects long radio-waves back to earth, which enables them to be received at a distance, rather than disappear into space. It allows the penetration of short radio-waves, which continue until they reach the APPLETON L., 240 km. (150 mi) up.

heavy industry The manufacture of articles of considerable bulk, using much steel and other materials; e.g. bridges, ships, locomotives, heavy machinery. Ct. LIGHT I.

hectare (*ha.*) A metric unit of areal measurement.

1 ha. = 2·47106 acres
1 acre = 0·4047 ha.
1 ha. = 10,000 sq. m.
100 ha. = 1 sq. km.

hekistotherm A plant such as reindeer moss or lichen, which can exist where the mean temperature of the warmest month is under 10°C. (50°F.).

Helada, Tierra The highest altitudinal climatic zone in Mexico, including the highest peaks and permanently frozen or snow-covered areas.

helicoidal flow A kind of 'cork-screw' motion in the current of a river within a bend, so tending to move material from its concave bank to its convex bank, thus emphasizing deposition on the inside and erosion on the outside of a MEANDER.

helictite A formation of calcium carbonate in a cave, of varied shape; it may be as thin as a thread, arranged in spirals or loops, or festooned with tentacles. It is formed by the deposition of calcite from solutions percolating down fine fissures on to the ceilings of caves in limestone country, usually where a strong current of air is present.

heliotropism The turning of plants towards the light.

Helm wind An E. or N.E. wind blowing strongly and gustily down the Crossfell escarpment in the N. Pennines into the Eden valley; with it is associated the *helm-cloud* or Helm, resting above the ridge (a type of BANNER-CLOUD), and the *helm-bar*, a parallel line of cloud a few km. to the W. These clouds are mainly the result of LEE-WAVES, eddies and turbulence, together with an INVERSION-layer at about 1800 m. (6000 ft.). [*f* LENTICULAR CLOUD]

hemera A palaeontological term for the period of max. development of an organism. The corresponding stratigraphical division is an *epibole*.

hemipelagic deposits Material deposited near the shore in the shallow water and bathyal zones.

hemisphere Half a sphere. The globe is divided into the N. and S. h.'s by the equator. There is also a 'land' and a 'water' h. The term 'W. H.' is sometimes used for the Americas.

hemp Used loosely for a number of fibre-producing plants, which may also yield oil from their seeds and drugs from the flowers and leaves. The plants are usually grown for a single purpose only, mostly as a source of fibre. (i) *True* or *soft h.* (*Cannabis sativa*), the only h. in a strict sense. It is botanically a member of the hop and nettle family, growing under the same conditions as flax, but is taller and yields coarser fibres. It is used for making string, and as a substitute for jute in sacking. It is chiefly grown in U.S.S.R. (which dominates world production), Italy, China, Yugoslavia, Roumania, Poland, Turkey. The leaves have narcotic properties, and hashish, bhang and marihuana are obtained therefrom. (ii) *Tropical* or *'hard' h.'s* are not strictly h., and the fibres are obtained from their leaves, not the stems. They include: (*a*) *sisal* (*Agave sisalana*), originating in Yucatan (Mexico) and introduced into Kenya and Tanganyika; and (*b*) *henequen* (*Agave fourcroydes*), which is closely related, and also grows in Yucatan. The plant forms a clump of sharp, sword-like leaves, from which is obtained a coarse fibre used for binder-twine, string and mats. (iii) *Manila* or *abaca h.*, a type of plantain, 90% of which comes from the Philippines. It produces a very strong fibre which resists water, and is used for hawsers and cables.

Hercynian Orig. used in Germany for any mountains produced by folding and faulting, of various ages; later limited to the late Palaeozoic (Upper Carboniferous-Permian) orogeny. This is given various names; sometimes the whole mountain system in central Europe is named H. (after the Lat. name, *Hercynia Silva*, of the Harz Mtns.), sometimes the mtns. are called the *Altaides*. The term VARISCAN (coined by E. Suess) is used by some as syn. with H., by others as the E. representatives of it. The mtns. of Brittany and S.W. Britain are of the same age, known specif. as ARMORICAN. In Europe the worn-down tilted blocks, the eroded remnant of the H. folds, survive as massifs S.W. Ireland, Brittany, the Central Massif of France, the Vosges, the Black Forest, the Ardennes, the Middle Rhine Highlands, the Bohemian 'Diamond', Upper Silesia, an the Donetz (a syncline). Remains of fold ranges of broadly similar age include the Urals, the Tien Shan and Nan Shan in C. Asia, the Appalachians of N. America, and the foothills of the Andes.

heterogeneous A specif. meaning applied to a gen. word, implying rocks of varying kind or nature in close juxtaposition, or to a rock which is composed of different materials, and therefore tends to weather unevenly; the contrast is with HOMOGENEOUS.

hiatus A gap in the geological succession, the result either of strata not being deposited during a period of earth-movement or denudation (e.g. the absent Miocene in Britain), or as the result of their removal by denudation. See UNCONFORMITY.

hierarchy of central places A series of distinctive levels or orders of C.P.'s, which at each level or order performs a similar set of functions that distinguish this from all others.

'high' Used in meteorology for an area of high atmospheric pressure, surrounded by closed isobars; an ANTICYCLONE.

highland High uplands, used (in pl.) specif. as a proper name; e.g. the H.'s of Scotland.

high seas The open s.'s beyond territorial waters, used both expressively and as a technical term in mercantile law.

high-speed steel Very hard s.-alloys, containing various amounts of tungsten, chromium, vanadium, molybdenum, used for very hard cutting-tools.

high water The highest point reached by the sea in any one tidal oscillation. *H.W.M.M.T.*, High Water Mark Medium Tides, is shown on British O.S. maps as the High Water Mark.

hill Uplands of less elevation than mountains in the same area; there is no specif. height-definition, because this is a matter of relative scale. The S. Downs, the Chilterns and the Cotswolds are h.'s. The Black H.'s of Dakota rise to 2207 m. (7242 ft.), the Nilgiri H.'s in S. India to 2694 m. (8840 ft.).

hill farming Gen., f. carried out in h. country, but in Britain the term is strictly applied for Government grant purposes to a specific category of farmer. It is partic. concerned with sheep-rearing and in recent years with beef-cattle.

hill-shading A method of relief depiction on maps, by assuming an oblique light from the N.W., hence the slopes facing E. and S. are shaded; the steeper the slopes the darker. Ridge-crests, plateaus, valley-bottoms and plains are left unshaded, thus giving the impression of a relief model. Printed topographical maps make effective use of h.-s. applied in a subdued stipple in a neutral tone, esp. in conjunction with contours; e.g. the Ordnance Survey 'Tourist' edition, 1 in. maps.

hill-station A settlement in the tropics deliberately sited at a high altitude among the hills to escape the summer heat of the plains, esp. in connection with Europeans administering colonial territories; e.g. a chain of h.-s.'s in Java formerly used by the Dutch; Simla 2204 m. (7232 ft.) and Darjeeling 2248 m. (7376 ft.) in India.

hindcasting In studying a sequence of events, it is possible to extrapolate either forwards (FORECASTING) or backwards (h.). The latter enables a probable sequence of events and processes to be more accurately worked out, using recorded events and processes in conjunction with those obtained by theoretical h. This may be based both on actual laws and also on tendencies and probabilities.

hinge The line along a fold where the curvature of the strata is at its maximum.

hinterland Orig. the 'back country' of a coastal settlement; now the sphere of influence of a port, the area from which it derives its exports, and within which it distributes its imports. Some use the term with respect to urban spheres of influence, the urban h.'s of both coastal and inland towns. Ct. UMLAND.

hirsel A large area of grazing for sheep in the S. Uplands of Scotland.

histogram A graphical representation of a frequency distribution, such as seasonal frequencies of rainfall. The amounts are plotted as abscissae, the scale of frequencies as ordinates. [*f* ALTIMETRIC FREQUENCY GRAPH]

historical geography No agreement exists as to the scope and content of this academic borderland between g. and history. Much that is included by some (e.g. the influence of the geographical environment on historical events, the evolution of states and their boundaries, the history of geographical exploration) is really geographical history. H. g. involves a reconstruction of past environments, either as a cross-section or a series of successive sections in time, or a sequential and retrospective appraisal of changes throughout time. Some would include only the sources of material available at the time for which the reconstruction is being made, others would allow all available material to be used (e.g. a later knowledge of geology, soils).

historical geology The time-record of changes in or on the earth. This is one of the 2 major divisions of g., the other being PHYSICAL G.

hoar-frost The deposition of tiny ice-spicules directly from water-vapour on to the surface of plants and objects on the ground, which have cooled by nocturnal radiation to a temperature below the DEW-POINT, which itself is below freezing-point. Sometimes called *white-dew* in U.S.A. See also DEPTH HOAR.

hogback, hog's back A form of CUESTA in which the dip of the strata is great, so that both front- and back-slopes are steep (rather than with one steep and one gentle slope), forming a narrow crested hill-ridge; e.g. the H.'s B., W. of Guildford; in Colorado, the h.'s of Dakota Sandstone (Lower Cretaceous) parallel to the Front Ranges of the Rockies. [*f*]

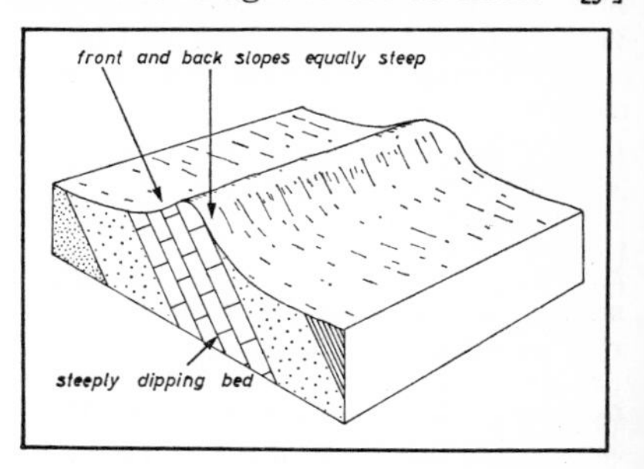

holistic App. to the view that in nature functional organisms are produced from individual structures which act as complete 'wholes'. Geography is said to have this h. approach, whereby phenomena are viewed not as individual entities but as interrelated complexes. Hence *holism*, the doctrine itself.

holm, holme (i) A small islet in an estuary or river; e.g. Steep H., Flat H. in the Bristol Channel. (ii) An area of level land liable to flood, along the banks of a river, esp. in N. England and S. Scotland, now a common valley place-name; e.g. Milnholm, Murtholm, Broomholm, in the valley of the R. Esk, Dumfriesshire; and Atholme, Lincs.

Holocene The most recent (*lit* 'wholly recent life') of the geological periods (the younger part of the Quaternary era), and the rocks laid down during that time; e.g. alluvium peat. It occupies the time since the termination of the Ice Age; the Upper H. is regarded as the last 5000 years. Many American authorities do not use the term, simply speaking of *Recent*.

holokarst An area of KARST, the features of which are developed to the fullest extent; used by J. Cvijić though by few other authorities.

holt A small wood, a wooded knoll; hence a widely used rural place-name.

homestead (i) An abode, more usually applied to a farm and its attached outbuildings. (ii) Specif. in U.S.A., a lot of 160 acres, granted for the residence and maintenance of a family under the H. Act (1862). (iii) Gen. a small rural settlement of scattered farms, without the identity of a village.

homoclimes Places with broadly similar CLIMOGRAPHS [*f*].

homocline Coined by R.A. Daly in 1916 for a succession of strata dipping gently and continuously in one direction. Cf. MONOCLINE.

homogeneous Of the same kind, character or nature, applied specif. to rocks, in ct. to HETEROGENEOUS.

Homolosine Projection See GOODE'S INTERRUPTED HOMOLOSINE P.

homoseismal line, homoseism A line on a map linking places affected simultaneously by an earthquake shock.

honeycomb weathering In some types of rock along the shore, the w. and erosion of small water-filled pools may produce distinctive features. The pools may become enlarged by chemical and physical w., leaving only sharp interconnecting ridges; the deeply pitted rock surface resembles hat of a h.

iorizon (i) The *visible* (local, apparent, natural, visible, geographical or ensible) h. is the boundary of the arth's surface as viewed from one oint, where earth (or sea) and sky ppear to meet. A near-by eminence hich interrupts the view is not part f the horizon. If the observer stands ith his eye-level at 10 ft. above the orizontal, his h. is 4·1 mi. away; at oo ft. it is 13·0 mi.; at 500 ft. 29·2 i.; at 1000 ft. 41·3 mi.; at 5000 ft. 2·4 mi.; and at 10,000 ft. 130·0 mi. if e visibility is unimpaired (E. Raisz). i) In geodetic surveying, a great rcle on the celestial sphere whose ane is at right-angles to a line from the zenith to the point of observation; this is the *true, astronomical*, or *celestial h.* (iii) In geology, the plane of a stratified surface, or a bed with a partic. series of fossils. (iv) Each main layer or zone within a SOIL PROFILE [*f*]; see A-, B-, C- H.'S.

horizontal equivalent The distance between two points on the land-surface, projected on to a h. plane, as on a map. [*f* SLOPE LENGTH]

Horn (Germ.) A pyramid peak, formed when several CIRQUES develop back to back, leaving a central mass with prominent faces and ridges. It is used commonly in the Alps as a suffix in a mountain name; e.g. Wetterhorn, Matterhorn. [*f* PYRAMIDAL PEAK]

hornblende A rock-forming mineral, occurring as a dark-coloured crystalline mass in acid and intermediate igneous rocks. It consists mainly of calcium, iron and magnesium silicate, and forms one of the amphibole mineral group.

Horse Latitudes The sub-tropical high-pressure belts of the atmosphere, about 30°N. to 35°N. and 30°S. to 35°S. (though interrupted by the pattern of distribution of land and sea); zones of calms and descending air, from which air-masses move poleward and equatorward. These belts may in part be the result of a gen. movement of air in the upper part of the TROPOSPHERE from the equator, coming under the influence of the CORIOLIS deflection so that an accumulation of air occurs in these latitudes. Possibly an equatorward movement of air in the upper troposphere from high latitudes tends to descend in the H. L., so increasing the air accumulation. They are said to be so called from the throwing overboard of horses in transport from Europe to America if the ship's passage were delayed by calms. [*f* ATMOSPHERIC CIRCULATION]

horse-power A British unit of power; it represents a rate of work of 550 ft.lb.wt. per sec. 1 h.p.=746 watts.

horst (Germ.) Introduced by E. Suess to indicate a block, usually with a level summit-plane, sharply defined by faults, and left upstanding by differential movement, either by the sinking of the crust on either side of a pair of faults, or by bodily uplift of a mass between these faults; e.g. the Vosges, Black Forest, Harz, Sinai, Korea, Morvan. The h. may be denuded so that although the structural pattern remains, the upstanding relief form may disappear. Cf. in this sense RIFT-VALLEY and GRABEN. [*f*]

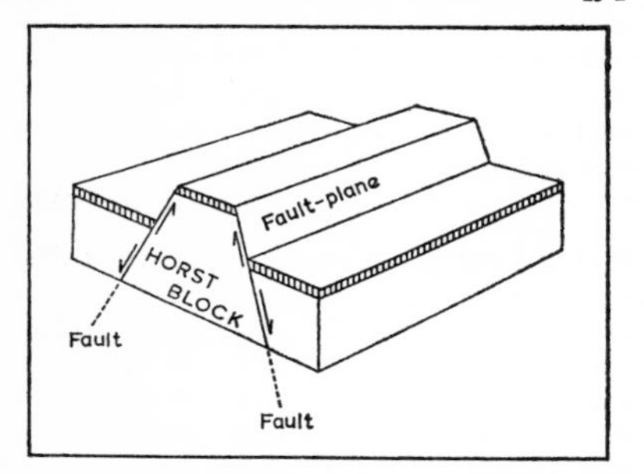

horticulture Orig. the cultivation of a garden; it is now used more widely to cover the intensive cultivation of vegetables, fruit and flowers on a small plot, including market-gardening, nursery-gardening and glasshouse cultivation.

hot spring (thermal spring) In some areas associated with past or present vulcanicity, hot water (21°C. to near boiling-point) flows out of the ground continuously, in contrast to the forceful periodic emission of a GEYSER; e.g. in Yellowstone National Park, Wyoming; Lassen Park, California; Iceland (where the water is piped and used for central heating and swimming-baths); N. I., New Zealand. H. s.'s do occur in some non-volcanic areas, e.g. Bath in England, Spa in Belgium.

how A low hill, notably in Cumberland, from Old Norse *haugr*; e.g. Pickett How in the Loweswater valley. The term is widely used in the W. Lake District as a suffix in farm names. *Note.* The name Torpenhow, a village in Cumberland, consists of 3 syllables, each meaning hill in a different language or dialect.

howe A depression, or 'hollow place', esp. in E. Scotland; e.g. H. of Fife, H. of the Merse.

huerta (Lat.-Sp.) An irrigated lowland in Valencia and Murcia, Spain. Sometimes the term is restricted to land which produces 2 or more crops a year. Ct. VEGA.

hum A residual hill-mass, shaped roughly like a haystack, in a limestone area, esp. in the KARST.

human geography Gen., that part of g. dealing with Man and human activities. It includes the study of Man in terms of ethnographic units; in terms of population: its distribution, density, sex and age structure, changes and trends, social and occupational structure, migration and movement; and in terms of Man's settlements, their origins, distributions, forms and patterns. Some authorities use h. g. as merely a composite term, in contrast to physical g. Most would deplore this arbitrary distinction, except as a matter of convenience, for Man is an integral part of the *Zusammenhang* between himself and his physical surroundings; the h. geographer must produce '. . . a well-ordered world-picture with man as an integral element' (W. G. East).

humic acid A complex organic a[cid] formed by the partial decay of vegetation through which passes percolating water. This may assist in the chemical weathering of rock, esp. of felspar, and also contribute to the formation o[f] certain soil-types.

humidity The condition of th[e] atmosphere with ref. to its water vapour content. See ABSOLUTE H[.,] RELATIVE H., SPECIFIC H., MIXIN[G] RATIO, SATURATION DEFICIT.

hummock A low rounded hilloc[k] of earth, rock or ice.

humus The remains of plants (an[d] to a less extent of animals) in the soi[l] which have decomposed into a darkis[h] amorphous mass through the action of bacteria and other organisms. Essentially the organic constituent of soil. See MULL, MOR.

hundred Formerly an administrative division of an English county, so called possibly because it contained 100 freemen or individual families, or possibly contained 100 hides (an area of arable land which was a unit of taxation assessment). In the 11th century, Staffordshire had 5 h.'s, Worcestershire 12. They survived as units of local government until the 19th century.

hurricane (i) An intense tropical storm in the W. Indies and the Gulf of Mexico, usually occurring in August and September, accompanied by winds of terrific force (160 km., 100 m.p.h. plus), torrential rain and thunderstorms. H.'s originate in mid-or W. Atlantic in latitudes 5°N. to 20°N. Their tracks are gen. to the W.N.W. over the Caribbean Sea and Florida, turning more in a N.E. direction from about 30°N. along the Atlantic coast of U.S.A. Their effects are sometimes felt as far as Maine and even the Maritime Provinces. They are usually indicated by a female code-name in alphabetical order as they occur. For details of a tropical storm gen., see CYCLONE. (ii) Force 12 (121 km., 75 m.p.h. plus) on the BEAUFORT WIND SCALE.

hurst Used commonly in place-names (esp. in the Weald of S.E. England) to denote a wood, thicket, knoll, or occas. a sand-bank in a river.

hydration A process of weathering, whereby minerals take up water and expand, causing stresses within the rock; e.g. anhydrite becomes gypsum ($CaSO_4.2H_2O$). This can cause SLAKING in a clay-rich sedimentary rock.

hydraulic force The eroding f. of water on rocks through its sheer power; e.g. a river surging into cracks, sweeping against banks on the outside of bends, with turbulence and eddying. Similarly, waves can exert h. f. as they pound the base of cliffs, esp. where air is compressed in cracks and fissures. No load of material is involved, or the erosive process becomes ABRASION.

hydraulic gradient The slope of the WATER-TABLE, expressed as the ratio between the HEAD (h) and the actual length of flow (l). Thus if $h=$ 6 m. and $l=60$ m., $h.g.=10\%$ or 0·1.

hydraulic radius The ratio between the area of the cross-section of a river channel and the length of its wetted perimeter, used by hydrologists and engineers.

hydraulic tidal current A c. which forms to compensate for the difference in height of the water when high tide occurs at different times at either end of a long narrow stretch of water; e.g. in the Menai Straits, the Pentland Firth.

hydrocarbon A compound of hydrogen and carbon: liquid, solid or gaseous. There is a large number of these, many of which are familiar in commercial use; e.g. coal-gas, petroleum.

hydro-electric power Electric p. generated in dynamos, the motive energy or prime mover for which is moving water, either a natural fall (e.g. Niagara) or one created by building a high dam across a valley (e.g. Grand Coulee Dam, Washington, U.S.A.; Génissiat Dam in S.E. France). See ELECTRICITY.

hydrogenation of coal The manufacture of mineral oil from coal by the action of hydrogen; used by Germany extensively during the war of 1939–45 to make up for the lack of imported oil, at such centres as Gelsenkirchen and at Merseburg (the Leuna plant, now in operation again, where brown-coal is used).

hydrogen-ion concentration A measure of the acidity of a solution, since the properties of acids are the result of the presence of hydrogen ions, expressed in terms of the negative index of the logarithm of the concentration. In soil science, this is given as pH VALUE.

hydrograph A graph representing stream DISCHARGE measured at a given point as a function of time.

hydrography The description, surveying and charting of the oceans, seas and coastlines, together with the study of tides, currents and winds, though mainly and essentially from the point of view of navigation; e.g. the Hydrographer to the Admiralty, the Hydrographic Office. Ct. HYDROLOGY.

hydrolaccolith A 'blister', approx. 6 m. high, raised by hydrostatic pressure in TUNDRA or PERIGLACIAL areas, when autumn freezing may trap a layer of water between the frozen surface and the underlying PERMAFROST. This may later collapse, leaving a hollow which may become a small pond. Ct. PINGO.

hydrologic(al) cycle The endless interchange of water between the sea, air and land: EVAPORATION from the oceans, movement of WATER-VAPOUR, CONDENSATION, PRECIPITATION, then some surface RUNOFF to the oceans, movement by GROUNDWATER, some EVAPORATION, and some EVAPOTRANSPIRATION. 'All the rivers run into the sea; yet the sea is not full; unto the place from whence the rivers come, thither they return again.' *Ecclesiastes*, i, 7.

hydralogy The scientific study of water, esp. (by usage) inland water, both surface and underground, including its properties, phenomena, distribution, movement and utilization. Ct. HYDROGRAPHY.

hydrolysis A form of chemical weathering, a process of chemical reaction involving water, strictly one in which a salt combines with water to form an acid and a base. E.g. the breakdown of felspar, whereby colloidal silica is removed in solution and clays are formed.

hydrometeor An ensemble of liquid in the atmosphere: RAIN, DRIZZLE, SNOW, SLEET.

hydrophyte A plant which lives in water or saturated soil, such as floating and submerged aquatic plants.

hydrosphere The waters of the surface of the earth, complementing the ATMOSPHERE and the LITHOSPHERE, though the h. and the atmosphere overlap by way of the HYDROLOGICAL CYCLE.

hydrostatic equation In meteorology, the basic relation in the atmosphere between density, pressure, gravity and altitude. This e. can be solved to produce a *barometer-height formula*, relating barometric pressure to temperature and altitude.

hyetograph (i) A type of self-recording RAIN-GAUGE, with a float attached to an inked pen which indicates the amount of water in the gauge; sometimes called a 'tipping-bucket' gauge. (ii) A columnar diagram which shows the max., min. and mean rainfall for each month, also the standard deviation and the probable deviation from the accepted mean. A chart record of the rate of rainfall is a *hyetogram*.

hygrograph An instrument which makes a continuous record of change of the RELATIVE HUMIDITY of the atmosphere by transmitting the record from a HYGROMETER to an inked pen on a rotating drum.

hygrometer An instrument whic measures the RELATIVE HUMIDITY c the air, using a human hair or lithium-chloride strip. The forme lengthens and shortens, and the resistance of the latter varies, these change being amplified and registered. A wet and dry-bulb thermometer (PSYCHRO METER) is another type.

hygrophyte A plant which lives an environment with a plentiful wat supply, such as species in the Tropic Rain-forests, usually with large bro leaves. Hence adj. *hygrophilous, hygr phytic*.

hygroscopic App. to a substan with a marked affinity for water; e particles of salt in the atmosphere which condensation may take pla

hypabyssal rock An intrusion of igneous r. along cracks and lines of weakness, which has solidified as a DYKE or a SILL; intermediate in physical form between deep-seated PLUTONIC r. and extrusive VOLCANIC r. Some of the types of this r. include well-formed crystals of different minerals embedded in a ground-mass of micro-crystalline glassy material; e.g. porphyries.

hypogene Proposed by C. Lyell to signify all rocks formed within the earth, including both plutonic and metamorphic, excluding sedimentary and those formed by the solidification of volcanic material on the surface. It is sometimes used as an adj. (*hypogenic*), in ct. to external or *epigenic* features. This word has not been gen. accepted or used.

hypsithermal Used in U.S.A. to denote the interval of mild climate which followed the last major phase of the Quaternary glaciation. Its temporal span has been defined by RADIOCARBON DATING as between 9000 and 3000 years before the present, though of course this varies with latitude (i.e. is *time transgressive*). In W. Europe the h. period corresponds broadly to the BOREAL and ATLANTIC climatic phases.

hypsographic (or **hypsometric**) **curve** A graph used to indicate the proportion of the area of the surface of an island, continent or the globe at various elevations above, or depths below, a given datum (usually sea-level). The vertical axis indicates heights, the horizontal axis the areas of land. A percentage h. c. uses percentages of the total area for the horizontal axis, instead of absolute areas. [*f*]

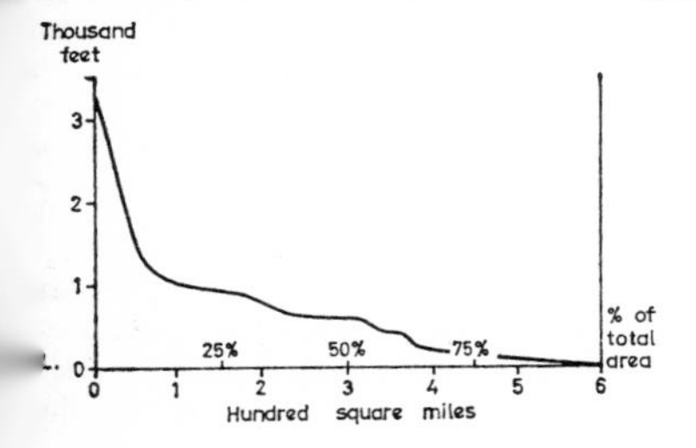

hypsometer An instrument which with the aid of tables, enables altitude to be determined by measuring the temperature at which water boils at that height, since this temperature varies with the pressure of the atmosphere. The boiling-point of water at 1013 mb. is 100°C., at 909 mb. is 97°C., at 728 mb. is 90°C.

hythe, hithe A small haven or mooring point, esp. in a river, commonly found as a suffix in place-names; e.g. Rotherhithe.

hythergraph A climatic diagram plotting temperature against humidity or precipitation, orig. used by T. Griffith Taylor. It summarizes broad climatic differences in relation to human activity, esp. settlement.

ice The solid form of water, formed by: (i) freezing; (ii) the compaction of snow; and (iii) the condensation (SUBLIMATION) of water-vapour directly into crystals. Density 0·9166; i.e. less than water, and thus ice floats.

ice age A geological period of widespread glacial activity, when ice-sheets covered large parts of the continents. In popular usage, the 'I. A.' or 'Great I. A.' signifies the latest (Pleistocene or Quaternary) glaciation. There is considerable argument as to its date of origin, with a disputed range of more than a million years. The 'LONG TIME-SCALE' puts this at about 1·8–2·0 million years ago, the 'SHORT TIME-SCALE' at *c.* 600,000 years ago. At least 3 others have occurred, the Huronian in Canada, the widespread Pre-Cambrian and early Cambrian, and the late Carboniferous which mainly affected the S. continents. The glaciation of the Pleistocene period took place in a number of distinct glacial phases, separated by interglacial periods of warmer conditions. The elucidation of these stages of the Quaternary glaciation in terms of climate and resultant deposits is receiving close attention in current research. As these ice-sheets at their max. covered about

8 million sq. mi., their impress on the landscape through erosion and deposition was considerable.
[*f* MINDEL]

ice-barrier The edge of the Antarctic ice-sheet; e.g. the Ross I.-B.

iceberg A large mass of ice which has been broken off from the tongue of a TIDAL GLACIER or the edge of an ice-barrier, then has floated away under the influence of currents and winds. The ratio of the volume of submerged ice to that above water depends on the relative density of ice and sea-water. It was thought that this ratio was about 6:1, some estimates as much as 9:1, but recent opinion believes that it is actually more like 4 or 3:1. The N. hemisphere i.'s are mostly *castellated*, towering high above the water; 10–15,000 each year are formed, mostly from Greenland, carried S. by the E. Greenland and Labrador Currents. They are rarely found in the Pacific Ocean, as the Bering Strait is shallow and narrow. The S. hemisphere i.'s are mainly *tabular*, large floating islands of ice up to 80–100 km. (50–60 mi.) in length. They break off from the edge of the Antarctic ice-barrier, and float N. as far as about 60°S. in the Atlantic, 50°S. in the Pacific. I.'s may move at about 6 km. per day, depending on the currents and the strength and direction of the wind.

ice blink The glare from the underside of a cloud-layer, produced by the reflection from an i. surface below, as in the case of an ice-sheet or of pack-ice. This may produce eye irritation and even snow-blindness.

ice-cap A permanent mass of ice covering plateaus and high-latitude islands, but smaller than an ice-sheet; e.g. Spitsbergen; Novaya Zemlya; Franz Josef Land; part of Iceland (where 37 named examples have been distinguished); the Jostedalsbre and other examples on the Norwegian plateau.

ice-dam A dam on a river caused by blocks of ice, which may result in widespread flooding in spring and early summer, as along the Canadian and Siberian rivers (also called an *ice-jam*). During the Quaternary glaciation, solid i.-d.'s ponded up water to form PROGLACIAL LAKES. Some present lakes result from ice-damming; e.g. the Märjelen See, Switzerland.

ice-edge The boundary between open water and a mass of floating sea-ice.

icefall A confused labyrinth of deep clefts and ice-pinnacles (SÉRACS), formed by the intersection of CREVASSES where a glacier steepens; e.g. the glaciers moving down from the flanks of Mont Blanc to the Chamonix valley, descending over 1800 m. (6000 ft.) in only 3 km. (2 mi.). The huge Khumbu i. was a major obstacle in the approach of climbers to the S. Col of Everest.

ice-field Strictly a large continuous area of PACK-ICE or sea-ice, by both U.S.A. and British definition. Gen. used much more widely; e.g. the Columbia I.-f. in Jasper National Park, Alberta, Canada, which is 340 sq. km. (130 sq. mi.) in area.

ice-floe See FLOE.

ice-fog A fog consisting of minute ice-crystals suspended in the air, in conditions of calm air and low temperature. If the sun is shining above the fog-layer, the effect is dazzling, and without dark goggles the risk of snow-blindness is great.

ice-front A cliff of ice, the seaward face of a floating mass of ice, such as an ice-shelf or tidal glacier.

ice-jam (i) A mass of broken ice fragments, esp. during spring-melting, jammed in a narrow channel, causing flooding. (ii) A mound of broken ice piled up along the shore of a lake, esp. the Great Lakes of N. America.

Icelandic 'low' The mean sub-polar atmospheric low pressure area in the N. Atlantic Ocean between Iceland and Greenland, most marked in winter. It is not a very intense stationary 'low', but an area of rapid moving individual 'lows', interrupted by occasional periods of higher pressure. Cf. ALEUTIAN LOW.

ice-rind A thin crust of sea-i., formed by the freezing of snow-sludge on the surface of a calm sea.

ice-sheet Any large continuous area of land-ice, incl. ice-caps (e.g. the Greenland Ice-cap); the *Ice Glossary* also includes floating sea-ice, such as ice-shelves. It is preferable to restrict the term ice-cap to smaller masses, and use i.-s. for: (i) the 2 very large existing masses: Antarctica (13 million sq. km., 5 million sq. mi.) and Greenland (1·6 million sq. km., 600,000 sq. mi.); and (ii) the i.-s.'s of the Quaternary glaciation, covering at their max. large areas of N. America and N.W. Europe.

ice-shelf A large floating ice-sheet attached to the coastline; e.g. the Ross I. S.

ice-wedge A wedge-shaped mass of ice in the ground, tapering downwards. If the ground freezes under PERIGLACIAL conditions to low temperatures (*c.* −21°C.), espec. where it is covered with a thick mantle of loose material, cracks will develop. In summer these fill with melt-water, which freezes to form an i.-w. Each winter the cracks and w.'s may enlarge and ultimately they may penetrate to a depth of 11 m. Repeated melting and re-freezing not only causes an increase in size, but also helps to shatter the surrounding material. The sites of former w.'s can often be recognized by curving infilling masses of material outlining their shapes. [*f opposite*]

ichor Derived from the ethereal fluid which was believed to fill the veins of the Gk. gods, and used to denote the MAGMA which penetrates and circulates through existing rocks, thus causing *granitization*. See also MIGMATITE and GRANITE.

iconic model A m. in which some aspect of the physical world is presented, showing the same properties, though reduced to scale; these m.'s may be either *static* or *working*. An i. m. cts. with an *analogue* or *simulation m.*, in which the actual properties are represented by different, though analogous, properties.

icing The accumulation of a thickness of clear ice on exposed objects, caused by the freezing of SUPERCOOLED droplets in a cloud, as on the leading edges of the wings of aircraft, or by rain falling on an aircraft flying in sub-freezing-point temperatures, though not in a cloud; ice can also form on branches, telegraph wires (which may be brought down by the weight), on the surface of a road, and on railway lines and points; the last is an espec. difficult problem if the i. occurs on the conductor rail of an electrified line. Cf. GLAZED FROST, RIME.

ideographic An approach to Geography whereby individual cases and situations are studied, rather than those of a gen. type; ct. NOMOTHETIC.

igneous rock A r. which has been formed by the solidification of molten r. material or MAGMA. Its character depends on: (i) the *chemical composition* of the magma, whether: (*a*) it is ACID (granite, rhyolite, obsidian); (*b*) BASIC (gabbro, dolerite, basalt); or (*c*) INTERMEDIATE (diorite, andesite);

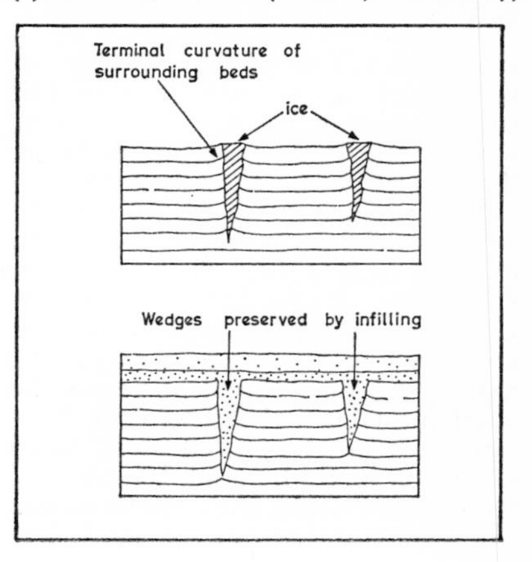

(ii) the *mode of cooling*, whether: (*a*) at depth within the crust, slowly and therefore large-crystalled, hence intrusive or PLUTONIC (granite, diorite, gabbro, peridotite); (*b*) on the surface, rapidly, and therefore fine-crystalled or glassy, hence extrusive or VOLCANIC (rhyolite, obsidian, andesite, basalt); or (*c*) intermediately, hence HYPABYSSAL (granophyre, porphyries, dolerite). Some i. r.'s may be formed from consolidated fragments of pre-existing ones (BRECCIA, TUFF).

IGY The most recent International Geophysical Year, which lasted from 1 July 1957 to 3 Dec. 1958.

Illinoian In America the 3rd main glacial advance, corresponding to the RISS in the Alps. First used by T. C. Chamberlin in 1896. Its deposits consist of a clay-till, widely spread over Illinois, with extensions into Indiana, Wisconsin and E. Iowa. Much of this is covered with LOESS. In N.W. Illinois and S. Wisconsin the drift is stony, with areas of KAME-MORAINE, and the much eroded remnants of TERMINAL MORAINE can be distinguished.

illuviation The redeposition at a lower depth (usually in the B-HORIZON), sometimes forming a HARD-PAN, of material removed from the upper soil horizon by ELUVIATION, including colloids, salts and mineral particles. [*f* SOIL PROFILE]

imbricate(d) structure Wedges of rock piled up on one another like roof-tiles, as a result of an extreme form of an OVERTHRUST; the rocks are thrust over in 'slices', without folding, each separated by a THRUST-PLANE, with which each 'slice' makes an oblique angle. E.g. in the N.W. Highlands of Scotland, where the Cambrian quartzites and limestones are sliced in this way, between the underlying 'Sole' thrust-plane and the overlying Glencoul thrust-plane. This can be seen very well near Loch Glencoul, in Sutherland. The 'slices' are inclined to the S.E., the direction from which the pressure came. I. s. is known in Germ. as *Schuppenstruktur*. [*f*]

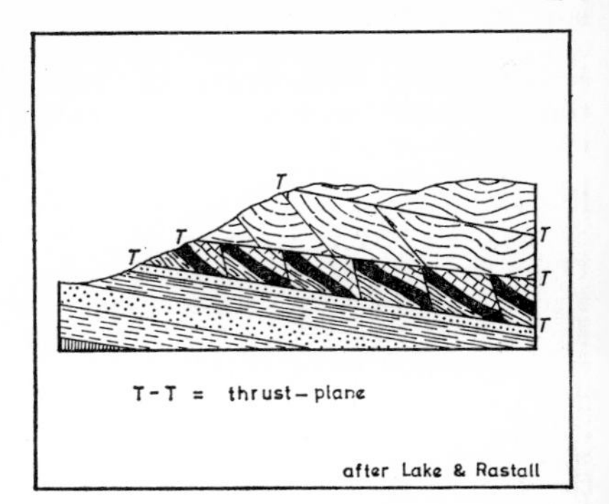

immature soil See AZONAL S.

immigrant A person (sometimes used of plants and animals) who moves into another country to settle. Some countries have been largely populated by immigration; e.g. the U.S.A., where 40 million people entered between 1820 and 1955, including 35 million from Europe. However, certain restrictions became desirable, esp. after the war of 1914–18, and a quota system was established. Immigration restrictions have also recently been established in Britain.

impermeable The quality of not allowing the passage or transmission of water through a rock, because it is either *non-porous* (with two exceptions), or IMPERVIOUS, or both. The exceptions are clay (which is porous but becomes i. because of its fine pores, which when wet are filled with water held by surface tension, so sealing the rock against the passage of water), and unjointed chalk.

impervious App. to a rock which is not pervious; i.e. it does not allow the passage of water because it has no joint-planes, cracks or fissures; e.g. slate, shale, gabbro, massive granite.

improved land An area of l. that has been made more productive by drainage, reclamation, fertilizing and marling.

inch (i) See LENGTH. (ii) (Derived from the Gaelic *innis*). A small rocky island; e.g. I. Marnock, in the Sound of Bute, W. Scotland. (iii) A flat area of alluvium in a river valley, which at times may be flooded; e.g. Inches of Perth in the Tay valley.

incised meander The down-cutting of a river due to REJUVENATION which maintains the pattern of a m. at a progressively lower level below the general surface; e.g. the rivers Wye, Dee, Meuse. A distinction is usually made between INTRENCHED and INGROWN M'.s, though most i. rivers show both types at different stages of their courses. [*f*]

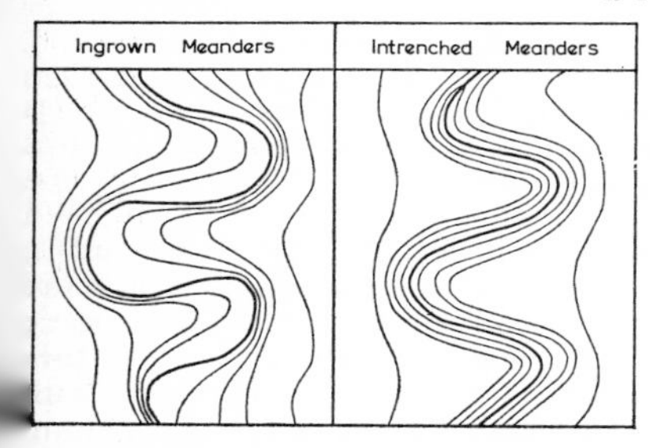

nclination-dip The angle of d. of VEIN, FAULT or STRATUM, as measured rom the horizontal.

nclosure The process of terminatng common rights over a piece of nd, and turning it into freehold land y means of a legal award, hence 'I. ct'. Strictly this may be distinguished om ENCLOSURE, which is merely the rocess of putting a fence round a iece of land without necessarily any gal justification or sanction.

competent bed A rock-stratum at is so weak that it would be comexly distorted by plastic flow and formation, rather than by bending, ring folding movements. Such a b. ay thicken at the hinge of a fold, d thin out in its limb; e.g. the pper Jurassic rocks along the Dorset ast, where these were folded to m the N. limb of the Ringstead ticline.

consequent drainage A d. patn which has not developed in a systematic relationship with, and is not consequent upon, the present structure; it is either ANTECEDENT or SUPERIMPOSED, and the river is said to be DISCORDANT; cf. INSEQUENT, where the drainage pattern appears to be independent of surface features. [*f* ACCORDANT DRAINAGE]

index fossil A f. characteristic of a special zone in the rocks, which is named after it.

Indian summer A spell of fine sunny weather which seems to occur with some regularity during October and early November in U.S.A., said to be used by the Indians for storing crops and making ready for the winter. The term is sometimes used in Britain for a much earlier spell of fine weather in early September, which frequently occurs as a result of an extension of the Azores anticyclone. According to C. E. P. Brooks, it could be recognized in 43 out of 52 years, the peak being 10 Sept.

indifferent equilibrium See NEUTRAL INSTABILITY.

indigenous Originating in a specif. place and remaining *in situ*, used of plants, animals, population. Sometimes applied to rocks, minerals and ores originating *in situ*, and not transported there from elsewhere.

induced expansion In town planning, a planned and directed e. of population in some specif. area, helped by the establishment of some specif. attraction or inducement.

induration Hardening, used of rocks so transformed by heat, pressure or cementation, and of soil HORIZONS hardened by chemical action to form a HARDPAN.

industrial archaeology The study on the ground of the remains of past i. activity; e.g. of Elizabethan mining in the English Lake District; the early I. Revolution.

industrial inertia The tendency for an industry to maintain its activity in a place where the reasons for its introduction are no longer significant or even no longer exist, as when sources of fuel, ore or other raw material are exhausted, or where they can now be obtained more economically from elsewhere; e.g. the Sheffield steel industry; the Liège zinc industry.

industrial linkage An aspect of the study of manufacturing location, involving all exchanges between i. plants and the factors, material and non-material, involving: (i) *process* (the movement of goods between different firms as stages in manufacturing); (ii) *service* (the supply of machinery, equipment and tools); (iii) *marketing* (links with firms concerned with the distribution and sale of goods); and (iv) *financial* (links with banks, stockbrokers, etc.). *Operational l.'s* are associated with sources of raw material, labour, marketing areas, land ownership, planning regulations, etc.

industrial location theory The study of the l. of industry, using DATA BANKS to provide the available information, which can then be processed (e.g. by linear programming) to find the 'best l.'s' (e.g. for a new factory). This involves such factors as labour supply, suitable site, supply of raw materials, nature and extent of marketing area, or (in an assembly industry) the availability of components.

industrial momentum The tendency for places with established industries to increase in importance after the initial conditions leading to their original establishment have changed, the result of their reputation and goodwill, specialized labour supplies, site values, and immovable large-scale equipment; e.g. the tobacco industry in Bristol. This is very similar to INDUSTRIAL INERTIA, but the term momentum implies an actual increase in importance, inertia a mere maintenance of activity.

industry In its widest sense, any work done for gain; in its narrow sense, mining and manufacture, as ctd. with agriculture, commerce and personal service. The Ministry of Labour in the U.K. distinguishes between 'Manufacturing I.'s' and 'Basic I.'s', the latter comprising Coal Mining, Quarrying, Public Utilities (gas, water, electricity), Transport and Communications, Agriculture and Fishing. Frequent sub-divisions are distinguished, according to their nature; e.g. 'servicing i.', 'extractive i.', 'light' and 'heavy i.', 'metallurgical i.', 'textile i.', servicing i.'s. It may also be classified as (i) PRIMARY; (ii) SECONDARY; and (iii) TERTIARY I.

infant(ile) mortality The average number of deaths of infants under 1 year of age per 1000 live births. A high i. m. is usually an indication of limited medical services, malnutrition and general underdevelopment.

infield Farmland lying around a farmstead, as opposed to the more distant 'outfield'. A characteristic infield-outfield system of agriculture has been developed in Scotland, with a group of small enclosed fields around a croft and extensive areas of rough grazing beyond; similarly in the hill country of N. England.

influent stream A river whos[e] channel lies above the water-table[,] e.g. a s. flowing in a chalk valley[,] which only maintains its flow becaus[e] its bed may be lined with fine sil[t], thus forming an impermeable laye[r]. Some of these rivers may lose part [of] all of their water down crevices or joint[s] in limestone and chalk; e.g. the [R.] Mole near Dorking; the Mymmsha[ll] Brook in Hertfordshire.

ingrown meander An INCISED M[EANDER] where the valley-sides are asym[metrical], the result of considerab[le] lateral erosion, so that one side of t[he] valley presents steep, even underc[ut] banks, while the other side is qu[ite] gentle; e.g. the lower Seine, whi[ch] flows 90 m. (300 ft.) below t[he] flanking chalk plateaus; the low[er] Ribble, Lancashire. Ct. INTRENCH[ED] M. [*f* INCISED MEANDE[R]

initial landform A l. where the original features, as formed when uplifted by various TECTONIC forces, have been only slightly modified, as opposed to SEQUENTIAL L.'s.

inland basin See BASIN.

inland drainage See INTERNAL D.

inland sea An isolated sheet of water of large extent, which has no link with the open s.; e.g. the Caspian, Aral, Dead S.'s.

inlet A narrow opening of the sea into the land, or of a lake into its shores.

inlier An exposed mass of rocks wholly surrounded by younger ones, commonly occurring when the crest of an anticline or an elongated dome is removed by erosion; e.g. a small i. of Jurassic rocks occurs in the centre of the Weald; an i. of Silurian rocks is at Woolhope, Herefordshire. [*f*]

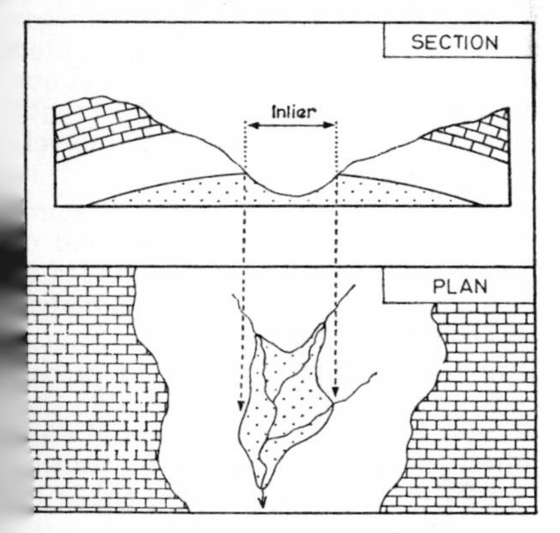

ıner lead A stretch of calm, ıeltered water between the coast of mainland and a string of parallel ffshore islands; e.g. the skerry-guard Norway; along the coast of British olumbia (the line of Georgia, Queen harlotte, Hecate, Clarence, Chatham raits).

put-output analysis An attempt trace in quantitative terms the w of goods between industrial ctors (an important tool used in the ıdy of economic regions), and to use employed labour (labour 'input') as a criterion of the concentration of manufacturing, taking into account different regional levels of labouring productivity (manufacturing 'output').

inselberg (Germ.) A prominent residual rock-mass in the desert, isolated by circumdenudation, usually smoothed and rounded, rising from a distinctive near-level erosion surface. It is believed to develop from the extension of PEDIMENTS into upland areas, in time reduced to PEDIPLAINS. The mountain-front on each side recedes through SLOPE RETREAT, ultimately reducing the intervening upland to an i. Some writers suggest that certain i.'s result from the exposure and reshaping of 'cores' of resistant rock revealed by the removal of a deep layer of weathered REGOLITH. An i. is a characteristic landform of semi-arid or SAVANNA landscapes in a late stage of the CYCLE OF EROSION.

insequent drainage A d. pattern which has developed on the present surface, but which appears to reflect chance development, and is not consequent upon it nor controlled by features of the surface; e.g. DENDRITIC DRAINAGE.

insolation The energy emitted by the sun which reaches the surface of the earth. The sun, a mass of intensely hot gases, with a temperature at the surface estimated to be 5700°C., and at the centre 45,000,000°C., pours out radiant energy in the form of waves. These consist of very short wave-length X-rays, gamma rays and ultraviolet rays; the visible light-rays; and the longer infrared rays. The earth receives only about one 2-thousand-millionth of the total i. poured out by the sun, but this is vital to it; the amount received at the outer limit of the atmosphere is the SOLAR CONSTANT. This enters the earth's atmosphere, which absorbs part; some is lost by 'scattering' caused by air molecules, dust particles and water-vapour; part is reflected back into space by clouds and dust. About 45% reaches the earth, of which nearly a ¼ is immediately reflected back into space, depending on the nature of the surface. The remainder is converted into long-wave heat-rays, which heat the surface of the earth and (by conduction) the layer of air resting upon it. See also ALBEDO.

instability, atmospheric The physical condition whereby a mass of air is warmer, and therefore less dense, than the air above, causing bodily rising and expansion. The vertical movement of 'parcels' of air of differing temperature and humidity, the result of i., is intimately associated with atmospheric disturbances, and is the most potent cause of precipitation.

installed capacity Used esp. of hydro-electricity plant, giving its total c. when it is fully utilized; the actual output varies as between peak and offpeak loads. I. c. is given in kilowatts or megawatts.

insular App. to an island; hence i. location, I. CLIMATE.

insular climate A climatic régime experienced by islands and coastal areas, characterized by equable c.'s, with a low seasonal range of temperature; e.g. Ocean I., 1°S. in the W. Pacific, has an annual range of only 0·01°C., and Jaluit, in the Marshall I.'s. (6°N.), of 0·3°C. Even in middle latitudes, such i. c.'s have relatively small ranges; e.g. Scilly I., Jan. 7°C. (44°F.), July 16°C. (61°F.), (range 9°C., 17°F.), and frost is almost unknown.

intaglio See PHOTOGRAVURE.

intake An area of land taken in from moorland or upland, usually fenced or enclosed, and improved by draining, liming and fertilizing.

intensify An increase in the amount of some physical quantity. In meteorology, it is used of the gen. strength of air-flow around a low pressure system; i.e. a depression intensifies.

intensity of rainfall The rate at which rain falls, ranging from a fine drizzle to a torrential downpour. The hourly i. of r. is obtained from $\frac{\textit{total rainfall}}{\textit{no. of hours of rain}}$. The less specific daily i. of r. is $\frac{\textit{total rainfall}}{\textit{no. of}\ \text{RAINDAYS}}$. The hourly i. of r. at Boston, U.S.A., is 9·1 mm. (0·36 in.); at Cherrapunji in N.E. India is 106·0 mm. (4·17 ins.). The Island of Réunion once experienced a total of 1168 mm. (46 ins.) in 24 hours.

intensive cultivation The application of capital and labour to a relatively small piece of land in order to obtain high yields, involving heavy fertilization, sometimes irrigation, and the production of several crops (including CATCH-CROPS) per season; e.g. market-gardening near a large city; bulb- and flower-c. in S. Holland; vineyards in France; c. of vegetables and fruit at a N. African oasis and in the Nile delta and valley. The term is also used in a wider sense for rice-c. in S.E. Asia.

interactance The activities which result from the contacts and relationships between 2 separate (i.e. divided by a political boundary) yet contiguous groups of people. S. C. Dodd (1950) produced a GRAVITY MODEL and developed the i. hypothesis.

interchange In a road system partic. of MOTORWAYS, the system o links at different levels and grades t provide access between the major roa and its connections.

interdependence Used in urba ecology to describe the adaptation tha occurs between units in any are These adaptations result in mutu benefit to the several units, either b SYMBIOSIS or COMMENSALISM.

interface A zone of sharp chan between 2 different substances, between 2 different characteristics the same substance; e.g. in a crystal.

interfluve The area of land betwe 2 rivers.

interglacial period A p. of tir between 2 advances of the co tinental ice-sheets during the Quate nary glaciation; commonly 'inte glacial' alone is used as a nou Between these advances, the clim was milder and the ice-sheets shrar from the evidence of plant remains (rhododendron, box) some i. p.'s were possibly warmer than at present. The 1st i. p. (Günz-Mindel in the Alps) lasted about 60–80,000 years, the 2nd or Great I. (Mindel-Riss) about 190–240,000 years, and the 3rd (Riss-Würm) about 60,000 years; this is on the 'SHORT TIME-SCALE'; see PLEISTOCENE. The corresponding periods in N. America are known as the *Aftonian*, *Yarmouth* and *Sangamon* i.'s respectively. [*f* MINDEL]

intergranular translation The movement of ice-grains within a mass of glacier-ice, so that they slide over each other (as in the case of a mass of lead shot) and thus contribute to the overall glacier-movement; one of the contributory categories to the gravity flow of a glacier.

interior drainage See INTERNAL D.

interlocking spur A projecting s. in a 'young' river valley, where a stream tends to follow a winding course round obstacles, each bend separated by an i. s.; viewed upstream, these s.'s 'interlock', 'overlap' or 'interdigitate' with each other. [*f opposite*]

Intermediate Area See GREY AREA.

intermediate layer Formerly given to, and still occasionally used for, the SIMA l. in the earth's crust.

intermediate rocks Igneous r.'s classified according to chemical composition as lying between the ACID and BASIC groups; i.e. those with a silica content of 52–65%; e.g. trachyte, andesite (EXTRUSIVE), diorite, syenite (INTRUSIVE).

intermittent saturation, zone of A layer, just below the surface soil, which may contain GROUND-WATER after long-continued rain (VADOSE WATER), but which will dry out after a short period of drought.

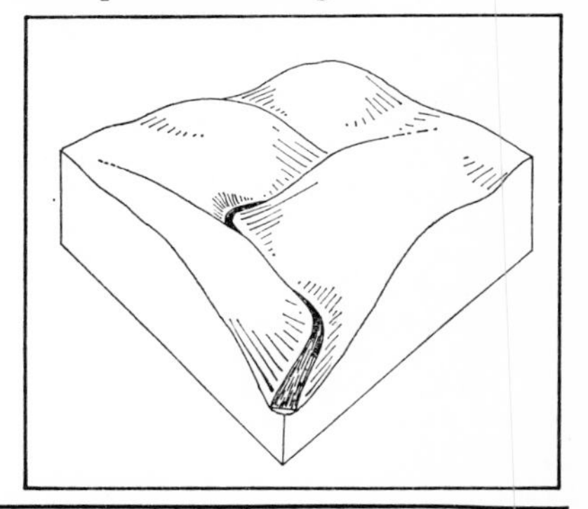

intermittent spring A s. whose discharge varies intermittently, usually the result of fluctuations in precipitation and therefore of the height of the WATER-TABLE. Sometimes it may be the result of the presence of a SIPHON in a cave-system, which opens out to the surface at the source of the spring; e.g. the Ebbing and Flowing Well, Settle, Yorkshire. [*f*]

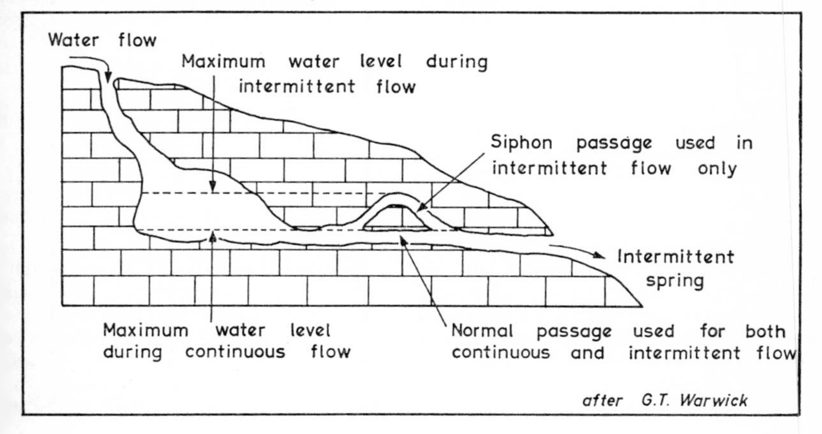

intermont, intermontane Between the mountains, as of a high plateau between ranges; e.g. the Bolivian and Peruvian Plateaus between the E. and W. ranges of the Andes; the Great Basin of Utah between the Sierra Nevada and the Wasatch; the Tarim Basin between the Altyn Tagh and the Tien Shan.

internal drainage A d. system whose waters do not reach the ocean. It is also known as *inland* or *aretic d.*; e.g. the R. Jordan and the Dead Sea; the R. Volga and the Caspian Sea; the Amu and Syr Darya and the Aral Sea; the Tarim R. in the Lop Nor; the many temporary streams converging on L. Eyre and L. Chad.

International Date-Line A line following approx. the 180° meridian (with sundry deviations to avoid some land areas), where, to compensate for the accumulated time-change of 1 hour in each 15° longitude time-zone, a ship sailing W. (e.g. San Francisco to Tokyo) omits a day (i.e. Monday is followed by Wednesday, the 27th by the 29th), while one travelling E. repeats a day (i.e. Monday is followed by another Monday, the 27th by the 27th). [*f opposite*]

International Million Map A series of m.'s on a scale of 1:1 million (15·782 miles to 1 in.), first proposed in 1891 by A. Penck; the principle was accepted in 1909. Each sheet covers 4° of latitude and 6° of longitude (except in high latitudes, where it is 4°×12°), reckoned from the Greenwich meridian and the equator, and is known by an index letter (alphabetically from the equator) and a number reckoned E. from Greenwich, prefixed by N. or S. according to the hemisphere. The m. is drawn on a modified POLYCONIC PROJECTION. The headquarters of the I. M. Bureau were established at Southampton. Progress was slow and sporadic; by 1931 only 89 sheets had been produced, though others on a comparable scale were published by private bodies (e.g. the American Geographical Society were responsible for 16 maps of S. America), and both the U.S. Army Map Service and the British G.S.G.S. published sheets, esp. during the war of 1939–45. After post-war discussions, the functions of the Central Bureau were transferred to the U.N. Cartographic Office, which gives a summary of progress in its annual report. Modifications have been made to the specifications of the m.

interpluvial periods Those stages in low latitudes (beyond the reach of GLACIATION) which correspond to, though are not necessarily synchronous with, the interglacial periods of higher latitudes. However, the correspondence between *pluvials* and *glacials* may be only approx.

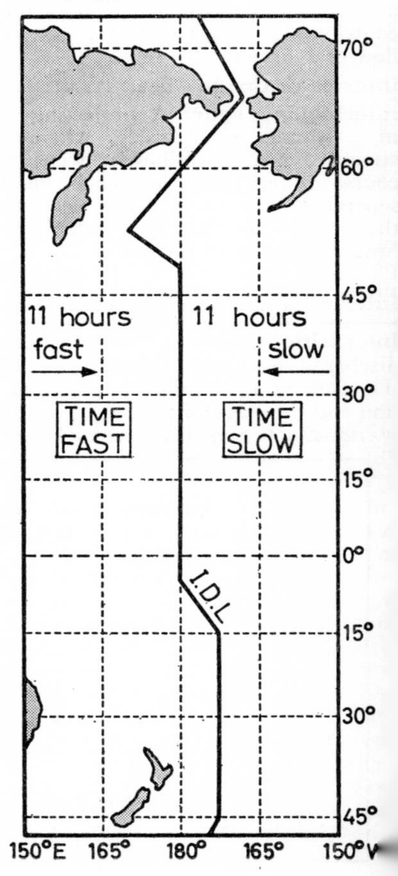

interruption of, interrupted, map projection A p. may be drawn with several standard meridians instead of a single one; each is centred over a continent, with true scale along it, and a section of the p. in a lobate form is plotted from there. Gaps are left occurring in the oceans between each re-centred section to accommodate the errors. This reduces the overall distortion of shape, esp. towards the margins, and the linear scale discrepancy, while retaining the equal area property. I. p.'s are widely used for world distribution maps. E.g. MOLLWEIDE, SANSON-FLAMSTEED, GOODE'S I. HOMOLOSINE P.'S. Several i. p.'s have been specially designed for *The Times Atlas*; e.g. *Bartholomew's Regional P.* (conformal and near equal area, designed to show continental relationships), and the *Lotus P.* (to show oceans, currents, etc.). [*f*]

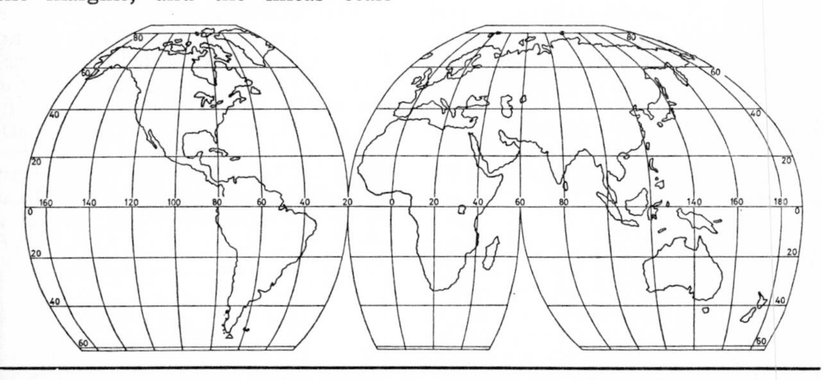

interstadial Lit. between stages, partic. used with respect to periods of time between 2 glacial stages or 2 phases of glacial retreat, in which case it is commonly used as a noun (an 'i.'). It is of a more minor character, or is less marked, or of briefer duration, than an INTERGLACIAL. Thus an i. oscillation occurred in each of the first 3 glaciations of Quaternary times, and at least 2 in the 4th (WÜRM) glaciation.

Inter-Tropical Convergence Zone A broad trough of low pressure, lying more or less along the axis of the equator, though moving slightly N. and S. with the seasons, towards which the trade winds blow, i.e. towards which Tropical Maritime (*Tm*) air-masses converge. It is usually abbreviated to *I.T.C.Z.* This boundary z. was formerly known as the Inter-Tropical Front (*I.T.F.*), but this has gen. been replaced, since the phenomenon is usually zonal rather than frontal in character, esp. over the oceans, where the *I.T.C.Z.* is only weakly defined. The air-masses may be virtually stagnant, i.e. in the DOLDRUMS, and the winds are usually light or variable. The air is very unstable, and convectional rainfall is common. In this region shallow, slow-moving 'LOWS' develop; sometime these move poleward out of the z., intensify and develop into intense tropical storms; see CYCLONE. Over a continental area, the front may be more sharply defined; temperature differences in the converging air-streams are small, but humidity may differ markedly; e.g. in W. Africa a dry continental air-mass from the Sahara may come into contact with a humid one from the Gulf of Guinea, causing a distinct front to develop.

intrazonal soil A s. type which is not restricted to a latitudinal zone, but is the result of partic. conditions or constituents. It includes: (i) *peat s.'s* (fen-peat, bog-peat, dry-peat, meadow-soils); (ii) *saline s.'s* (e.g. SOLONCHAK, SOLONETZ); (iii) *calcareous s.'s* (e.g. RENDZINA, TERRA ROSSA).

intrenched meander An INCISED M. where the valley sides are steep and symmetrical, where vertical erosion has been dominant; e.g. the R. Wear near Durham. Ct. INGROWN M. [*f* INCISED MEANDER]

intrusion, vb. **to intrude,** adj. **intrusive** (i) The penetration or injection of molten rock (MAGMA) into existing rocks, usually along a line of weakness. (ii) The particular form of the mass of solidified igneous rock thus formed; see BATHOLITH, DYKE, LACCOLITH, LOPOLITH, PHACOLITH, SILL. [*f in each case*]

intrusive rock A category of igneous r.'s (as opposed to *extrusive r.*), which solidified at depth among the pre-existing r.'s.

invar An alloy of steel containing 36% nickel, invented by C. E. Guillaume. It has an extremely low, almost undetectable, coefficient of thermal expansion, and is therefore used for surveyors' tapes, esp. for measuring base-lines of triangulation systems.

inversion of temperature An increase of air t. with height, so that warmer air overlies colder, contrary to the normal LAPSE-RATE. It can occur both at the surface of the earth (*ground-i.* or *surface-i.*) and at high altitudes. (i) A *ground i.* may be the result of rapid heat-loss from the ground by radiation at night, esp. when the air is calm and the sky clear, or by warm air advection over a cold surface. I. may be partic. marked in a valley or basin, for radiation of heat into space on the upper slopes is rapid, and the cold dense air then drains down into the hollows. (ii) A *high altitude i.* may occur when a cold air-mass undercuts a warm one (at a COLD FRONT), or a warm air-mass overrides a cold one (ahead of a WARM FRONT), or develops into an OCCLUSION. The subsidence and ADIABATIC heating of a high altitude mass of air may cause an i. This is esp. important in ANTICYCLONES; e.g. the Trade Wind i.

inverted (inversion of) relief The result of prolonged denudation, by which SYNCLINES may ultimately form high ground and ANTICLINES form low ground. The crest of an anticline is commonly structurally weak, having been subject to tension, and a stream may develop along it, forming a valley with infacing steep slopes; these will gradually recede outwards, so reducing the area of the orig. adjacent synclinal valley. In due course the anticlinal valley will lie at a level below the synclinal valley, and the remains of the syncline will remain as a ridge or peak; e.g. the mountains E. of the Bow R. in Alberta (such as Mt. Eisenhower), which are the remains of a syncline; Snowdon in N. Wales is the remnant of a great SYNCLINORIUM; the Great Ridge on Salisbury Plain lies on a synclinal axis. [*f*]

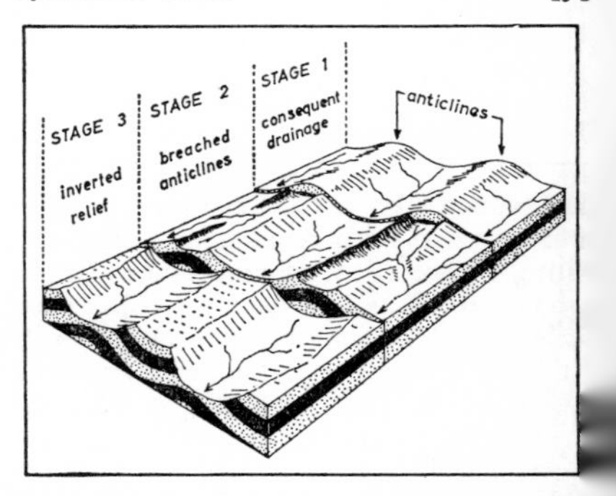

inverted (inversion of) structure When stratified rocks have been acutely overfolded or overthrust, the orig. older (lower) beds may now rest upon the younger (upper) beds; e.g. in the Ardennes, where in places Cambrian rocks lie on top of Devonian; in the Meuse valley in Belgium where Devonian, Carboniferous Limestone and Lower Coal Measures have been thrust over younger coal-bearing Upper Coal Measures; in Glacier National Park, Montana, U.S.A. where Pre-Cambrian rocks rest on Cretaceous rocks, the result of the great Lewis overthrust from the W.

invisible exports Credit values in

country's balance sheet, not derived from e.'s of actual goods, but from payments for services such as banking, insurance, shipping, tourism and interest on investments.

involution (i) The evidence of frost-heaving and disturbance in the upper layers of the soil. Broadly syn. with CRYOTURBATION. (ii) The modification of NAPPES after their formation, either by the re-folding together of 2 nappes, or the penetration of the older by the younger nappe.

ion An atom or group of atoms in an electrically unbalanced form, with either an excess (*positive i.* or *cation*) or deficiency (*negative i.* or *anion*) of electrons; e.g. the hydrogen atom without its electron is a hydrogen ion. Almost all matter owes its existence to i.'s, held together by an ionic bond, as in the case of compounds, formed by combinations of elements. I.'s are partic. important in soil chemistry, since COLLOIDS are able to attract and hold i.'s of calcium, magnesium and potassium (bases), and i.'s of hydrogen. The degree to which the hydrogen i.'s are held by the soil-colloids is the pH VALUE.

ionosphere (syn. with *thermosphere*) The portion of the atmosphere above the STRATOSPHERE characterized by distinctive layers (indicated by letters D to G), which reflect electromagnetic waves (including radio signals) back to earth, and the zone in which the AURORA occurs. In this zone the atmospheric gases are ionized by incoming solar radiation. The i. includes the HEAVISIDE-KENNELLY LAYER at 100–120 km. (60–75 mi.) and the APPLETON LAYER (*c.* 240 km., 150 mi. up). The lower level of the i. sinks during the daytime to about 56 km. (35 mi.), rising to about 100–105 km. (60–65 mi.) at night. Recent practice avoids the term, substituting the lower EXOSPHERE and the upper MAGNETOSPHERE.

Iowan In N. America the earliest sub-stage of the Wisconsin or 4th main glacial advance, corresponding to Würm I. in the Alps and to the Brandenburg stage of the Weichsel in N. Europe. The term was first used by T. C. Chamberlin in 1894. The drift consists of a stony yellowish-brown clay, and has been identified in N. Iowa; it is a thin deposit, usually less than 6 m.

irisation Colour sometimes seen on clouds, mingled or banded, through which the sun may be partially seen.

iron ore I. is the 2nd most wide-spread metal (after ALUMINIUM), estimated to make up 5·06% of the earth's crust by weight. The chief o.'s are: (i) *haematite* (red ferric oxide, Fe_2O_3), found in the 'i. ranges' near L. Superior (U.S.A.), in a slightly different form at Krivoi Rog (U.S.S.R.), in Spain (Bilbao), and (nearly exhausted) in W. Cumberland; (ii) *limonite* (hydrated ferric oxide, $2Fe_2O_3.3H_2O$), mainly of Jurassic age, found in French Lorraine; (iii) *magnetite* (black oxide of i., Fe_3O_4), in N. Sweden; and (iv) *siderite* (ferrous carbonate, $FeCO_3$), in the Coal Measures and Jurassic limestones of England. The o. is smelted in a blast-furnace to produce pig-i., and is turned into steel in either: (*a*) a *Bessemer converter*; or (*b*) an *open-hearth furnace*; or (*c*) an *electric furnace*. C.p. = U.S.S.R., U.S.A., China, Canada, France, Sweden, India, Brazil, Venezuela, Liberia, Australia.

ironpan See HARDPAN.

irrigation The artificial distribution and application of water to the land to stimulate, or make possible, the growth of plants in an otherwise too arid climate. I. may be: (i) *basin* or *flood*, in which water brought by a river in flood is held on the land in shallow basin-shaped fields surrounded by banks; e.g. in Egypt, along the banks of the Nile and in the delta; (ii) *perennial*, or 'all-year' i., where water can be lifted on to the fields from a low-level river, a well, a 'tank' or small reservoir. Primitive devices (*shaduf*, *sakiyeh*) have been long used, operated by man- or animal-power, later by windmills, and recently by steam, petrol or diesel pumps. Perennial i. is esp. possible where a supply of water can be obtained from the mountains, esp. from snow-melt. Modern perennial i. involves large dams or barrages to hold back a great volume during river-floods, which can be released through aqueducts and canals as required; e.g. the Aswan Dam on the Nile, the Sennar Dam on the Blue Nile, the Lloyd Barrage on the Indus, the Hoover Dam on the Colorado, Grand Coulee on the Columbia. Many of these are multi-purpose; i.e. are also concerned with hydro-electricity production, flood control and navigation.

isallobar A line on a map joining places where the same change in atmospheric pressure has taken place during a specific time. They are plotted in order to reveal to the forecaster how a pressure system is developing and moving.

isanomal, isanomalous line, A line on a map joining places with an equal difference (or *anomaly*) from the mean or normal of any climatic element; specif. the difference between mean temperatures for individual stations and the average of all available stations in the same latitude. Tables of standard distribution of temperature with latitude are available, or the following formula (by J. D. Forbes) can be used: $t = -17{\cdot}8 + 44{\cdot}9 \cos^2 (\theta - 6{\cdot}5)$°C., where θ is the latitude and t is the standard temperature. Anomalies are plotted and i.'s are interpolated. Areas of high positive anomaly are *pleions* or *thermopleions*, of high negative anomaly are *meions*, *antipleions* or *thermomeions*. On a map, areas of positive anomaly are conventionally tinted red, negative ones blue. The map emphasizes the winter cold of continental interiors, the summer heating of land-masses, and the effects of the oceans.

island A piece of land surrounded by water, smaller than a continent; the diminutive or poetic is isle, sometimes islet. The largest i.'s are Greenland (2·2 million sq. km., 840,000 sq. mi.), New Guinea (820,000 sq. km., 317,000 sq. mi.), and Borneo (750,000 sq. km., 287,000 sq. mi.).

iso- Derived from the Gk. word meaning 'equal', and used as a prefix to various lines on maps linking points with similar values or of similar quantities; they are known gen. as *isopleths*, *isarithms*, *isontic lines*, *isograms* or *isolines*. A formidable terminology has developed, which has by no means received universal acceptance. The following terms may be regarded as standard: ISALLOBAR, ISANOMAL, *isobar* (pressure), *isobase* (elevation or depression of land), *isobath* (depth), *isobront* (places having thunderstorms at the same time), *isocheim* (winter temperature), *isochrone* (travelling time), *isocryme* (coldest period of time), *isogon* or *isogonic line* (magnetic variation) [*f* AGONIC LINE], *isohaline* (salinity), *isohel* (sunshine), *isohyet* (rainfall), *isohypse* (contour), *isoikete* (degree of habitability), *isokinetic* (equal wind speed), *isomer* (the mean monthly rainfall as a percentage of the average annual amount), *isoneph* (cloudiness), *isonif* (amount of snow), *isophene* (flowering dates and other botanical and biological occurrences), *isophyte* (heigh of vegetation), *isopore* (annual chang of magnetic variation), *isoryme* (frost) *isoseismal* (earthquake intensity), *isostade* (significant dates), *isotach* (equa distance travelled in a period of time) *isotherm* (temperature). As man stations as possible are plotted, the values noted, an isopleth interval i chosen, and lines are drawn throug stations with the selected values or ar interpolated proportionally betwee them.

isocline, isoclinal folding A seri of tightly packed parallel overfold all limbs dipping at approx. the san angle, and in the same direction; e. in the S. Uplands of Scotland. If th folding is still more intense the folds a

ruptured and become IMBRICATED. [*f*]

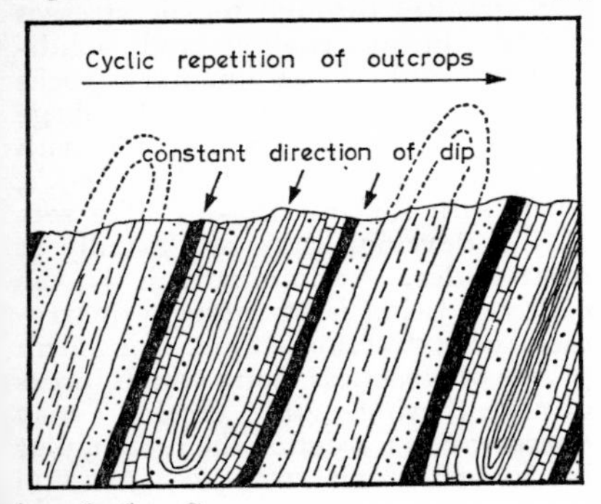

isopleth See ISO-.

isometric A type of graph paper with 3 principal axes (vertical and 2 diagonals), thus dividing the surface into equilateral triangles. This can be used in producing simple pseudo-perspective block-diagrams. [*f*]

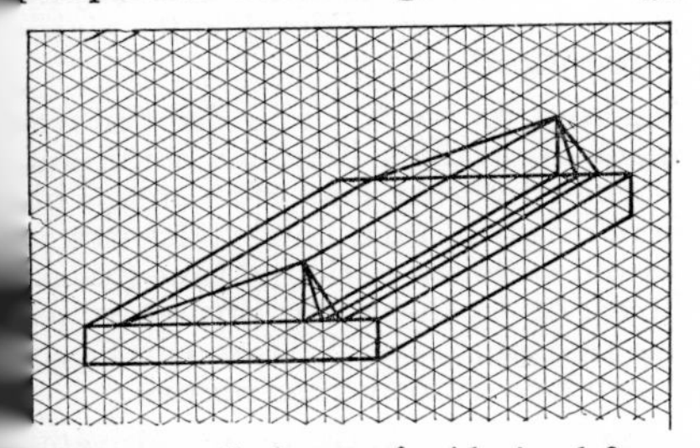

isostasy, adj. **isostatic** (derived from Gk. words meaning 'equal standing') Coined by C. E. Dutton in 1889, signifying a state of equilibrium or balance in the surface crust of the earth, the result of the necessity for equal mass to underlie equal surface area. Thus where mountains (of less dense SIAL) rise high above the mean surface of the GEOID (i.e. excess mass), their roots of similar, less dense, sial must penetrate more deeply into the denser SIMA, thus exercising i. compensation. Under the oceans the deficiency of mass is compensated by the higher density of the sima. It is as though the continents are lighter 'rafts' of sial 'floating' on the dense underlying sima. I. involves: (i) vertical adjusting movements of sections of the crust; e.g. as an ice-sheet melts, the continent rises; as a mountain mass is eroded, the continent rises; and as a great weight of sediment is deposited on a delta, there is a sinking movement; and (ii) horizontal movement of sub-surface material to balance the differences in mass. [*f*]

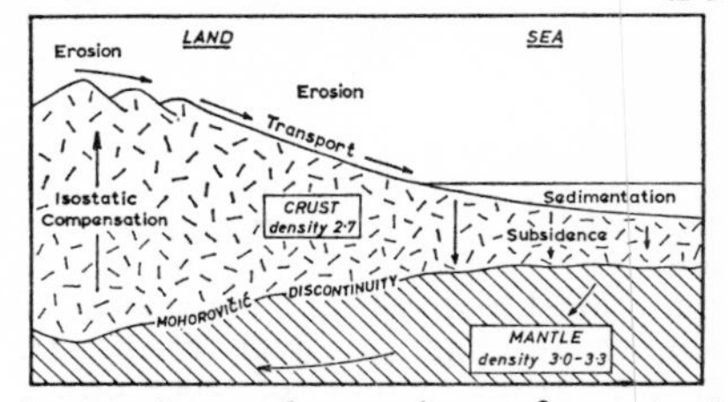

isosteric surface A s. of constant density in the atmosphere.

isothermal layer A l. of the atmosphere in which the temperature remains the same for a considerable height.

isotope The alternative form of an element, though with virtually identical chemical properties, having a different number of neutrons in its nucleus. The atomic number of the element and its i.'s is the same, but their atomic weights differ. Thus oxygen has three naturally occurring i.'s, $_8O^{16}$, $_8O^{17}$ and $_8O^{18}$. *Radiogenic i.'s* are those formed from the breakdown of a radioactive parent element, and this is of great importance in finding the age of certain rocks; see LEAD-RATIO and RADIOACTIVE DECAY. The radioactive i. of carbon, carbon 14 ($_6C^{14}$), can be used for dating organic material less than 50,000 years old (RADIOCARBON DATING).

isthmus A narrow neck of land between two seas or oceans; e.g. the I. of Panama.

jade A hard translucent compact nephrite (a variety of AMPHIBOLE) or a variety of PYROXENE (a bisilicate of calcium, magnesium, aluminium, etc.), greenish to whitish in colour, used for carving jewellery and ornaments.

jet A hard, dense, compact, black form of LIGNITE, capable of taking a brilliant polish. Found in the U.K. in the Upper Lias, notably in Yorkshire.

jet stream (i) A high-altitude (about 12,000 m., 40,000 ft. in the TROPOSPHERE) W. air movement, consisting of relatively strong winds concentrated in a narrow belt. In summer its mean speed is about 50–60 knots, but in winter it may be much stronger (95–120 knots); on occasion winds of 200–250 knots have been recorded. Several such streams are distinguished, including a *polar-front j. s.*, a *polar night j.s.*, an *arctic j.s.*, a *subtropical j. s.* [*f below*], between 20° and and 30° latitude N. and S. (ii) A more local high-altitude air-stream of almost tubular or oval form. In recent years, study of the upper atmosphere has shown that the position and nature of j.s.'s have important climatological effects, esp. on FRONTOGENESIS (a j. s. may discourage CONVERGENCE and encourage DIVERGENCE), on the bursting of the MONSOON, and on the formation of a TORNADO.

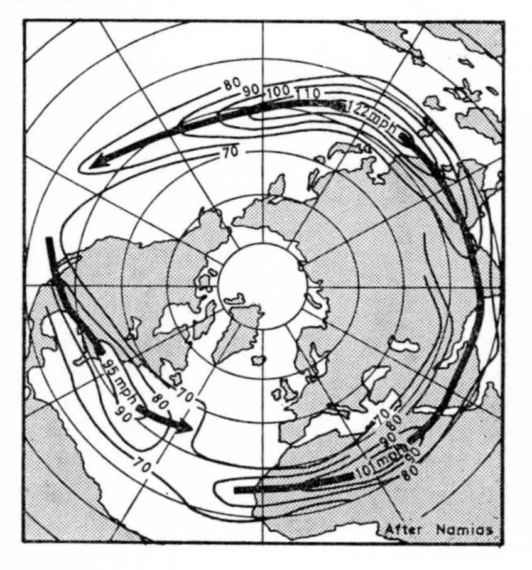

joint A surface of division or fracture in a mass of rock, the result either of tearing under tension, or of shearing under compression, but involving little or no actual movement or displacement of the rocks (ct. FAULT). A j. is usually transverse to the bedding; e.g. Checkerboard Mesa, Zion Canyon, Utah, where well-marked vertical j.'s cross well-defined bedding-planes, causing a 'checkerboard' effect. Other types of j. occur in igneous rocks (e.g. granite) through tensile stresses caused during cooling and solidification, and in sedimentary rocks during consolidation (shrinkage cracks). When the j.'s coincide with the STRIKE, they are called *strike-j.'s*, when with the DIP, *dip-j.'s*. If 2 sets of j.'s develop, they are usually at right-angles. A dominant j. is the *master-j.* J.-ing is of value to quarrymen in extracting stone. J.'s form planes of weakness along which forces of weathering and erosion may operate; e.g. j.'s define the rectangular blocks of a granite TOR. [*f* GRIKE]

joint-plane The plane of a joint; when 2 sets of vertical joints are at right-angles to the BEDDING-PLANE, the rock is divided into rectangular masses.

jökull (Icelandic) A small icecap, derived from the numerous examples in Iceland; e.g. Vatnajökull, Heinabergsjökull.

journey to work The daily movement of workers between home and work in different parts of an area. This is normally a 'tidal flow' concentrated within a limited period of time ('peak hour' or 'rush-hour'), during which the transport system has to carry very much heavier loads than normal.

jungle Orig. waste or uncultivated ground (specif. in India); now used popularly for any type of tropical forest with thick undergrowth.

Jurassic The 2nd (middle) of th geological periods of the Mesozoi era, and the system of rocks laid dow during that time, extending fro about 180 to 135 million years ag the age of the great reptiles. The nan is derived from the type-region in tl Jura Mtns., to the N.W. of the Alp Sediments were laid down in fluctua ing, shallow seas, including clays a sands, and coral reefs were also wid spread, with some periods of estuari and fluviatile deposition. The resu is a varied system of rocks, traceal in England in a diagonal band acro the country from the Yorkshire co (S. of the mouth of the Tees) to t

Dorset coast. The series are as follows:

Purbeck	limestones, clays
Portland	limestone and sandstone
Kimmeridge Clay	dark grey or black shaly clay
Corallian	limestone, with some clay
Oxford Clay	heavy clays
Great Oolite	oolitic limestone, fuller's earth
Inferior Oolite	oolitic limestone, with beds of clay and sandstone
Lias	mainly grey or blue clays

jute A coarse fibre derived from 2 species of tall plant, 4–5 m. high, *Corchorus capsularis* and *C. olitorius*, mainly grown in the Ganges–Brahmaputra delta, esp. in E. Pakistan. It is manufactured into hessian and sacking in mills in E. Pakistan, along the R. Hooghly in India, and in Dundee, Scotland.

juvenile water Hot, mineralized w. liberated during igneous activity, which has come from the depths of the earth, not from the surface or the atmosphere. Syn. with *magmatic w.*

kame Derived from an old Scots word signifying an elongated steep ridge (from 'comb'). Applied to a hummocky deposit of sand and gravel aid down by melt-water along the edge of an ice-sheet, gen. aligned parallel to the ice-front. Much confusion has resulted in the terminology. If k. is used at all, it is now in an unspecific and gen. sense, relating to any hummocks or mounds of poorly mixed fluvioglacial materials, about whose formation there is some doubt. . K. Charlesworth terms all these deposits ESKERS, and divides them nto: (i) OS (pl. ÖSAR); (ii) K.-MORAINE; nd (iii) K.-TERRACE [*f*].

ame-and-kettle country A landscape of undulating KAME-MORAINES, nterspersed with shallow hollows or ETTLES; e.g. W. of L. Michigan, .S.A., in the state of Wisconsin.

kame-moraine Undulating mounds of bedded sands and gravel, deposited unevenly from melt-water along the face or front of a stagnant, slowly decaying ice-sheet. Their inner faces represent 'moulds' of the former ice-contact slope. They are in effect groups of ALLUVIAL CONES. If the deposition took place in a PROGLACIAL LAKE, the materials are more clearly bedded and sorted, with a distinct flat top and FORESET BEDS; this variety is known as a *k.-delta.* K.-m.'s are found widely on the formerly glaciated lowlands of N. America and N.W. Europe. A belt has been traced in loops from Long I. as far W. as Wisconsin, and in Europe more or less parallel to and S. of the Baltic Heights. They are widespread among the bogs of central Ireland; one line can be traced from near Dublin into Galway, though cut through by the Shannon near Athlone.

kame-terrace A t.-like ridge of sand and gravel along the side of a valley, laid down by a melt-water stream occupying the trough between a glacier and its enclosing valley-wall; it is sometimes called an *ice-contact t.* K.-t.'s look like the shorelines of former glacial lakes, but they are much less regular, and have numerous depressions where blocks of ice lingered before melting. They are found commonly in the valleys of the Appalachians in New England. [*f*]

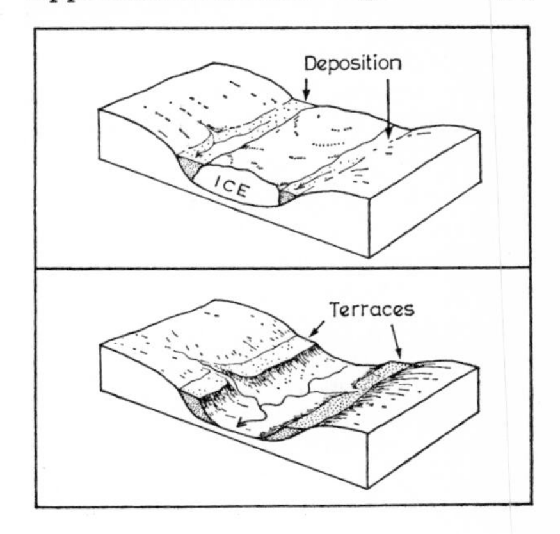

kampong A patch of coconuts, fruit-trees and vegetable gardens, interspersed with houses, and with ponds breeding fish and growing water-hyacinth. A k. has a considerable density of population and intensity of cultivation, and is a land-use category typical of Malaya and Singapore.

kanat See QANAT.

Kansan In America the 2nd main glacial advance, corresponding to the Mindel in the Alps. First used by T. C. Chamberlin in 1894. Its deposits consist of extensive, near level TILL, consisting almost entirely of clay, with little evidence of stones, sands and gravel. It covers much of Iowa, Nebraska and Missouri, though it has been much dissected by the development of post-glacial river valleys. TERMINAL MORAINES at its S. margin and some RECESSIONAL MORAINES have been distinguished; i.e. there was a steady recession of the ice, with no prolonged halt. The till was mainly brought from the Keewatin centre of ice-accumulation.

Kant's index of concentration, of rural settlements Designed by E. Kant for use in reducing a map showing distribution of habitations by means of non-quantitative dot-symbols, to one in which the dispersion and concentration of settlement is more precisely reflected in terms of the distance between the habitations.

$X = \frac{1}{M}\sqrt{\frac{A}{D}}$; where X = the interval between two settlements, $1/M$ = scale of the map, A = area under consideration, and D = density of habitations.

kaoliang A type of MILLET grown in N. China, esp. Manchuria, where it forms a staple food.

kaolin A whitish clay produced by the complex chemical breakdown of aluminium silicates, very largely the felspars and esp. orthoclase, partic. in granite, the result both of weathering (HYDROLYSIS) and ascending gases and vapours from a magma reservoir (PNEUMATOLYSIS); the process is known as *kaolinization*. Strictly, *kaolinite* is the mineral, hydrated silicate of alumina ($2H_2O.Al_2O_3.SiO_2$); k. is the clay, including also particles of quartz and mica from the decomposed granite. Its name originates from the Kauling ('high hill') range in Kiangsi province, China, where it has been worked for many centuries for pottery, hence the alt. name 'china-clay'. Most k. deposits are found *in situ*, and esp. where PNEUMATOLYSIS is the chief agent are of great thickness; e.g. in the neighbourhood of St. Austell in Cornwall, and gen. around Dartmoor and Bodmin Moor. Some k. is transported by running water, and redeposited in thin beds as 'residual k.' or 'sedimentary k'.; e.g. the 'ball clay', near Wareham, Dorset, though this type is of lower grade. The k. is washed out in Cornwall from open pits by high-pressure jets of water, allowed to settle and dry, and removed in large lumps, the waste forming conical spoil-heaps. K. is used for pottery, as a filler and in paper-making.

kar (Germ.) See CIRQUE.

karaburan A strong dust-laden N.E. wind in the Tarim Basin of central Asia; *lit.* '*black buran*', in ct. to the '*white buran*' of winter, which blows clouds of snow.

karez (Baluchi) An underground irrigation channel, bringing water from the foot of hills to an otherwise arid basin or plain; found in Baluchistan and Iraq. Syn. with FOGGARA, QANAT.

karren (Germ.) A limestone surface intersected with furrows; syn. with LAPIAZ.

Karroo, Karoo (i) The Gondwana series of rocks in S. Africa, ranging from the Upper Carboniferous to the Lias in age. (ii) A flat-topped landform, which may rise in a series of terraces, called after the proper name for such a feature in the S. of the Republic of S. Africa.

karst (from It. *Carso*, Serbo-Croat *Kars*) (i) The proper name of an area of rugged limestone plateaus an ridges near the Adriatic coast of Yugoslavia. (ii) Used (with lower-case k) for any area of limestone or dolomite with the associated phenomena of SINKS, underground drainage and cave-systems, the result of carbonation-solution; e.g. parts of the N. Pennines (Craven), S.W. France (the Causses), parts of Indiana and Kentucky, Yucatan, Jamaica. It must be remembered that the original k. phenomena were described in an area where the features are emphasized by the long aridity of summer and the intensive but short-lived autumn and winter rain. In the Pennines these features are modified by a more constant humidity, and also in many areas by a cover of till. A full terminology of k. phenomena was put forward by J. Cvijić, with Serbo-Croat words now mixed with English ones, which sometimes creates confusion. The International Geographical Union has established a Commission on Karst Phenomena, and various terms have been discussed and recommended; the more important are listed in this *Dictionary*.

katabatic wind A cold w. blowing downhill. Frequently it occurs at night in valleys, caused by the gravity flow of dense air chilled by radiation on the upper slopes. The effect is emphasized when the cold air blows down from an ice-cap; e.g. the nevados from the Andes of Ecuador. Ct. ANABATIC.

katafront A COLD FRONT in which the warm air-mass is for the most part descending over the cold wedge. Ct. ANAFRONT.

kavir Used in Iran as syn. with a PLAYA.

Keewatin (i) The oldest Pre-Cambrian rocks in the vicinity of L. Superior. (ii) An ice-accumulation centre (or group of centres) in N. America, W. of Hudson Bay, from which the Quaternary ice-sheets moved S. At the glacial max., the Keewatin ice-sheet probably merged with the ice of other centres (*Patrician* and *Labradorian*) to form a single *Laurentide* ice-mass.

Kelvin scale, of temperature Syn. with the scale of ABSOLUTE T. 0°K. = −273·16°C (usually approximated to −273°C.), or −459·4°F.

Kelvin wave A tidal system which develops in a more or less rectangular area of the sea (e.g. the English Channel), in which the tidal range increases on the right of the direction of the PROGRESSIVE W., and decreases on the left. This largely explains why the tidal range is much less on the S. coast of England (about 4 m.) than on the N. coast of France (11–13 m.)

kettle, kettle-hole A circular hole in glacial drift, commonly water-filled, caused by the previous presence of a large detached mass of ice which has subsequently melted; e.g. around Brampton to the N.E. of Carlisle; near Lancaster; in the Vale of Pickering. Numerous examples occur in the Kettle Moraine country of Wisconsin, U.S.A. [*f*]

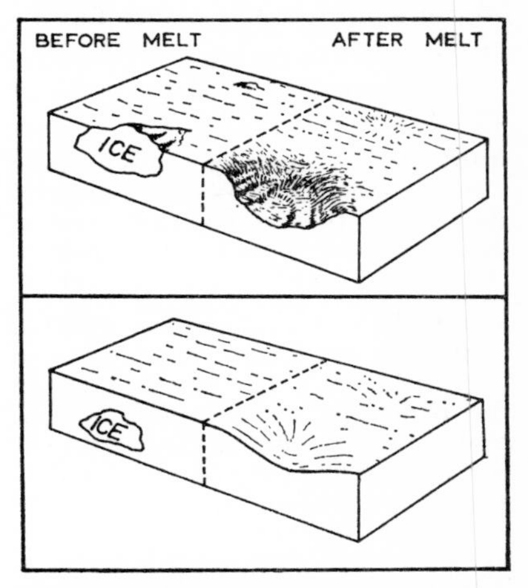

kettle-moraine An uneven area of morainic hills with numerous KETTLE-HOLES, so called after the Kettle Range in Wisconsin, U.S.A.

Keuper A series of rocks, Upper Triassic in age, consisting mainly of red clays and marls, with some bands of more compact sandstone. Beds of common salt and gypsum sometimes occur. Outcrops of the K. are found in the Midlands of England, Cheshire (where the salt is extensively worked) and Lancashire.

Kew barometer A standard mercury b., which requires no adjustment for the level of the mercury in the cistern (as does the FORTIN type), so that the pressure can be read directly from the top of the mercury column in the graduated glass tube.

key See CAY.

khamsin (Arabic) A hot S. SIROCCO-wind blowing in Egypt and the S.E. Mediterranean area. These winds are commonly preceded by a heat-wave and followed by a dust-storm. The wind is usually very dry, often dust-laden, but it is sometimes humid when it has blown over the sea. It is associated with the front of a depression moving E. through the Mediterranean Sea.

kilogram A metric unit of mass = 1000 grammes (gm.) = 2·2046 lb.

kilometre Metric unit of distance: 1 km. = 1000 m. = 1093·61 yds. = 0·621372 mi.

kilowatt A unit of power = 1000 watts. *k.-hour* (*kWh*) = 1 k. expended for 1 hour, the normal unit used in indicating the output of a power-station.

kimberlite An ultrabasic material (including varieties of MICA and OLIVINE), which fills volcanic PIPES in the Kimberley district of S. Africa, sometimes containing diamonds. At the surface it is weathered ('yellow ground'), passing downward into weathered but less oxidized 'blue ground', then into unweathered k. The pipes and the k. are probably the result of FLUIDIZATION.

Klimamorphologie (Germ.) Coined by Germ. geographers to denote the study of the relationship between climate and morphological features.

klint (Swedish or Danish) A steep cliff around the shores of the Baltic Sea, esp. in Denmark and Sweden; e.g. Möns K., a chalk cliff on the island of Mön, rising 130 m. (420 ft.) from the sea.

klippe (Germ.) Part of an overthrust rock-sheet, isolated by denudation; a special case is a *nappe-outlier*. E.g. (i) the zone of k. in Switzerland, esp. the Chablais Pre-Alps to the S. of L. Geneva, now eroded into a maze of limestone ridges and peaks over 2400 m. (8000 ft.), of great structural complexity. (ii) Hills in Assynt (Sutherland, N.W. Scotland), consisting of k. of Lewisian Gneiss, Torridonian Sandstone and Cambrian sediments, lying W. of the main edge of the Ben More Thrust which carried the nappes W.-ward. (iii) The mountains in Glacier National Park, Montana, of Pre-Cambrian rocks, remnants of nappes carried E. by the Lewis Overthrust of LARAMIDE age, and now resting as isolated fragments on Cretaceous rocks (e.g. Chief Mountain). [f.

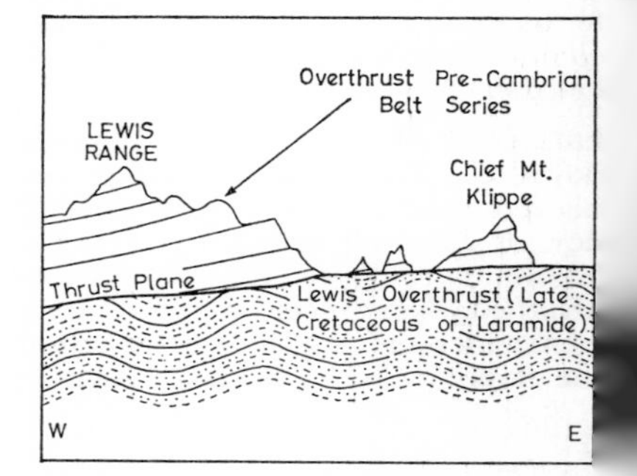

kloof (Afrikaans) A gorge or ravine sometimes a pass, in S. Africa.

knap A hill-crest, or rising ground

knickpoint (from Germ. *Knickpunk*
A break of slope in the long-profile of a river valley, where a new curve of erosion, graded to a new sea-level (after a relative lowering of former sea level), intersects an earlier one; the recedes upstream as erosion proceeds Sometimes the Germ. word is r served for a rejuvenation featu formed in this way. The Englis word may refer to any break of slo in the long-profile of a stream, whi

may also result from an outcrop of resistant rock (a 'hard bar'), or from causes other than rejuvenation. [*f* REJUVENATION]

knoll A small rounded hill. See REEF K.

knot (i) A speed of 1 nautical mi. (6080 ft., in U.S.A. 6076 ft.) per hour, so called because the speed of a ship used to be measured by counting the knots, as a knotted rope (LOG-line) ran out, against a period of time indicated by a sand-glass. (ii) A structural junction-area of 2 or more ridges of fold-mountains; e.g. the k. of Pasco in the Andes, the Pamir k. in central Asia.

kolkhoz See COLLECTIVE FARMING.

kopje (Afrikaans) A small, but prominent, isolated hill in S. Africa, frequently an erosion remnant forming a castellated pile of rocks (*castle-k.*).

Köppen climatic classification An empirical c. of climatic types, first devised by W. Köppen in 1918, and several times revised, esp. in collaboration with R. Geiger. 5 major climatic groups were devised, based on the climatic requirements of certain types of vegetation, lettered *A.* to *E.* These are sub-divided according to features of the rainfall régime and certain thermal characteristics, using various lower case letters; e.g. *Csa*, a warm temperate climate, with dry hot summer and warm moist winter (*humid mesothermal*), gen. known as the Mediterranean type.

koum, kum (Turkestan) The sandy desert, with dunes and 'sand-seas', equivalent to the ERG of the Sahara.

kraal (Afrikaans) A term, cognate to 'corral', used for an African village, sometimes for a cattle-pen or enclosure surrounded by a thorn-fence.

kratogen(ic) A rigid, relatively immobile part of the earth's crust, which was later modified to *kraton* and then to CRATON. The last is occas. used, but the term SHIELD is more popular.

kyle (Gaelic) A channel, sound or strait between the mainland and an island or two islands; e.g. the K. of Loch Alsh, the K.'s of Bute.

laccolith, laccolite Coined by G. K. Gilbert in 1877 from 2 Gk. words meaning 'rock-cistern'. It is the result of an intrusive mass of MAGMA which has forced up or domed the overlying strata. *Note:* It is now thought that a l. is not necessarily fed from below by a PIPE, but sideways from a BYSMALITH, thus forming a tongue-shaped bulge. In the Henry Mtns. the l.'s cluster around the bysmalith '. . . like the petals of a lop-sided flower . . .' In its simplest form, the magma solidifies as a cake-like mass, but several l.'s may be formed one above another from a single intrusion (see CEDAR-TREE L.). E.g. the Henry Mtns. of S. Utah, where 5 groups of trachytic l.'s were injected among sandstones of age from Carboniferous to Cretaceous; some are simple forms (Mt. Ellsworth), others are complex (Mt. Ellen is thought to be a cluster of 30 individuals, Mt. Hillers of one large and 8 small). These l.'s have been subject to much denudation, and they reveal various stages of destruction and exhumation; the highest is Mt. Ellen (3497 m., 11,473 ft.). The La Sal and Abajo Mtns. are other l.'s in Utah, and there are several in Wyoming and S. Dakota (Belle Fourche valley). [*f*]

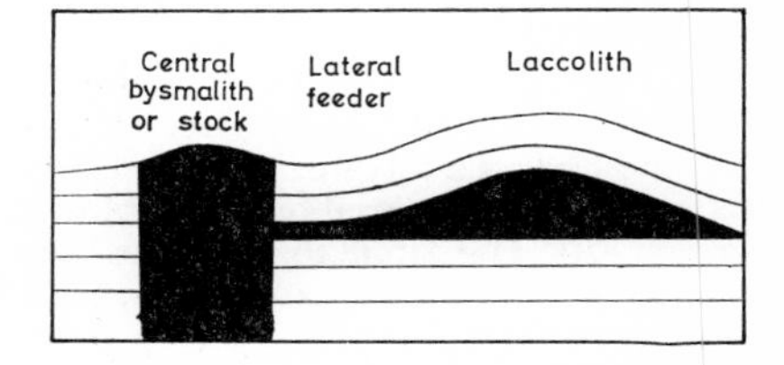

lacustrine App. to a lake, esp. to: (i) deposits of sediment therein; and (ii) terraces around the margins of a lake whose surface was formerly at a higher level. L. deposits often reveal clear layering or banding, each corresponding to one season's deposition. See L. DELTA, L. PLAIN, VARVE.

lacustrine delta A d. built out by a stream into a lake; e.g. the d. of the Rhône into L. Geneva, of the R. Kander into L. Thun, Switzerland.

lacustrine plain The floor of a former lake, filled in by the deposition of material by inflowing streams. The soils developed on these p.'s are usually favourable for agriculture, since they consist of fine, well-sorted materials, and the surface is level. Sometimes artificial drainage is necessary, because of the low-lying swampy character of the l. p. L. p.'s are widespread because of the former numerous PROGLACIAL LAKES, which have now disappeared or are represented only by fragments. E.g. L.'s Winnipeg, Winnipegosis and Manitoba are the remnants of 'L. Agassiz', leaving extensive l. silts; Airedale in Yorkshire, which now contains a l. p. of at least 30 m. thickness of deposits; the l. p. between Derwentwater and Bassenthwaite in the English Lake District.

ladang A system of agriculture in Indonesia, Malaysia and the Philippines, whereby a forest is felled, usually leaving the roots *in situ*, burning all the vegetation in the dry season, digging in the ashes, and planting crops at the start of the next wet season. After 2 or 3 years, when much of the fertility has gone, the area is abandoned for another. The cleared land soon becomes covered with secondary jungle (*bloekar* in Indonesia).

lag (i) Coarse residual material (sometimes called *l. gravel*) left behind on the bed of a stream, or in a desert after the finer material has been blown away. (ii) The time that elapses between a change in conditions, and the recording of the same by an instrument. Measuring instruments, esp. in meteorology, do not respond immediately to changed conditions.

lag-fault A low-angled fault formed when a series of rocks is moved over one another, the uppermost of which lags behind the lower.

lahar A mud-flow associated with volcanic activity in Java.

lagoon (i) A sheet of salt water separated from the open sea by sand- or shingle-banks; e.g. the Venetian l.'s. (ii) The sheet of water between an offshore reef, esp. of coral, and the mainland. (iii) The sheet of water within a ring- or horseshoe-shaped ATOLL.

lake A water-filled hollow, more or less extensive, in the earth's surface. Some are actually saline inland seas; e.g. Caspian Sea (437,000 sq. km., 169,000 sq. mi.), Aral Sea (63,300, 24,500), Dead Sea (1048, 405). The largest named l.'s are: (*a*) *in N. America:* Superior (82,410, 31,820), Huron (59,600, 23,010), Michigan (58,020, 22,400), Great Bear (30,200, 11,660); (*b*) *in Africa:* Victoria (67,860, 26,200), Chad (fluctuating, about 51,800, 20,000), Nyasa (36,780, 14,200), Tanganyika (Malawi) (32,890, 12,700); (*c*) *in Asia:* Baikal (29,990, 11,580), Balkhash (18,260, 7050); (*d*) *in S. America:* Maracaibo (21,487, 8296), Titicaca (8290, 3200); (*e*) *in Europe:* Ladoga (18,130, 7000), Onega (9840, 3800); (*f*) *in Australia:* Torrens (6216, 2400), Eyre (fluctuating). The largest in England is Windermere, in Britain Loch Lomond, in the British Isles Lough Neagh (381, 147). The longest in the world, excl. the Caspian, is Tanganyika (675 km., 420 mi.). At the other end of the scale are small sheets of water, *meres*, *ponds*, *tarns* (English Lake District), *llyns* (Wales), *loughs* (Ireland), *lochs* and *lochans* (Scotland), *étangs* (France), *stagni* (Italy). L.'s may be classified according to the mode of origin of the hollows in which they lie: (i) *erosion:* glaciation ('finger' l.'s and tarns), solution (salt-meres), wind ('shotts' in N. Africa); (ii) *deposition* or *barrier l.'s*, enclosed by rock-falls, sand bars, deltaic deposits, morainic deposits, ice-dams, vegetation dams (l.'s in the W. European heaths and moorlands), calcareo dams (a barrier of deposited calcite, as L. Plitvicka in Yugoslavia); (iii) *structural:* rift-valleys, hollows formed by crustal sagging, down-faulted basins; (iv) *volcanic:* crater-l.'s, lava-dams. The largest man-made l.'s are Kariba (Zambia–Rhodesia) formed behind the Kariba Dam across the R. Zambezi, and Mead behind the Hoover Dam across the R. Colorado, U.S.A. The deepest lake is Baykal in the U.S.S.R. (1940 m., 6365 ft.), the lowest surface the Dead Sea (395 m., 1296 ft. below sea-level), the highest named surface L. Titicaca (3811 m., 12,506 ft. above sea-level).

lake dwelling A d. erected on piles in standing water; e.g. at Meare in Somerset; along the shores of L. Constance, W. Germany. Recently doubt has been expressed as to whether the remains of the many types of settlement in bog and lakeside situations do in fact represent former settlements built above the surface of l.'s. The piles may simply have been driven into wet or marshy ground. However, in some cases it seems clear that l. d.'s, *sensu stricto*, did exist.

lake rampart A ridge of material on a lake-shore, formed when the ice of the frozen l. exerted lateral pressure against the shore; these are clearly shown around the shores of the Great Lakes in N. America. It is not to be confused with a lacustrine terrace, representing a former level of the l.

lalang (Malay) Thick coarse grass which rapidly covers abandoned plantations and clearings.

Lambert's Conformal Projection A type of CONIC P. in which parallels are concentric circles, meridians are radiating, equally spaced, straight lines, with the scale true on 2 standard parallels, increasing N. and S. from them. This p. gives true direction at every point, and is used for aeronautical charts, weather maps, and any type of distributional map where correct direction is important. It is a popular p. for maps of U.S.A., since there is little distortion for this shape and area, esp. when using parallels 29°N. and 45°N. The calculation of this p. is complex, and it is usually drawn from tables.

Lambert's Zenithal Equal Area Projection (L.'s 6th) In its polar form, meridians are represented by straight lines at their true angles to each other, while the parallels are circles of which the meridians are radii. The radius of each parallel is calculated from the formula $2a \times \frac{\sin z}{2}$, where z = the co-latitude and a = the earth's radius to scale. The p. can also be applied to any part of the earth's surface, for deformation is symmetrical around the central point from which GREAT CIRCLES radiate. The equatorial and oblique cases are plotted from the co-ordinates of grid intersections given in tables. Directions are correct at the central point, distances are true from the central point, it is equal area, and there is but little deformation for 30° from the central point.

laminar flow (i) The movement of a glacier, produced by a thrust along the line of slope caused by pressure resulting from the weight and solid character of the ice higher up. The ice is fractured and sheared, and thrust along glide-planes. This thrust may even push the 'snout' of the glacier uphill for a short distance. (ii) Non-turbulent flow in a fluid.

lamination, lamina, *adj.* **laminar** The finest scale of stratification or sheeting in rocks, esp. metamorphic varieties; sometimes defined as less than 1 cm. in thickness, though there may be 40 or more layers in a cm. Each l. may be regarded as representing a layer of original deposition in sedimentary rocks; each period of flood deposited a thin layer of mud, which compacted, and the next flood added another layer, or the wind added successive layers of sand to dunes. In such rocks as shale, the micas and clay minerals lie parallel with the bedding. E.g. the Green River Shales of Wyoming, 800 m. (2600 ft.) in thickness, laid down in layers 0·17 mm. thick; it is estimated that these took 6·5 million years to be deposited.

land breeze A cool night b. blowing from the l., which has been cooled by radiation, towards the warmer sea, occurring partic. near the equator, and elsewhere in calm, settled weather. During the night pressure is slightly higher over the l. than over the sea, so there is an airflow towards the latter, reinforced by the outward spread of cool denser air. In some islands these winds are so regular that fishing-boats go out at night with the l. b., returning next afternoon with the SEA B. [*f*]

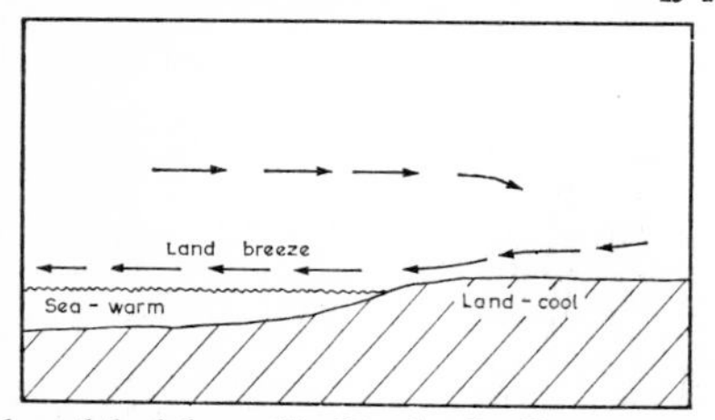

land-bridge (i) Geologically, a past or present land-link between continents; e.g. (past) across the Bering Strait; (present) the isthmus of Panama. (ii) By biologists and anthropologists, former migratory routeways by which terrestrial animals, including man, were dispersed over the earth's surface.

land classification A c. of l. into categories, which can be used 'for a broad national policy of land-use planning and conservation of land resources' (L. D. Stamp). Professor Stamp proposed 3 major categories and 10 types, based on: (*a*) the nature of the *site* (elevation, slope and aspect); and (*b*) the nature of the *soil* (its depth, texture and water conditions). The major categories are: I. good; II. medium; and III. poor. The 10 types are: (i) first-class l.; (ii) good general-purpose farmland; (iii) first-class l., with grass; (iv) good but heavy l.; (v) medium-quality light l.; (vi) medium-quality general-purpose farmland; (vii) poor-quality heavy l.; (viii) poor-quality mountain and moorland; (ix) poor-quality light l.; (x) the poorest l.

land competition A situation whereby a floating demand for alt. uses settles on a partic. area of l. In a free market this is resolved for the most part economically. The function of planning is to evaluate these demands in the interests of the community, which may result in social or aesthetic considerations overriding purely economic reasoning, though the reverse may occur; e.g. the Festival of Britain site on the S. Bank of the R. Thames, intended for an open-air theatre or something of that nature, was occupied by large office blocks.

landes (Fr.) (i) The proper name for the sandy lowlands ('*Les L.*'), with dunes and lagoons, bordering the S.W. shores of the Bay of Biscay. (ii) With lower case, used for other wastelands, as in the heathlands of Belgium.

landform The shape, form and nature of a specif. feature of the earth's land surface, the study of which was formerly called PHYSIOGRAPHY, in its modern scientific concept known as GEOMORPHOLOGY.

landscape Used orig. by artists to denote rural scenery (Dutch, *Landschap*), now a general term for the sum total of the aspect of any area, rural or urban. 'L. Geography' developed as part of the regional view of the subject, stimulated by the work of P. Vidal de la Blache in France, which examined both the 'natural' and 'cultural' aspects of the face of the earth in terms of concrete reality. See CULTURAL L., NATURAL L.

landscape evaluation An effort to establish a scale of l. (and TOWNSCAPE) values as a contribution to planning, development and CONSERVATION. The problem is to replace pure subjective judgement and personal bias by means of some scale of values, though of course personal observation from a chosen series of viewpoints ('view evaluation') is required, supplemented by colour photographs. One suggested scale (by K. D. Fines) is a series from 0 to 32, with 6 categories: *unsightly*, 0–1; *undistinguished*, 1–2; *pleasant*, 2–4; *distinguished*, 4–8; *superb*, 8–16; *spectacular*, 16–32; e.g. 18 is the highest in Gt. Britain (a view of the Cuillins of Skye across Loch Coruisk), 12 is the highest in Lowland Britain. A l. e. map has been made of E. Sussex, using a series of tracts (shaded on a CHOROPLETH basis), each largely related to geological and physiographical features. In this e., 9 lowland l. types, 3 highland l. types and 6 townscapes were suggested.

Landschaft (Germ.) The German word for landscape, which has come to have a specialized, though often diverse, meaning, sometimes in the sense of 'scene', sometimes with the character of a unit region. Translating the word simply as 'landscape' has led to confusion in certain theoretical discussions, and it is well to remember that it cannot be simply regarded as a cognate.

landslide, landslip A fall of earth and rock-material down a slope or mountain-side, the result of gravity and rain-lubrication; see MASS-MOVEMENT. The two words really denote the same process; the former is more customary in U.S.A., the latter in Britain.

Land Survey System (U.S.A.) A s. by which much of U.S.A. was surveyed and divided into congressional TOWNSHIPS of 6 mi. square. A series of 32 major areas was demarcated, each with a principal N.–S. meridian crossed by an E.–W. base-line corresponding to a parallel of latitude. Within each of these, 6-mi. units (each known as a RANGE) were measured off both to W. and E. along each base-line from the principal meridian, and 6-mi. units, each known as a TOWNSHIP, were measured off along the principal meridian both to N. and S. from each base-line.

[*f opposite*]

land use, utilization The use made of the land by Man, as surveyed and mapped in a series of recognized categories. In Gt. Britain the first l. u. survey was carried out in the 1930s, under the direction of L. D. Stamp. The basic classification for mapping on the 1-in. scale was as follows: (i) forest and woodland (dark green); (ii) arable land (brown); (iii) meadowland and permanent grass (light green); (iv) heath and moorland (yellow); (v) houses with gardens, orchards, nurseries (purple); (vi) land agriculturally unproductive, close building, pits, quarries, cemeteries and industrial premises (red). Various symbols, shading and stipples were superimposed in black to indicate certain sub-divisions. In 1950 a classification for a *World Land-Use Survey* was drawn up under the auspices of *Unesco*: 1. settlements and associated non-agricultural land (dark and light red); 2. horticulture (deep purple); 3. tree and other perennial crops (light purple); 4. crop-land: continual rotation cropping (dark brown), land rotation (light brown); 5. improved permanent pasture (light green); 6. unimproved grazing: used (orange), not used (yellow); 7. woodlands: dense (dark green), open (medium green), scrub (olive green), swamp forests (blue-green), cut or burnt-over forest (green stipple), forest with subsidiary cultivation (green with brown dots); 8. swamps and marshes (blue); 9. unproductive land (grey). A new l. u. survey of England and Wales, on a scale of 1:25,000, has been recently undertaken under the direction of Miss A. Coleman, and numerous sheets have appeared.

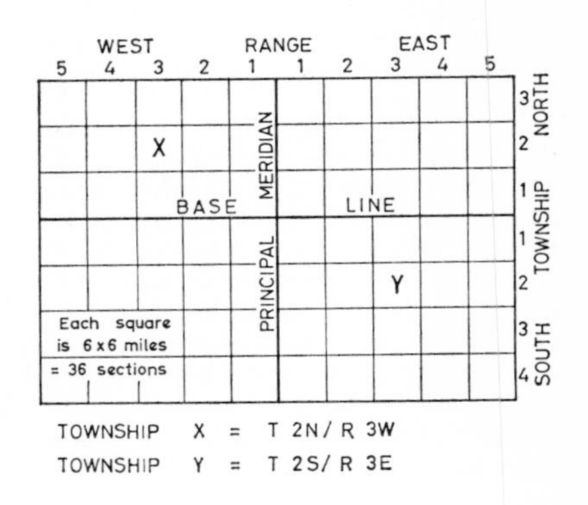

lane (i) A commonly used route by sea or air. (ii) A channel of clear water through an ice-field. (iii) A minor, narrow, usually unpaved or unmetalled route, esp. in rural areas.

lapiaz, lapiés (Fr.) A 'pavement' of limestone, fretted by carbonation-solution along well-marked joints, with ridges and furrows (equivalent to CLINTS and GRIKES respectively). The term *lapié* (plur. *lapiés*) is sometimes used for the individual furrows within the whole limestone pavement.

lapilli (It.) Small fragments of cinder, about the size of a pea, ejected from a volcano.

lapse-rate The rate of temperature decrease (or *vertical temperature gradient*) in the atmosphere with increasing altitude. The ENVIRONMENTAL L.-R. is about 0·6° C. per 100 m., or 1°F. for every 300 ft. of ascent. This l.-r. seems to continue steadily as far up as the TROPOPAUSE, beyond which the temperature remains fairly constant at about −90°C. (−130°F.) over the equator, −48°C. (−65°F.) in mid-latitudes, and −45°C. (−50°F.) over the Poles. This normal l.-r. may be interrupted by inversion, and it is affected markedly by the vertical ascent of air 'parcels' of different temperature and humidity (PROCESS L.-R.) See DRY ADIABATIC L.-R., SATURATED ADIABATIC L.-R.

Laramide orogeny A mountain-building movement responsible for the folding of the main Rocky Mtns., which started in late Jurassic times and continued into the early Tertiary.

latent heat The amount of heat-energy expended in changing the state or phase of a body, without raising its temperature, expressed in calories per gm. In meteorology, the l. h. of evaporation indicates the energy required or heat absorbed to convert water to water-vapour. When CONDENSATION occurs, the heat given out is the *l. h. of condensation*. The *l. h. of fusion* represents the change from solid to liquid.

latent instability In climatology, the state of the upper part of a conditionally unstable column of air, lying above the level of free CONVECTION. See CONDITIONAL I.

latent potential, in movement of population Before any movement of people takes place between areas in response to some attractive force, a sorting-out process occurs among the people who are able to move. From all those people who satisfy the conditions necessary to move, only a certain number actually do so, depending upon the requirements of the system at each level of attraction. At any point in time, therefore, in any system of movement a l. p. of movers exists. This cannot be released, or satisfied in its ambition to move, because of the inherent or 'built-in' limitations of the system.

lateral dune A d. flanking a main DUNE within a pattern formed in a desert where some obstacle occurs. [*f* DUNE]

lateral erosion The e. performed by a stream on its banks, in ct. to vertical erosion on its bed. This is most marked on the outside of a bend or meander, so that the current impinges there, resulting in undercutting and BANK-CAVING. Gradually a river extends its amount of l. e., widening the swing of each meander and so its overall valley; each meander gradually moves downstream. The valley-floor is bordered by a BLUFF, the outer limit of l. e. The creation of a broad, nearly flat, surface in this way is often termed *l. planation*.

lateral moraine Rock debris lying on the surface of a glacier, forming a low ridge or band more or less parallel to and near its edge. When the glacier melts, the debris may remain as a distinct embankment along the valley side; several of these, broadly parallel, will indicate stages in the shrinkage of the ice. In the case of a big glacier, a l. m. may be 30 m. or more in height. [*f* MORAINE]

laterite A porous layer of reddish material formed by weathering from the breakdown of rocks, partic. igneous types, chiefly under humid tropical conditions. The material is heavily leached of silica and alkalis, leaving a concentration of sesqui-oxides of aluminium and iron (i.e. a hydrated form of aluminium oxide). When formed it appears to be quite soft, and has a plastic quality when wet, but on exposure to the atmosphere it becomes extremely hard. It is thus used for building material (Lat., *later*, a brick) and road construction. L. and lateritic soils have received much scientific attention. Some writers define the term more closely than that in gen. use as above, and others have abandoned it. See LATOSOL. L. occurs widely in Brazil, the W. Indies, W. and E. Africa, S. India and Ceylon.

lateritic soils Zonal s.'s developed on LATERITE, not usually of good agricultural value because of their porosity and poverty in minerals as a result of leaching.

laterization The process of weathering under hot damp conditions within the tropics to form LATERITE.

latex The milky fluid exuded from the laticiferous cells of various trees; e.g. *Hevea Brasiliensis* (rubber-tree), from which by *tapping* (a pattern of cuts through the bark of the trunk) the l. is obtained, the raw material of manufactured natural rubber.

latitude The angular distance of any point along a meridian N. or S. of the equator (which is 0°), the Poles being 90°. A circle drawn around the earth parallel to the equator is a parallel of l. and every point on it has the same l. 1° of l. at the equator = 110·551 km. (68·704 mi.), at 45° = 111·130 km. (69·054 mi.), near the Poles 111·698 km. (69·407 mi.), as based on the Clarke spheroid. All parallels of l. intersect all meridians of LONGITUDE at right-angles. [*f opposite*]

latosol Prospsoed in U.S.A. to replace LATERITE, on the grounds that the latter is not sufficiently precise and definitive. It is not much used outside U.S.A.

Laurasia The N. primeval land-mass; see PANGAEA.

laurence Used in U.S.A. for the shimmering effect on a road-surface on a hot day.

lava Molten rock (MAGMA) extruded on to the surface of the earth before solidification. It is classified as: (i) ACID L.; (ii) BASIC L.; and (iii) intermediate l. See AA, PAHOEHOE and PILLOW L.

lavant See BOURNE.

lava tunnel, tube If the surface and sides of a narrow flow of basaltic l. solidify to form a thick outer crust, the hot l. inside continues its forward movement, ultimately draining out of this crust to leave a tunnel. This may later be occupied by an underground stream. If this water freezes in winter, it may not thaw out in summer because the basalt crust forms an insulating layer; e.g. in the Modoc L. Beds National Monument, California.

law A rounded hill in Scotland, esp. near the Firth of Forth; e.g. Traprain L.

lawn (i) An open area of grassland among woodland, notably in the New Forest of Hampshire. (ii) The surface of terraces on the limestone of the Isle of Portland, Dorset.

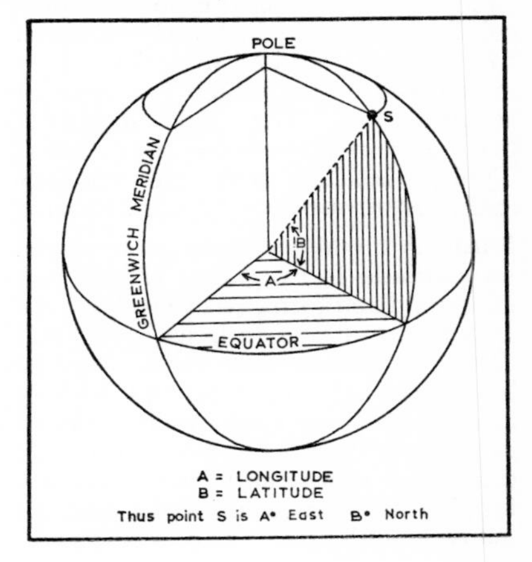

layer-tinting The distinctive shading or colouring of a map between partic. pairs of contours in order to reveal the pattern of distribution of high and low land at a glance. The CONTOUR-INTERVAL at which each colour change is made must be carefully chosen. One alternative is to use a sequence of greens, yellows and browns, perhaps leading to red, purple and even white in high country. Another method is to use one colour only, ranging from the faintest to the darkest possible density, as in the purple-tinted R.A.F. maps. A 3rd method is to merge or grade successive tints so as to avoid the stepped appearance inherent in contour-filling. Line-shading in black can be used, but it should be avoided if possible, owing to its obscuring other detail and its general lack of quality.

'lazy bed' A form of potato cultivation in the W. Highlands of Scotland and the Hebrides, where the seed is placed in a row on the ground, and the soil or turf from either side is placed over it. Thus the potato grows on a raised bed, and the trench between helps drainage, with a min. of labour. In areas of deserted crofts, the pattern of l. b.'s, overgrown with grass and heather, can be clearly seen; e.g. on the Isle of Soay, near Skye.

leaching The removal of soluble salts, esp. bases, from the upper layer (the A-HORIZON) of the soil by percolating soil-water in humid climates. Much of this leached material is held in the B-HORIZON, which is a zone of accumulation. The soils gradually become mineral-deficient and 'sour'. Typical leached soils are PODZOLS and LATERIC SOILS [*f*] SOIL PROFILE

lead A metal occurring as an ore in its sulphide form, *galena* (PbS), also in an oxidized form, *cerussite* ($PbCO_3$). It is mostly found in lodes, veins and 'flats' in sedimentary rocks, esp. in limestone. Silver and zinc are usually associated. L. is easily reduced from galena, and has been used since early times for pipes, and more recently for paint, storage batteries and in many alloys. C.p. = Australia, U.S.S.R., U.S.A., Canada, Mexico, Peru, Yugoslavia.

lead-ratio In a piece of rock containing uranium and/or thorium, the l.-r. is the ratio of lead^{206} to uranium^{238}, of lead^{207} to uranium^{235}, or of lead^{208} to thorium^{232}, formed by the radioactive breakdown of some of the uranium/thorium. As the rate of breakdown is known (e.g. the HALF-LIFE of $U^{238} = 4{\cdot}51 \times 10^9$ years, of $U^{235} = 7{\cdot}13 \times 10^{18}$), an estimate of the age of the rock may be obtained. See ISOTOPE, RADIOACTIVE DECAY.

league An obsolete British unit of length = 3 nautical mi.

leat (i) An open watercourse, commonly conveying water to a mine. (ii) A ditch which contours a hill-side, picking up minor streams, to feed a reservoir; e.g. a l. contours the side of the Ogwen valley in N. Wales, delivering its water into Llyn Cowlyd.

Lebensraum (Germ.) Lit. living-space; a term formerly used by German geopoliticians to justify aggression and expansion of their State, esp. in the Hitlerian era.

ledge (i) An underwater ridge of rocks. (ii) A projecting horizontal, shelf-like mass of rock on a mountain- or cliff-side.

lee, leeward The side away from, or protected from, the wind; e.g. the lee-side of a range of mountains.

lee depression A low pressure system formed in the l. of a mountain range, the result of a dynamically created eddy as an air-stream crosses it; e.g. along the E. margins of the Canadian Rockies; to the E. of the S. Alps in New Zealand. In the W. Mediterranean basin, a cold Polar Maritime air-stream, associated with a COLD FRONT, flows S. over the Maritime Alps towards the sea, where the low pressure is intensified to form a distinct d.

lee-shore The s. towards which a wind is blowing; the danger s. for shipping.

lee wave A w. form in air motion caused by a relief barrier. When a gentle wind crosses a barrier, the effect on the leeward side is limited to simple LAMINAR FLOW or *streaming*. With a stronger wind, a STANDING W. forms, and with still stronger winds a l. w. is formed which may develop into complicated TURBULENCE, sometimes called *rotor-streaming*. This may produce striking BANNER, LENTICULAR [and *f*.], and arched clouds.

legend An explanation of the symbols, shading and colours used on a map, put either as an underline or in a panel on the map.

leiotrichous The quality of having 'straight hair', hence in some classifications of mankind (e.g. by A. C. Haddon) one category is of the *Leiotrichi*. Ct. CYMOTRICHOUS, ULOTRICHOUS.

length A basic unit of l. is the *Ångström Unit* (Å. U.) $= 10^{-10}$ m., or $3{\cdot}937 \times 10^{-9}$ in., used in atomic dimensions (an atom of most elements has a diameter of about 2 Å). See METRE.

1 m. = 1·09361 yds.; 1000 m. = 1 km. = 0·62137 mi.; 1 yd. = 0·9144 m.; 1760 yds. = 1 statute mi. = 1·6093 km.; 1 cm. = 0·3937 in.; 1 in. = 2·5400 cm.; 1 ft. = 0·3048 m.; 1 m. = 3·281 ft.

lenticular cloud (*lenticularis*) A lens-shaped c., often associated with wind eddies over hills and mountains; e.g. the Helm-c. over the Crossfell escarpment in the N. Pennines, lying along or just above the ridge; the 'Long White Cloud' or 'North-West Arch' of S. Island, New Zealand. See HELM WIND. [*f*]

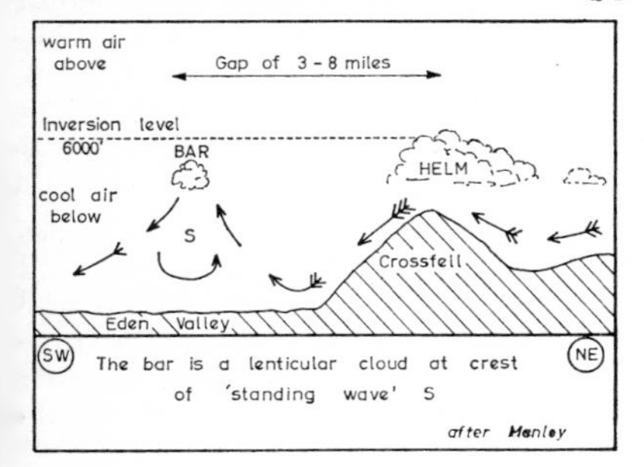

letterpress A method of reproduction of maps (and other material), in which the detail to be printed is raised mirrorwise above the level of the printing plate (raised metal type or blocks in which the detail stands up). When the plate is inked, a sheet of paper will reproduce this detail.

levante, levanter (Sp.) An E. or E.N.E. wind blowing between S.E. Spain and the Balearic I., and in the Straits of Gibraltar, when a depression is in the W. Mediterranean basin. It is gen. a mild and humid wind, and can cause heavy precipitation and floods.

leveche (Sp.) A hot, dry S. wind of SIROCCO-type, blowing from Morocco across to the coast of S. Spain. It occurs in the warm sector of a depression moving E.-ward through the W. Mediterranean basin.

levee, levée (orig. Fr.) (i) A natural bank built up by a stream along the edges of its channel, esp. during flood; when the water subsides the banks remain. The bed of the river thus tends to rise above the surrounding FLOOD-PLAIN. If the l.'s are breached during a subsequent period of river high-water, widespread floods may occur; e.g. the Mississippi, Hwang-ho. (ii) An artificial embankment constructed along the course of a river to check flooding; e.g. along the lower Mississippi.

level Used in a variety of senses where a horizontal line or plane surface is implied. Specif. in a geographical sense it includes: (i) A surveying instrument (ABNEY L., DUMPY L.). (ii) A large tract of l. land, usually drained; e.g. the Bedford L. (E. Anglia). (iii) A horizontal tunnel, passage or gallery in a mine; e.g. the Nentforce L. in a lead-mine near Nenthead, Cumberland; cf. ADIT.

levelling In surveying, the operation of finding the difference in height between successive pairs of points, by sighting through an instrument with a peep-sight or a telescope on which is mounted a spirit-level, on to a graduated measuring-rod. A line of levels can thus be run from the starting height or DATUM, and the heights of points on the line are thus obtained. *Geodetic l.* is of a high order of accuracy, used as the basis of a topographical survey. See ABNEY LEVEL, DUMPY LEVEL. [*f*]

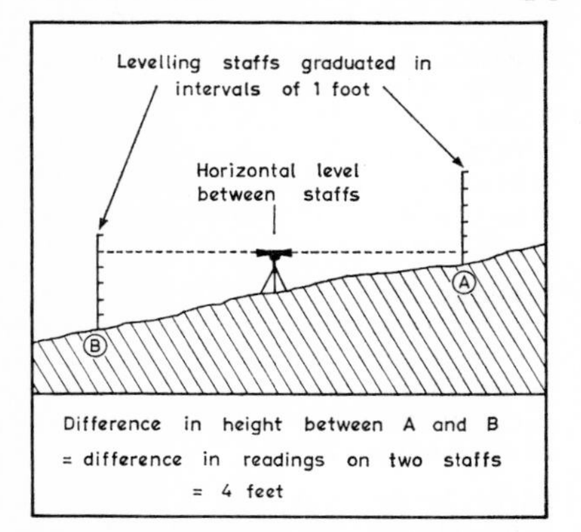

ley An area of cultivated grasses or clover within an arable rotation, remaining down for a few years ('*short ley*') or up to 20 years ('*long ley*') before the field is ploughed.

Lewisian A group of Pre-Cambrian rocks in N.W. Scotland, esp. in Sutherland and Ross, and in the Outer Hebrides. The name refers specif. to the lower of the 2 Pre-Cambrian systems (the other being the TORRIDONIAN), and consists mainly of gneiss, with some altered schists. The L. occurs both as part of a BASAL COMPLEX and as a series of DYKES and SILLS; it has been much altered by earth-movements and metamorphism.

liana, liane A woody climbing plant with roots in the ground and its stem around a tree, found esp. in Tropical Rain-forests; e.g. the rattan in Indonesia, so long and strong that it can be used as a rope.

Lias A Lower Jurassic formation, consisting of clays, shales, limestones and sandstones, conspicuous by the ammonite fossils. It can be traced across England from the Dorset to the Yorkshire coasts in a diagonal outcrop, and occurs also in S. Wales. The term is also used for limestones typically interbedded with shale or clay in other formations.

libeccio (It.) A strong W. or S.W. wind blowing towards the W. coast of Corsica. It has its max. frequency in summer, when it is sometimes boisterous and invigorating, although when it crosses the narrow Cap Corse it may reach the E. coast as a hot, desiccating squall. In winter the wind may bring rain or snow to the W. slopes of the mountains.

lichenometry A method of time-estimation based on the rate of growth of lichen on a stone, thereby giving some indication as to how long the stone (e.g. in a MORAINE) has been lying in that position; this has been used for periods up to 3 centuries. The method must be used with discretion, since the growth rate is affected by so many variables, esp. of climate.

lido (It.) A BARRIER-BEACH, so named after the L. at Venice; now used in the sense of a sophisticated bathing-beach.

light industry The manufacture of articles of relatively small bulk, using small amounts of raw materials; e.g. tools, instruments, radio and television sets. Ct. HEAVY I.

lightning A visible electrical discharge in the form of a flash; see THUNDERSTORM. Many types are defined: *forked*, *sheet*, *ball*, *streak*. It may be within a single cloud, cloud to cloud, or cloud to ground.

light-year The distance l. travels in 1 y. at 186,326 mi. per second ($3{\cdot}04 \times 10^8$ m. per second), i.e. 5,878,310,400,000 mi., or approx. $9{\cdot}7 \times 10^{12}$ km. (6×10^{12} mi.). The nearest star (*Alpha Centauri*) is 4·29 l.-y.'s away from the earth.

lignite Used gen. to denote types of low-grade coal, in structure and carbon-content between PEAT and sub-bituminous coal, as in the production figures below. Strictly, a distinction should be made between l. and BROWN-COAL; l. is stonier, darker in colour, with a slightly higher carbon content, less moisture, and a less evident vegetable structure. L. is mainly burnt to produce heat for thermal-electric generators. C.p. = E. Germany, U.S.S.R., W. Germany, Czechoslovakia, Yugoslavia, Hungary.

limb, of fold The rock strata on either side of the axis, or central line, of a f. [*f* ANTICLINE]

limestone A rock consisting mainly (sometimes defined as 80%, by others 50%) of calcium carbonate ($CaCO_3$); i.e. it is *calcareous*; there is a large number of varieties, defined by texture (e.g. *oolite*), mineral content (e.g. *dolomite*), origin (e.g. *coral*), age (e.g. *Carboniferous*, *Jurassic*). Chalk is also a l., though as the latter is thought of as a resistant hard rock and chalk is manifestly different in appearance, this is often forgotten. See individual types of l.

limnology The scientific study of fresh-water lakes and ponds, dealing with the various physical, chemical and biological conditions and characteristics.

limnoplankton Microscopic organisms, both vegetable and animal, that live in still water (lakes and ponds).

limon (Fr.) A superficial fine-grained deposit, akin to LOESS, from which brown, loamy soils have developed. It is probably of wind-blown origin, laid down during dry, steppe-like interglacial, or immediately post-glacial, times. In Europe wind blowing outwards from high-pressure areas over the N. ice-sheets exercised its sorting effect and removed the finer elements from the vast mass of materials deposited by fluvioglacial action beyond the ice-margins, and transported them to the W. It occurs at all heights in Belgium and the Paris Basin, and even in Brittany, on the highest interfluves and plateaus; thicknesses of 21 m. have been found. A distinction is made between the l. thought to be wholly wind-blown and wind-deposited, and that which may have been reworked and redeposited by later river action.

limonite A yellowish brown ore of iron, hydrated ferric oxide ($2Fe_2O_3 . 3H_2O$), found notably in Lorraine.

limpo (Portuguese) A type of SAVANNA in Brazil, dominated by grasses, with relatively few bushes or trees.

linear eruption Syn. with FISSURE E.

linear town A long, drawn-out urban settlement, orig. formed either because of site restrictions, or as the result of clustering along a main routeway. This urban form is becoming popular with some planners; by focusing along a routeway, rapid transportation between points is aided, and the congestion problems associated with the more normal circular city are avoided.

line-squall A sharply-defined COLD FRONT, associated with very stormy conditions, along a line as much as 480 km. (300 mi.) in length, with dark rolls of cloud, downpours of rain and hail, and violent gusty winds, sometimes of gale force.

linkage analysis The quantitative reduction of a mass of data into a number of interconnected classes, using a selected series of factors; e.g. the reduction of the number of regions into homogeneous groups, inevitably losing detail at each stage of the grouping. This technique has only become really practicable with the availability of computers. See also INDUSTRIAL L.

links Gently undulating sandy ground, with dunes, coarse grass and shrubs, near the sea-shore in Scotland and N.E. England; e.g. The L.'s, Holy I., Northumberland.

Linmap A computer-mapping routine (*line-printer mapping*), using a standard computer line-printer to produce diagrammatic maps of census and other data. It will accept magnetic tape input, and works from a system of data identification based on grid co-ordinates.

linseed Seeds of the flax plant from which oil is produced by crushing (hence 'linen-seed'). It forms a drying-oil used in paint, and for making linoleum. C.p. = U.S.S.R., U.S.A., Canada, Argentina, India.

lithification The conversion of an accumulation of sediments into rock, through compression, compaction and cementation.

lithogenesis The accumulation of sediment, sand and mud in the sea (notably in a GEOSYNCLINE), later compacted to form solid rocks.

lithography The art of drawing mirrorwise on stone in a greasy ink and transferring the image to paper, formerly used extensively in reproducing maps, invented in 1798. The present offset PHOTOL., using specially prepared grained aluminium or zinc plates, are direct descendants of the original process, as are various other methods of reproduction.

lithology The study of rocks in connection with their physical, chemical and textural character (hence *lithological*).

lithometeor An ensemble of solid, non-aqueous particles in the atmosphere; e.g. HAZE, DUST, smoke, DUST-STORM, SAND-STORM.

lithosol Used in U.S.A. for an AZONAL soil consisting of stony unweathered or part-weathered rock-fragments, scree, boulder-clay; in Britain called *scree-soils*.

lithosphere Syn. with the CRUST of the earth.

litre A metric unit of volume; 1 kilogram of water at 4°C. and 760 mm. pressure = 1000·027 c.c., usually accepted as 1000 c.c. 1 l. = 0·219976 Imperial gallons.

'Little Ice Age' (i) A name for the period (*c.* 3000–500 B.C. in E. Anglia, though this varies with latitude, i.e. is *time-transgressive*), equivalent to the SUB-BOREAL stage, when the climate became cooler than in the previous few millennia. Glaciers which had disappeared from parts of Alaska and the Sierra Nevada once again developed. It has been proposed to drop this somewhat journalistic name in favour of NEOGLACIAL. (ii) Used for a much more recent advance of the glaciers in Switzerland during the 16th and 17th centuries. Records show that villages were overwhelmed by i., summer pastures were no longer usable, and former passes were blocked. Latterly there has been a gen. overall, though fluctuating, shrinkage. In Alaska there seems to have been a maximum advance *c.* A.D. 1850.

littoral (i) App. gen. to the seashore. (ii) The zone between high- and low-water marks, sometimes termed the *enlittoral zone*. Other authorities, esp. in connection with living organisms, extend the zone more widely, some to 200 m. depth, though the area between 60 and 200 m. depth is usually specif. denoted as *sub-littoral*. [*f* ABYSSAL ZONE]

littoral deposit A d. of sand, shells and shingle between high- and low-water marks, though sometimes it includes all shallow-sea d.'s, including offshore mud.

llano (Sp.) Tropical grassland or SAVANNA on the Guiana Plateau in S. America. Initially it was a term simply for 'plain'; by extension, the word has come to mean the vegetation found in such a situation. This vegetation reflects the summer rainfall régime (April to Oct.).

load The material transported by a natural agent of transportation, one of the integral parts of the process of denudation, used specif. of a river. Material is carried: (*a*) in *solution*; (*b*) in *suspension*; (*c*) by SALTATION (*bed-l.* or *traction l.*). The COMPETENCE or ability of a stream to move particles of a certain size is thought to be proportional to the 6th power of its velocity (SIXTH POWER LAW). A stream can carry a much larger l. of fine material than of coarse. The term l. is sometimes also used in respect of the material carried by a glacier, the wind or moved by waves, tides and currents.

loam soil A permeable, friable mixture of particles of different size, forming: (i) *sandy l.*; (ii) *silty l.*; (iii) *clay l.*, according to the proportion of constituents. The term has been given a more precise definition in U.S.A. (where it contains 7–27% clay, 28–58% silt, and 30–52% sand); particles of sand are 1·0–0·05 mm. in diameter; silt, 0·05–0·005 mm.; clay less than 0·005 mm. (U.S. definition).

lobe (i) A rounded tongue-like mass, used esp. of: (*a*) ice projecting from a larger sheet; (*b*) a mass of wet clay moving down a steep slope on to a beach; and (*c*) a tongue of DRIFT projecting further than the main mass. (ii) In U.S.A. it refers to the land enclosed by an acute meander of a river.

local climate The c. of a small area which possesses marked contrasts with other areas within a short distance, resulting from minor differences of slope and aspect, colour and texture of the soil, the proximity of a water-surface, the nature of the vegetation cover, and the effects of buildings. A large number of carefully sited recording stations is needed over the area under consideration.

local relief The difference in altitude between the highest and lowest points in a limited area under consideration. Syn. with *relative relief*.

local time Syn. with APPARENT T.

location factor Introduced by P. Sargent Florence in 1937, indicating a comparison between the proportion of occupied persons in a given industry in a given area, and the corresponding proportion for the country as a whole.

locational (localizational) coefficient A statistical measure used to determine the rel. amount of any function, or of any population, present in the sub-area of a large region. Various formulae and methods are available. E.g. P. S. Florence obtained a l. c. by summing the positive or negative deviations of the percentage of all workers in a particular industry in each sub-area from the total occupied population in the whole region as a percentage of all workers. A c. of 0·0 would represent complete coincidence of the selected industry with all occupations, and 1·0 would represent complete differentiation.

locational theory An approach to geography, esp. developed by those with interests in economics; e.g. J. H. von Thünen (1875) on agricultural location, A. Weber (1909) on industrial location. More recently it has been developed through concepts derived from topological mathematics and geometry, as by W. Bunge, B. J. L. Berry and W. L. Garrison.

loch (Scottish) (i) A lake in Scotland; e.g. L. Lomond, Rannoch, Laggan. (ii) A long narrow arm of the sea on the coast of Scotland; e.g. L. Linnhe, Fyne, Hourn.

lochan (Scottish) A small lake in Scotland, usually lying in a coire (CIRQUE); e.g. L. Meall an t-Suidhe on the slopes of Ben Nevis; L. Lagan in the Black Cuillins of Skye.

lode (i) An artificial channel or watercourse, usually embanked (E. Anglia). (ii) A mineral vein, or closely parallel veins of ore, a term used esp. in Cornwall. It would seem to orig. from the fact that the miner could be led or guided in his search for ore by following a l.

lodgement till Sometimes used as syn. with GROUND-MORAINE.

loess, löss (Germ.) Derived from the name of a village in Alsace, used to indicate local deposits by farmers and brick-makers and adopted as the name of a fine-grained, coherent, friable, porous, yellowish dust. It was probably initially removed by the wind, either from desert surfaces where the loose material was unprotected by vegetation, or during dry inter- and post-glacial periods from unconsolidated glacial and fluvioglacial materials. L. was studied by v. Richthofen in N.W. China, where a sheet, covering nearly 650,000 sq. km. (250,000 sq. mi.), swathes the landscape to a depth of from 90–300 m. (300 to 1000 ft.). It occurs here at all elevations, from near sea-level to 2500 m. (8000 ft.). It reveals innumerable vertical tubes, lined with calcium carbonate, thought to be the 'casts' of grass stems. It was deposited where vegetation had some binding effect, and possibly where increased rainfall helped to wash it down from the air to the ground. It is also found in Europe S. and W. of the terminal moraines of the glacial advance, from E. Germany, through W. Germany (the Bördeland), Belgium and France (see LIMON). In Europe much may have been reworked and redeposited by running water. It is also found in the W. states of U.S.A. in the Mississippi–Missouri River valleys and in Argentina. From it develop fine-textured, easily worked, deep, well drained soils, good for wheat and sugar-beet.

loessoïde Given by Dutch geologists to deposits in S. Limburg believed to be of loessic origin, but reworked and redeposited by stream-action, possibly with an admixture of residual materials formed by the decomposition *in situ* of the Upper Chalk. The resultant soils are referred to as *loessleem*, and are extremely fertile.

log An apparatus for measuring the speed of a ship, which came into gen. use at the beginning of the 17th century. The *common l.* was replaced by a mechanical or *patent l.* at the beginning of the 20th century. The common l. consisted of a *l.-ship*, a flat piece of wood weighted with lead, secured to a *l.-line* by a *bridle*, the l.-line fastened to a *reel*. The l.-line was divided into equal parts by knotted cords. The l.-ship was thrown overboard, and a *l.-glass* (a sand-glass) was used to measure the number of knotted cords, hence the term KNOT.

logan A rocking-stone, after a granite block in the Land's End area of Cornwall, very delicately balanced on its base. These rocks have become individual masses by chemical weathering along the joints in the granite. See TOR.

logarithmic scale A s. in which an increase of one unit represents a power increase in the quantity involved. When it is desired to plot rates of change (output, population), a vertical log-s. is used, with a horizontal linear s. for the time intervals; such semi-logarithmic graph-paper is available. Log log graph-paper, with both horizontal and vertical log-scales, is used for some frequency graphs.

longitude The angular distance of a place E. or W. of the prime (0° or Greenwich) MERIDIAN, measured in either direction to 180°. A circle drawn around the earth through each Pole is a meridian of l., and cuts all parallels of latitude, including the equator, at right-angles. All places on the same meridian have the same l. As meridians converge towards the Pole, the length of 1° of l. becomes less. 1° of l. at the equator = 111·320 km. (69·172 mi.); at latitude 45° = 78·848 km. (48·995 mi.); at 70° = 38·187 km. (23·729 mi.); at the Poles = 0. 15° of l. = 1 hour in local time. [*f* LATITUDE]

longitudinal coast See CONCORDANT C. [*f*].

longitudinal crevasse See CREVASSE.

longitudinal valley A term sometimes used when a v. is parallel to the gen. trend of mountain ranges (cf transverse v.'s). E.g. the *vaux* (sing *val*) in the Jura Mtns. Strictly and correctly, the term should be reserved for a v. developed along the STRIKE of the strata.

longitudinal wave A form of shock w. produced by an earthquake, which travels along the surface of the ground its nature controlled by the elasticity of the strata. It therefore arrives after the *P*- and *S*-*w.'s*, which have taken more direct courses through the earth's mass. They are subdivided into L_Q (*Love w.'s* or *Querwellen*) and L_R (*Rayleigh w.'s*), now often referred to as *Q* and *R* waves. [*f* EARTHQUAKE

Longmyndian A series of mudstones and sandstones, once thought to be Cambrian, now definitely known to be of Pre-Cambrian age. They outcrop typically in the Longmynd, in S. Shropshire.

long-profile, of a river The PROFILE of a r. bed, drawn from source to mouth.

long-range forecast A f. for a period greater than 5 days. These have been issued by the U.S. Weather Bureau for some time; public forecasts of this kind for a month ahead were started in Britain in Dec. 1963.

longshore drift A d. of material along a beach as a result of waves breaking at an angle. A breaker (the swash) sweeps material obliquely up the beach, the backwash drags some of it down again at right-angles to the shoreline, thus there is a net movement along the beach; e.g. to the E. along the S. coast of England, since the dominant winds and waves are from the S.W. A l. current will also help this movement. [*f*]

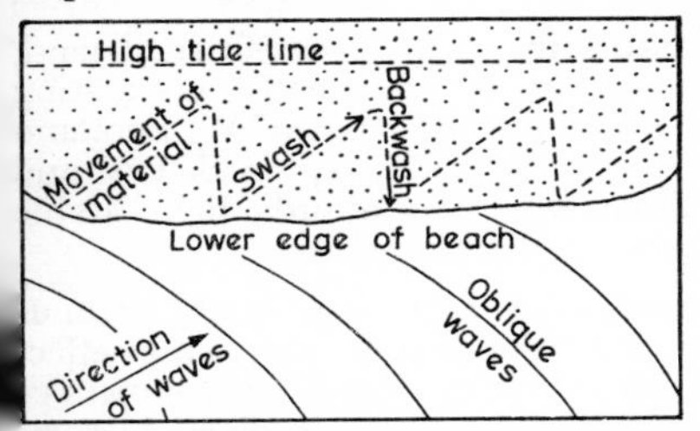

'long time-scale' A t.-s. of the PLEISTOCENE glaciation, in which its onset is pushed back to 1·8 to 2 million years ago. Ct. 'SHORT T.-S.'. Much of the evidence for this proposed lengthened span of the Pleistocene is based on the radiochemical analysis of CORE SAMPLES from the deep-sea sediments of the Atlantic Ocean. One such study indicated that the average rate of accumulation of the sediments was 2·5 cm. per millennium. The whole Pleistocene section, its start delimited by an abrupt change from warm water to cold water planktonic foraminifera, totalled [illegible]0 m.

lopolith A large-scale saucer-shaped intrusion of igneous rock lying CONCORDANT with the strata, and forming a shallow basin; e.g. the Bushveld L. in the Transvaal, more than 480 km. (300 mi.) across; the rhyolite plateau of Yellowstone National Park between the Gallatin and Absaroka Mtns. [*f*]

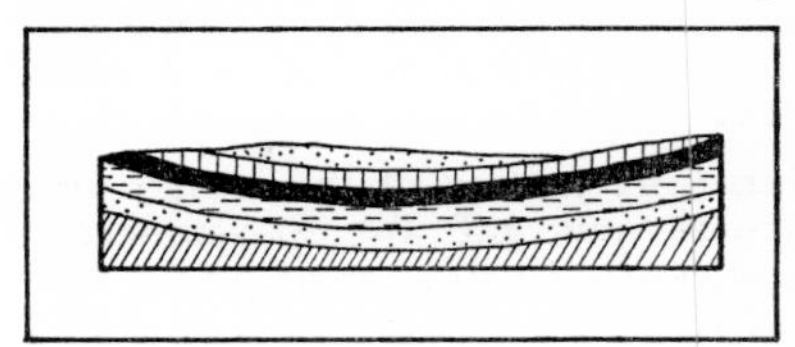

Lorenz Curve A c. drawn on squared graph paper, using % values on each axis, designed to show, by means of its degree of concavity, to what extent a particular distribution (e.g. a concentration or clustering of population) is uneven in cf. a uniform distribution (which would be shown by a straight line crossing the graph at an angle of 45°).

lough (Irish) A lake or arm of the sea (equivalent to Scottish loch); e.g. L. Neagh, Derg.

'low', atmospheric A DEPRESSION, or low pressure system in the atmosphere, shown on a chart by closed isobars of diminishing values towards the centre. [*f* DEPRESSION]

low, beach A long shallow depression on a b., roughly parallel to the shoreline, separating 2 ridges or FULLS. This is also known as a *swale*. E.g. on the Lancashire coast near Formby; on Dungeness.

lowland A vague word with no precise meaning, gen. referring to land below 180 m. (600 ft.), though distinguished rather by its contrast with adjacent higher land.

loxodrome See RHUMB-LINE.

lucerne See ALFALFA.

lunar day The period of time taken by the earth in rotating once in respect to the moon, i.e. between 2 successive crossings of the same meridian, which is 24 hours, 50 mins. This is because while the earth is rotating once in 24 hours, the moon has its own orbital motion around the centre of gravity of the moon and earth, and so crosses each meridian 50 minutes later. This is the cause of the interval of about 12 hours, 25 minutes between 2 successive high tides; i.e. any high tide is 50 minutes later than the corresponding tide of the previous day.

lune A portion of the surface of a sphere cut off by 2 semi-GREAT CIRCLES (i.e. half a great circle, with spherical distance of 180°). Area of l (where lunar angle is θ) =

$$\frac{\theta}{360} \times 4\pi R^2 = \frac{\theta . \pi R^2}{90}$$

lutite A SEDIMENT or SEDIMENTARY ROCK consisting entirely of fine clay-particles (each less than 0·002 mm. in diameter); i.e. ARGILLACEOUS. E.g. MUDSTONE.

L-wave See LONGITUDINAL W. (of an earthquake).

lynchet A man-made terrace on a hill-side, usually parallel to the contours. It is ascribed to ancient cultivation practice (from Iron Age or earlier), and was constructed to provide a level, well-drained strip of land with a S. aspect, and also to check soil-erosion. L.'s are found esp. on the chalk country in S. England, and also on the N. sides of the Yorkshire Dales. [*f*]

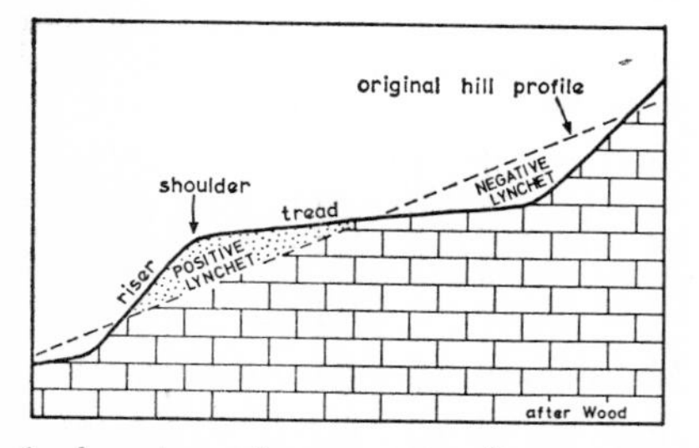

lysimeter An apparatus for measuring the quantity of water percolating through the soil; it consists of a container filled with the material under examination, with measuring devices to assess the amount of water and dissolved materials which have passed through.

Maar (Germ.) A small, near circircular sheet of water situated in an explosion-vent, the result of an eruption which has blown a hole in the surface rocks, surrounded by a low crater-ring of fragments of the country rock, but accompanied by no extrusion of igneous rock. The term is derived from the Eifel Mtns., W. Germany; e.g. Merfeldmaar, Pulvermaar, Lachermaar.

macadam Road-building method named after J. L. McAdam, who found that certain types of angular stones, roughly of uniform size and preferably calcareous, would form a hard and permanent surface if rolled or compressed. Orig. merely water was sprinkled before rolling, hence 'water-bound'; now the stones are usually coated with tar, hence *tar-m.*

macchia (It.) See MAQUIS.

machair A fine whitish shell-sand, found in broad, gently undulating tracts along the coast of W. Scotland and the Hebrides, esp. in S. Uist and Tiree, where it affords light arable soils for the crofters.

mackerel sky See CIRROCUMULUS CLOUD. Sometimes the same effect appears with small ALTOCUMULUS CLOUDS. The name is given because the cloud-pattern resembles the scales of a m.

macroclimate A broad large-scale climatic region, as distinguished orig from a MICROCLIMATE.

maelstrom (Dutch) (i) A whirlpool. (ii) A powerful eddy in a tidal current in a restricted irregular channel; e.g. the famous M. of tidal origin in a channel between the Lofoten I.'s, Norway.

maestrale (It.) The It. name of the wind known as the MISTRAL.

mafic mineral Syn. with FERROMAGNESIAN M.

magma Molten rock material under the surface of the earth at a very high temperature, charged with gas and volatile materials, and under enormous pressure. The m. is probably formed in local concentrations at a depth of 16 km. or more, and cannot be regarded as a continuous layer; the fusion of its constituents may be due to the local accumulation of radioactive heat. It consists chemically of a solution of a wide range of elements, mainly in oxide form, including silica and basic oxides, the relative proportions of which determine whether it is an ACID or BASIC m. When it solidifies under the surface, INTRUSIVE (PLUTONIC) rocks are formed. If it reaches the surface, much of its gas and water is lost and it becomes LAVA, from which EXTRUSIVE, *eruptive* or *volcanic* rocks are formed upon solidification. Hence *magmatic differentiation* or *segregation*, the process by which different individual igneous rocks are formed from a single m. See also STOPING.

magmatic water See JUVENILE WATER.

magnesian limestone A l. with a certain proportion of magnesium carbonate. With capitals, it is a series of Permian age. See DOLOMITE.

magnesite Magnesium carbonate ($MgCO_3$), an ore of magnesium. It is one of the sources of the metal magnesium, others being magnesium chloride and sea-water. M. is also used for making refractory bricks for lining basic open-hearth and electric steel-converters, and for making special cements. C.p. = Austria, Czechoslovakia, Yugoslavia, U.S.A., Greece, India.

magnetic declination (or **variation**) The angle at any point on the earth's surface between the m. needle of a compass (pointing to the Magnetic Pole and indicating the m. meridian) and True North (or the geographical meridian), expressed in degrees E. and W. of True North. M. d. varies considerably in different parts of the world, and also with time. In Gt. Britain it was about 8°W. in 1968, decreasing by ½° every 4 years; i.e. if the rate is sustained m. d. will be zero in about A.D. 2033. At present it is zero on a meridian passing approx. through Cincinnati, Ohio, the AGONIC LINE [*f*].

magnetic pole One of the 2 poles of the earth's m. field, situated in N. America and Antarctica, and indicated by a free-swinging m. needle in a horizontal plane. The locations of these shift in a complex way, the N. one in 1968 being situated in Canada near Prince of Wales I. (about 73°N., 100°W.), the S. one in S. Victoria Land in Antarctica. Nor are these poles at the extremities of a diameter of the earth, for a line joining them misses the centre of the earth by about 1200 km. (750 mi.).

magnetic storm A sudden and temporary, sometimes worldwide, disturbance of the earth's m. field, which can have serious effects on m. surveys and short-wave radio. The cause is not fully known, but it seems to be associated with an active period of sun-spots and solar flares, and with the AURORA.

magnetite An ore of IRON (Fe_3O_4); this is a valuable source of iron in N. Sweden and the Urals. Known as *lodestone*.

magnetosphere The outermost part of the earth's atmosphere, from about 2000 km. (1200 mi.) upwards, within which lie the Van Allen radiation belts (at about 4000 km., 2400 mi.).

magnitude, of an earthquake A scale, devised by C. F. Richter in 1935, of earthquake shocks based on instrumental records. It is an index of the earthquake's energy at its source. The numbers range from 1 upwards, the largest earthquake so far recorded being 8·6. The m. differs from the *intensity* of an earthquake, which is related to the surface effects of the waves; see MERCALLI SCALE.

maidan An open space in or near a town in India.

maize A cereal crop originating in the New World, growing in a sub-tropical or warm temperate climate, and producing a tall stalk with a large 'cob' or ear. Modern hybrids have been developed, esp. in the U.S. 'Corn Belt'. The bulk of the m. crop is fed to cattle and pigs ('marketed on the hoof'), though in Africa much is eaten by Man as 'mealies'. C.p. = U.S.A., China, Brazil, Mexico, U.S.S.R., Roumania, Yugoslavia, Argentina.

mallee A dense scrubby thicket of dwarf eucalyptus, growing to about 2 m. in height in arid parts of S.E. and S.W. Australia.

malpais (Sp.) An area of rough, barren lava-surface, so called because of the difficulty of crossing it.

Malthusianism A doctrine derived from the writings of T. R. Malthus (1766–1834), esp. in his *Essays* of 1798 and 1803. His basic idea was that population increases in a geometric ratio, while available resources for subsistence increase only in an arithmetic ratio. Population thus grows up to and beyond the limit of subsistence, and can be checked only by famine, war and pestilence. He was writing chiefly in respect of the unemployment and economic distress in England of the early Industrial Revolution. 19th century-colonialism and the opening up of new areas of arable land prevented the development of a Malthusian situation in many parts (e.g. in W. Europe), but it is a permanent problem in many parts of Asia (e.g. India), esp. where the death-rate has been decreased by medical science and hygiene, while the birth-rate continues to increase. This is sometimes referred to as *Neo-M*.

mamélon (Fr.) Syn. with CUMULO-DOME

mammatus App. to breast-shaped clouds, usually associated with convectional conditions and the piling-up of thunderclouds.

mammilated surface Rock s.'s which have been smoothed and rounded by various agencies, esp. glacial action; e.g. in the Canadian Shield, Finland, Sweden.

mandate, mandated territory An area established under Article 22 of the League of Nations covenant, whereby the former German colonies and parts of the old Turkish empire were placed under the tutelage and administration of various Allied powers, appointed by and responsible to the League. They were then divided into 3 groups: A. those to attain full independence in the near future (Iraq and Palestine); B. those to be administered as colonies (Tanganyika); C. those to be regarded as an integral part of the mandatory power's territory (W. Samoa to New Zealand, S.W. Africa to the Union of S. Africa, German New Guinea to Australia). Other m.'s created were Syria (France), Ruanda (Belgium), some Pacific islands (Japan), Cameroons and Togoland (jointly France and Gt. Britain). The m. was replaced by the TRUST TERRITORY under the charter of the U.N.O., and most have now achieved independence.

manganese A metal found mainly in concentrations of m. oxide in bands and nodules in clay. Its chief importance is in steel manufacture; it is added in the furnace in the form of ferro-m. (80% *Mn*) or *spiegeleisen* (a ferro-alloy with 20% *Mn*), to act both as a cleanser of impurities and to harden and toughen the steel. C.p. = U.S.S.R., S. Africa, India, Brazil, Gabon, China, Ghana, Morocco.

mangrove (i) A collective name for some genera and species (*Rhizophora* and *Bruguiera*), which have the ability to grow on tide-washed mud-flats in the tropics. They have short stumpy trunks, sometimes supported by a maze of aerial roots, or by roots which bend at right-angles ('knees') others send out horizontal roots from which other vertical ones grow up through the mud. These root-systems both anchor the m. in the mud, and act as aerating organs. (ii) A swamp or swamp-forest composed of a dense growth of such m.'s; e.g. along the coast of S. America near the Amazon delta; along the edge of the Niger delta; along the coasts of Sumatra and Borneo.

manor From the Lat. *manerium* and Fr. *manoir*, introduced into England after the Norman Conquest to denote a feudal unit of a house and land, as recorded in *Domesday Book*. Part of the estate was retained by the lord (the *demesne*), the rest was let to various grades of tenants on differing terms of rents and services. M.'s became very diverse in terms of size, land-use, obligations, agricultural methods, and esp. after the Agricultural Revolution their importance dwindled, the last legal survivals being abolished in 1926. The 'm. house' and the courtesy title of 'lord of the m.', however, still survive in places.

mantle (i) A layer of ultrabasic rocks, density 3·0–3·3, 2900 km. (1800 mi.) thick, lying between the CRUST and the CORE of the earth (formerly called the 'lower layer'). Its upper surface (at about 32 km. (20 mi.) under the continents, 6–10 km. (4–6 mi.) under the oceans) is the MOHOROVIĆIĆ DISCONTINUITY, its lower surface is the GUTENBERG DISCONTINUITY (at about 2900 km. down). (ii) The surface accumulation of soil and weathered rock; the equivalent of REGOLITH. [*f* ISOSTASY]

manufacturing industry The conversion of a raw material into fabricated articles; one branch of i. in its broad sense. The opposite of the implied lit. meaning ('made by hand') is usually to be inferred; i.e. the making of articles on a large scale by machines.

map A representation on a plane surface (paper, card, plastic, cloth or some other material) of the features of part of the earth's surface, drawn to some specific scale. It obviously involves certain degrees of generalization, selective emphasis and conventionalization, according to the scale and the detail involved. A m. on a large scale, on which everything is drawn exactly to scale, is usually called a PLAN.

map projection The representation of the earth's parallels and meridians as a net or GRATICULE on a plane surface. Some p.'s are theoretically constructed on a developable surface, i.e. cone, cylinder or plane, capable of being laid out as a plane on to which the graticule is 'projected' geometrically or by calculation, though many conventional p.'s are not constructed in this way. See under the names of individual p.'s.

maquis (Fr.) A low scrub of evergreen aromatic plants, found in the Mediterranean lands; it includes oleander, rosemary, heath, arbutus, lavender, myrtle, a profusion of interlacing creepers, vines, herbaceous and bulbous plants. It is partly a reflection of the Mediterranean climatic régime of summer aridity, partly the result of the felling of former evergreen oak forests and therefore a degeneration of the vegetation cover. It is usually found on siliceous soils, being replaced by GARIGUE on limestone. This scrub formed helpful cover for the French Resistance Movement during the war of 1939–45; hence the name was assumed by the people so engaged.

marble A crystalline limestone which has been metamorphosed by pressure and heat, so forming a hard, patterned shiny rock; it is used for decorative purposes. Sometimes the name is given loosely to any sort of decorative stone that will take a polish.

march, marchland A frontier zone or debatable land between 2 states; e.g. the parts of England bordering Wales and Scotland: the Welsh and Scottish M.'s. In another form, *mark;* e.g. the mark-state of Austria. Note also 'the E. Marchlands of Europe' (H. G. Wanklyn). [*f*]

'mares' tails' Long drawn-out wispy CIRRUS CLOUDS, indicating strong winds in the upper atmosphere.

marginal deep A long narrow trough in the ocean floor, parallel and close to an island arc; e.g. the Sunda D. parallel to Sumatra and Java.

marginal land L. which is barely worth cultivating, or which may or may not be cultivated according to changes in economic conditions, or according to the length and nature of wet and dry seasons. Some l. (e.g. bordering a desert) may be fluctuatingly marginal, other l. (e.g. an altitudinal zone above a valley-floor on a hill-farm) may be permanently marginal.

marginal sea A semi-enclosed s. that borders a continent, and lies on a submerged portion of a continental mass, rather than within an ocean basin; e.g. S. of Okhotsk, S. of Japan, Yellow S., Sulawesi (Celebes) S., Baltic S., North S., Mediterranean S., Red S., Persian Gulf.

marin A moist warm S.E. wind, blowing esp. in spring and autumn across the coast of S. France, the result of a depression in the Gulf of Lions.

maritime air-mass See AIR-MASS.

maritime climate A climatic régime experienced on islands and near coasts (particularly W. coasts in mid-latitudes), usually with a small seasonal and diurnal temperature range, and with appreciable cloud and precipitation. It can occur in any latitude; e.g. Tropical M. (Köppen's type *Af*), Humid Sub-tropical (*Cfa*), Cool Temperate M. (*Cfb*, *Cfc*).

market-gardening A form of intensive cultivation of vegetables, fruit or flowers, comparable to the American *truck-farming*, although the British farmer usually grows a greater range of crops. However, in England recently there has been a decline in traditional methods, with its replacement by intensive production in glass-houses, and by more extensive cultivation of vegetables using farm methods.

market-town A large settlement which grew up at a convenient centre of communications where people could buy and sell goods. It is therefore a common place-name element, esp. of small country towns; e.g. M. Drayton, M. Harborough, M. Rasen.

marl (i) A clay with an admixture of at least 15% calcium carbonate. (ii) A term used loosely for any friable clay soil. (iii) In geology, the proper name of several particular types of rock; e.g. Keuper M., Chalk-m.

marling The application of mar (or even just clay) to a loose sandy soil to help its consistency and moisture-retaining capacity.

marsh (i) *Coastal*: see SALT-M. (ii *Inland*: an area liable to temporar inundation, but the land is usuall wet and ill-drained; rivers frequent ly overflow, and much silt is de posited. There are extensive sheet of shallow water, with rushes, reed and sedges, and occasional wate tolerant trees (e.g. alder). Whe drained, a m. is usually fertile, wit its mineral-rich soil; e.g. the sil lands of the N. Fen District, Englan the Pontine M.'s, Italy (both no largely drained). Ct. BOG, FEN, SWAM

marsh gas Methane (CH_4), the chief constituent of natural g., also resulting from the decaying of vegetation in marshes. The same gas may occur in coal-mines as *fire-damp*.

mascaret (Fr.) See BORE.

massif (Fr.) A compact plateau-like mass of uplands, usually with clearly defined margins; e.g. the M. Central of France. The term is applied to most of the Hercynian blocks in Europe. It is also used of a distinct mountain group; e.g. the Mont Blanc m.

massive App. to a rock which is markedly free from stratification, bedding-planes, jointing and cleavage; a thick uninterrupted stratum.

mass-movement, mass-wasting The downward movement of material on a slope under the influence of gravity, usually lubricated by rain-water or snow-melt. The movement may be: (i) *slow*: soil-creep, rock-creep, scree- (or talus-) creep, SOLIFLUCTION; (ii) *rapid*: EARTH-FLOW, EARTH-SLIDE, SLUMP, ROCK-SLIDE, ROCK-FALL. A classification of such m.-m. by C. F. S. Sharpe (1938) has been gen. accepted. See individual types.

mass production The large-scale, continuous p. of a specif. manufactured item, using the principle of the 'conveyor-belt'.

'Master Plan' A gen. strategic p. showing basic land-use zones and communications networks over a planning area; e.g. Abercrombie's *Greater London Plan* (1944). Now superseded by STRUCTURE PLAN under the Town and Country Planning Act (1968).

mature An advanced stage in the development of a landscape, river, shoreline or soil. This term is much used for partic. types of landscape, following the writings of W. M. Davis, though there has been considerable criticism of the use of such terminology.

maximum thermometer A t. which registers the max. temperature recorded over an interval of time. Various types are used. One consists of a metal rod within the sealed glass tube of the t., resting on the mercury meniscus. As the temperature rises this is pushed upwards, but it so fits into the tube that when the mercury falls again (with a drop in temperature), the marker is left at the highest point reached. This can be read later.

Mayen (Germ.) An intermediate stage in the seasonal movement of cattle from a valley floor to high pastures (see TRANSHUMANCE) in the Swiss and French Alps.

Meade's Ranch A geodetic station in central Kansas, U.S.A., approx. in the centre of U.S.A., the latitude and longitude of which have been precisely fixed in relation to the Clarke Spheroid of 1866 (lat. 39°13′ 26″·286, long. 98° 32′ 30″·506). It is the starting point for the geodetic triangulation of U.S.A., Canada and Mexico. The positions of all other primary control points were related to M.'s R., a system known as the *North American Datum of 1927*. Topographical surveys on all scales are related to this.

meadow An area of grassland which is mown for hay, in ct. to that grazed by animals (PASTURE). M. grasses are usually taller than turf grasses. American usage often implies marshy conditions, while occasionally English usage is restricted to land near rivers or on valley-floors (see WATER-M.). It is often used for grassland among wooded mountains, or for a grassy glade.

meadow soil An INTRAZONAL S. formed in the flood-plains of rivers where flooding occurs for a short part of the year, resulting in the deposition of a thin layer of silt and mud, but with considerable growth of vegetation at other times. The A-HORIZON is dark, with much organic material, with an underlying GLEI horizon, where waterlogging prevents oxidation and the presence of ferrous salts gives the s. a bluish-grey appearance, with occasional iron concretions.

meander A curved, loop-like bend or sinuosity in the course of a sluggish river, or of a valley (*valley-m.*);

derived from the Maiandros R. in Asia Minor. Various types of m. have been distinguished, such as INCISED, INTRENCHED and INGROWN. As the current flows round a bend, the curve is accentuated, since the water impinges most strongly on the concave side (the outside of the curve), causing max. erosion there, even undercutting (forming a *bluff* or *river-cliff*), while there is little erosion and usually deposition in the slack of the current on the inside of the bend (*slip-off slope*). Sometimes an initial slight bend in a river may be transformed into a m. M.'s may develop until only a thin neck separates the stream on each side, and a 'cutoff' or OXBOW lake may then be produced, with a remnant of the spur left as a M.-CORE. Gradually the m.'s will move downstream. However, many m.'s are not initiated in this way. The size (or wavelength) of m.'s and the width of the m.-belt appear to be related to such factors as the discharge and bedload of the river, and to varying depths as the stream crosses an uneven surface, or from non-resistant to resistant rocks. [*f*]

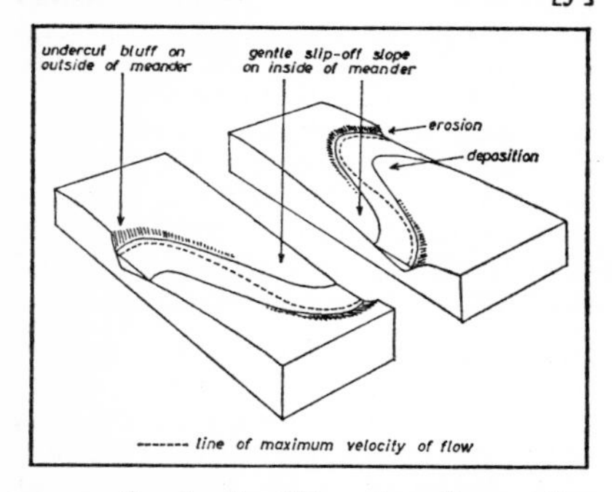

meander-belt The flat floor of a valley between the outside limits of successive m.'s, within which the migration of a stream and its channel occur. Gen. with an increase in the discharge of a stream, or a reduction in its load, the wider the m.-b. becomes. [*f opposite*]

meander-core The area of land almost surrounded by the nearly complete circle of a river in an INCISED M. E.g. Palace Green, Durham, nearly surrounded by the R. Wear; there are many examples in the Moselle valley, where a castle sometimes crowns a m.-c. Some definitions limit the term m.-c. to the isolated hill left when the river completely breaks through the 'neck of' the m.; e.g. in the Wye valley near Redbrook; the Dee in N. Wales to the N. of Llantisilio.

meander-scar A m. abandoned by a river (see CUTOFF), filled in by deposition and vegetation, but still traceable.

meander-terrace An unpaired river-t., formed when a river is meandering freely, though eroding vertically to some extent. As it swings across the valley, it removes part of a fomer higher level of its flood-plain, thus leaving a portion as a higher t.; e.g. in the Val d'Hérens, Switzerland. [*f*]

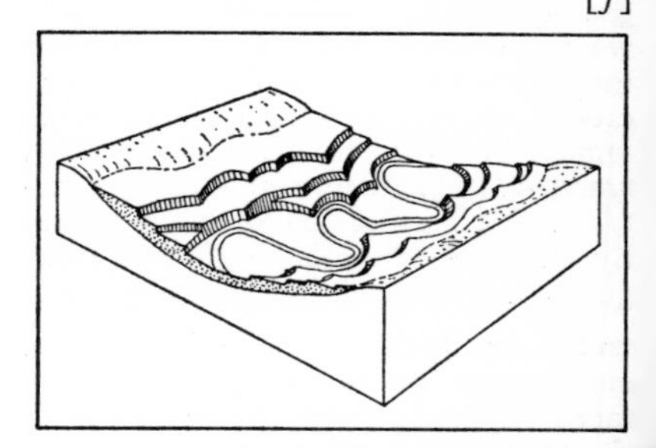

mean sea-level An average level of the sea, calculated from a long series of continuous records of tidal oscillations (see ORDNANCE DATUM)

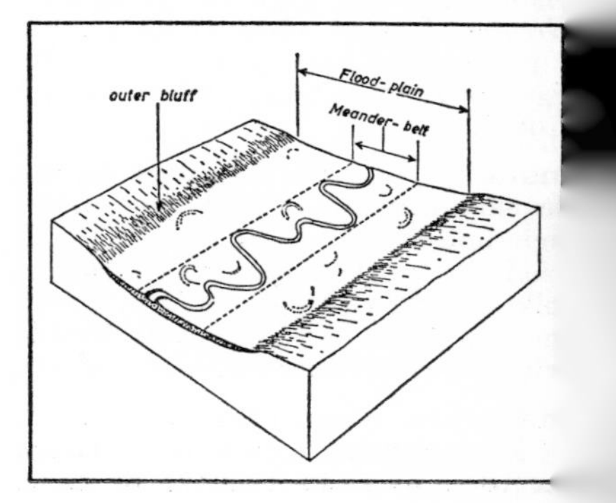

M.S.L. is calculated for Gt. Britain at Newlyn, Cornwall.

Mean Solar Time An average or *mean solar day* of 24 hours, used because the interval between successive transits of the sun over the meridian (APPARENT T.) is not uniform. See EQUATION OF T.

mechanical (or **physical) weathering** The disintegration of rock by agents of the weather (frost, temperature change), but without involving chemical change.

medial, median moraine A line of debris down the centre of a glacier, formed by the merging of the LATERAL M.'s of two confluent glaciers. M.'s occupying a central position on a glacier may also occasionally be produced below a rock-step or an ice-fall along its course. Following the shrinkage of the glacier, the m. m. may be left as an irregular ridge along the centre of the valley. E.g. the Aletsch Glacier, Switzerland, where several m. m.'s follow the curve of the valley. [*f* MORAINE]

median mass A high intermontane area within a zone of fold-mountains. As the two forelands of a GEOSYNCLINE approached, the result of orogenic pressure from either side, their bordering portions may have been overthrust on to the margins of each foreland, so forming fold-mountain ranges, but with a lofty, relatively unfolded part of the geosyncline remaining between them as the m. m. E.g. the plateau of Tibet, lying between the Himalayas on the S. and the Kuen Lun on the N.; the Hungarian plain between the Dinaric Alps and the Carpathian Mtns.; the Persian plateau, between the Elburz Mtns. on the N. and the Zagros Mtns. on the S. [*f*]

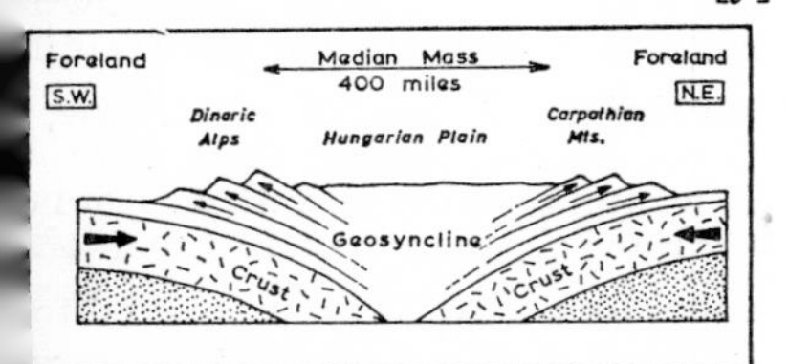

medical geography The study of the distribution of disease on the earth, together with the causative factors in their geographical setting; the study of the environment as it affects human health. An important aspect is the portrayal of MORBIDITY and MORTALITY statistics in cartographical form; e.g. G. M. Howe, *National Atlas of Disease Mortality* (1963).

Mediterranean climate Sometimes called Western Margin Warm Temperate c. (*Csa* and *Csb* types on W. Köppen's system). A warm temperate c., occurring on the W. margins of continents in latitudes 30–40°, characterized by hot, dry, sunny summers, and moist warm winters, the result of an alternation between gen. high pressure conditions in summer, and the passage of depressions associated with moist winds from the oceans in winter. It occurs in the New World as a relatively narrow coastal area, sharply defined inland by mountain ranges, in central California and central Chile; in the S.W. of Cape Province (S. Africa); S.W. and S.E. Australia; and around the Mediterranean Sea, where the total rainfall decreases progressively E. (Gibraltar 914 mm. (36 ins.), Athens, 406 mm. (16 ins.)). A wide variety of c.'s is prevalent within the general type-area of the Mediterranean; ct. S.W. Spain with Israel.

megalith A large stone used as a monument, or as part of one, or as part of a burial chamber, esp. in Neolithic times; e.g. the megalithic monuments of Stonehenge and Avebury in Wiltshire.

megalopolis Derived from a city-state in the Peloponnesus (Greece), which still survives as a small town. J. Gottman has applied the term to the almost continuous extent of densely populated urban and suburban area from S. New Hampshire to N. Virginia, and from the Atlantic coast to the Appalachian foothills.

megatherm One of a category of plants with the temperature requirement of each month with a mean of more than 18°C. (64·5°F.); typical of Tropical Rain-forest.

Megathermal Stage See ATLANTIC STAGE, of climate.

melt-water Water formed by the melting of snow and ice; e.g. a m.-w. stream issuing from the snout of a glacier.

Mercalli Scale (modified) App. to earthquake intensity, as given on a numbered scale ranging from I (only detectable by a seismograph) to XII (catastrophic, involving the total destruction of buildings). Ct. RICHTER SCALE.

Mercator Projection A CYLINDRICAL P. with CONFORMAL properties, which Gerhard Mercator used for his world-map of 1569. The parallels are straight lines, drawn the same length to scale as the equator, equally divided for the meridians intersecting at right-angles. The distance between the parallels increases from the equator to preserve the correct ratio between latitude and longitude, thus giving great and increasing distortion in high latitudes. The parallel intervals are most conveniently obtained from tables; their computation involves the use of calculus. Its chief advantage is that lines of constant direction are straight lines, so that it is used widely for navigation, when it is useful to be able to plot bearings (LOXODROMES) as straight lines (see GREAT CIRCLE [*f*]). One min. of latitude is given on the side of a chart, as a scale-line equal to a nautical mile. See also TRANSVERSE M. P. [*f*]

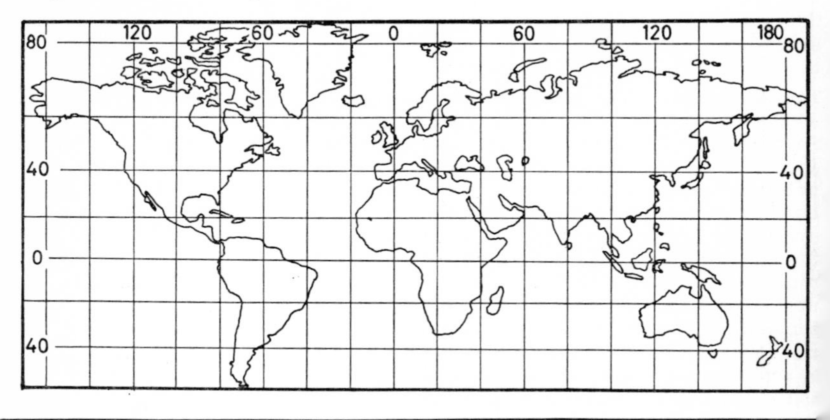

mercury The only metal which is liquid at ordinary temperatures; it is obtained mainly from red mercuric sulphide (or *cinnabar*). It is used in the 'amalgam' process of the extraction of gold from crushed auriferous quartz, in the electrical industry, in barometers because of its great density, in thermometers, and in various compounds for a wide range of industrial uses. C.p. = Spain, Italy, U.S.S.R., China, Mexico, U.S.A., Yugoslavia.

mere A small, usually circular lake esp. on a clay-covered plain; e.g Breckland. It occurs specif. in Cheshire the result of subsidence caused by th removal of underground salt deposits

meridian A line of LONGITUDE, Great Circle passing from Pole t Pole, numbered to E. and W. from ((Greenwich, the Prime M.) to 180°.

meridian day The term used fo (i) the d. on which the INTERNATIONA

DATE-LINE is crossed; (ii) the actual d. which is repeated on a ship crossing the International Date-Line when sailing in an E. direction.

meridional flow A type of atmospheric circulation in which N. to S. (meridional) movement of air is dominant. Ct. ZONAL F.

mesa (Sp.) (i) A flat-topped eminence (*lit.* tableland), commonly capped with a resistant rock-stratum, the remnant of the denudation of a plateau in a semi-arid area; it is more extensive than, though similar to, a BUTTE. (ii) A tableland extending back from a scarp-edge; this use is now less common. E.g. in Arizona, Utah, and S. California in S.W. U.S.A., where it forms a common place-name, as Checkerboard M. (Zion National Park, Utah), and M. Verde (Colorado). [*f*]

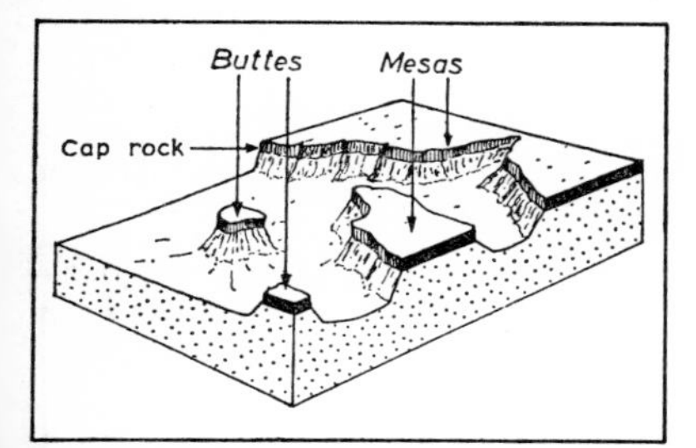

mesocephalic A head-form that is neither long nor broad, with a CEPHALIC INDEX between 75 and 83·1.

Mesolithic period (lit. 'middle stone', from the Gk.) A culture p. in Britain from the 8th to the 4th millennium B.C. During this time England was separated by sea from the continent of Europe (*c.* 5000 B.C.), and the moister ATLANTIC climatic stage began.

mesophyte One of a category of plants requiring a moderate or average amount of moisture, including most trees; adj. *mesophytic*. Ct. XEROPHYTE, HYGROPHYTE, HYDROPHYTE.

mesosphere Sometimes used for the zone of the atmosphere between about 50 and 90 km. (30 and 55 mi.) in altitude, including the upper part of the STRATOSPHERE and the lower part of the IONOSPHERE.

mesotherm One of a category of plants with mean temperature requirements of the warmest month over 22°C. (72°F.), of the coldest month between 6°C. and 18°C. (43°F. and 65°F.); e.g. the olive, the 'type-tree' of the Mediterranean climate.

Mesozoic Derived from the Gk. 'middle life', app. to the 2nd (hence sometimes Secondary) main division (era) of geological time subsequent to the Pre-Cambrian, and to the rock-groups deposited during that time, lasting from *c.* 225 to 70 million years ago. It is divided into three periods: (i) *Triassic;* (ii) *Jurassic;* (iii) *Cretaceous.*

mesquite A deep-rooted, drought-resistant leguminous shrub or stunted spiny tree, growing in the arid parts of S.W. U.S.A., Mexico and as far S. as Peru. It often forms dense thickets.

mestizo The offspring of a union between a European and an American Indian, found mainly in S. America.

metacartography Used by W. Bunge to define the study of the way maps fulfil the function for which they are intended.

metamorphism, adj. **metamorphic** The processes by which an already consolidated rock undergoes changes in or modifications of texture, composition or structure, either physical or chemical. These changes may be brought about by: (i) the *pressures* involved in earth-movements (DYNAMIC or regional M.); (ii) *heat*, caused by the intrusion of a mass of molten rock (THERMAL or contact M.). Hence *m. rocks*, which tend to be compact and resistant to denudation, and form masses involved within areas of mountain-building; e.g. much of the Highlands of Scotland, the Hercynian uplands of Central Europe. A special case is *impact m.*; see COESITE.

metasomatism Changes in a rock-substance as a result of processes in which a mineral is replaced by another introduced from external sources by percolating solutions, or by penetrating vapours of high chemical activity (PNEUMATOLYSIS).

métayage (Fr.) A system of land tenancy, esp. in France, by which the landlord furnishes seed and stock and takes a proportion, sometimes as much as half, of the produce. This system makes for wasteful cultivation, since the landlord puts in as little and takes out as much as possible, and the insecurity of the tenant makes for bad farming. The *métayer* is usually a man with little capital, but with a large family available for labour. The system has declined markedly in France; only in the *départements* of Landes and Allier is much land still cultivated thus.

meteor A body of matter travelling through space, which becomes incandescent and visible when heated by friction with the atmosphere; hence its popular name of 'shooting star' or 'falling star'. It usually becomes luminous at heights between about 145 km. (90 mi.) and 113 km. (70 mi.), disappearing at about 80 km. (50 mi.). If it is not wholly consumed, it reaches the earth's surface as a METEORITE.

meteoric water W. on the earth's surface derived from the atmosphere in the form of rain or snow. Ct. JUVENILE W.

meteorite A mass of either nickel-iron or silicate minerals from outer space, which has survived its passage through the earth's atmosphere and has landed on its surface; ct. METEOR. Several large and many small m.'s have been recorded as falling on the surface of the earth; e.g. the Great Meteor Crater in Arizona, a hole 1251 m. (4150 ft.) across, 5 km. (3 mi.) in circumference and 174 m. (570 ft.) deep, was probably made by the impact of a m. about 50,000 years ago. Recent gravimetric surveys have shown that below the floor of this 'crater' lies a mass of metal weighing about 3 million tons, 250–275 m. (800–900 ft.) down. The Tunguska Crater in Siberia was made by a m. in 1908. There are nine other authenticated m. craters in the world.

meteorology The scientific study of the physical phenomena and processes at work in the atmosphere. This is partic. applied to forecasting, the basis of which is the construction of a SYNOPTIC CHART from simultaneous observations of atmospheric phenomena at a large number of stations.

metre A unit of length in the metric system, orig. taken to be 1/10 millionth of the quadrant of the meridian through Dunkirk. The standard m. is the distance between two lines on a platinum-iridium alloy bar at 0°C. (kept in Paris); it is now precisely defined in terms of the wave-length of the orange light emitted by krypton^{86}. 1 m. = 10 decimetres = 100 centimetres (cm.) = 1000 millimetres (mm.); 1000 m. = 1 kilometre; 1 m. = 39·3701 ins. = 3·281 ft., = 1·094 yds.; 1 cm. = 0·394 ins.; 1 sq. m. = 1550 sq. ins. = 10·7639 sq. ft. See LENGTH.

metrication The system of measurement based on a rationalized metric system, which the U.K. is in process of adopting. See S.I.

metropolis A town or city which predominates as a seat of government, of ecclesiastical authority, or of commercial activity. It is used as an adj. (*metropolitan*) in several senses, but it gen. denotes the activities on state-wide or even world-wide scale. Hence *metropolitan area*, *— region*, *— centre*. *Note:* (i) Boroughs within London were officially termed *metropolitan boroughs* until April, 1965. (ii) Metropolitan France implies the whole of mainland France, Corsica and formerly Algeria. See also STANDARD METROPOLITAN AREA.

mica A group of silicate minerals having a perfect cleavage, and splitting into thin tough lustrous plates. The main m.'s are *biotite* (dark), *muscovite* (light) and *phlogopite* (yellowish) (which is magnesium-rich).

microclimate The climate of the immediate surroundings of some phenomena on the surface of the earth, partic. around plants and groups of plants. The dimensions of the space considered varies with the object; thus a larger scale is involved in studying trees than grass. *Urban m.'s* are also receiving much attention, including the effects of buildings on temperature, and also air pollution. Hence *microclimatology*, the scientific study of m.'s. Ct. MICROMETEOROLOGY, LOCAL CLIMATE.

microgranite A type of granite with medium-grained rather than coarse-grained texture, occurring in minor INTRUSIONS.

micrometeorology The detailed scientific study of the lowest layer of the atmosphere, esp. from ground-level up to about 1·2 m. (4 ft.).

micron A unit of length, 1-millionth of a m., denoted by the symbol μ.

microtherm One of the category of plants with mean temperature requirements of the warmest month between 10°C. (50°F.) and 22°C. (72°F.), and the coldest month above 6°C. (43°F.); e.g. most temperate deciduous trees, such as oak.

mid-latitude A latitudinal zone, in its broadest sense between 23½° and 66½° (in both the N. and S. hemispheres). It is increasingly used as a more specif. term than TEMPERATE.

midnight sun In latitudes higher than parallel 63½°, the s. does not sink below the horizon during the period from mid-May to late July in the N. hemisphere, and from mid-November to late January in the S. hemisphere.

mid-ocean ridge An elongated arch or swell rising from the o. floor, notably in the Atlantic o. (the *Mid-Atlantic R.*, known as the *Dolphin Rise* in the N. and as the *Challenger Rise* in the S.), and in the Indian o., where a r. can be traced from S. India to the Antarctic o. The r.'s are covered by 3000–3600 m. (1700–2000 fthms.) of water, and from them rise occasional islands (Azores, Ascension, Tristan da Cunha). The r.'s seem to be basaltic, probably volcanic in origin. Another theory believes that the Mid-Atlantic R. is actually a double r. separated by a gaping crack or rift, a possible indication of the moving apart of the continental mass on either side, i.e. of CONTINENTAL DRIFT.

migmatite A rock so called from the Gk. *migma*, mixed, formed by the cooling of a mixture of MAGMA injected into metamorphosed unmelted rock, part of the process of *granitization* (see GRANITE). The m. represents a stage in this process, whereby mica-schists have been changed via m.'s into granite over a very long period of time. Sometimes alt. layers or 'lenses' of schist and granite can be distinguished; e.g. in the islands off S.W Finland. See also ICHOR.

migration The movement of people, either *internal* (within a country) or *external* (to or from a country); in the latter case, it can be movement out (*emigration*) or in (*immigration*). M. is a result of two complementary forces: *expulsion* (for political, religious or economic reasons), and *attraction* (cheap empty land, minerals (esp. gold), employment). The term can be used of creatures other than people; e.g. eels, lemmings, caribou, reindeer, formerly bison.

migration of divide The change in position of a d. or watershed, the result of a more active river (with a greater runoff or a steeper slope) on one side cutting back more rapidly, and capturing an area of land formerly drained by the weaker stream. Also in the case of a CUESTA, the steeper slope will usually erode more rapidly than, and thus at the expense of, the back-slope; the watershed of the cuesta at its crest is therefore continually shifting down-dip.

mile (i) *Statute mi.:* a linear measurement of 1760 yds. = 5280 ft. = 63,360 ins. = 880 fthms. = 80 chains = 1609·3 m. (ii) *Nautical mi.:* the length of 1 minute of arc, or 1/21,600 of a mean Great Circle; i.e. 1 minute in latitude $\theta = 6076{\cdot}8 - 31{\cdot}1 \cdot \cos 2\theta$ ft. This is standardized (in Great Britain) at 48°N. = 6080 ft. = 1·1516 statute mi. = 1853·25 m.; the U.S.A. (since 1954) and many other countries use the *International Nautical mi.* = 6076·1033 ft. = 1852 m. 1 knot = 1 nautical mi. per hour. (iii) *Geographical mi.:* strictly the length of 1 minute of arc measured along the equator = 6087·2 ft.; in practice, it is also taken as 6080 ft. Orig. the mi. was the Roman measurement of 1000 paces; hence the name (from *milia*).

milieu (Fr.) Sometimes used for Man's environment or surroundings.

military grid A g. system used by the British War Office until the NATIONAL G. was introduced for all official British maps. The U.S.A. has a m. g. system; see U.T.M. and U.P.S. G.'s.

mille map A quantitative distribution m. on which there are 1000 dots, representing the sum total of the quantity depicted; each dot therefore represents 1/1000 of the total, and is located as accurately as possible according to the available evidence. E.g. if 2760 cattle are to be located, each dot will represent 2·76 animals. These m.'s are esp. useful for comparing the distribution of the same item at different dates; the disadvantage is the usually irregular value ascribed to each dot.

millet A variety of small edible grains grown widely, mostly in tropical and sub-tropical areas, esp. where the climate is too dry for rice or the soil too poor for wheat, gen. as a SUBSISTENCE CROP. Outside the subtropics it is gen. grown as a fodder crop, but KAOLIANG (sorghum m.) is a staple in N. China. Each stalk produces a head of tightly packed seeds. The main type is sorghum (Great M., Indian M. or Guinea Corn); there are many varieties and local names.

'millet-seed' sand Wind-borne s.-grains are in a state of constant movement, impacting against each other and the rock-surfaces they meet; hence by ATTRITION each particle becomes progressively more rounded. Not only are the desert-s.'s of today like this, but the grains in the red Triassic sandstones of the English Midlands and the Pre-Cambrian sandstones of Charnwood Forest have a similar rounded character.

millibar A pressure unit of 1000 dynes per sq. cm., used in recording atmospheric pressures as indicated by a barometer. 1000 mb. = 1 bar. There is no universal formula for conversion, except at a constant temperature (0°C.) and latitude (45°); 29 ins. of mercury = 982 mb.; 30 ins. = 1016 mb.; 31 ins. = 1049 mb.; 1000 mb. = 29·531 ins. (750·1 mm.).

million-city A c. with a population of a million or over; about 1965 there were 113.

Millstone Grit A hard, coarse-grained sandstone, sometimes in massive beds, which occurs under the Coal Measures at the base of the Upper Carboniferous. Occasional layers of shale, thin seams of coal, and bands of ironstone are included. It was laid down under shallow-water marine conditions, probably as a delta-deposit. It is found in the central Pennines (where it forms gritstone moorlands), and in Northumberland; in places it is up to 1500 m. (5000 ft.) thick, and forms some prominent edges, notably around Kinder Scout (between Manchester and Sheffield).

Mindel The 2nd of the four periods of fluvioglacial deposition on the Alpine Foreland, which A. Penck and E. Brückner correlated with periods of glacial advance during the Quaternary Glaciation. The Younger Deckenschotter (or upper outwash terrace) on the Alpine Foreland is associated with this glacial advance. This fourfold concept of glacial advances has been accepted elsewhere, although later research has now shown the picture to be more complicated in many areas, sometimes with more, sometimes with fewer, glacial periods. The M. probably corresponds with the Elster glaciation in N. Europe and with the Kansan in N. America. [*f*]

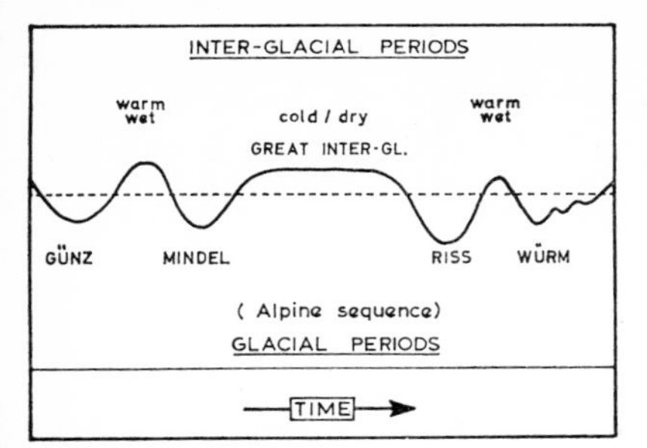

mineral (i) In gen. terms a substance obtained by mining: coal, oil, a metallic ore. (ii) Scientifically, an inorganic substance with a specif. chemical composition; mixtures of m. particles comprise rocks. Nearly all m.'s are crystalline. Some are simple in composition, consisting of a single element (e.g. diamond, carbon), most of two (e.g. pyrites, of iron and sulphur, FeS_2). M.'s have several properties: crystal form, hardness, specific gravity, colour, lustre and transparency, streak, cleavage, fracture, striations.

mineral spring A s. of water containing an appreciable amount of m. salts in solution, including iron compounds (Harrogate), hydrogen sulphide, magnesium chloride and sodium chloride (Droitwich). It may sometimes be of medicinal value, and its presence may lead to the development of a spa, as at Bath, Carlsbad (now Karlovary), Vichy.

mineral water Orig. w. impregnated with some m. substance, often drunk locally for medicinal purposes or bottled for export; e.g. the alkaline Vichy w. from France. This word now covers artificial imitations of such natural w., and in gen. drinks of a similar nature.

minette (Fr.) An iron ore, either a carbonate or a silicate, with a metal content of 24–40% and with 1·7 to ·9% of phosphorus, occurring in strata of Middle Jurassic age in French Lorraine and S. Luxembourg.

minimum thermometer A t. that automatically records the lowest temperature attained over a given period of time. One type consists of an alcohol t. fixed horizontally, with a small dumb-bell-shaped marker kept just beneath the meniscus of the alcohol by surface tension. When the temperature falls, the marker is drawn to the level of the lowest temperature attained, where it remains, even when the temperature subsequently rises.

minority A group of people living in a country, different in any aspect of race, religion, language, social customs and national sympathies from the majority of the people. This was esp. the case after the war of 1914–18, when the creation of new national states inevitably left m.'s within the new boundaries; e.g. Sudeten Germans in Czechoslovakia; Austrians in N. Italy (S. Tirol); Hungarians in Roumania; Bulgars in Greece. Today there are still such problems as Turkish Cypriots in Cyprus.

minute A unit equivalent to: (i) 1/60 of an hour; (ii) 1/60 of a degree of latitude and longitude; (iii) 1/60 of an angular degree, horizontal or vertical.

Miocene The 3rd of the geological periods of the Cainozoic era (4th in U.S.A., where it is classed as an epoch) from approx. 25 to 11 million years ago. No rocks of proven M. age have been found in Britain, where it was a time of earth-movements and erosion, though the Molasse of M. age is widespread in Central Europe. The max. of the Alpine orogeny occurred during the M.

miogeosyncline A GEOSYNCLINE which experienced little VULCANICITY during its infilling by sedimentation. Ct. EUGEOSYNCLINE.

mirage An optical illusion caused by the refraction of light by the atmosphere, partic. in hot deserts, when the layer of air near the ground is greatly heated by conduction, hence becomes less dense, so that rays of measured off along the continuation of light from the sky may be bent upwards; thus the sky may be seen by refraction, giving the impression of a shimmering sheet of water. A m. may be seen over a road-surface on a hot day; this is known as an *inferior m.* In a *superior m.*, where the light rays are bent down from a warm layer of air which is resting on a cold one (e.g. in high latitudes), a sharply defined double or inverted image of a distant object, such as a ship, may be seen.

mire An area of soft, spongy waterlogged ground. It is used in place-names; e.g. Great Close M., W. Yorkshire; Foxton M., Dartmoor.

misfit river A r. obviously much too small (*underfit*) for its present valley, because: (i) its headwaters have suffered capture and its size has been thereby reduced; e.g. the R. Meuse in Lorraine; (ii) a change of climate has occurred and its volume has been reduced; e.g. most English rivers; (iii) a valley has been enlarged by glaciation to a broad U-shape; e.g. the small R. Ogwen in the wide Nant Ffrancon, N. Wales.

[*f* CAPTURE, RIVER]

Mississippian Formerly in U.S.A. the lower of the two divisions of the Carboniferous period (and system); now it is regarded in U.S.A. as a period and system in its own right, the 5th in order of age, and characterized by widespread deposits of Carboniferous Limestone. It was succeeded by the PENNSYLVANIAN.

mist An obscurity of the ground layers of the atmosphere, the result of the condensation of water droplets, with visibility between 1 and 2 km. It is really a form of FOG, with not such restricted visibility. Ct. HAZE.

mistral (Fr.) A strong, cold, dry N.W. or N. wind, blowing from the Massif Central of France towards the Mediterranean Sea, felt esp. over the Rhône delta and the Gulf of Lions. The cold air is funnelled down the lower Rhône valley from the winter anticyclone over Central Europe towards a low pressure area over the W. Mediterranean basin. The wind averages about 60 km. (40 mi.) per hour, but over 137 km. (85 mi.) per hour has been recorded. See also BORA.

mixed farming Agriculture involving both crops and livestock. This is not to be confused with 'm. cultivation', implying merely a series of different crops.

mixing ratio The r. of the weight of water-vapour in a 'parcel' of the atmosphere to the total weight of the air (excluding the water-vapour), stated in gm. of water-vapour per kilogram of dry air. Ct. SPECIFIC HUMIDITY.

mizzle Very fine rainfall, in the form of a misty drizzle. Cf. Dutch dialect form, *miezelen*.

model (*a*) In its simple conventional form, a m. can provide a 3-dimensional reproduction of the landscape, with length and breadth to scale, though of necessity altitude is usually exaggerated. The material may be of superimposed layers of paper, card, pulp, hardboard or wood, moulded plaster or potter's clay, or various plastics (which may be produced by vacuum-forming); these are *hardware m.'s*. The m. may be prepared from topographical maps and/or air photographs. Apart from showing land-forms, m.'s are widely used in civil engineering projects, town-planning, geological studies etc. A *working scale m.* enables processes to be reproduced, to enable their results to be studied; e.g. a tidal-m. (as of Southampton Water at the University of Southampton), of the Mersey at the Hydraulics Experiment Station at Wallingford), a river m. (at the U.S. Corps of Engineers Research Establishment at Vicksburg, Tennessee) (*b*) 'The identification and association of some supposedly significant aspects of reality into a working system which seems to possess some special properties of intellectual stimulation (R. J. Chorley). This brings together certain aspects of reality into a clear cut theoretical m., forming a bridge between observation and theory, and providing a working hypothesis against which reality can be tested. They may be divided into m.'s of (i) physical systems; (ii) socio-economic systems; (iii) mixed systems; and (iv) information systems. They can be created in any aspect of geography, e.g. in geomorphology, climatology, industrial location, or in the form of regional m.'s. They may be mathematical, theoretical (or conceptual), experimental or natural; they may be *iconic* (presenting the same properties, though reduced to scale), *analogue* or *symbolic* (in which the actual properties are represented by different, though analogous, properties), according to the degree of abstraction, generalization and presentation of information. See also SIMULATION.

model building The abstraction by a research-worker of those parts of the complex reality which are relevant to the problem under discussion, their sorting, sifting and presentation to form a m. of reality.

mofette (Fr.) A small hole in the earth's surface, from which issues carbon dioxide, with some oxygen and nitrogen, and occasionally water-vapour; e.g. in the Phlegraean Fields near Naples; in Auvergne in central France; and in Java. It indicates a late stage in minor volcanic activity.

Mohorovičić Discontinuity The d. between the earth's crust and MANTLE, so-called after the scientist A. Mohorovičić, who discovered it in 1909 while studying a Balkan earthquake. The discontinuity greatly affects the speed at which earthquake waves travel. American scientists made preliminary trials towards drilling a hole through the crust into the mantle ('Operation Mohole', off W. Mexico, abandoned 1966). Soviet scientists made similar drillings off the Kuril I. Both U.S. and Russian drills reached a layer of BASALT. The M. D. lies at a depth of about 32 km. (20 mi.) under the continents, but at only 6–10 km. (4–6 mi.) under the oceans. [*f* ISOSTASY]

Mohs' Scale See HARDNESS S. A s. named after F. Mohs (1773–1839), a German-Austrian mineralogist.

Molasse A soft greenish sandstone, with conglomerate and marl, mainly of Miocene age. The materials which constitute it were worn from the Alpine ranges of Europe during and after the orogenic max. (ct. FLYSCH, laid down before the max.), and deposited under continental or shallow fresh-water conditions. Used with l.c., the term is increasingly applied to all deposits of similar origin and character.

mole A protective wall or breakwater designed to improve or safeguard the facilities of a harbour; e.g. the m. at Zeebrugge, Belgium.

mollisol See ACTIVE LAYER.

Mollweide Projection An EQUAL-AREA P., in which the central meridian is a straight line half the length to scale of the equator. If this central meridian is 0° longitude, the area bounded by the parallels 90°E. and W. will represent a hemisphere πr^2, with a radius $(r) = \frac{1}{2}$ central meridian. Area of hemisphere $= 2\pi R^2$, where R is the radius of the earth, therefore $\pi r^2 = 2\pi R^2$, therefore $r = \sqrt{2}R$. If R is unity (i.e. the scale is 1/250 millions), $r = 1{\cdot}414$; length of the central meridian $= 2r = 2{\cdot}828$; and length of equator $= 4r = 5{\cdot}657$. The parallels are drawn as straight lines at right-angles through the central meridian; their distances apart are obtained from the following table, and they are spaced more closely towards the Poles:

(Pole to Equator = 1)
Distance from Pole

10°	=	0·137
20°	=	0·272
30°	=	0·404
40°	=	0·531
50°	=	0·651
60°	=	0·762
70°	=	0·862
80°	=	0·945
90°	=	1·000 (i.e. equator)

Each parallel within the hemispherical circle is divided equally according to the meridian interval (e.g. by 9 to give 20° intervals), and the same divisions, the parallels outside the circle, will give the points of intersection of the outer meridians. Smooth ellipses are drawn from Pole to Pole, through each parallel–meridian intersection. The p. is equal area, since the distance apart of each parallel was calculated to make this so. The linear scale is true only on the parallels 40° 40′ N. and S., increasing poleward and decreasing equatorward. The p. is quite good for a world map of distributions centred on longitude 0°, but if drawn with a central meridian through N. America, Asia is cut in two. Within the inner hemisphere shape is good, but distortion increases towards the margins in high latitudes, though less so than on the SINUSOIDAL P. The Mollweide P. is sometimes used in INTERRUPTED [*f*] form, and its high latitude portions in GOODE'S HOMOLOSINE P.

molybdenum A metal occurring in an ore *molybdenite* (MoS_2), and also in *wulfenite*. It is used similarly to TUNGSTEN as an alloy in high-speed tool steels, and increasingly as a lubricant in suspension. C.p. = U.S.A., Canada, U.S.S.R., Chile, China.

monadnock A residual hill or erosional survival standing above the general denuded level, named after Mount M. (965 m., 3165 ft.) in New Hampshire, U.S.A.

monoclinal fold An asymmetrical f., the result of compression in the crust, with one limb markedly steeper than the other. Very often it is simply a pronounced bend in the strata, on each side of which the rocks are either horizontal or lie at the same angle.

monocline The bending or flexing of strata along a line, through tension in the crust; the strata are near horizontal (though at different levels), except along the line of flexure. The term should never be used for a MONOCLINAL FOLD, which is a compressional feature. An American usage is for beds dipping in one direction (known in Britain as a HOMOCLINE). [*f*]

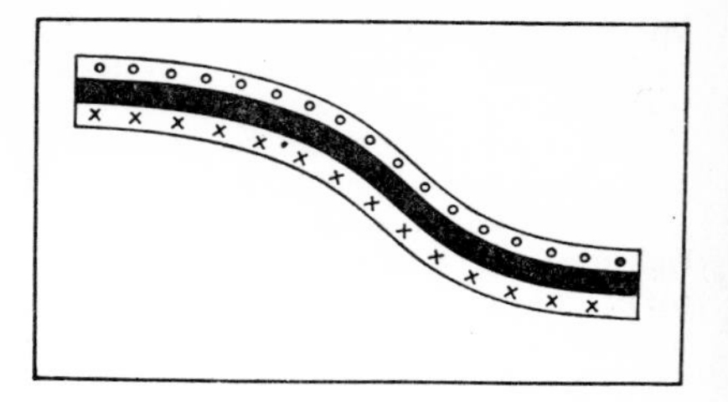

monoculture An emphasis on one dominant crop, either large-scale and extensive (e.g. wheat in parts of the Canadian prairies), or small-scale and intensive (e.g. rice in the Monsoon Lands, the vine in parts of S. France, tobacco in Virginia). M. is not usually sound practice, partly because of the dangerous dependence on one crop, its yield and price, partly because the soil usually becomes impoverished.

monoglaciation A glacial period in which only one major advance of an ice-sheet occurred, in ct. to the 4-fold (at least) advances of the Pleistocene ice-sheet.

monolith A single block of stone, often used of rock buttresses on mountains; e.g. the M., Cwm Idwal (N. Wales).

monsoon (derived from the Arabic word '*mausim*' for 'season') The seasonal reversal of pressure and winds over land-masses and neighbouring oceans. The term orig. referred merely to winds, specif. in the Arabian Sea, but it is now popularly used both for 'the rains' accompanying inflowing moist winds, and for various climatic types. In part, it is the result of the much more marked differential heating and cooling of land by comparison with neighbouring oceanic areas, in part the result of the poleward shift of the hemispheric wind-belts in summer. The Asiatic M. is considerably affected by the interrupting effect of the Plateau of Tibet on the Upper Westerlies (incl. the sub-tropical JET STREAM) in summer these shift N. of the Plateau, permitting the N.-ward surge of the SW. M.

monsoon forest A tropical f. which experiences a marked seasonal drought. It may occur in true monsoonal lands (Viet-Nam, Burma, Indonesia, India and N. Australia), and also on the margins of the equatorial climatic belt, where it forms a transition zone between the rain-f. and the tropical grasslands. Luxuriant growth occurs during the wet season, and a markedly deciduous habit of leaf-shedding during the dry season. The dominant trees are teak, bamboo (Asia), acacia (Africa) and eucalyptus (Australia).

montaña (Sp.) The forested slopes of the eastern Andes in equatorial latitudes.

monte An area of low XEROPHYTIC scrub in the foothills of the Andes in Argentina.

Moon The earth's sole satellite, revolving around it in a plane inclined at approx. 5° 9′ from the plane of the ECLIPTIC, in a period of 27 days, 7 hours, 43·25 minutes (a *sidereal month*). It also has an apparent motion in the Celestial Sphere, relative to the stars, from W. to E., and so completes one revolution relative to the sun in a mean time of 29 days, 12 hours, 44 minutes (a *lunar* or *synodic month*, from one new m. to the next). A LUNAR DAY is of 24 hours, 50 mins. This time varies slightly because of the eccentricities in the orbits of both the m. and the earth. The m.'s min. distance from the earth is 354,000 km. (220,000 mi.), max. 407,000 km. (253,000 mi.). The m.'s mass is approx. 1/81 of that of the earth, its diameter approx. ¼. It has no water or atmosphere. The phases of the m. are: (i) *1st quarter* (a semi-circle with the bow facing W.); (ii) *full m.;* (iii) *3rd* or *last quarter* (a semicircle with the bow facing E.); and (iv) *new m.* (invisible). The m. is *gibbous* when more than a half-circle is visible. It rotates on its axis once during each revolution in its orbit, and therefore the same face is always turned to the earth. A major result of the m.'s presence is its contribution to the gravitational forces mainly responsible for the TIDES. *Note.* Man first landed on the m. on July 20–21, 1969.

moor, moorland Strictly an upland area of siliceous rocks, such as Millstone Grit, where acid peat (derived from sphagnum, cotton-grass, purple moor-grass) has accumulated to an appreciable thickness, under damp conditions. This is the lit. sense of the Germ. *Moor* or *Hochmoor*. In more gen. terms, it includes: (i) any area of unenclosed upland waste ('the moors'); (ii) undulating 'upland heaths', dry and covered with ling and heather (as the N. York M.'s, the Scottish grouse-m.'s); (iii) an area of lowland marsh (Sedgemoor in Somerset). These three types are ecologically incorrectly so termed.

moorpan See HARDPAN.

mor A 'raw' humus, markedly acid (with a pH less than 3·8), formed on heaths, moors and in pine-woods, the result of the slow and only partial decay of vegetation because of the cool, moist conditions and the poverty of soil organisms.

moraine (Fr.) (i) Masses of boulder-clay and stones carried and deposited by a glacier; (ii) the arrangement of this material to form a partic. land-form. M. was a word used by Fr. peasants in the Alps in the 18th century for any bank of earth and stones, and gradually it became an accepted term in Alpine literature. Frost action is potent on rock-buttresses and mountain-slopes above a glacier, and blocks of all sizes fall on to the ice, some being carried on the surface, others sinking in. Material is also picked up from the underlying valley-floor through plucking and abrasion by the ice. Ultimately much of the material is deposited in well-defined humps and mounds, known variously as LATERAL, MEDIAL, PUSH,

RECESSIONAL, STADIAL and TERMINAL M.'s. [f]

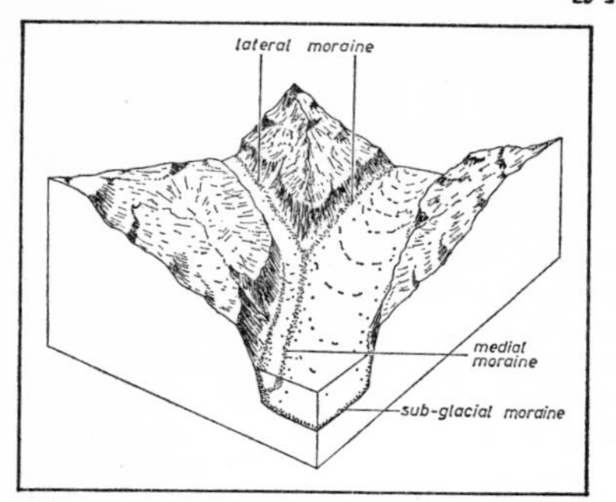

morass A vague and gen. term, with no specif. connotation, for an area of waterlogged ground, swamp, marsh or bog.

morbidity The liability to a partic. disease, not necessarily fatal. The analysis and mapping of m. statistics is an important part of MEDICAL GEOGRAPHY.

morphochronology The dating of MORPHOLOGICAL features.

morphogenesis, adj. **morphogenetic** The origin of forms, applied partic. to landforms. Thus a m. description of a landform is one in which origin, and probably development, is considered. A *m. region* is one in which partic. climatic and erosional conditions predominate, thus giving the area a regional character which cts. with those of other areas subjected to contrasting processes and conditions.

morphographic map A small-scale m. showing physiographic features by means of standardized pictorial symbols, based on the appearance such features would have if viewed obliquely from the air. Such m.'s have become well-known through the work of A. K. Lobeck and E. Raisz in U.S.A.

morphological region In geomorphology, a distinctive unit, of various scales or orders, demarcated according to its form, structure and evolution. D. L. Linton has considered the delimitation of such r.'s and has suggested a hierarchy of increasing size and complexity, each with a distinctive name: *site*, *stow*, *tract*, *section*, *province*, and *continental subdivision*. On a m. map, m. types are distinguished by means of a range of symbols and tints.

morphology The scientific study of form in various connections; e.g. landforms, towns.

morphometry The exact measurement of the external features of landforms; a mathematical treatment of them. This basis of exact measurement, whether of stream characteristics, drainage basins or slopes, is becoming much more prevalent in geomorphology. Relationships between precisely measured aspects of the landscape are being revealed, partic. by the use of statistical methods. While some authorities doubt the real value of several of these techniques, it is true to say that a morphometric approach does ensure an objectivity which hitherto was lacking, and the increased precision of current research work is a useful corrective to the subjective knowledge of the landscape on which many older theories and ideas were based.

morphostasis Used in GENERAL SYSTEMS THEORY to describe a counteracting process to DEVIATION. Any tendency to change a system is counteracted, and the steady state characteristics are perpetuated.

mortlake See OXBOW.

mosaic A composite vertical photograph of an appreciable area, made by joining individual overlapping photographic prints.

moss An area of waterlogged boggy land, orig. derived from the extensive development of sphagnum and other m.'s. The term is used in several ways: (i) an area of siliceous rocks, with layers of peat, characterized by sphagnum and cotton-grass; e.g. Featherbed M. in the S. Pennines, N.E. of Glossop; (ii) a lowland m. of basic fen-peat; e.g. the S. Lancashire

M.'s (Chat M., Risley M.); (iii) coastal marshes; e.g. Solway M.

moss forest A type of f. found esp. on mountains in the tropics (Ceylon, E. Africa), and in warm temperate regions (Olympic F., N.W. of Washington State, U.S.A.), where the trees are covered with thick layers of dripping m. It is the result of very heavy precipitation throughout the year.

mother-of-pearl cloud A somewhat rare, high altitude (about 24 km., 15 mi.) c.-form, delicate in form, usually lenticular in structure, occasionally visible in winter in high latitudes during low pressure atmospheric conditions. It reveals an iridescence which can persist after sunset. It is scientifically called a *nacreous c.*

motorway A road with separate carriageways in each direction, with limited access, used by high-speed motor transport; e.g. the *autobahnen* in Germany, *autoroutes* in France, *autostrade* in Italy (notably the *Autostrada del Sole*, 1600 km., 1000 mi. long, from Milan to Naples), motorways in Britain, freeways in U.S.A.

moulin (Fr.) A circular sink-hole ('glacier mill') in the surface of a glacier, sometimes also continuing into the bed-rock beneath, worn by swirling melt-water falling down a CREVASSE. There is commonly a loud roaring noise, probably the origin of the name. Some authorities have, however, doubted whether a hole in a glacier would remain stationary for a sufficiently long period for a cavity to be worn in the rock floor beneath.

mound A gen. term for a low hill, which may be either natural or artificial. The latter is common in areas which were, or still are, liable to flooding, hence refuge-m.'s; e.g. *terpen* in Friesland, *wierden* in Groningen, *vliedbergen* in Zeeland (all in the Netherlands).

mountain A markedly elevated landform, bounded by steep slopes and rising to prominent ridges or individual summit-peaks. There is no specif. altitude, but it is usually taken to be over 2000 ft. (about 600 m.) in Britain, except where eminences rise abruptly from the surrounding lowlands, e.g. Conway M. In such a case, the term Mount is sometimes used; e.g. Mt. Caburn in Sussex, an outlier of the S. Downs. The world's highest m.'s are: Everest (8848 m. ±3 m., 29,028 ±10 ft.), K2 or Godwin-Austen (8610 m., 28,250 ft.), Kanchenjunga (8578 m., 28,146 ft.). The highest in N. America is Mt. McKinley (6194 m., 20,320 ft.); in S. America, Aconcagua (7003 m., 22,976 ft.); in Africa, Kilimanjaro (5895 m., 19,340 ft.); in Europe, Mont Blanc (4811 m., 15,782 ft.); in Australia, Mt. Kosciusko (2227 m., 7308 ft.); in Gt. Britain, Ben Nevis (1343 m., 4406 ft.); in England, Scafell Pike (978 m., 3210 ft.). in Wales, Snowdon (1085 m., 3560 ft.).

mountain-glacier See GLACIER.

mountain sickness The physiological effects of the increasing shortage of oxygen with increasing altitude. In untrained or unaccustomed people, this usually begins at about 2000 m. (7000 ft.) (varying with the individual), resulting in lassitude and headaches, shortness of breath and dizziness. After a short period of acclimatization this usually wears off, but some people seem to have a low 'ceiling' and never really acclimatize. This is one of the problems of high-altitude mountaineering, as in the Himalayas; the development of efficient, lightweight oxygen apparatus has made a great contribution to the successful ascents of the world's highest peaks in recent years. Sir E. Hillary and Tensing Norkay reached the summit of Everest in 1953; each used an open-circuit oxygen apparatus, from which they inhaled 3 litres of oxygen per minute.

mountain wind See ANABATIC W.

mouth, of river The junction of a tributary with its main stream; the outfall of a main stream into the sea or a large lake.

moving average See RUNNING MEAN.

muck (i) The rural word in Britain for animal manure. (ii) In U.S.A. a soil with a high % of organic material, though less than peat.

mud, oceanic Bathyal deposits of Blue, Coral, Green and Red M.'s deposited on the CONTINENTAL SLOPE, derived from clay particles worn from the land, larger than the very fine ooze on the ocean-floor.

mud-flat An area of fine silt, usually exposed at low tide but covered at high, occurring in sheltered estuaries or behind shingle-bars and sand-spits. In British waters these m.-f.'s are colonized by eel-grass (*Zostera*), marsh-samphire (*Salicornia*), perennial rice-grass (*Spartina townsendii*). Some plants help to trap particles of mud and bind it with their roots. In tropical waters, m.-f.'s are commonly colonized by MANGROVES.

mud-pot A pool of boiling m., usually of a sulphurous quality, sometimes brightly coloured, which bubbles away in an area of minor volcanic activity. They may be up to 10 m. in diameter. E.g. in Yellowstone National Park, Wyoming; Iceland; N. Island, New Zealand.

mudstone Used to describe argillaceous (so called by R. I. Murchison) non-fissile strata, initially of the Silurian system, but also used of similar rocks elsewhere.

mud-volcano The ejection of hot water and m. from a volcanic vent, thus building a small short-lived cone; e.g. near Paterno in E. Sicily; at Krafla in Iceland; and in the S. of the N. Island of New Zealand. Sometimes natural gases, associated with oil-deposits, escape through soft water-logged deposits, building up small cones; e.g. near Baku, Caucasus, U.S.S.R.

mulatto The offspring of a union between white and negro parents, esp. in S. America.

mulga A dense thicket of spiny acacia-scrub (*Acacia aneura*) on the margins of the desert of central Australia.

mull (Swed.) (i) A mild HUMUS (pH 4·5–6·5) derived from leaf-mould, found in deciduous forests as a surface layer. (ii) The surface horizon where the soil is mixed with humus, giving a granular crumby texture. (iii) A headland in Scotland; e.g. M. of Galloway

multivariate analysis Quantitative methods of examining and evaluating the variables in any problem, espec. where more than 2 variables are involved. E.g. the relationship between DISCHARGE and LOAD of a river involves the dimension and nature of the load, water temperature, volume, velocity, etc. If sufficient statistical information is available, such methods as multiple correlation and multiple regression may be used with the aid of a computer.

Muschelkalk (Germ.) The middle of the 3 series of the TRIASSIC system, its strata consisting of a fossiliferous limestone (containing ammonoids and crinoids). It occurs in central Germany; e.g. in the Teutoburger Wald, and in Thuringia where it forms a low out-facing CUESTA; also in Luxembourg and French Lorraine.

muskeg Waterlogged depressions in the subarctic zone of Canada, largely filled with sphagnum moss, with scattered lakes, and with groups of tamarack and fir-trees on slight eminences. There are festoons of meandering, though virtually stagnant, streams. Sometimes pools are covered over with moss, with an appearance of solidity, but are liable to collapse under the unwary. It is very difficult country to cross, and in summer is mosquito ridden.

nab (i) A spur, in North England, esp. in the North York Moors; e.g. N. End, Glaisdale; N. Ridge, near Helmsley. (ii) A headland; e.g. Long N., White N., Cunstone N. near Scarborough.

nacreous cloud See MOTHER-OF-PEARL C.

nadir The point on the CELESTIAL SPHERE opposite to the ZENITH; sometimes applied gen. to 'the lowest point'.

nailbourne See BOURNE.

nappe (Fr.) (i) An overthrust mass of rock in a near-horizontal fold, in which the reversed middle limb has been sheared out as a result of the enormous pressure; it is actually the hanging wall of a very low-angled THRUST-FAULT. As a result, the rocks have been forced for many miles (some American definitions stress more than 1 mi.) from their 'roots', so covering the underlying formations. Some writers use the term simply to denote an unbroken recumbent anticline, but this is not correct. A portion of a n., surviving after denudation, is a KLIPPE. E.g. in the Alps, a whole series of n.-remnants may be distinguished, now forming distinctive relief regions: (*a*) the *Pre-Alps*, S. of L. Geneva; (*b*) the *Helvetic* (Helvetian) n.'s, Helvetides or High Calcareous Alps, six individual n.'s; (*c*) the six n.'s of the *Pennine Alps*: the three Simplon–Ticino n.'s, the Grand St. Bernard, Monte Rosa and Dent Blanche n.'s. Remnants of the last two form the high peaks of the Pennine Alps. (ii) Esp. in France, the term is used more widely for any overlying, covering sheet; e.g. a lava flow. This is equivalent to the Germ. *Decke*. [*f*]

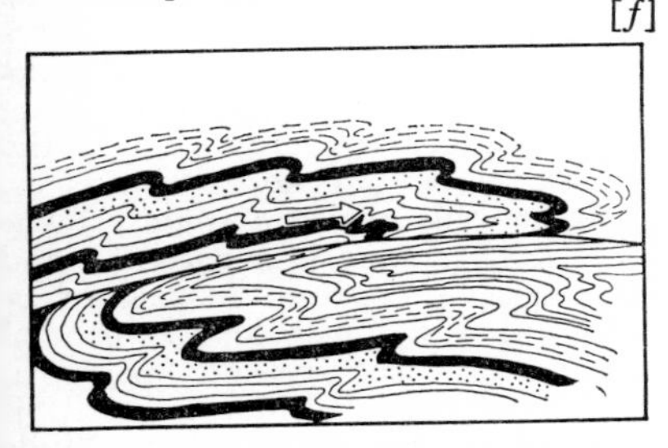

narrows A constricted section of a river, a strait, valley or a pass. 'The N.' is a common place-name; e.g. The N. between Staten I. and Long I., separating the Lower and Upper Bays in New York Harbour.

nation, nationality A group of people associated with one another by ties of history, sentiment, descent and (sometimes) language, frequently organized as a separate political unit. It is common practice to equate a n. with a separate independent state (as in the United N.'s), but a state may in fact constitute two or more national groups (e.g. Cyprus). This term has been the object of close and detailed attention, and the concept of n. is a complex one. Ct. NATIONALISM.

National Grid The g. based on the TRANSVERSE MERCATOR PROJECTION, used on all current British Ordnance Survey maps. The axes of the N. G. are 2°W. and 49°N., intersecting at the True Origin, from which the FALSE ORIGIN is transferred 400 km. W. and 100 km. N. The N. G. is drawn on the metric system, with: (i) 500-km. squares designated by a letter; (ii) within these are 100 km. squares designated by a second letter (A to Z, excl. I); (iii) within these are 1-km. squares, with every 10 km. accentuated, which appear on 1-in. maps. Thus a single reference system is provided for the whole of Britain, correct on the 1-in. scale to 100 m., with still greater precision on large-scale plans. Thus the N. G. reference of Southampton University is SU 427153. G.-lines correspond with sheet-lines, so that the 10 km. g. on the 1-in. map is an index of the 1 : 25,000 series. Each of these, together with the 6 in., 1/2500 and 1/1250, is numbered and identified by the N. G. reference of its S.W. corner. [*f* GRID]

nationalism A sentiment associated with membership of a nation, though in its extreme modern forms it is apt to be identified with anti-colonialism and anti-Europeanism.

nationalization The bringing of some form of activity under state control, administered by the central government; e.g. the National Coal Board in Britain; *Électricité de France* and *Charbonnages de France*. N. has become a political issue of some magnitude. Under Communist and dictatorial régimes, most aspects of economic life are nationalized.

national park An area set aside for the preservation of scenery, vegetation, wild-life and historic objects, 'in such manner and by such means as will leave them unimpaired for future generations', both for scientific purposes and public enjoyment. The status of n.p.'s varies considerably from country to country. In U.S.A. the concept began with the designation of Yellowstone in 1872; the N.P. Service under the Department of the Interior was established in 1916. By 1968 there were 29 n.p.'s, 86 n. monuments and 61 'preserved places', totalling 23 million acres, and including Yosemite, Katmai in Alaska with its active volcanoes, Death Valley, Grand Canyon, and the Statue of Liberty. Canada also has spectacular n.p.'s (e.g. Banff and Jasper), and Japan has Mt. Fuji. In Africa the main problem has been the preservation of animal life, as in Kruger (Transvaal) and Serengeti (Tanzania). In Gt. Britain n.p.'s were first established in 1949, and now comprise ten areas of natural beauty; i.e. Lake District, Snowdonia, Exmoor, Brecon Beacons, Yorkshire Dales, N. York Moors, Peak District, Dartmoor, Northumberland and Pembrokeshire coast. While these include inhabited countryside, some safeguards and amenities are provided.

national plan A comprehensive p. for an entire country, drawn up by its government to establish machinery to implement and integrate policies over a period of time; e.g. the 5-Year Plans of the U.S.S.R., the overall series of p.'s (*La Planification française*) of France, beginning in 1946; the U.K. Economic Plan (1965) under the Department of Economic Affairs for the period 1966–70.

natural gas See GAS, NATURAL

naturalize (i) To confer the rights and privileges of citizenship of a specif. country upon an alien. (ii) The introduction of animals and plants into an area where they formerly did not exist, often under different physical circumstances; e.g. rhododendron and azalea into Gt. Britain.

natural history An old term, popular in the 19th century, wide in its scope and application, implying the study of n. objects: animal, vegetable and mineral. Some would restrict the term to living creatures, but a n. historian is interested in all his physical environment (incl. rocks and minerals; e.g. A. Harker, *The Natural History of Igneous Rocks*). The term seems to be growing in popularity again, implying a science student with wide interests, rather than a narrow specialist.

natural landscape The l. as unaffected by man, in ct. to the CULTURAL L. It is regarded as syn. with *physical l.*, i.e. concerned with relief and n. vegetation. But so little of the earth's l. has been unaffected by man that it is better not to make any distinction, and to refer to n. and cultural elements in the l. as a whole.

natural region (i) Unit-area of the earth's surface, with certain uniform and distinctive physical characteristics (structure, relief, climate, vegetation), the basis of regional division. (ii) A geographical r., a sum total of all significant characteristics, in contrast to political or administrative r.'s. The concept of a r.—natural or otherwise—has been the subject of much debate, for while r.'s delimited in some way are the basis for much geographical teaching, many writers have wondered whether they exist other than as the result of applying subjective criteria formulated in the minds of geographers themselves. Whether this criticism is fully justified or not, there is a current trend against using the term 'n.' in connection with r.'s, unless in very general terms. The concept of 'major n. r.'s' was formulated in 1905 by A. J Herbertson, revised in 1913. These divisions have been the basis of many textbooks and much teaching in schools.

natural vegetation The primeval plant-cover, unaffected either directly or indirectly by Man. Recent investigations have shown that very little of the earth's present v. cover remains unaffected in this way; some writers have doubted whether the concept is of any value at all, in view of its present small extent and doubtful definition. The term is usually taken more widely to mean all v. not deliberately organized or included in farming activity, and includes 'wild' or SEMI-NATURAL V.

nature reserve An area preserved so that botanical and zoological communities may survive. This does not necessarily mean an untouched WILDERNESS, but may involve careful control to maintain a partic. environment; e.g. Wicken Fen in Cambridgeshire, where a high water-table is maintained. Regulations are usually imposed to limit public access or disturbance. The N. Conservancy in the U.K. has demarcated numerous n. r.'s; e.g. Old Winchester Hill, Hants., Cwm Idwal, N. Wales.

nautical mile See MILE.

naze A promontory or headland, cognate with 'nose', NESS, NAB and Nase; e.g. the N., Essex; Carr Nase, near Filey, Yorkshire.

neap tide When the earth, sun and moon are in quadrature (i.e. at right-angles, with the earth at the apex), the t.-producing forces are in opposition and the tidal range is reduced, producing t.'s of low amplitude (i.e. high low t.'s and low high t.'s), about the time of the first and last quarters of the moon.

Nebraskan In U.S.A., the first main advance of the continental ice-sheet during the Pleistocene glaciation, corresponding to the GÜNZ in the Alps. The term was first used by B. Shimek in 1910. The ice moved from the Keewatin accumulation centre, and advanced to a point just S. of the present Mississippi–Missouri junction. The N. drift has been identified in Nebraska, S. Iowa (where in places it is 46 m. (150 ft.) deep), N. Missouri and W. Illinois, but it lies on the surface only in part of Nebraska, and is elsewhere identified only when it has been revealed by post-glacial denudation. The N. glacial period was followed by the Aftonian interglacial period.

neck (i) A mass of solidified lava which fills the pipe or vent of a volcano; this may be later exposed by the removal of the surrounding material by denudation; e.g. Castle Rock, Edinburgh. (ii) An isthmus or promontory, a 'n. of land'.

needle-ice A small spike of ice, often known as a PIPKRAKE, formed just below the surface of the soil or of loose weathered material. Its presence contributes to the downhill movement of particles of material.

neese A prominent ridge or spur, cognate with NOSE, NESS, NAB and NAZE; e.g. Gavel N., Great Gable, Cumberland.

negative anomaly, of gravity A g. measurement below that computed for an ideal globe, implying the downward penetration of the less dense SIALIC rocks to a greater depth than elsewhere, as beneath the rift-valleys of E. Africa and beneath the deeps parallel to the Indonesian arcs. It is sometimes called a *Meinesz zone*, after V. Meinesz, who studied n. a.'s from a submarine in 1926.

negative landform A relatively depressed or low-lying l., incl. valleys, basins, plains and ocean-basins, in ct. with upstanding or POSITIVE L.'s.

negative movement, of sea-level A change in the relative l. of land and s., causing the net lowering of actual s.-l. This may be the result of either: (i) a world-wide or EUSTATIC fall in the l. of the water (e.g. during the Quaternary glaciation, when the abstraction of water in the ice-sheets was equivalent to a 90 m. fall); or (ii) a more local uplift, warping, tilting, or ISOSTATIC recovery of the land.

'negro head' Given on the Great Barrier Reef of Australia to a mass of coral, darkened by the growth of lichens, which has been broken off the outer part of a coral reef and hurled over on to the reef-flat behind by storm-waves.

Nehrung (pl. **Nehrungen**) (Germ.) A long sand-spit on the S. coast of the Baltic Sea, formed across the seaward side of a shallow embayment (HAFF), the result of the deposition of sand by longshore currents moving E. in the Baltic; e.g. the Kurische N., Frische N. (their names are now altered to Russian and Polish forms), 80 km. and 56 km. (50 and 35 mi.) long respectively, though only 0·4 km. to 3·2 km. (0·25 to 2 mi.) wide. These are lined by almost continuous tracts of sand-dunes up to 60 m. (200 ft.) high, mostly 'fixed' with marram-grass or planted with pines. [*f*]

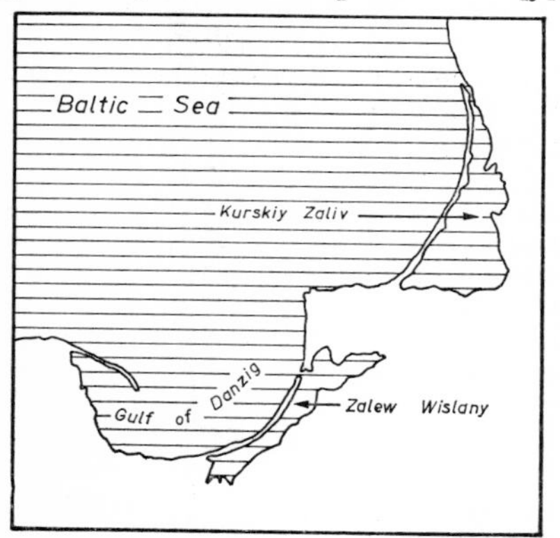

neighbourhood unit A part of a town which is virtually self-contained, with shopping and other service facilities. This is a concept of modern town-planning, and in the design esp. of new towns the site may be divided initially into a number of n. u.'s which form the basis for planning.

nekton Plant and animal organisms that actively swim in the surface waters of the ocean, as distinct from merely floating and drifting. Ct. PLANKTON, BENTHOS.

Neogene The younger of the two divisions of the Cainozoic era, as defined by the International Geological Congress, and used by European geologists. It has not found favour in Gt. Britain and U.S.A., and there is some doubt as to the time-period and rocks it covers. Some use the term to indicate all time and rocks from the Miocene onwards, others just the Miocene and Pliocene. In Gt. Britain the Miocene is gen. regarded as absent, and the Pliocene is represented only by the Coralline Crag of E. Anglia, but this is a convenient term for a period of erosional geomorphology.

Neoglacial, Neoglaciation Used increasingly in U.S.A. in place of 'LITTLE ICE AGE' to refer to a renewal of glacier growth in the Sierra Nevada and Alaska after their shrinkage or complete disappearance during the milder HYPSITHERMAL phase.

Neo-Malthusianism See MALTHUSIANISM.

Neolithic (lit. 'new stone', from the Gk.) A culture period in Britain from the latter part of 4th millennium B.C., lasting until the BRONZE AGE, *c.* 2000–1400 B.C. It was characterized by finely made flint (at Grimes Graves, Norfolk) and stone (at Langdale, Westmorland) implements, with great progress in the domestication of animals, cultivation of crops and the making of pottery, and the construction of long barrows, megalithic tombs and several great religious sanctuaries; e.g. Woodhenge and the first part of Stonehenge.

Neozoic (after the Gk. *neos*, *zoe*, 'new life') A term not gen. in use, but sometimes denoting a combination of Mesozoic and Cainozoic, as ct. with Palaeozoic ('old life').

nephoscope An optical instrument for measuring the direction and speed of movement of clouds.

neritic Related to shallow water, hence *n. deposits*, organic material derived from the remains of shell-fish, sea-urchins and coral, found in the LITTORAL and shallow water zones along the coast.

ness Syn. with NAZE; e.g. Orford-ness, Dungeness.

nested sampling A statistical technique used in regional science, whereby the region is divided into equal major areas, several of which are chosen at random, again broken down, and so on as far as data are available. Cartographic techniques can be applied to present the resultant s. designs, using an appropriate type of variance analysis.

nesting In urban geography, the pattern shown when low order trade areas lie within the boundaries of high order trade areas.

net national product The GROSS N.P., less depreciation and other capital consumption allowances. In U.S.A. the n. n. p. is about 90% of the g. n. p.

net register tonnage The GROSS R. T. of a ship less deductions for crew living space, stores, engines and ballast; i.e. it indicates the passenger- and cargo-carrying and earning capacity of the ship. Harbour and docking dues are assessed on n. r. t.

net reproduction rate In population studies, the n. r. r. affords a more exact indication of whether a generation can reproduce itself than is given by the GROSS R. R. Allowance is made for deaths of girls before attaining child-bearing age, later deaths, non-marriage and infertility, and applied as a correction to the gross r. r.; e.g. in France in 1935–6, the gross r. r. was 1·004, but the n. r. r. was only 0·88. A n. r. r. of unity indicates an ability to maintain a population, and one exceeding unity indicates an increase. Recent refinements have been introduced, using both female and male n. r. r., since alteration of the sex-age distribution (e.g. by war, as in the case of France in 1914–18) will cause discrepancies.

network A pattern of interconnected lines: (*a*) *material:* a railway, road or pipeline system; (*b*) *conceptual:* a n. summarizing a relationship, which may not exist in a material form; e.g. between units in an administrative hierarchy. A n. may be described verbally (e.g. as TRELLISED or DENDRITIC DRAINAGE, a grid-iron or radial road pattern), or in terms of topological ratios. The creation of n. MODELS, involving a number of points, their possible links, and the flows between them, is a valuable analytical tool in the study of transport and communications.

neutral coast A coastline where there has been no change in the relative level of land and sea; the term includes stationary c.'s of sedimentation (MUD-FLATS, MARSHES and DELTAS), CORAL c.'s and LAVA c.'s.

neutral stability An unsaturated column of air has n. s. when its ENVIRONMENTAL LAPSE-RATE is equal to the DRY ADIABATIC LAPSE-RATE, and a saturated column has n. s. when its environmental lapse-rate is equal to the SATURATED ADIABATIC LAPSE-RATE; i.e. the 'parcel' of air is in equilibrium with its surroundings. Syn. with *indifferent equilibrium.*

Nevadian (Nevadan) orogeny A period of earth-movement in N. America, in late Jurassic–early Cretaceous times, incl. folding movements in the W. part of the N. American cordilleras, together with the emplacement of massive BATHOLITHS of granitic rock in the Sierra Nevada. It seems to have lasted until, or even overlapped, the LARAMIDE O. in the E. part of the Cordilleras of N. America.

nevados, nevadas (S. Am.) A cold down-valley wind blowing from the Andean snow-fields to the high valleys of Ecuador. It is a type of ANABATIC WIND, strengthened by the chilling of air in contact with a snow surface, resulting in a gravity flow downhill.

névé (Fr.) Usually regarded as syn. with FIRN (Germ.); the latter is preferred by most glaciologists. *Note:* Some differentiate between the two terms, the firn being regarded as the actual snow, the n. the *accumulation area* of the firn, or the *firn-field.*

New Red Sandstone S. rocks deposited during Permian and Triassic times; the line between the Palaeozoic and Mesozoic rocks is actually drawn in the middle of the N. R. S. The formation occurs widely in N.W. England (St. Bees Head and the Eden valley); Devonshire; Durham; and various parts of Scotland (incl. the S. part of the Isle of Arran).

newton The unit pressure in the S.I. 1 n. per sq. m. = $1 \cdot 4504 \times 10^4$ lb. per sq. in.; 1 lb. per sq. in. = 6894·8 n. per sq. m.; 1 n. = 10^5 DYNES; and 10 n. = 1 BAR.

new town A t. deliberately designated by government planning, to relieve population pressure in large cities (OVERSPILL) and to provide more pleasant surroundings in which to live. Since 1945, the following n.t.'s have been created in Britain: Crawley, Bracknell, Stevenage, Hemel Hempstead, Welwyn, Hatfield, Harlow, Basildon (all around London); Corby (Lincs.); Newton Aycliffe, Peterlee, Washington (Durham); Skelmersdale (Lancs.); Runcorn (Cheshire); Redditch (Worcs.); Killingworth, Cramlington (Tyneside); Dawley (Shropshire); Cwmbran (S. Wales); East Kilbride and Cumbernauld (Glasgow); Livingston (W. Lothian); Glenrothes (Fifeshire); Irvine (Ayrshire); Newtown, Montgomery (Mid-Wales); Peterborough (county of Huntingdon and Peterborough); and, under the New Towns Act of N. Ireland (1967), Craigavon and Antrim/Ballymena. In 1967 Britain's first 'new city' was scheduled: Telford (Shropshire). It will include the present towns of Wellington, Oakengates and the n. t. of Dawley.

niche-glacier A very small CIRQUE G., lying high in the mountains in a steeply sloping hollow, bench or gully, probably developed from a compacted snow-patch through NIVATION.

nickel A metal occurring as a sulphide-ore in conjunction with copper and iron. It is resistant to corrosion and non-tarnishable. Its chief use is in steel-alloys, which are hard, strong, tough and non-magnetic. '*Invar*' is a 36% n. alloy with steel, with virtually no expansion or contraction with temperature change, and is used for surveying tapes. C.p. = Canada, U.S.S.R., New Caledonia, Cuba, U.S.A.

nick-point See KNICKPOINT.

nife The mass of nickel-iron (*Ni-Fe*) of density about 12·0, which is believed to comprise the CORE of the earth, lying beneath the MANTLE below a depth of about 2900 km. (1800 mi.) from the surface.

nimbostratus (*Ns*) A low thick cloud, from which continuous rain falls; it is dark grey in colour and nearly uniform in texture, and occurs partic. above a warm frontal surface in a low-pressure system, having progressively thickened from ALTOSTRATUS.

nimbus A gen. term for clouds from which rain is falling, attached to the names of other cloud-types; e.g. CUMULONIMBUS, NIMBOSTRATUS. It is not used by itself in the international system of cloud classification.

nip A distinct break of slope at the higher edge of a beach, clearly related to high-water mark, and usually just a few feet above the level of the highest tides.

nivation (i) The rotting or disintegration of rocks beneath and around the margins of a patch of snow by chemical weathering and alternate FREEZE-THAW, sometimes called 'snow-patch erosion'. Melt-water by day trickles into cracks and re-freezes at night, causing shattering. The sligh hollow in which the initial snow-patch accumulated is thus progressively enlarged, representing an early stage in the creation of a CIRQUE. (ii) In a mor gen. sense, the term includes the work of snow and ice beyond or outside th true glacial limits.

noctilucent cloud A luminous bluish or silvery-white c. seen (though only rarely) after sunset on summer evenings in the STRATOSPHERE at a altitude of from 50–80 km. (30–5 mi.), believed to consist of either du (some believe cosmic dust) or ice-crystals. It has been most frequently recorded in mid to high latitudes.

nocturnal A form of star-clock, developed at the beginning of the 15th century for determining time during the night at sea. By measuring the angle between the plane of the local meridian and a line through any circumpolar star (usually the 'pointers' of the Plough) and the Pole Star, SIDEREAL TIME can be measured. By means of another dial, sidereal time can be converted to APPARENT TIME. The medieval n. was frequently a decorative work of art.

nodal region A r. organized with respect to a single node or focus by means of patterns of circulation.

node A central point or focus in any complex, system or pattern of distribution; in topological terms a point of no dimension. In HUMAN GEOGRAPHY it can denote a CORE area, an agglomeration where activity is concentrated. See also NODAL REGION. N.'s may be linked in various ways on a 2-dimensional surface to form a *topological graph*.

nomad A member of a social group which leads a life of constant movement in search of pasture for their herds of animals; e.g. formerly the Kirghiz in Turkestan; the Bedouin Arabs; the Masai in E. Africa with their cattle. Hence the noun *nomadism*, adj. *nomadic*. The life is well described in the Old Testament. *Note:* the term does not include casual wanderers and collectors (e.g. Bushmen and Australian aborigines), nor those who practise SHIFTING CULTIVATION or TRANSHUMANCE (as in the Alps) who have permanent homes.

nomograph A graphical method of solving the functions of 3 or more variables. The graph consists of a series of three or more related scales, plotted on some type of graph-paper, so that a value can be immediately read off graphically by lining up the variables concerned. In the case of the fig. below, the 'dial' is moved until the 10 on it is aligned between 6400 acres on the left scale and 100 persons on the right scale, because 100 persons per 6400 acres is equal to 10 persons per sq. mi. Once set the n. may be used to calculate densities and reduce acres to sq. mi. in one operation. [*f*]

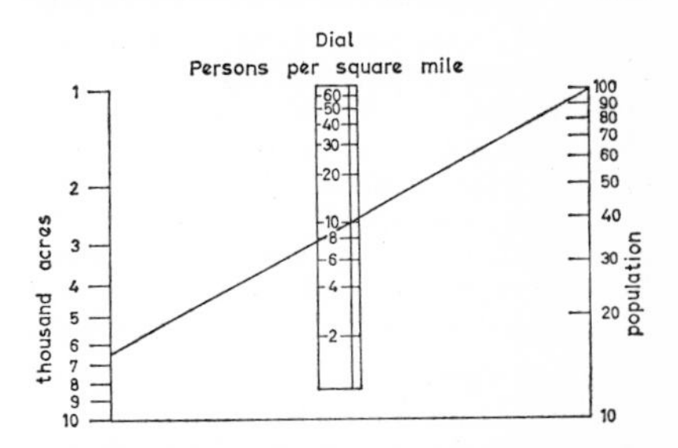

nomothetic An approach to geography whereby situations of a general type are studied, rather than individual cases. Ct. IDEOGRAPHIC.

non-basic function (or **service function**) In urban geography, those economic activities within a town that serve surrounding communities. Ct. BASIC F.

non-conforming use In town-planning, an unsympathetic land-u. established within an area allocated on a town map primarily to another use.

nonconformity, adj. **nonconformable** A type of UNCONFORMITY where igneous rocks are denuded and then overlain by material which gradually compacts to form sedimentary rocks; e.g. near Cape Town, a dark sandstone overlies nonconformably a mass of pale-coloured granite.

non-sequence In a CONFORMABLE series of rocks, a brief gap in the normal sequence, during which no deposition took place.

Nordic Projection A map p. designed by J. Bartholomew, 'an oblique area-true p. designed to give optimum representation to Europe, and to routes across the Atlantic, Arctic and Indian Oceans'. The major axis is a GREAT CIRCLE touching 45°N., the lesser axis the Greenwich meridian.

Norfolk rotation A r. of crops evolved in E. Anglia at the beginning of the 18th century during the Agrarian Revolution: wheat, a root crop, barley and a legume (beans or clover).

normal erosion Orig. used for river e. under conditions of a temperate climate, in contrast to desert and glacial e., which were termed 'special e.' It is now realized that e. as found in temperate areas may in fact be abnormal, certainly when considered in respect of the earth as a whole, and in relation to past geological conditions.

normal fault A f. resulting from tension, when the inclination of the f.-plane and the direction of the downthrow are in the same direction; i.e. the HANGING WALL [*f*] has been depressed relative to the FOOT WALL. A vertical f. is usually regarded as a n. f. The term can be misleading if it is thought to imply that this kind of f. is the one most commonly found; this is far from the case. [*f*]

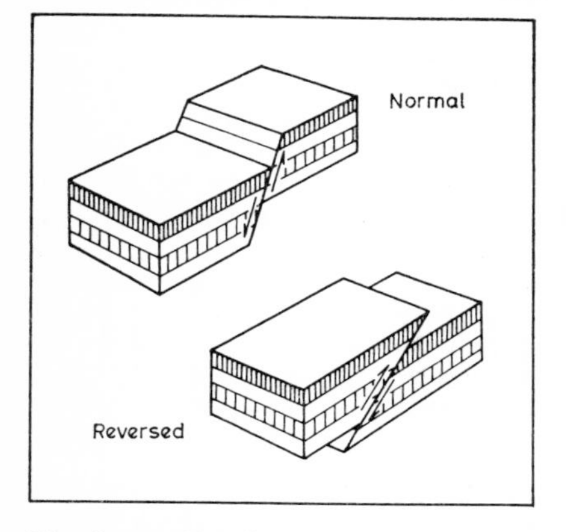

Northeast Trades The TRADE WINDS of the N. hemisphere.

norther A cold dry N. winter wind bringing low temperatures to Texas and the Gulf coast of U.S.A. It is a form of polar outbreak, whereby a cold air-mass moves S. across the N. American continent, unimpeded by any transverse relief barrier. A similar wind, the *norte*, affects the coasts of Mexico and Central America. The term n. is used for cold polar winds in many parts of the N. hemisphere (California, Portugal).

northing The distance on a map from the origin of a GRID in a northerly direction, providing the second half of a grid-reference; ct. EASTING, which provides the first half.

Nor'wester A FÖHN-like wind in the S. Island of New Zealand.

nose A prominent spur or ridge, cognate with NESS, NAZE, NAB, NESSE; e.g. Wrynose Fell, on the N. side of Wrynose Pass, Westmorland.

nosology (from Gk., *nosos*, disease) The scientific classification of diseases, hence *nosography*, the scientific description of diseases. The construction of *nosological maps* is an important part of MEDICAL GEOGRAPHY.

notch (i) The undercutting of a cliff at high-watermark, which produces a n. at its foot. (ii) A pass, esp. in the Appalachians; e.g. Pinkham N., White Mtns., New Hampshire.

nuclear power The use of radioactive energy created in the process of n. fission, in this sense for peaceful ends. In the n. reactor, or atomic pile, uranium is allowed to disintegrate, releasing energy in the form of heat, which is applied to water circulating through and around the pile, so producing steam which drives a turbine, which in turn drives a dynamo. The U.K. Atomic Energy Authority was formed in 1954, and the first reactor for generating electricity was Calder Hall in W. Cumberland opened in 1955. The first American station began operation in 1958.

nucleated settlement, nucleation A group or cluster of houses in the countryside, as ctd. with isolated individual or DISPERSED S.; e.g. in Belgium in the Hesbaye, the Ardennes and Belgian Lorraine.

nuée ardente (Fr.) A mass of hot gas, superheated steam and incandescent volcanic dust, which may move down the slopes of a volcano during an eruption as a glowing avalanche; the particles slide over each other, with a high internal velocity. E.g. during the eruption of Mont Pelée in the W. Indies in 1902, the town of St. Pierre was obliterated, destroying its 25,000 inhabitants. Another n. a. rolled down the flanks of Lassen Peak in 1915, where a 'Devastated Area' was created as the forests were swept away by 'the Great Hot Blast'.

nullah (Indian) The bed of a stream which flows only occasionally, following sporadic though intensive, downpours of rain.

numerical forecasting The f. of weather by solving, usually with the aid of a computer, sets of equations relating to associations of observed atmospheric phenomena.

nunatak (Eskimo) A rock-peak projecting prominently above the surface of an ice-sheet; orig. used in Greenland, where they are numerous along the coastal rim; they are also very common in Antarctica.

oasis A place in a desert where sufficient water is available for permanent plant growth and human settlement, varying in scale from a small group of palms around a spring to an area of some hundred sq. km. An o. may occur wherever the WATER-TABLE approaches or reaches the surface, as in a DEFLATION hollow (e.g. the district of the Shotts, in Algeria and Tunisia), and esp. in an ARTESIAN BASIN (Libya). Sometimes the water must be pumped from great depths; e.g. in S. Algeria, in the Mzab, where there are 3300 wells, from which water must be laboriously raised to supply the irrigated gardens.

oats A cereal (*Avena sativa*) preferring a rather cooler, damper climate than others, mainly grown as fodder for animals, though eaten by humans, esp. in Scotland. A bulky grain, usually consumed on the farm where it is grown. C.p. = U.S.A., U.S.S.R., Canada, Poland, France, W. Germany.

oblate A flattening or depression of opposite parts of a sphere, hence an *o. spheroid*, the shape of the earth.

oblique aerial photograph A p. taken from an aircraft, with the camera pointing down at an angle, not vertically. It combines the familiar ground-view with the pattern obtained from a height. In low undulating areas, it may be used for map-making (e.g. the Canadian Shield), but since the scale decreases away from the bottom (foreground) of the p., this presents considerable difficulty.

oblique fault A FAULT whose strike is oblique to the strike of the beds it traverses.

obsequent fault-line scarp A f.-l. s. facing in a direction opposite to that caused by the initial earth-movement. After the f.-l. s. has developed, denudation may progress so far that the resistant strata on the higher side may be removed, exposing less resistant underlying strata. The strata on the DOWNTHROW side may now be the more resistant, and will increasingly stand out as the gen. surface is lowered, so that a f.-l. s. will develop along the line of the same fault, but facing in the opposite direction; e.g. the Mere Fault on the N.W. of the Vale of Wardour, Wilts. [*f* FAULT-LINE SCARP]

obsequent stream Introduced by W. M. Davis to describe s.'s which flow in a direction opposite to the original CONSEQUENT drainage to join a SUBSEQUENT S. The term is often used now for a *scarp s.* or *anti-dip s.*, one flowing in the opposite direction to the dip of the strata. This transference of meaning can be confusing, esp. if consequent drainage, initiated perhaps on a marine platform, does not conform to the dip of the rocks. There is still a difference of opinion about the usage; many French and American geomorphologists use the orig. Davisian concept as 'anti-consequent', while

some English authorities prefer the 'anti-dip' meaning.
[*f* SUBSEQUENT STREAM]

obsidian An extrusive igneous rock, with 65% or more silica, in appearance like a mass of dark glass, and rhyolitic in composition.

occlusion, occluded front The overtaking of a WARM F. by a COLD F. in an atmospheric depression, which ultimately lifts the WARM SECTOR off the surface of the earth, so forming an occluded front. Thus the cold air in the rear of the depression has now come up against the cold air against which the warm front was orig. formed. If the overtaking air is colder than the air in front, it is a *cold o.*, if not as cold it is a *warm o.*, as in the [*f*] below. If there is no marked difference in temperature, it is a *neutral o.* [*f*]

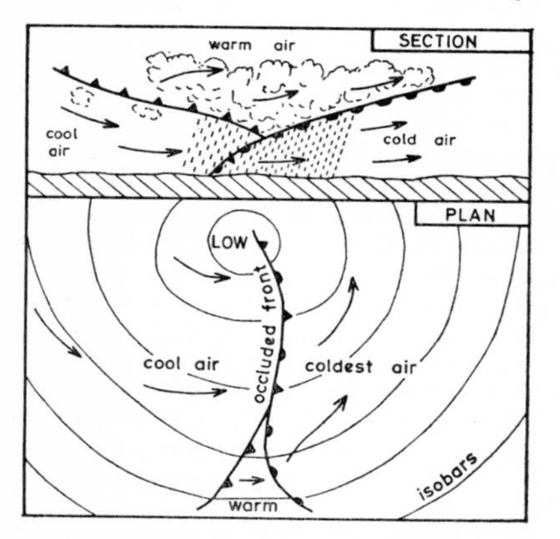

occupancy rate A measure of the degree of overcrowding of people in any area, calculated as

$$\frac{\textit{population}}{\textit{number of 'habitable rooms'}}$$

In gen., living conditions are comfortable up to an occupancy rate of about 1·0, and overcrowding occurs above a rate of about 1·5. (Habitable rooms are defined as all rooms other than kitchen, bathroom, etc.).

ocean One of the main water-areas of the globe, lying in a basin; o.'s occupy in all 71% of its total surface.

	Area 1000 *Sq. km.*	*Area* 1000 *Sq. mi.*	*Greatest depth* *m.*	*Greatest depth* *ft.*
Pacific	165,384	63,855	11,033	36,198
Atlantic	82,217	31,744	8,378	27,498
Indian	73,481	28,371	8,047	26,400
Arctic	14,056	5,427	5,449	17,880

This total excludes shallow marginal seas, such as the Caribbean and Mediterranean. The first three include the various sectors of the Southern O., sometimes defined as the o. areas S. of 40°S. Some authorities speak of six oceans: the N. and S. Atlantic, the N. and S. Pacific, the Indian and the Arctic.

oceanic climate See MARITIME C.

oceanography The scientific study of phenomena associated with the oceans. It includes: (i) *physical o.*, the extent and shape of ocean basins, the structure and relief of their floors, the movements of sea-water, its temperature and salinity; (ii) *biological o.*, the study of life in the oceans.

odograph A recently devised automatic instrument which is carried in a vehicle, and draws a continuous map of the route taken, using a mileometer for distances and a magnetic compass for directions. The route is plotted directly on paper by electronic and photo-electric methods. This instrument has been used experimentally in U.S.A.

odometer A device for measuring distance, based on the principle of a revolution counter attached to a wheel. A simple o. was used by such cartographers as Christopher Saxton. One type, attached to the person, will measure the distance walked by a pedestrian.

offshore App. to movement away from the shore, as o. wind, o. current or to accumulation of material away from the shore, as o. BAR [*f*].
[*f* COASTLINE]

ogive One of a series of bands of light and dark ice, arcuate downstream, on the surface of a glacier or within its mass, and similar patterns of 'waves', depressions and dirt-bands. The o. shape results from the more rapid flow of the centre of the glacier by comparison with that at the edges, where it is retarded by friction. The dirt-bands represent the refreezing of meltwater containing much dirt. The light bands contain numerous air-bubbles, and this white ice was probably formed from winter accumulation of snow, when melting and refreezing were negligible. The bands of dark ice are virtually bubble-free, and represent partial melting and refreezing. The surface waves are due to the fact that the white bubbly ice reflects more insolation; the dark bands suffer greater melting because of their increased conductivity, thus become troughs, and accumulate more meltwater and dirt in summer. In U.S.A. they are sometimes called *Alaskan bands* or *Forbes bands*.

oil-pool An accumulation of PETROLEUM held in the pores of a sedimentary rock. Though o. is widespread, it occurs in concentrations only in favourable structural locations; e.g. above a salt-dome (in Texas and N. Germany), in a symmetrical anticline (o.-dome) above water, in any stratigraphic trap.

oil-sand S. which is impregnated with hydrocarbons (bitumen); e.g. the tar-sands in N. Canada, which cover a large area of the Athabasca valley 160 km. N. of Edmonton. These are believed to comprise the largest o. reserves in the world, though only in recent years has an economic process been devised to extract the o.

oil-shale S.'s containing hydro-carbons, which can be distilled to give o. These occur in the Carboniferous rocks of the Midland Valley of Scotland in West Lothian, and were first worked commercially in the middle of the 19th century, yielding for a time 30 gallons per ton. During the war of 1939–45, output was stimulated to about 149,000 tons of o. (in 1941), since when the annual average has been about 110,000 tons until 1963, when production ceased. The s. was processed at Pumpherston.

okta The proportion of the sky which is cloud-covered, expressed in 8ths of the total, and indicated on a weather map by a proportionally shaded disc. 0 okta = clear sky, 8 oktas = total cloud cover.

Older Drift The older of the two main groups of Pleistocene glacial deposits, which is more weathered than, and is overlain in part by, the YOUNGER D. More complex subdivisions have been made in recent years, with several periods of glacial and interglacial deposition being distinguished. In Europe the term O. D. denotes that laid down during the earlier glacial advances, lying S. and W. of the Baltic Heights, the main terminal moraine of the 4th (Weichsel) glaciation. The O. D. has been much weathered, eroded, resorted and redeposited by post-glacial rivers.

'oldland' An area of ancient rocks reduced by denudation to subdued relief.

Old Red Sandstone A series of red, brown and white s.'s, with some layers of conglomerate, marl, shale and limestone, of Devonian age. These rocks are non-marine, and were probably deposited in large inland lakes. They are found in Herefordshire, S. Wales, the Midland Valley of Scotland, N.W. Scotland (Caithness), and the Orkneys.

Oligocene The 2nd of the geological periods of the Cainozoic era (3rd in U.S.A., where it is classed as an epoch), and the system of rocks laid down during that time, from about 40 to 25 million years ago. In England, O. rocks occur only in the Hampshire Basin, the result of deposition under freshwater or brackish conditions, incl. limestones, clays and marls, and in Devonshire between Newton Abbot and Bovey Tracey (clays, gravel, sands and lignite). The O. is much more widespread in Europe, occurring in the Paris Basin, N.E. Belgium and N. Germany.

olive The fruit of an evergreen tree (*Olea Europea*), grown in the Mediterranean lands since ancient times, with narrow leathery leaves. It can grow on poor gravelly soils and steep limestone slopes, in areas with low rainfall. From its fruit oil is expressed. C.p. = Spain, Italy, the Mediterranean lands gen., California and Argentina.

olivine The chief constituent of a group of rock-forming minerals, basic and ultrabasic in character, consisting of silicates of magnesium and iron, usually dark green in colour. It forms much of the SIMA layer in the earth's crust. An older name is *chrysolite*. Common rocks in which o. is the most important mineral include *peridotite* and *serpentine*.

onion-weathering See EXFOLIATION.

onset and lee The result of the abrasive effect of a glacier on the upstream side of a rock, and its plucking effect on the downstream side. See ROCHE MOUTONÉE.

oolite A sedimentary rock, usually calcareous, consisting of small, rounded grains (*ooliths*), from 0·25 to 2·00 mm. in diameter. These are often likened to fish-roe, hence the name. A grain commonly has a nucleus of a quartz or shell particle, with a surrounding concentric calcareous structure. O. as a formation name refers to the Inferior and Great O. outcrops of Jurassic age, which can be traced across England from Dorset to N. Yorkshire.

ooze A very fine-textured PELAGIC DEPOSIT, which accumulates on the ABYSSAL zone of the ocean-floor. Types include: (i) *calcareous o.* (PTEROPOD, GLOBIGERINA); (ii) *siliceous o.* (RADIOLARIAN, DIATOM). O. accumulates very slowly, at the rate of about 2·5 cm. in 20,000 years in the Pacific Ocean. The total thickness is great; echo-sounding indicates a layer of about 3700 m. (12,000 ft.) thickness, which must have taken 300–400 million years to accumulate.

opencast mining A form of extensive excavation, only practicable where the mineral deposits lie near the surface, whereby the overlying material (*overburden*) is removed by large-scale machinery, and the seams or deposits thus revealed are then quarried. The equivalent American terms are '*strip-m.*' or '*open-cut*'. It is used for working coal, brown-coal (e.g. in the Rhineland and Upper Saxony), and iron-ore (e.g. in the L. Superior ranges in U.S.A. and the E. Midlands of England).

open-field Before enclosure, the land of an English village was cultivated in two or more large common f.'s, one of which usually lay fallow for a year. Each member of the community held strips scattered over the several f.'s, and had rights of common pasture at certain times of year.

open system An approach to a GENERAL SYSTEMS THEORY that is being increasingly applied to geography. The o. s. is characterized by the supply and removal of energy and material across its boundaries (ct. CLOSED SYSTEM). Under such conditions the s. regulates itself by homeostatic adjustments and attains a *steady state*. Throughout time the s. attains a constant magnitude and is said to behave equifinally, for different initial conditions can produce similar end-results. Application of the o. s. concept enables geomorphology to free itself from many constraints, and helps to solve the problem of GRADE. In a closed s. grade may be regarded as a hypothetical end-stage in the denudational process, but an o. s. approach demonstrates how a provisional equilibrium can be established at a much earlier stage. In human geography, an example of an o. s. is a NODAL REGION, with settlements, roads, land-use zones, etc., involving movement of people, goods and money, the concept of the URBAN HIERARCHY, and in fact all CENTRAL PLACE theory.

opisometer A small serrated wheel linked to a calibrated dial, used for measuring distances on a map (along a road, railway or river) by running the wheel along the partic. line.

opposition The position of 2 heavenly bodies when their longitude differ by 180°. When the moon and the sun are in o. to the earth, the result is the phase of full moon, resulting in SPRING TIDES. Ct. CONJUNCTION and QUADRATURE.

optimum population The number of people who can live most effectively in any area in relation to the possibilities of the environment, with a reasonable standard of living.

orbit, of the earth The path through space of the e. in its annual journey around the sun. This path is an ellipse, with a max. distance of 152 million km. (94·5 million mi.) on 4 July (APHELION), and a min. distance of 147 million km. (91·5 million mi.) on 3 Jan. (PERIHELION). The mean velocity at which the e. travels round this orbit is about 106,000 km. (66,000 mi.) per hour, but this varies according to the partic. portion of the orbit.

order, of a good, function, central place In urban geography, o. relates to the partic. level of a good (or function of place) in the hierarchy of goods (or functions of places). 'High o.' is used to distinguish those which are comparatively rare, 'low o.' is used for those which are rel. numerous; cf. in respect of their functions a jeweller and a butcher.

order, of streams See S. ORDER.

Ordnance Datum (O.D.) A mean sea-level calculated from hourly tidal observations at Newlyn, Cornwall, taken between 1915 and 1921, from which all heights on official British maps are derived.

Ordovician The 2nd of the geological periods of the Palaeozoic era, and the system of rocks laid down during that time, lasting approx. from 500 to 440 million years ago. Formerly it was regarded as the Lower Silurian (hence the index letter *b* is still used for both O. and Silurian on British Geological Survey maps), but in 1879 a separate system was proposed by C. Lapworth. In Britain the rocks occur in S., central and N. Wales and in neighbouring parts of Shropshire (an area once inhabited by a tribe, the *Ordovices*, hence the name); in the English Lake District; the S. Uplands of Scotland (a narrow belt continued into Ireland from near Belfast Lough); and S.W. Ireland from Dublin to Waterford. The rocks include shales, flagstones, grits and slates, with interbedded layers of volcanic tuff and lava, which compose mountainous country in Snowdonia and in the English Lake District (the Borrowdale Volcanic Series).

Series	*N. Wales*	*English Lake District*
4. Ashgillian	Blue-grey slates	Coniston Limestone
3. Caradocian	Rhyolitic lavas, with slates	
2. Llandeilian	Volcanic ashes, agglomerates, limestones, flags	Borrowdale Volcanic Series
1. Arenigian or Skiddavian	Arenig Series (flags, grits, shales)	Skiddaw Slates

Note: A series, the Llanvirnian, is sometimes intercalated between 1. and 2., assimilating the upper Arenigian and the lower Llandeilian.

---

ore A mineral containing metal, in sufficient concentration to be economically workable.

organic (i) In one scientific sense, the term refers to compounds containing carbon and hydrogen. (ii) In another sense, it app. to living organisms and their remains. Hence o. deposits: shelly limestone, pelagic oozes, coral, coal and lignite, chalk.

organic weathering The breakdown of rocks by plants such as moss and lichen, thus causing humic acids to be retained in contact with the rock; a tuft of moss may lie in a small hollow, which is slowly enlarged by rotting beneath it. Vegetation can also have a mechanical effect, the result of the penetrating and expanding force of roots. Worms, rabbits and moles have a considerable effect on the surface soil.

orientation The setting of a map or of a surveying instrument in the field so that a N. to S. line on the map or drawing sheet is parallel to the N. to S. line on the ground. Medieval maps were drawn with the E. at the top, hence the term.

origin, of a grid A point from which a g. system is laid out, at the intersection of the central meridian and a line drawn at right-angles (*true o.*). To avoid negative values, this is transferred to a point beyond the gridded area (FALSE O.). [*f* GRID]

origin, seismic The point in the earth's crust at which an earthquake shock originates as a result of tectonic movements. It is also known as the S. FOCUS. [*f* EARTHQUAKE]

orocline A bent or flexed (in plan) orogenic belt; e.g. the Alaskan O., the Baluchistan O.

orogeny A major phase of fold-mountain building. Hence *orogenesis,* the process of fold-mountain building; *orogenic,* the forces which cause this process. The main o.'s which affected Europe are: (i) Charnian (late Pre-Cambrian); (ii) Caledonian (Silurian-Devonian); (iii) Hercynian (Carboniferous-Permian); (iv) Alpine (mid-Tertiary). The chief American o.'s are: (i) Laurentian; (ii) Algomanian; (iii) Killarnean (all Pre-Cambrian); (iv) Taconian (Ordovician-Silurian); (v) Acadian (Devonian-Lower Carboniferous); (vi) Appalachian (Permian, early Triassic); (vii) Nevadian (late Jurassic-early Cretaceous); (viii) Laramide (late Cretaceous-early Tertiary); (ix) Cascadian (Pliocene-Pleistocene). Nos. (vii) and (ix) are not o.'s in the true sense of folding, since they involved mainly emplacement of batholiths, *en masse* uplift and vulcanicity (Sierra Nevada, Cascades).

orographic precipitation Caused by the ascent (hence cooling) of moisture-laden air over a mountain range. Heavy p. falls partic. on the windward slopes of mountains facing a steady wind from a warm ocean; e.g. Vancouver airport on the flat delta of the Fraser R. receives 635 mm. (25 ins.) of p., Vancouver city 1499 mm. (59 ins.), the mountains behind Vancouver at 1200 m. (4000 ft.) altitude about 2286 mm. (90 ins.). But Kamloops, E. of the Coast Ranges in a valley, receives only 254 mm. (10 ins.); see RAIN SHADOW. Gen. the o. factor merely emphasizes the amount of rainfall caused by other means; a relief barrier will cause an intensification of depressional p. (as in W. Britain), since the ascent 'triggers off' CONDITIONAL INSTABILITY, causes convergence and uplift, and retards the rate at which a depression moves (thus prolonging the period of actual p.).

orography In gen., the description or depiction of relief; more specif. the relief of mountain systems.

orrery A working model of the sun, moon, earth and planets, which when motivated illustrates the relative movements of these bodies.

orthodrome A section of a GREAT CIRCLE on the earth's surface; i.e. the shortest distance between any two points. All o.'s are straight lines on a GNOMONIC PROJECTION. Ct. LOXODROME. [*f* GREAT CIRCLE]

Orthographic Projection A category of p. in which a global hemisphere is projected on to a perpendicular plane as if from an infinite distance. Only at the centre of the p. is the scale accurate, and the error increases rapidly outwards. It is little used except for astronomical charts or for pictorial world maps. It can be drawn in a polar, equatorial or oblique plane but the largest area that can be shown is a hemisphere. [*f, page 25*

orthomorphic projection A category of p., syn. with CONFORMAL meaning lit. 'true shape'.

os, (pl. **osar**) (from Swedish *ås*, plur. *åsar*) An ESKER *sensu stricto*, a winding ridge of coarse sand and gravel, lying parallel to the gen. direction of movement of a glacier. It was formerly thought to represent the 'cast' of a stream-course within the ice, which survived when it melted. More probably it was formed as a continuous delta at the edge of a rapidly receding ice-sheet or glacier, deposited by a heavily laden ENGLACIAL or subglacial stream; e.g. in Finland, Sweden, S. Norway, the former E. Prussia.

oscillatory wave theory of tides The concept that the ocean surface can be divided into 'tidal units', each with its node or centre. In each unit the water is set oscillating, varying with the relative positions of the earth, moon and sun, together with the shape, size and depth of the water body within the unit; a gyratory movement is also imparted by the rotation of the earth. From the nodal points, where the height of the water remains virtually level, the height of the tidal rise increases outward. See also AMPHIDROMIC POINT, CO-TIDAL LINE. [*f* AMPHIDROMIC POINT]

outcrop That part of a rock occurring at the surface of the earth, though covered with superficial soil, vegetation and buildings. It is also used as a vb., *to o*. Ct. EXPOSURE.

outfall The narrow outlet of a river or drain; e.g. of the Fleet into the Thames.

outfield Land in the outlying parts of a farmstead, which may be moorland or fell-grazing, in ct. to the cultivated INFIELD.

outlet glacier A g. emerging from the edge of an ice-cap on a high plateau, rather than from a CIRQUE; e.g. in N. Norway, Iceland.

outlier (i) *Erosional:* an outcrop of newer rocks surrounded by older ones, the result of its separation by erosion from the main mass of which it now forms a detached portion, as in the *f* below; e.g. a *butte témoin.* (ii) *Structural:* an outcrop of newer rocks which has been let down between faults, or in a syncline, which has therefore survived when the adjacent portions of the same rock have been removed by denudation. Ct. INLIER. [*f*]

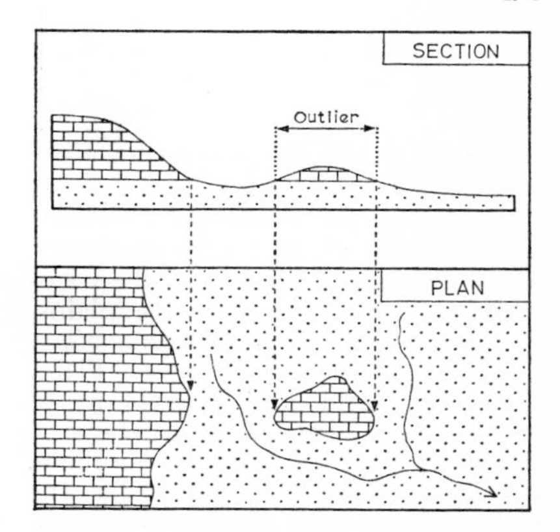

outport A subsidiary port at the mouth of an estuary, up which the main port may lie at an inconvenient distance. They have developed: (i) as shipping has grown in size; e.g. Avonmouth for Bristol, St. Nazaire for Nantes, Pauillac and Le Verdon for Bordeaux; (ii) as estuaries on which ports are situated have become progressively silted up; e.g. Zeebrugge for Bruges; (iii) as passenger traffic required speedy transport by rail to its destination, instead of proceeding slowly by ship further up the estuary; e.g. Tilbury for London, Cuxhaven for Hamburg, Wesermünde for Bremen. [*f*]

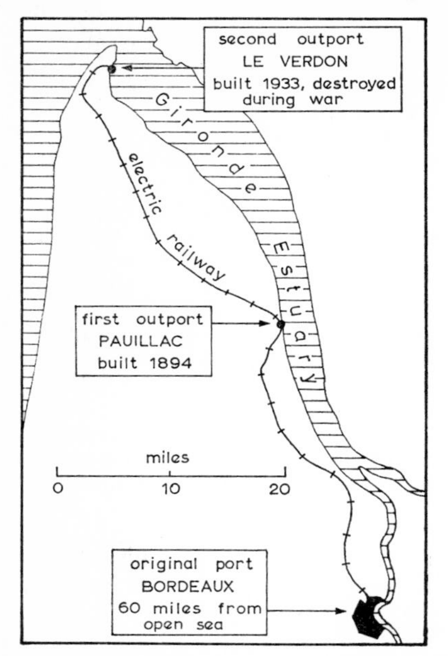

outwash Clay, sand and gravel washed out by melt-water streams flowing from glaciers and ice-sheets. These deposits beyond the ice-margins may form extensive plains, or more limited fans. Hence *o. plain*, *o. apron*. In the N. European Plain, sheets of sand and gravel up to 75 m. thick were laid down beyond the main lines of terminal moraine, cut into individual blocks by the valleys of the post-glacial rivers, and now forming heathlands. Most glacial forelands, and the terraces of rivers crossing them, are covered with layers of o., resorted and redeposited. [*f* ESKER]

overburden The overlying soil and rock that covers a deposit of ore or coal. It is stripped off before open-cast mining or quarrying begins.

overcast Used as a noun for a complete cloud cover, esp. in aviation.

overflow channel A c. by which a lake has overflowed during a former period of high water-level. The term is applied partic. to a c. draining a lake dammed up between a pre-glacial watershed and an ice-sheet; e.g. Newton Dale, cutting across the N. York Moors, which drained water from the PROGLACIAL LAKES on their N. flanks S.-ward into the Vale of Pickering; Forge Valley, through which the Derwent flows S., eroded when an ice-sheet blocked its orig. E. course to the N. Sea. When L. Bonneville occupied much of the Great Basin in U.S.A., and its height was 1000 ft. above the present Great Salt L., water spilled over through the Red Rock Pass into the Snake R. valley. The Great Basin is now an area of inland drainage. [*f*]

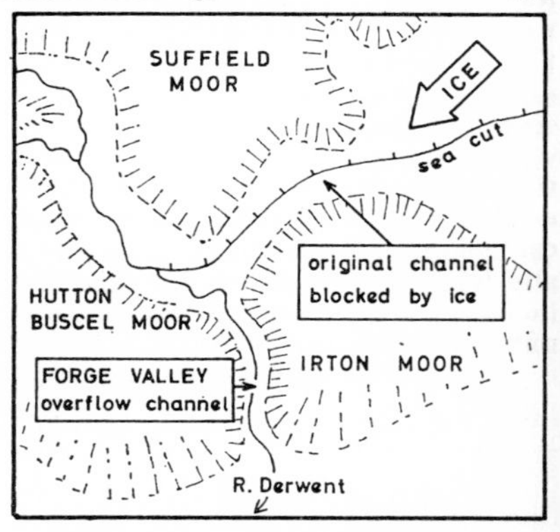

overfold An asymmetrical ANTICLINE which has been completely pushed over, or overturned, by compressional forces. [*f*]

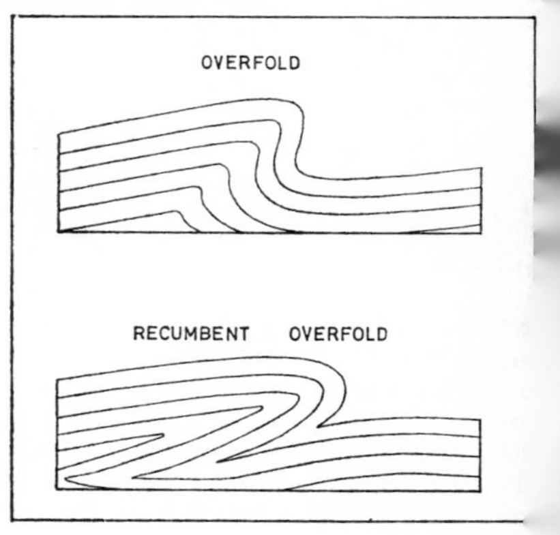

overgrazing, overstocking In farming, the putting of too many animals on to the land to graze, so that the vegetation cover is gradually destroyed. In an area of uncertain or short seasonal rains, this removal of the protective mat of vegetation can lead to serious SOIL EROSION.

overland flow A term used in U.S.A. for surface runoff, not concentrated in individual channels. Cf. SHEETFLOOD.

overpopulation A condition in which the population of a state, or any part of it, exceeds the optimum, and thus cannot be adequately supported by the available resources. It is not necessarily equated with dense population; some rural, mountain or desert areas may be overpopulated with a very low density.

overspill Surplus population from a densely crowded urban area, which can no longer be effectively housed and maintained there, and so is compelled to move. This may be the result of the actual growth of the population; or of slum-clearance and better housing which can accommodate fewer people. O. may result in expanding suburbs, or in the creation of 'new towns', as in the case of London.

overthrust The result of very intensive folding, causing the forcing of the upper limb of a fold along a horizontal or gently inclined plane over the lower limb (cf. NAPPE). This often involves fracturing of the rocks. E.g. the Lewis O. in Glacier National Park, Montana, where Pre-Cambrian rocks were thrust to the E. over the Cretaceous rocks; the 4 thrust-planes of the Caledonian folding in N.W. Scotland: Glencoul, Ben More, Moine and Sole. [*f opposite*]

oxbow The surviving portion of a former acute MEANDER, formed by the river current ultimately cutting through the meander-neck, thus abandoning a small crescent-shaped lake. It is also called a 'cutoff' or 'mortlake'; e.g. found on most rivers in their flood-plain tracts, as the Mississippi, the Trent. [*f*]

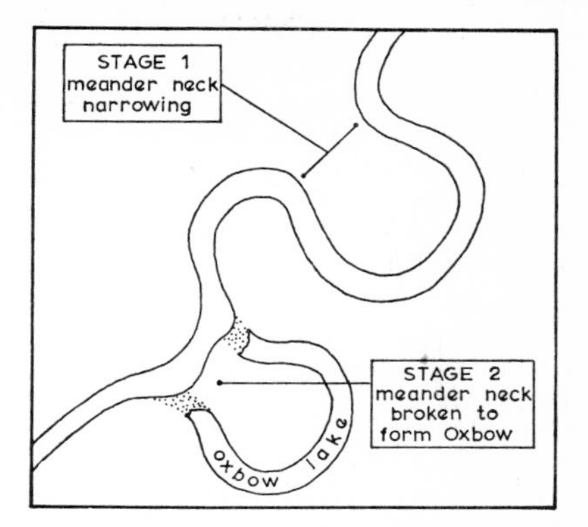

oxidation One of the processes of chemical weathering, involving the combination with oxygen. The results are most commonly shown when the rocks affected contain iron. The ferrous state changes into the oxidized ferric state, forming a yellow or brown crust which readily crumbles.

ozone An allotropic form of oxygen (O_3), very faintly blue in colour, found in minute quantities in the earth's atmosphere, with some concentration at about 19–26 km. (12–16 mi.) altitude. The concept of health-giving o. at the seaside is fallacious. An important result is that this layer absorbs most of the sun's ultraviolet rays, forming a screen (or 'hot layer') over the lower stratosphere.

Pacific type, of coast See CONCORDANT C.

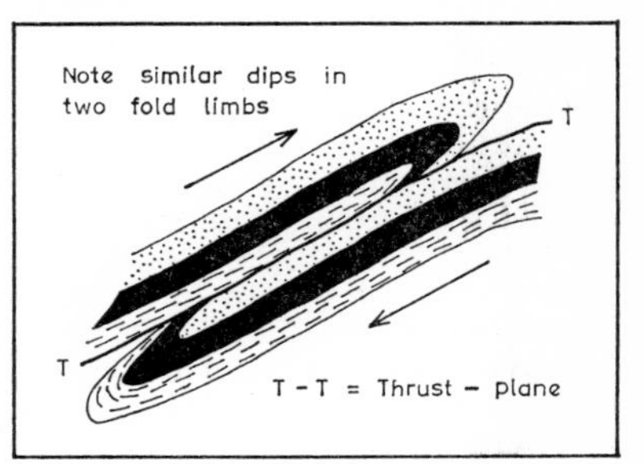

packet-boat A small steamer travelling between ports along the coast or on either side of a relatively narrow sea, such as the English Channel, N. Sea and Irish Sea, carrying mails, p.'s and passengers. In the past it was referred to as 'the p.'. It is now rather an obsolete term, but p. is still applied to some ports; e.g. Harwich, Dover, Newhaven, Calais, Boulogne, Fishguard.

pack-ice Masses of i. floating on the sea, specif. when floes have been driven together to form a mass, more (*close p.*) or less (*open p.*) coherent, but covering the entire sea-surface, with little or no open water. Frequently the individual floes are driven together, their margins buckled up, and irregular PRESSURE RIDGES are formed.

padang Used of land in S.E. Asia that is gen. treeless, with scrub and heath-type vegetation, commonly developed on sandstones with leached soils.

paddock A small field or an enclosure of pasture land, adjoining a farm or a stable.

paddy, padi Rice, whether referring to the whole plant, or just the grain while still in the outer husk. Hence *p.-fields*. The term rice should not be used until this husk has been removed.

pahoehoe (Hawaiian) A newly solidified lava-flow with a wrinkled, 'ropy' or 'corded' surface. The liquid lava is at a high temperature, the gases escape quietly, and the flow congeals smoothly; ct. the jagged 'block' lava (AA), PILLOW LAVA.

palaeobotany The study of plant life of past ages, mainly from the evidence of fossils.

Palaeocene Used in 3 contexts. (i) The earliest part of the Eocene, lasting from about 70 to 60 million years ago. (ii) Strata intermediate in age between the Cretaceous and Tertiary systems. (iii) In A. Holmes's time-scale (as revised in 1959), the P. is classified as the first of the epochs of the Tertiary period, lasting from 70 million to 60 million years ago.

palaeoclimate, palaeoclimatology The climate, or the study of a climate, of some period in the geological past. Climates have varied considerably in time, with major glacial, desert and pluvial episodes.

Palaeogene Used in Europe, though rarely in Gt. Britain and U.S.A., for the two earliest of the Tertiary periods, i.e. Eocene and Oligocene, in ct. to the NEOGENE. Syn. with EOGENE.

palaeogeography The reconstruction of the pattern of the earth's surface during its geological history.

Palaeolithic age (from Gk. *palaios* ancient, *lithos*, a stone) The Old Stone A., the earliest period of human pre-history, lasting from about 500,000 to 8000 years ago, coinciding with the greater part of the Pleistocene Ice-age. A series of culture-periods has been distinguished, related to the various glacial and interglacial phases: *Pre-Chellean*, *Chellean* (or *Abbevillian*), *Acheulean*, *Mousterian*, *Aurignacian*, *Solutrean*, *Magdalenian* and *Azilian*, named after places in France. The earliest implements were roughly fashioned palaeoliths of flint, but gradually these became more highly fabricated ('blade industries'), and at some stages bone was used. Cave-paintings are a significant feature, esp. in S.W. France, as at Les Eyzies in the Vézère valley.

palaeomagnetism Igneous rocks when cooled preserve a certain magnetization, the result of the presence of iron oxides, which is accordant with the direction of the earth's magnetic field at the time. This fossil magnetism thus provides a record by which the position of the magnetic poles at various times may be located, and has strengthened in partic. arguments that CONTINENTAL DRIFT has occurred. P information from rocks of the same age indicates that the continents must have moved relative to one another in order to harmonize the direction of magnetization in the differing areas.

palaeontology The scientific study of FOSSILS, the preserved traces of life of past geological time.

Palaeozoic Derived from Gk. 'old life', the first (hence 'Primary') of the main divisions (eras) of geological time subsequent to the Pre-Cambrian, and the rock-groups deposited during that time. It is divided into six periods (seven in U.S.A.), with corresponding systems of rocks, and lasted from *c.* 600 million to *c.* 225 million years ago. See individual periods.

Paleocene See PALAEOCENE.

paleoecology Used esp. in U.S.A. to study the ECOLOGY of past geological times, partic. in relation to the changing associated climates, using RADIOCARBON DATING, POLLEN ANALYSIS, the analysis of fossils and the landforms with which they are associated.

Palisadian In U.S.A. the series of block-faulting movements which affected the E. parts of that country in late Triassic–early Jurassic times.

palm oil, palm-kernel oil Derived from the o.-palm (*Elaeis guineensis*), a tree which flourishes in equatorial climates, and produces a large bunch of plum-like fruits; from their flesh p. o. is expressed; from the crushed kernels p.-k. o. is obtained. Both are used for the manufacture of margarine, cooking-fat and soap. C.p. = Nigeria, Malaysia, Congo (Kinshasa), Indonesia.

palsa (Swedish) A small hillock of peat, 3–6 m. high, found in TUNDRA areas. Hence *p. bog.*

palynology The science of pollen analysis. Pollen grains, blown by the wind and deposited in peat-bogs, are partic. resistant to decay. By counting and identifying these grains, a picture of the type of vegetation during past ages may be obtained. If a boring is made in a peat-bog, a core of peat can be brought up, and samples are studied under the microscope. The various pollen grains have characteristic and recognizable shapes, and the plants responsible for them can thus be identified. A 'pollen count' is made, and the contributory proportion of various plants, esp. trees, can be calculated.

pamir Poor grassland on the high plateaus of central Asia; now used as a proper name for a mountain complex in Tadzhikistan, where meet the Hindu Kush, Karakoram and Tien Shan.

pampa (Sp.) Orig. given by Sp. settlers to the extensive, monotonous plain in Argentina and adjacent parts of Uruguay. It comprises a basin of sedimentary deposits, with much superficial loess and alluvium. At the arrival of the Europeans, the p. was covered with tall bunch-grass called *pasto duro*. Gradually the term p. was transferred to this vegetation; p. (or pampas) is now sometimes used as syn. with temperate grassland. Much has been ploughed up and replaced by European meadow-grasses (*pasto tierno*), alfalfa and wheat, and a large part is occupied with the rearing of cattle on large *estancias*.

pampero (Sp.) A dry, bitterly cold wind, bringing esp. low temperatures in winter to parts of the Argentine and Uruguay, the result of a movement N. of a polar air-mass. It is associated with the cold front of an E.-moving low-pressure system, and blows from a S. or S.W. direction.

pan See HARDPAN.

pancake ice Small near-circular pieces or 'cakes' of newly formed ice on the surface of the sea.

pandemic With ref. to an infectious disease of large-scale, world-wide or continent-wide distribution; e.g. the cholera p.'s of 1816–23, 1826–37, 1842–62, 1865–75. Ct. ENDEMIC, EPIDEMIC.

panfan A term invented by A. C. Lawson (1915) to describe the end-stage of desert geomorphological development, when ridges have been worn down and destroyed, PEDIMENTS have become extensive, and basins filled in.

Pangaea A concept put forward by A. Wegener of a single large land-mass of SIAL in Pre-Cambrian times, surrounded by a primeval SIMA-floored ocean (*Panthalassa*). He suggested that the land-mass, P., broke up to form a S. (GONDWANALAND) and a N. (LAURASIA) land-mass, between which lay TETHYS.

panhandle A narrow projection of the territory of a state between that of others, as a strip or along a coast-line (specif. in U.S.A.); e.g. Texas, Alaska. [*f*]

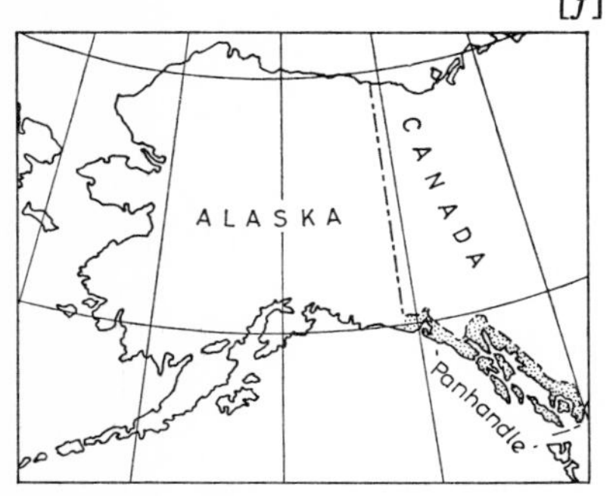

pannage The feeding, partic. of swine, in woodland on acorns and beech-mast. Rights of p. still exist for many areas of commonland; e.g. the New Forest. In some cases payment is made in respect of each animal by its owner.

panorama An outline sketch of a piece of country as viewed from some prominent point, covering a considerable horizon distance, and emphasizing foreground, middle-ground and background detail. It is an essential part of field-sketching. Various geometrical methods can be used. A p. can be drawn in the field (preferably), or from a contour map.

panplain Coined by C. H. Crickmay in 1933 to denote a plain formed by the coalescence of several flood-plains, each created through lateral erosion by its respective river. Hence *panplanation*. It was introduced in ct. to the PENEPLAIN of W. M. Davis, which is formed by the degradation of divides between the rivers.

pantanal A category of SAVANNA along the flood-plain of the Paraguay R. and its tributaries in Brazil, flooded for several months during summer, but with continuous drought for the rest of the year. It comprises a varied vegetation complex, with grasses on lower areas, and clumps of trees on the higher mounds.

pantograph An instrument for enlarging or reducing a map. It consists of four metal arms of equal length, loosely jointed, with one end fixed to a weighted stand. A cross-bar is moved along parallel to two of the sides, in order to fix the scale-factor. Tracing points are attached at the opposite end to the fixed one and in the cross-bar. If the line-work is carefully traced round by the point on the cross-bar, the outer point will draw the same pattern on an enlarged scale, and vice versa.

papagayo (Sp.) A dry, strong, cold N.E. wind, bringing low temperatures and clear weather in winter to the coastlands of Mexico. It is caused by air moving from the high plateau towards low pressure over the Caribbean Sea and the Gulf of Mexico.

parabolic dune A sand-d., curved in pattern, but with the curve in the opposite direction to that of a BARKHAN; i.e. the curve is convex downwind, and the 'horns' trail upwind. These occur esp. on sandy shores, after a 'blow-out' has been formed by deflation caused by eddying; the face of the d. is steep, and it may migrate inland. E.g. in the Landes of France and W. Denmark. P. d.'s are also found under similar 'blow-out' conditions in arid interior plateaus; e.g. Mongolia, Tarim Basin. If the d. moves rapidly forward, its 'horns' will be drawn out parallel to each other, forming a 'hairpin d'.

paradigm A type of large-scale MODEL, pattern or exemplar, stable in character, enabling the effective and selective presentation of theoretical and methodological belief in some aspect of scientific enquiry. This research involves the handling of a vast field of data matrices (storage, retrieval, analysis and display), possible only with an electronic computer.

parallel drainage A pattern of d. in which streams and their tributaries are virtually p. to one another. [*f*]

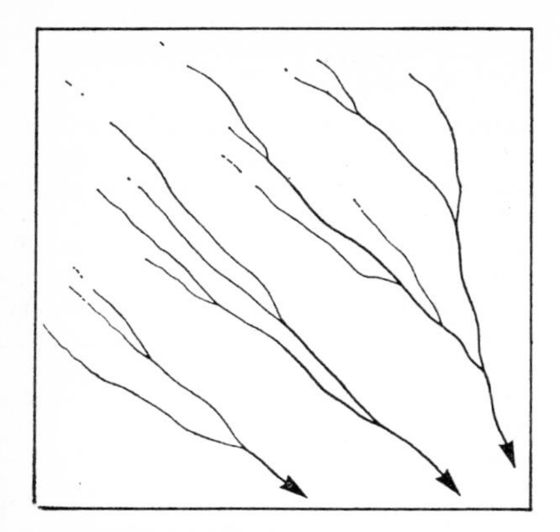

parallel of latitude A line on a map joining all points of the same angular distance N. or S. of the equator, and parallel to it. See LATITUDE. All p.'s are 'small circles', except for the equator, which is a GREAT CIRCLE.

parallel retreat of slope The r. of a s. under the attack of weathering and erosion without any appreciable change of gradient. The concept of p. r. is partic. associated with the name of W. Penck, and the idea that s. forms may remain fairly uniform throughout long periods of erosion has become increasingly accepted. P. r. is a fundamental process in L. C. King's theory of the formation of a PEDIPLAIN and in the development of INSELBERGS. [*f*]

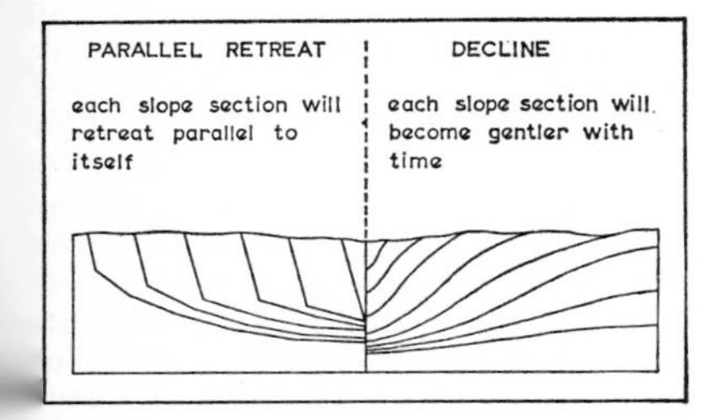

parallel roads' A series of horizon-al terrace-like features occurring p. to one another on each side of a valley. It is thought that these p. r.'s mark former shorelines of lakes, formed by ice-damming at several levels; e.g. Glen Roy, near Fort William, Scotland.

parameter Used in statistics, a value or quantity which characterizes, defines or delimits any complete set of data.

paramo (Sp.) A high, very bleak plateau in the Andes, between the tree-line and the permanent snow-line, covered with pasture and tundra vegetation.

parasitic city A c. that has a detrimental effect upon the economic development of an area. Commonly colonial c.'s have this effect because they act as NODAL POINTS in the break-down of the existing native economy and society. Since subsequent economic growth is focused upon these new nodes, the p. effect is usually only of short duration.

parent material The weathered rock-m. upon which the soil-forming processes operate to create soil. This is the C-horizon of an *in situ* soil.

parish A small administrative unit in Gt. Britain, usually consisting of a village and its outlying hamlets. Initially this was merely an ecclesiastical unit with a church and a clergyman, but it has also become a civil unit, though the boundaries of civil and ecclesiastical p.'s do not always coincide. In Tudor times the p. was made a unit of local government for poor law and highways administration, under the justices of the peace.

park (i) An area of land for ornament, recreation or pleasure, usually with grass and clumps of trees, sometimes (as in a town) with lawns and flower-beds of formal lay-out. (ii) See NATIONAL P. (iii) In the mountains of W. America, a high enclosed grass-floored valley, usually the location of a ranch. Hence *parkland*. See also P. SAVANNA

park savanna(h) S. grassland, with trees (acacia, baobab, palm) scattered sporadically among the tall grasses, and along the lines of water-courses or around water-holes.

paroxysmal eruption A volcanic e. with violent explosive activity, usually after a long period of quiescence of a volcano, during which subterranean pressures accumulated. It is sometimes referred to as a 'Vesuvian' or 'PLINIAN' E. after that of Vesuvius in A.D. 79. A p. e. occurred in 1883, when the island of Krakatoa, in the Sunda Straits between Java and Sumatra, blew up.

partial drought In British climatology, a period of 29 consecutive days, some of which may have slight rain, but during which the daily average does not exceed 0·25 mm. (0·01 in.). Ct. ABSOLUTE D., DRY SPELL.

pass (i) A col, notch or gap in a mountain range, affording a route-way across; e.g. Great St. Bernard P., Khyber P. (ii) The channel of a Mississippi distributary. (iii) A narrow channel through a coral BARRIER REEF; e.g. on the Great Barrier Reef of Australia.

pastoral farming A type of f. concerned mainly with the rearing of animals, whether for meat, milk, wool or hides. It may be large-scale (e.g. the sheep-farms of Australia, the ranches of U.S.A., the *estancias* of Argentina), or small-scale (a small Danish dairy-herd of a few animals). It may be primitive (carried on by NOMADS), or highly scientific, as in Denmark and the Netherlands (with careful breeding, milking techniques, butter-fat records). See also TRANSHUMANCE.

pasture An area of land covered with grass used for the grazing of domesticated animals, as distinct from that which is mown for hay (MEADOW), although the same field may be used for both purposes at different times of year. Some p. may be 'natural'; e.g. an ALP or mountain p., but most areas of grazing are usually improved in some way: by liming, fertilizing, draining, and periodic reseeding.

patana (Sinhalese) A coarse grassland on the uplands of Ceylon above about 1800 m. (6000 ft.); e.g. the Horton Plains.

paternoster lake One of a string or a series of l.'s in a glaciated valley, separated by morainic dams or rock bars, partic. where the valley is stepped, with the steps, the result of glacial plucking, facing down-valley. The name is derived from their resemblance to a string of beads. E.g. the series of l.'s in many of the parallel valleys of Sweden, draining E. to the Baltic Sea. [*f*]

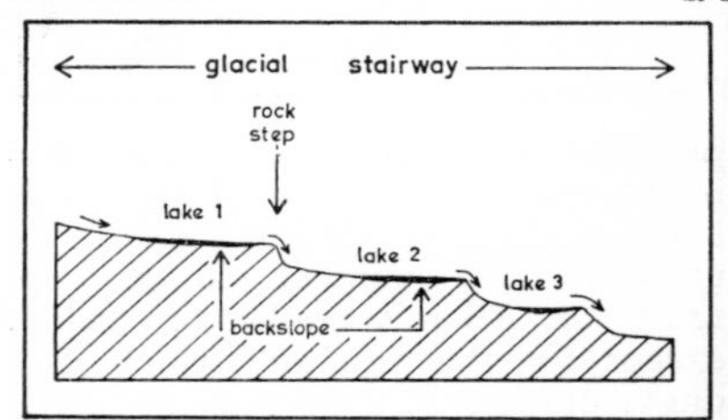

patina A coloured film on the surface of a rock or a pebble, which results from exposure to weathering.

patterned ground Certain well defined features, such as circles, polygons, nets, steps and stripes, characteristic of areas now or at some time subject to intensive frost action. This may produce features in which material is sorted either into polygonal forms of varying dimensions, with stones round the perimeter and finer material in the centre, or into 'stone stripes'. However, p. g. also includes non-sorted forms, such as polygons outlined by vegetation zones, and step-like forms on slopes. There is no certainty as to the mode of origin of these forms; A. L. Washburn has reviewed 19 possible hypotheses relying on different freeze-thaw processes, contraction following temperature change, moisture controlled movements and solifluction. In fact p. g. features probably have several origins, and exactly what these are is the object of much present research both in the field and in the laboratory. [*f*]

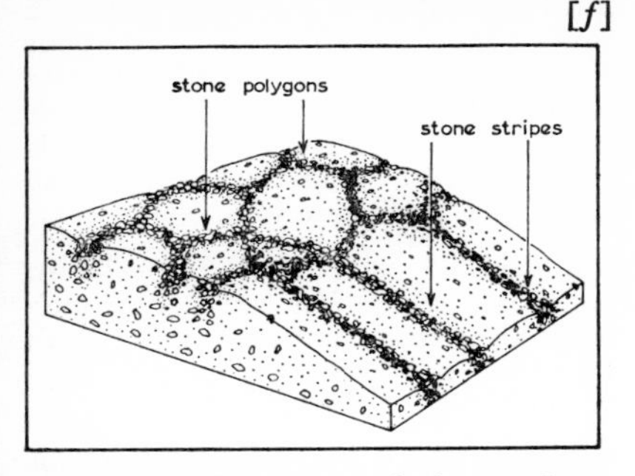

pavement In geomorphology, a bare rock surface, produced by weathering, wind erosion, or glacial scouring. See DESERT P., GRIKE. [*f* GRIKE]

pays (Fr.) A small region in France with a distinctive unity based on features of geology, relief and land-use which distinguish it from its neighbours, though gen. it has no administrative significance. Some p.-names are derived from feudal administrative units or families (e.g. *Valois*); some from striking physical characteristics (e.g. *Champagne humide*); some from types of land-use (e.g. *Ségalas, P. noir*); some from a near-by urban centre (e.g. *Bordelais, Lyonnais*). The origin of many is lost in antiquity, while others have been coined recently, sometimes by geographers. Hence *paysage*, the whole landscape.

peak A prominent pointed mountain summit, often incorporated in the proper name; e.g. Pike's P., Long's P. in the American Rockies. But the P. in S. Derbyshire, hence the P. District, is a flat-topped plateau.

peasant A word of very wide and generalized usage, but indicating an agricultural worker, usually (though not always) living at little more than subsistence level. The term may indicate the rice-growers of the Monsoon Lands, and the workers on collective and state farms in E. Europe. It is frequently, though quite wrongly, used in a disparaging sense.

peat Partially decomposed vegetable matter, dark brown or black, accumulated under waterlogged conditions. It may be either acid in reaction (*bog p., moss p.*), or neutral or alkaline (*fen p.*); see also BOG, FEN. It is formed chiefly under temperate or cold conditions, and obviously has accumulated more rapidly in the past (e.g. in the ATLANTIC STAGE) than at present. Acid p. is mainly used for fuel (e.g. in Ireland), both in domestic hearths and in thermal generators, or for animal litter; fen p. is used for horticultural purposes, and for dressing lawns and race-courses.

pebble A small water- or wind-worn stone, larger than GRAVEL, smaller than a COBBLE; it is sometimes defined as being 10–50 mm. (0·4 to 2 ins.) in diameter; another definition is 4–64 mm.

pedalfer A leached soil from which base compounds (esp. calcium) have been removed, leaving compounds of aluminium and iron, found in areas of humid climate (with over 24 ins. of precipitation), esp. when associated with high temperatures. The term is used broadly of one of the two major divisions of ZONAL SOILS, esp. in U.S.A., the other being PEDOCAL. A line drawn N. to S. through U.S.A. along the Mississippi broadly separates p.'s to the E. from pedocals to the W.

pediment A gently sloping rock-surface, bare or with but a thin veneer of debris, stretching away from the foot of a ridge or mountain in a semi-desert or desert region. There is much controversy about its origin, but it is usually regarded as the product of denudation under arid, sub-arid or savanna conditions. Its upper edge runs abruptly, with a marked change of slope (though sometimes masked by a fan), into the mountain-front; its lower edge slopes very gently away under the accumulation of sands and gravels found in a desert basin. One school of thought regards it as the product of lateral planation by streams, SHEETFLOOD, rills and downwash, resulting from episodic rainstorms. A second school regards a p. as resulting from SLOPE RETREAT; a steep mountain-front, with an angle of slope of 30° or more, retreats under the attack of weathering and erosion without having its gradient altered to any marked extent. The p. therefore is a low-angle (6°–7°) slope developing independently at the foot of a steep, parallel-retreating slope, commonly of granite; it is a basal slope of transport over which weathered material derived from the steep slope above is carried by occasional rainstorms. Some refashioning of the p. by lateral planation may occur, but only as a subsidiary process moulding a feature whose origin is the result of other factors. Some authorities claim that the production of features which are genetically p.'s is a much more widespread process. P.'s are widely found in S.W. U.S.A. (esp. Arizona), in the Kalahari Desert in S.W. Africa, and in the savanna lands gen. of Africa. The term *pedimentation* is used for the collective processes which produce a p. See INSELBERG, PEDIPLAIN. [*f*]

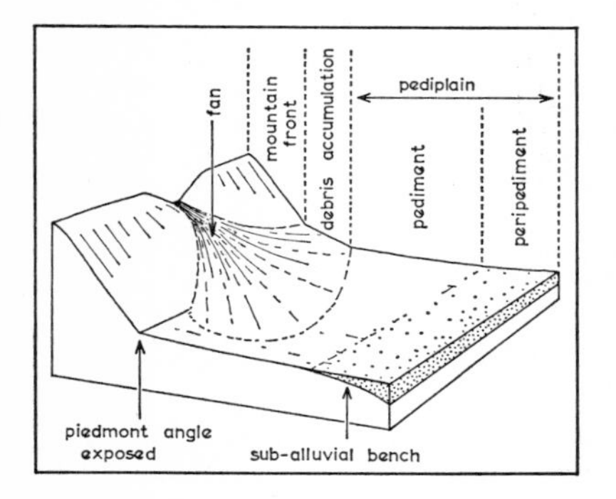

pediocratic period A p. of crustal calm during geological time between 2 OROGENIC p.'s; e.g. most of the MESOZOIC.

pediplain A multi-concave surface resulting from the coalescence of several large-scale adjacent PEDIMENTS [*f*]. Widespread erosion surfaces are thought to have been formed in this way; e.g. in Africa, the 'Gondwana p.' at over 1220 m. (4000 ft.) above sea-level probably dates from Jurassic times, and the 'Africa p.', at about 610 m. (2000 ft.,) is possibly of early Tertiary age. The word is alt. spelt 'pediplane', but it is best to reserve this for the vb. 'to pediplane' and for the process of 'pediplanation'.

pedocal A soil rich in calcium carbonate which has not been much leached, characteristic of areas with less than 610 mm. (24 ins.) of precipitation. Ct. PEDALFER.

pedogenesis The development of soil.

pedology The scientific study of soils, their origin and characteristics, and their utilization.

pegmatite An igneous rock characterized by large coarse crystals, tightly interlocking, commonly found as dykes or veins, esp. on the margins of BATHOLITHS. The individual crystals may be several inches or even feet in size; a crystal of a lithium mineral 12 m. in length has been found in the Black Hills of S. Dakota (according to L. D. Leet and S. Judson). The commonest type is a granite-p., chiefly of quartz and felspar. Some p.'s contain various rare minerals, concentrated during the final stages of magmatic cooling; e.g. of such elements as boron, fluorine, uranium, niobium, tantalium, lithium, thorium. P.'s may be simple or very complex in composition.

pelagic App. to the water of the sea, and to organisms living there, independent of either the shore or the sea-bottom.

pelagic deposit Material deposited as an ooze on the abyssal floors of the oceans, derived from floating organisms which die and sink. See individual oozes.

Peléan eruption A volcanic e. accompanied by clouds of incandescent ash (see NUÉE ARDENTE), named from the type-example of Mont Pelée which erupted in 1902, wiping out the town of St. Pierre, Martinique

'Pelé's hair' Glass-like 'hairs' or 'threads', extrusions from basaltic lava, produced by explosions or by the bursting of bubbles of contained gas in a pool of liquid lava; named after Pelé, the Hawaiian goddess of fire.

'Pelé's tears' Small drops of basaltic volcanic glass, thrown out and solidifying during eruptions of lava.

pelite A rock made of fine particles of clay; e.g. mudstone; i.e. an ARGILLACEOUS rock. Cf. PSAMMITE, PSEPHITE.

peneplain An almost level plain, the product of long-sustained denudation in the 'old age' stage of the cycle of landform development. The vb. is 'to peneplane', and the process is 'peneplanation', never 'peneplaination'. The term was introduced by W. M. Davis in 1889, and seems quite adequate for planation surfaces of whatever origin. However, because Davis associated particular processes with the production of such features, other authorities with different processes in mind have introduced other terms, notably PEDIPLAIN and PANPLAIN. Some authorities would exclude a surface of marine, as distinct from sub-aerial, planation.

peninsula An elongated projection of land into the sea or a lake; e.g. the S.W. P. of England, the Dingle P. of S.W. Ireland, the Iberian P. The term may be used adjectivally, as Peninsular Italy, P. Europe.

Pennsylvanian Formerly in U.S.A. it was the upper of the two epochs of the Carboniferous, equivalent to the British Upper Carboniferous (Coal Measures). It now has the status of a period in U.S.A., and is sixth in order in the Palaeozoic.

perambulation Lit. walking around a specif. piece of territory, usually by an official party to assert, preserve and record its defined boundaries. E.g. an annual ceremony around the boundaries of the Borough of Poole, Dorset, 'beating the bounds'. The word p. is used for the actual boundary of the New Forest, Hants.

percentage dot map A type of D.M. in which each located d. represents 1% (1/100th) of the total value involved in the particular distribution. Cf. MILLE M.

perched block A boulder perched in a delicate state of balance, usually left by an ice-sheet or glacier, or having fallen by gravity (e.g. the Bowder Stone, Borrowdale, Cumb.), or having been thus left by weathering *in situ*. Sometimes called a *pedestal rock*.

perched water-table An independent and isolated area of ground-water, above the w.-t. proper and separated from it by unsaturated rocks; i.e. it occurs in the VADOSE zone. [*f*]

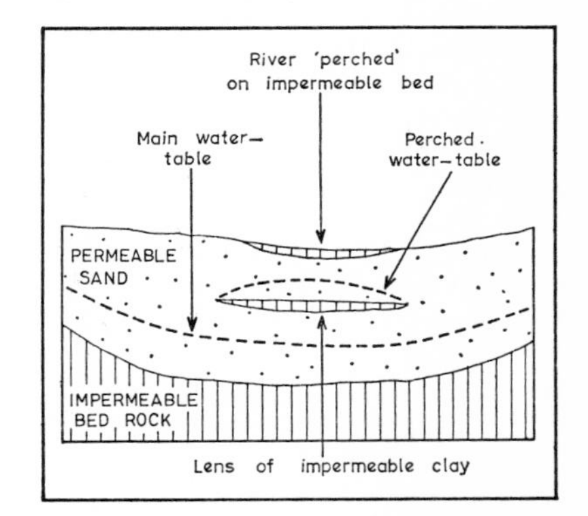

percolation The downward movement of water through the pores, joints and crevices within the mass of soil and rock. It can carry down soluble minerals, hence LEACHING.

perennial irrigation I. schemes whereby the land can be watered throughout the year. This may involve: (i) Lifting water on to the land from the low-level stage of a river, by means of primitive devices (SHADUF, SAKIYEH) or modern pumps; (ii) Lifting water from wells, similarly; (iii) The creation of dams or barrages to pond up sufficient water at the flood or high-water level on a river to last the year; e.g. on the Indus, Ganges, Tigris-Euphrates, Nile, Colorado, Columbia. The ct. is with 'basin' or 'flood' i., when a river overflows its banks and irrigates naturally the flood-plain for a short period each year.

perennial stream A s. that maintains its flow throughout the year, in ct. to an intermittent one.

pergelation The formation of PERMAFROST, or, in a more limited sense, the local freezing of the soil.

pergelisol See PERMAFROST.

pericline A small ANTICLINE which pitches along its axis in each direction from a central point, forming a type of elongated dome; e.g. the Kingsclere and Shalbourne p.'s to the W. of Basingstoke; the Pays de Bray in N. France. From their nature, p.'s are partic. vulnerable to breaching along their axes, as has happened in these three cases. Used commonly in Gt. Britain, rarely in U.S.A. (where it is sometimes called a *centrocline*).

peridotite A coarse-grained, plutonic igneous rock, consisting almost entirely of the mineral olivine, which is composed of ferromagnesian silicate. It is commonly dark green in colour.

perigean tide When the moon is at its nearest position to the earth (PERIGEE), high t.'s are higher and low t.'s are lower than usual.

perigee The point in the orbit of a planet when it is at its minimum distance from the earth. The term is now used strictly with ref. to the moon; when in p. it is about 354,000 km. (220,000 mi.) from the earth. Ct. APOGEE.

periglacial App. to an area bordering the edge of an ice-sheet, to the climate of that area, to the physical processes involving FREEZE-THAW activity dominant there, and to their results. Some confusion may be caused by the fact that climates along the edge of ice-sheets vary considerably, nor are all present or past ice-fronts marked by what are now gen. understood to be p. conditions or processes.

perihelion The nearest point of a heavenly body in its orbit around the sun; the earth at p. (3 Jan.) is at a distance of 147·3 million km. (91·5 million mi.) Ct. APHELION.

period A time-interval in the geological record, the corresponding division of the rock strata being a SYSTEM. A p. forms one division of an ERA, and itself is divided into EPOCHS. There are two usages of the term: (i) The general British practice is to recognize 15 periods of time since the beginning of the Palaeozoic era: Cambrian, Ordovician, Silurian, Devonian, Carboniferous, Permian, Triassic, Jurassic, Cretaceous, Eocene, Oligocene, Miocene, Pliocene, Pleistocene, Holocene (or Recent). (ii) The usual American (and sometimes British practice) is to demote the last 7 of these to epochs; thus the U.S. order is Cambrian to Devonian (as above), Mississippian, Pennsylvanian, Permian, Triassic, Jurassic, Cretaceous, Tertiary, Quaternary (the last 2 being within the Cainozoic era), hence 12 periods in all.

periodogram A form of deviational graph invented by A. Schuster in connection with the determination of climatic cycles; amplitudes of temperature in the form of deviations are plotted as ordinates and time-units as abscissae.

peripediment An area of alluviation in a desert basin, adjacent to a PEDIMENT [*f*].

permafrost The permanent freezing of the soil, sub-soil and bedrock, sometimes to very great depths; borings indicate that p. may exist in parts of the Siberian TUNDRA to 600 m. from the surface. Thawing may occur at and near the surface during summer, while the frozen ground below still forms an impermeable layer.

permanent hardness, of water The result of the presence of dissolved magnesium and calcium sulphate, which cannot be removed by boiling.

permeable rock A r. which will allow the passage of water, either because of its POROSITY or of its being

PERVIOUS (chalk, Carboniferous Limestone, quartzite). Hence *permeability*[1] (e.g. sand, sandstone, gravel, oolitic limestone, but not clay).

Permian The 6th of the geologica periods of the Palaeozoic era (the 7th in U.S.A.) and the system of rocks laid down during that time, lasting from about 270 to 225 million years ago. Until recently it was dated much later (235–200 million years ago). Sandstone strata were laid unconformably upon the worn-down Carboniferous rocks, except possibly in the coalfields of the English Midlands. The lower part of the Permian contains fossils which are related to Palaeozoic fauna, while the upper part in places passes indistinguishably into TRIASSIC rocks. Hence the term Permo-Trias is often used of the transition system of rocks (the New Red Sandstone, of desert origin). Other rocks of P. age include breccias, Magnesian Limestone in N.E. England, and beds of marls, sometimes containing salts. The P. period was probably a time of widespread glaciation in the S. continents. The name is derived from the province of Perm in the U.S.S.R., where the marine rocks of this age are well developed.

persistence When meteorological conditions exist for more than the normal length of time, they are said to be persistent; e.g. a persistent anticyclone. Some conditions are notably persistent; once the conditions exist, they are likely to remain for some time.

perspective projection A category of MAP P. based on geometrical p. from a given point, the point of p., through the surface of a globe. The category includes ZENITHAL (azimuthal), CONICAL and CYCLINDRICAL P.'s.

perturbation In meteorology, any disturbance in the steady state of a system. In partic., this is applied to departures from ZONAL FLOW and to similar wave disturbances in the atmospheric circulation.

pervious rocks One of the 2 groups of PERMEABLE R.'s. They are traversed by joints, cracks and fissures through which water can flow; e.g. chalk, quartzite, Carboniferous Limestone. *Note:* Some writers regard permeable and p. as syn., others refer to primary and secondary degrees of permeability respectively.

petrifaction The conversion of matter of vegetable or animal origin into stone by percolating solutions of silica or calcium carbonate, which slowly replace the original tissue and structure; e.g. the Petrified Forest, Arizona, where tree-trunks were buried under sediments including a layer of siliceous bentonite. The silica was carried down in solution and the trunks were converted into crystalline silica in the form of agate and other minerals.

petrogenesis A branch of PETROLOGY which deals with the origin of rocks, specif. of igneous rocks.

petroglyph A drawing made on a rock or cave-wall by prehistoric man.

petroleum In its broad sense, a mixture of hydro-carbons in the solid, liquid and gaseous states, though strictly it denotes the liquid form ('rock oil'). It is obtained by drilling, usually down into an anticline or 'oil-dome' below the surface found by a seismic survey. The crude oil is split into its 'fractions' (hydrocarbon groups) by distillation, using catalyst cracking plant at a refinery to produce the desired grades, which range from heavy fuel-oils to high octane aviation spirit. Output is measured in tons or in barrels: 1 ton = 7–8 barrels, according to its specific gravity. C.p. = U.S.A., U.S.S.R., Venezuela, Saudi Arabia, Kuwait, Iran, Libya, Iraq, Canada, Algeria, Indonesia.

petrology The scientific study of the chemical and mineral structure and composition of rocks.

phacolith, phacolite A concordant intrusion of igneous rock, lying near the crest of an anticline or the base of a syncline in folded strata, in the shape of a lens; e.g. Corndon Hill, Shropshire, where the lens of igneous rock, forced into an upfold of Ordovician rocks, has been exposed by denudation. [*f*]

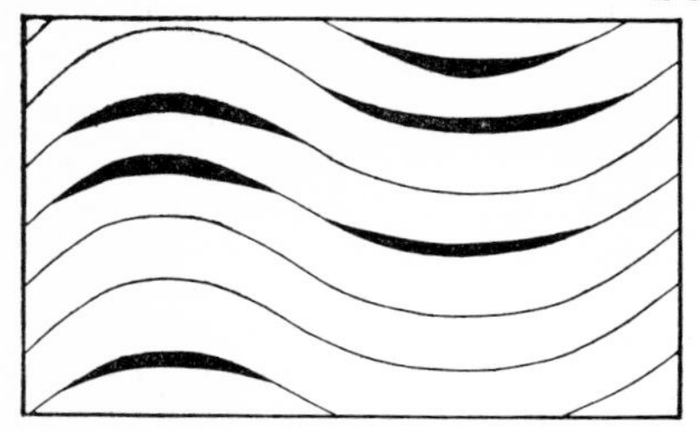

phase An aspect, appearance or stage of development, esp. used in the social sciences, notably in planning. In the case of the latter, there may be 3 p.'s: the *macrophase* (e.g. a NATIONAL PLAN), the middle phase, and the *microphase* (e.g. a town planning map, a planning permission, an investment appraisal).

phenocryst A large, conspicuous crystal in a porphyritic igneous rock, sometimes in a ground-mass of glassy or fine-crystalled material; e.g. felspar in the Shap Granite, and in the granite of Haytor, Dartmoor.

phenology The scientific study of the effects of seasonal climatic change upon recurring natural phenomena, such as bird migration and the flowering of vegetation.

phonolite A fine-grained extrusive igneous rock, consisting essentially of alkaline felspar, with the mineral nepheline. It was so named because of its ringing note when struck by a hammer. The Devil's Tower, Wyoming, is a great mass of p., 213 m. (700 ft.) high, with the rock in large pentagonal columns.

phosphate rock A variety of p., including phosphorite, phosphatic limestones and leached guano which has accumulated from sea-bird droppings (e.g. in Nauru). Phosphorus is also obtained from the mineral apatite and from basic-slag out of blast-furnaces. It is one of the 3 main elements needed by plants, and is used as a fertilizer in the form of superphosphates; i.e. p.'s treated with sulphuric acid. C.p. = U.S.A., U.S.S.R., Morocco, Tunisia, Nauru.

photoengraving A method of reproducing maps (and other material). The map, drawn on paper, card or plastic, or scribed on glass, is photographed to the desired size, and the negative is exposed over a sensitized zinc plate. The lines on the plate are strengthened with an acid-resistant resin, and the intervening areas are etched out with acid. Ct. PHOTOGRAVURE. The plate is mounted on a block of wood as a *line-block* for delivery to the printer. *Half-tones* (using a screen of fine dots) can be similarly produced on a copper plate for reproducing photographs.

photogrammetry The science of transforming an aerial photograph into a topographical map, using various plotting instruments. During the last few decades, the theory and practice of p. has developed so rapidly that it has revolutionized surveying and map-production.

photogravure A method of reproducing maps (and other material), whereby the image is incised below the surface of the printing plate, which is coated with ink; this is removed from the higher portions of the plate, and when it is pressed against the paper an image is transferred to it. Syn. with *intaglio*.

photolithography A method of reproducing maps (and other material). The map, drawn on paper, card or plastic, or scribed on glass, is photographed to the required size, and the negative is exposed over a sensitized zinc plate. After washing, the image remains on the plate, though it is usually chemically strengthened in various ways. The plate is then curved around the cylinder of a rotary offset printing machine.

photometeor A luminous phenomenon in the atmosphere produced by the reflection, refraction, diffraction and interference of light; e.g. BROCKENSPECTRE, HALO, IRISATION, RAINBOW, FOG-BOW.

photo-relief Shading on a map which gives the impression of a photograph of a relief model, usually with the appearance of the source of light from the N.W. (in U.S.A., *plastic relief*).

photostat A copy of a map or diagram, made from a paper negative on to paper.

phototypography The lettering of a map using a photo-typesetting machine; several different kinds have been developed. Each letter in turn (from a master matrix or disc) is exposed on a film, to the correct size, which is automatically developed and fixed, thus producing a name, in either positive or negative form, which can be stuck on the drawn map or plate. Some machines can also directly position the names on the map or plate.

phreatic water See GROUND W.

pH value The quantitative degree or scale of soil acidity and alkalinity, measured in terms of the negative index of the logarithm of the hydrogen-ion activity in soil-colloids. In pure water, one part in 10 millions (10^{-7}) is dissociated into hydrogen ions, and the pH is 7; this is a neutral state in the scale of acidity. If a strong alkali, such as caustic soda, is dissolved in water, the solution is markedly alkaline, and an infinitesimal part (10^{-14}) is dissociated into the hydrogen ions, hence the pH is 14. With hydrochloric acid, 1 part in 1000 is dissociated, i.e. 10^{-3}, a pH of 3. A neutral soil has a pH value of about 7·2, an acid soil of less than 7·2, and an alkaline soil of more than 7·2. Some very acid soils may have a pH of 4·5 or less.

phylloxera An aphid, of American origin, which first appeared in Languedoc in 1863, and had spread to the Bordeaux district by 1868. It multiplies rapidly, living in gails on the leaves and roots of the vines, where it cannot be reached by spraying; the plants become stunted and die. By 1884 every vine-growing area in France was affected. In 1891 it was discovered that vine-stocks from the E. U.S.A. were almost immune, and stocks from there were used in European vineyards; on to these, scions of the European varieties were grafted. While not wholly resistant, these vines are affected much less seriously.

physical geography Those aspects of g. which are concerned with the shape and form of the land-surface, the configuration, extent and nature of the seas and oceans, the enveloping atmosphere and the processes therein, the thin layer of the soil, and the 'natural' vegetation cover; i.e. Man's physical environment, though normally independent of the effects of Man. The term is gen. used in ct. to HUMAN GEOGRAPHY.

physical geology 'The nature and properties of the materials composing the earth, the distribution of materials throughout the globe, the processes by which they are formed, altered, transported and distorted, and the nature and development of the landscape' (L. D. Leet and S. Judson). This definition covers one major division of g., the other being HISTORICAL G. P. g. is sometimes divided into *structural g.* and *dynamic g.* It obviously includes much of the content of GEOMORPHOLOGY, but is wider in scope.

physical landscape See NATURAL L.

physical planning The allocation of land to specif. uses on a comprehensive territorial basis related to geographical criteria, with consideration given to social criteria. Ct. ECONOMIC PLANNING.

physical weathering Disintegration of rock by agents of the weather (frost, temperature change), but without involving chemical change.

physiographic pictorial map The depiction of relief on a m. by the systematic application of a standardized set of conventional pictorial symbols, based on the simplified appearance of the physical features they represent, as viewed obliquely from the air at an angle of about 45°. The American cartographer, Erwin Raisz, is a supreme exponent of this art.

physiography Orig. 'a description of nature', or of natural features in their causal relationships; then the term became almost syn. with physical geography; gradually, esp. in U.S.A., it became limited to the study of landforms. By some authorities the term has now been superseded by the more scientific GEOMORPHOLOGY; by others, it is still regarded in wider terms as an integration of geomorphology, plant geography and PEDOLOGY.

phytogeography Plant geography.

phytoplankton Microscopic floating plant-life in the sea.

pictogram A map of distributions (esp. commodities), in which small pictorial representative symbols (e.g. sacks, bricks, barrels) are located over the area of production. They may be drawn to scale to give some impression of the quantities involved, or as groups of repeating units.

piedmont An area near the foot of a mountain range; it is used widely as an adjective, as with p.-*plain*, -*fringe*, GLACIER. As a proper name, it denotes: (i) the plateau lying E. of the Appalachians; (ii) a province in N. Italy at the foot of the Alps.

piedmont angle The sharp angular junction between the mountain front and the PEDIMENT [*f*] at its foot.

piedmont glacier A mass of ice on the flanks of a mountain range, formed by the coalescence of a number of parallel valley-g.'s. P.g.'s were common in the Quaternary Ice-Age, when they spread down on to the Bavarian and Swiss Forelands, the N. Italian Plain, and the Lannemezan in the N. Pyrenees. Present examples are the Malaspina G. in S. Alaska, covering 1500 sq. mi.; the Frederikshaab G. in W. Greenland; and the Butterpoint G. in S. Victoria Land (Antarctica). [*f*]

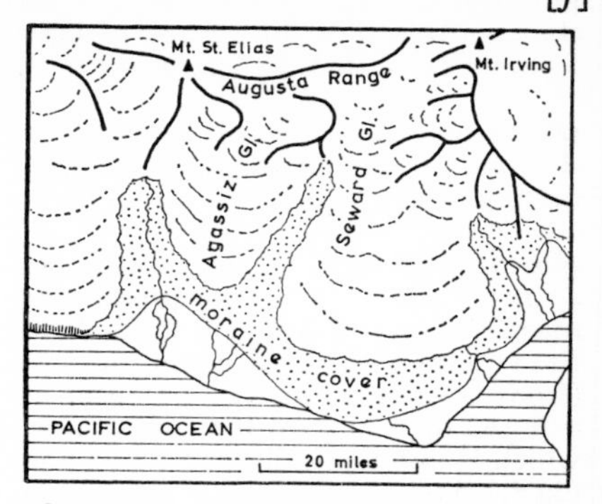

pie-graph A popular descriptive name for a DIVIDED CIRCLE g.

pig-iron The impure iron resulting from the smelting of iron ore, using coke and limestone as a flux, in a blast-furnace. In this process it absorbs 3–5% of carbon, giving it a brittle quality; this is removed when converting it to steel. As the molten metal flows from the blast-furnace, it is run into parallel channels, known as 'pigs', hence the name. In most modern integrated steel-works, the molten i. is passed directly to the steel-convertor, thereby avoiding loss of heat.

pike A mountain peak, esp. in the English Lake District; e.g. Scafell P., the Langdale P.'s; and in the Pennines; e.g. Buckden P.

pilang An area of close acacia-forest in Indonesia.

pillow lava L. which has solidified in the appearance of a pile of pillows, probably under water, partic. in the case of basic l. as basalt.

pinch-out Where a rock-stratum thins out in a horizontal plane and disappears.

pingo (Eskimo) An isolated dome-shaped or conical mound of earth or gravel found in Alaska, Arctic Canada, Greenland and Siberia. Autumn freezing traps a layer of water between the newly frozen surface and the underlying PERMAFROST, and hydrostatic pressure resulting from the expansion of this water on freezing may raise a 'blister' on the surface. Many have been recently formed, though some large ones, from the evidence of RADIOCARBON DATING, have been formed some 5–6000 years ago. American workers distinguish between *'closed system p.'s.'*, where the layer of water thus trapped is an isolated body [*f*], and *'open system p.'s'*, where the hydraulic head is related to an adjacent slope. The latter are esp. common in eastern Greenland, where plentiful water from summer snowmelt on the mountains moves downwards towards the coast over the zone of permafrost. P.'s range in size from small mounds of 6 m. (20 ft.) to hills of over 90 m. (300 ft.) in height with a basal diameter of 0·8 km. (0·5 mile). The top of the p. commonly collapses, leaving a 'crater' partly filled with a shallow pond. Other workers have sought to call these features *hydrolaccoliths* or *cryolaccoliths*, restricting the term p. to a small isolated mound with an existing ice-core, but there appears to be no unanimity about this usage.

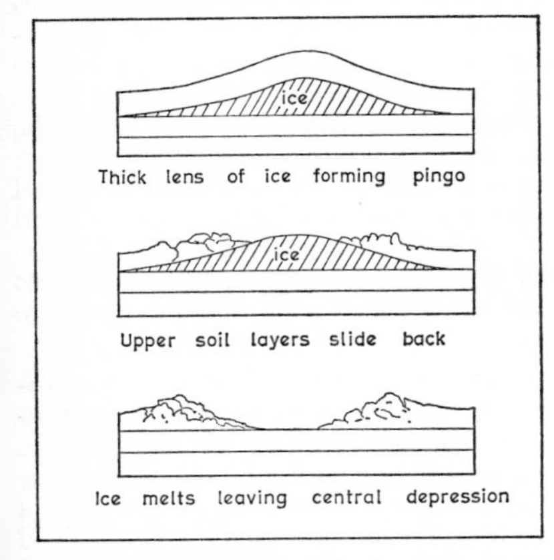

pinnate drainage A d. pattern in which the main tributaries receive a large number of closely spaced affluents at acute angles, like a feather. [*f opposite*]

pioneer An early leader, explorer, hunter or settler, esp. in a 'new' country, and taking a major part in its opening-up; hence the term *p.-fringe*, the zone beyond the present settled area.

pipe (i) The vent of a volcano which leads to its crater. (ii) In chalk, vertical joints enlarged by carbonation-solution, and filled with sand and gravel. (iii) A cylindrical mass of mineral ore or diamantiferous rock; e.g. the diamond-bearing p.'s of the Kimberley (S. Africa) district, containing an ultrabasic rock (KIMBERLITE) in which the diamonds occur. These p.'s. were probably formed by FLUIDIZATION.

pipeline A tube, usually of steel, sometimes of plastic, which can be used to transport liquids (water, oil, chemicals, even milk), gas, and solids (in the form of 'slurries'). P.'s are esp. used for petroleum and natural gas. The U.S.A. was a pioneer, and that country is covered with a network of crude and refined oil p.'s and gas p.'s. Canada has p.'s from the Alberta fields running W. over the Rockies to Vancouver, E. to the Lake Peninsula and St. Lawrence valley. In the Middle East, p.'s cross the desert from Iraq to Tripoli and Baniyas, and from Ras Tanura (Arabia) to Sidon (*Trans-Arabian P.* or '*Tapline*'). More recently, many European p.'s have been opened to supply inland refineries; e.g. from Wilhelmshaven and Rotterdam to the

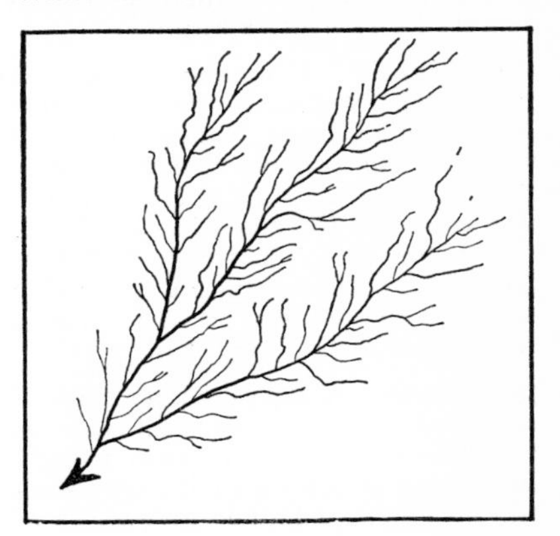

Ruhr and Cologne; the S. European P. from Lavéra (near Marseilles) to Strasbourg and Karlsruhe; E.N.I.P. from Genoa to Aigle (Switzerland); an E. European p. from the Volga to Schwedt-an-der-Oder (E. Germany); a short p. from Fawley (Southampton) to near London Airport. Gas-p.'s in Europe include one from Lacq in S. France to Paris, Lyons and Nantes; the 'Methane Line' from Canvey Island through the Midlands to Leeds. Conveyance through p.'s of solids in 'slurry' (i.e. suspended in water) is used for coal, china-clay (Cornwall), chalk, waste ash from power-station boilers, wood pulp, wood chips. The economics of 'solid p.'s' depends mainly on alternative means of transport, the cost of reducing the medium to suspendable form, and the cost of drying at its destination.

pipkrake (Swedish) A needle, wedge or lens of ground-ice in the soil, formed under PERIGLACIAL conditions.

piracy, of streams See CAPTURE, RIVER.

pisolith, pisolite A coarse-grained oolite, consisting of rounded grains of about the size of a pea (1–10 mm.), hence the name; e.g. the Jurassic Pea Grit.

pitch (i) The direction in which the AXIS of a fold dips. (ii) A solid hydrocarbon, formed in the distillation of coal-tar, though sometimes used loosely for asphalt or bitumen.

pitchblende A complex oxide of the metal uranium (*uraninite*), black, with a pitchy lustre, containing small amounts of other elements—lead, thorium, radium, helium. It is found in the Congo, Canada (Gt. Bear Lake, Gt. Slave Lake districts), U.S.S.R. The other main uranium mineral is *carnotite*, found esp. in the Jurassic sandstones of Utah and Colorado.

pitchstone A waxy-looking mass of solidified lava. When acid lava is extruded on to the surface, usually forming a glassy mass, it may sometimes contain a higher proportion of water than usual, so losing its glassy lustre. E.g. in the islands of W. Scotland, esp. Arran, Mull, Eigg. The highest point on the last island, the Sgurr of Eigg, is made of p.

piton (Fr.) (From a Fr. word for a metal spike or bolt with a ring-head, used in mountaineering) (i) Applied to a sharp-pointed peak. (ii) Used in KARST terminology (*Karst à pitons*) for sharp projections, the result of the acute dissection of limestone under sub-tropical or tropical conditions.

pitted outwash plain A p. of fluvioglacial deposition, dotted around with KETTLES; found extensively in Wisconsin, Minnesota and adjacent parts of Canada.

placer A mass of sand or gravel containing particles of gold, tin or platinum, which have been eroded from exposed veins, washed down by a river, and laid down as an alluvial deposit; e.g. tin in Malaya, gold in S.W. U.S.A.

plage (Fr.) A sandy beach, esp. at a seaside resort; syn. with LIDO.

plagiosere A series of communities which forms through the disturbance of natural CLIMAX VEGETATION by human interference leading to a *plagioclimax*. Many English uplands, formerly wooded, have been turned into grassland as a result of initial clearing, and have been so maintained by subsequent grazing. The decline of grazing intensity on hill-slopes has led to bracken colonization and the re-establishment of SUBSERES.

plain A continuous tract of comparatively flat country, though sometimes gently rolling or undulating, with no prominent elevations or depressions. There is, however, a wide variation in usage; some would restrict it to an area of horizontal structure, others to any flat area at a low elevation. The term is used in various contexts; see COASTAL P., UPLAND P., FLOOD-P. It is a common place-name element; e.g. Salisbury P., the Great P.'s of N. America, the N. European P.

plan (i) A large-scale map on which everything, even widths of roads, is drawn to scale. In Gt. Britain the standard series of p.'s are the 25-inch series (actually 1/2500, or 25·344 ins. to 1 mi.) and the 50-inch series (actually 1/1250, or 50·688 ins. to 1 mi.). (ii) A large-scale survey or basis for economic and social development; see NATIONAL P., REGIONAL P., DEVELOPMENT AREA, PLANNING REGION, PHASE.

planation The denudation of rocks to produce a fundamentally flat surface. Various types of p. have been distinguished, such as peneplanation (see PENEPLAIN), pediplanation (see PEDIPLAIN) and panplanation (see PANPLAIN). The term *p. surface* is to be preferred to erosion surface when used to describe this type of feature, since an erosion surface can be of any form, flat or otherwise.

plane of the ecliptic The p. of the apparent path of the sun. See ECLIPTIC. [*f* AXIS, OF EARTH]

plane-table A small drawing-board fixed on a tripod, used for topographical surveying in conjunction with an ALIDADE. See RESECTION.

planetary winds A popular name for the general ATMOSPHERIC CIRCULATION [*f*], comprising a series of latitudinal wind-belts (equatorial easterlies or Trades, mid-latitude westerlies, and polar easterlies). The zonal longitudinal component of the circulation, and the vertical movement, are brought about mainly by a thermally controlled pressure distribution (see HADLEY CELL); the latitudinal meridional component is the result of deflection due to the rotation of the earth (see CORIOLIS FORCE), producing WESTERLIES, which intensify aloft into JET STREAMS. The idealized pattern is further complicated by the imposition of perturbations on all scales, from major 'waves' in the Westerlies (with wavelengths measured in thousands of miles), to smaller travelling depressions and local convection 'cells', valley w.'s., etc. These disturbances of the broad pattern result from: (i) *thermal effects:* differential heating and cooling of land and sea, and seasonal variations in heating, responsible for some MONSOON w.'s and local w.'s such as sea breezes, and also often considered as the main cause of major waves in the broad pattern of upper w.'s; (ii) *orographic effects:* the influence of relief on local w.'s, and the influences of major mountain belts on the general pattern of latitudinal w.'s, which is held by many to be of greater importance than the thermal effects.

Planetesimal Hypothesis A theory put forward in 1904 by F. R. Moulton and T. C. Chamberlin to account for the origin of the earth and other planets. The main principle is that they were formed from the coalescence of small particles (planetesimals), rather than from a gaseous, then molten, mass, which slowly cooled. This theory has been revived in a modified form by some modern cosmologists.

planetoids See ASTEROIDS.

planets The nine solid heavenly bodies revolving in the same direction around the sun; these are:

		mean distance from sun		*time of 1 revolution in orbit*	
		(million km.)	*(million mi.)*	*earth-years*	*earth-days*
(i)	Mercury	58	36	–	88
(ii)	Venus	108	67	–	225
(iii)	Earth	150	93	1	0
(iv)	Mars	228	142	1	322
(v)	Jupiter	777	483	11	315
(vi)	Saturn	1426	886	29	167
(vii)	Uranus	2869	1783	84	6
(viii)	Neptune	4495	2793	164	288
(ix)	Pluto	5900	3666	247	255

planèze (Fr.) A triangular wedge of lava on the slopes of a dissected volcano, narrowing upwards to culminate in a broken projection on a crater-rim. These occur where the lava flows protect an otherwise unresistant cone; e.g. around the Plomb du Cantal (1858 m., 6096 ft.) in the Central Massif of France, where many of the p.'s have proper names (P. de St. Flour, Limon, Carlades).

planimeter An instrument for measuring areas on a map. Several types are available, varying in complexity from a simple form of tracer-bar (*hatchet-p.*) to delicate instruments fitted with recording dials (*wheel-p.* or *polar-p.*). In each case, the method is that a pointer is traced round the perimeter of the area to be measured. Some models record the area in sq. ins., which must be converted according to the scale of the map; others have variable tracer-arms which may be set to any scale-value desired. Hence *planimetry:* the measurement of areas.

planina, pl. **planine** (Serbo-Croat) Long limestone ridges, with comparatively level flattened crests, in Yugoslavia, trending broadly N.W. to S.E. in conformity with the geological 'grain' of the Dalmatian area. The high p. rise to 900–1500 m. (3–5000 ft.).

plankton Minute plant (not including seaweed) and animal organisms that float or drift in the waters of the ocean. Ct. NEKTON, BENTHOS.

planning region A specif. unit area, for which a co-ordinated plan is made, with 2 main aspects: (i) A city, urban r. or CONURBATION, largely social in concept (redevelopment of city centres, housing, suburbs, OVERSPILL, NEW TOWNS, etc.); in the U.K. this comes under the Town and Country Planning Acts (1947, 1962, 1968), the New Towns Act (1946), the Town Development Act (1952). Much planning is based on such specif. studies as *Traffic in Towns* (Buchanan, 1963), *S. E. Study* (1964). (ii) A r. suffering from unemployment and industrial stagnation, gradually fostered since the SPECIAL AREA legislation of the 1930s and subsequent to then. Reports include Government 'White Papers'; e.g. *Central Scotland* and *N.E. England* (1963). In 1964, the Department of Economic Affairs was created; England was divided into 8 economic p. r.'s, together with Scotland, Wales and N. Ireland, each with a Planning Council (of laymen) and an Economic Planning Board (of officials, controllers, representatives of nationalized industries, etc.).

planosol A soil in which a clay HARDPAN has developed midway between the surface and the bedrock, commonly found on an upland surface in a humid climate. The shallow A-horizon is usually leached, but as the claypan develops the soil becomes more or less permanently saturated and reveals GLEI features.

plantation (i) An estate on which the large-scale production of CASH-CROPS (rubber, tobacco, cacao, sugar-cane, coffee, cotton, tea, bananas, usually on a monocultural basis) is carried on, gen. by scientific, efficient methods, under a manager with an often large force of paid labour, involving a considerable community organization and administration. The concept originated during the colonial system, using European organization, skill and capital, in some cases in early days with slave- or forced labour, later paid, sometimes native, sometimes imported; e.g. negroes from Africa as slaves in the cotton-p.'s of the 'Deep South' of U.S.A.; Tamils from India as paid labourers in the tea-p.'s of Ceylon. Some processing of the product is usually required, involving 'factory' buildings; e.g. the drying and maturing of tea, the making of crêpe rubber from latex, the expressing of palm-oil. The p. system is found mainly in the tropics and sub-tropics, esp. where the use of white labour (other than managerial or directive) was thought to be impossible for climatic and health reasons. Some p.'s in N. Queensland operate entirely with white labour.

(ii) An area of trees, usually quick-maturing conifers, planted (e.g. by the Forestry Commission in England) to provide supplies of softwood, pulp, often as part of land-reclamation.

Plastic deformation, of ice Under certain conditions, the i. near the base of a thick glacier becomes p., as a result of pressure causing inter-molecular movement. This is an important contributory process to glacier flow.

plastic relief The depiction of r. by means of HILL-SHADING, so producing something of the effect of a 3-dimensional r. model.

plateau (Fr.) An upland with a more or less uniform summit level, sometimes bounded by one or more slopes falling steeply away, sometimes rising on one or more sides by steep slopes to mountain ridges. They include: (i) *tectonic p.'s* (Arabia; Deccan; Meseta); (ii) *residual p.'s*, formed from ancient fold-mountains by denudation and sometimes later uplifted (Middle Rhine P.; Allegheny-Cumberland P. of Appalachians); (iii) *intermont p.'s* enclosed within fold-ranges (British Columbia; the Great Basin of Utah and Colorado; the p.'s of Ecuador and Bolivia; Anatolia and Armenia in S.W. Asia; Tibet); (iv) *volcanic p.'s* (Antrim; Abyssinia; Columbia-Snake; N.W. Deccan). Some p.'s may be extensively dissected, leaving ACCORDANT SUMMITS as indications of the orig. p.-surface. The term is used very widely, both in gen. terms and as part of a proper name.

plateau-gravel A sheet of small stones, sand and grit, often compacted, capping a plateau or its dissected remnants. Poor acid soils develop on p.-g. Gravels of several different origins are marked as such on the maps of the Geological Survey of Gt. Britain. They are found over parts of S. England up to about 120 m. (400 ft.). E.g. in the Southampton area between the Test and Itchen valleys.

Plate Carrée Projection See EQUIRECTANGULAR P.

platform (i) Used for any level terrace or benchlike surface; specif. it refers to the product of wave erosion; as WAVE-CUT P. (ii) Sometimes used for a more extensive block forming a continental basement; e.g. the Russian P., which extends eastward from the Baltic Sea, buried progressively more deeply.

platinum A rare metal with a white lustrous appearance, normally found in nature alloyed with other rare metals (osmium, iridium, palladium). Used in the chemical, electrical and metallurgical industries, as a catalyst in petroleum refining (in 'cat-crackers'), for making crucibles to withstand high temperatures, and as a setting for jewellery. Produced at Rustenburg in the Transvaal in S. Africa, Sudbury in Canada, the Urals in the U.S.S.R., and in U.S.A.

playa (Sp.) (i) A basin of inland drainage containing a shallow, fluctuating lake, usually saline, surrounded by sheets of saline and alkaline crust or mud. (ii) The lake itself in such a basin; e.g. in Nevada and Arizona. [*f*]

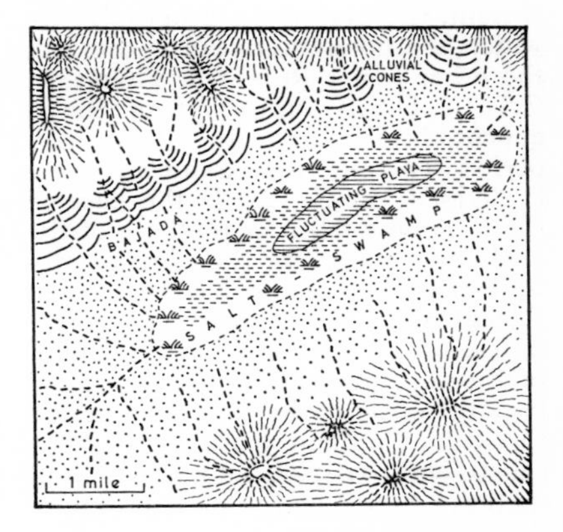

Playfair's Law Enunciated by J. Playfair in 1802, concerning rivers and their valleys. In areas of uniform bedrock and structure which have been subject to river erosion for a long period of time: (i) Valleys are proportional in size to the streams which they contain; and (ii) Stream junctions in these valleys are accordant in level.

plebiscite A direct vote of the whole body of adult citizens of a state (male only in some cases, as in Switzerland) on some specif. issue; a referendum. E.g. p.'s are frequently held in Switzerland should the representatives of 8 cantons or 30,000 citizens desire one; p.'s were held in the Saar in 1935 and 1955, in Luxembourg in 1919, in Gibraltar in 1967.

pleion An area with a high positive ANOMALY in respect of some climatic element, notably temperature (THERMOPLEION).

Pleistocene Derived from Gk. words, meaning lit. 'most recent'. It is, however, used with several meanings: (i) The last million years (or by some authorities, the last 600,000), including the Recent or Holocene, and coinciding with Man's appearance. (ii) The last million years (or 600,000), but excluding the Recent or Holocene (post-glacial); i.e. coinciding with the last Great Ice Age. (iii) In Gt. Britain, the first of the 2 periods into which the Quaternary era is divided, the other being Recent or Holocene. (iv) In U.S.A., the first of the 2 epochs into which the Quaternary period (within the Cainozoic era) is divided. In E. Anglia the base of the P. is now taken as the base of the Red Crag, reflecting the climatic deterioration before the Ice Age. *Note:* Recent work has shown very great difference of opinion as to the length of the P. Some workers have pushed back the onset of the 1st glacial period to 1·8 to 2 million years ago (the so-called 'LONG TIME-SCALE'), others limit its onset to 600,000 years ago (one of the 'SHORT TIME-SCALES').

Plimsoll Line A series of l.'s named after Samuel Plimsoll (1824–98, promoter of the Merchant Shipping Act of 1876), required to be placed on a British ship, marking the level to which she is submerged with a specific amount of cargo. A circle is crossed by a l. indicating the Lloyd's Register Mark, while further l.'s indicate the levels under Tropical Fresh Water, Fresh Water, Tropical, Summer, Winter and Winter N Atlantic conditions.

Plinian eruption A PAROXYSMAL volcanic E., so-called because it was observed and described by the younger Pliny, during the Vesuvius e. of A.D. 79.

Pliocene In Britain, the 4th of the geological periods of the Cainozoic era, lasting about 11 million years, and the system of rocks laid down during that time; i.e. the shelly sands and gravels known as the Coralline Crag in E. Anglia. In U.S.A. it is the 5th epoch of the Cainozoic. Until recently the boundary of the P. in Britain was taken to be the top of the Cromer Forest Bed Series, but this has been shown to be an interglacial deposit. The underlying Weybourne Crag may thus be representative of the 1st glacial period. The base of the P. in England is the base of the Coralline Crag, its top is the base of the Red Crag. There is still much argument about the position of this P.-Pleistocene boundary.

plottable error The smallest distance that can be shown on any map, depending on its scale. Based on the fact that the finest possible line that can be shown on a map is 0·25 mm., the p. e. on a map of scale 1 : 10,000 is 2·5 m.

plucking The pulling away of masses of rock on the valley-floor of a glacier, by means of ice freezing on to protuberances and detaching them as it moves on; one of the two main processes of glacial erosion. It is partic. marked in the case of well jointed rocks.

plug A more or less cylindrical mass of acid lava, occupying the vent of a dormant or extinct volcano, commonly exposed by denudation; e.g. Arthur's Seat and Castle Rock, near Edinburgh; the Rocher d'Aiguilhe and Rocher Corneille in Auvergne (Central Massif of France).

plug-volcano A mass of viscous acid lava, squeezed out of a vent in a molten form, and solidifying as dacite, rhyolite, trachyte; e.g. the Mont Pelée 'spine', formed in 1902; the Santiagito plug in Guatemala, formed in 1922–4; Mt. Lassen, Cascades, a dacite plug with an estimated volume of 1 cu. mi. *Note:* The last is not typical of p.-v.'s, since it had an eruption in 1914, and new craters were formed near its summit.

plum-rains See BAI-U.

plunge-pool A hollow eroded by the force of the falling water at the base of a waterfall, partic. by the eddying and CAVITATION effect; e.g. under the Horseshoe Falls, Niagara.

plutonic rock Intrusive igneous r., which has cooled slowly at considerable depth in the earth's crust, of large-crystalled coarse texture; e.g. granite, syenite, diorite, gabbro, peridotite. Hence the word *pluton,* used esp. in U.S.A. for any large intrusive igneous mass.

pluvial, pluviose Rainy, or app. to rain. Used in connection with a past rainy period, in ct. to dry periods; e.g. the ATLANTIC was a p. period between 5500 and 3000 B.C., with the drier preceding BOREAL and succeeding SUB-BOREAL.

pluviometric coefficient A value arrived at by expressing the mean monthly rainfall total for a given station as a ratio of the hypothetical amount equivalent to each month's rainfall were the total rainfall to be equally distributed throughout the year. E.g. suppose the mean annual rainfall = 31 ins., and the mean Jan. rainfall = 4·0 ins. If it were uniformly distributed, Jan. average $= \frac{31}{365}$ of 31 ins. = 2·63 ins. Then p. c. $= \frac{4 \cdot 00}{2 \cdot 63} = 1 \cdot 52$. See EQUIPLUVE.

pneumatolysis The chemical changes produced in rocks by the action of heated gases and vapours (notably when MAGMA is cooling and solidifying), both within the igneous material itself and in the surrounding country-rock. Usually new minerals are formed; e.g. tourmaline, kaolin, fluorite, topaz.

pocket beach See BAY-HEAD B.

podzol, podsol (Russian) A soil formed under cool, moist climatic conditions with a vegetation of coniferous forest or heath, esp. from sandy parent materials, resulting in the intense leaching-out of base-salts and iron compounds. The thin A_0-horizon consists of raw humus, beneath which is the partly leached A_1-horizon, and the heavily leached, ash-grey A_2-horizon; the B_1-horizon contains an accumulation of the partly cemented leached materials; the B_2-horizon contains an accumulation of sesquioxides, clay and humus. These soils are widely spread in the heathlands of W. Europe and in the coniferous forests of N. Europe. The process is *podzolization.*

point A narrow projection of land into the sea, usually low-lying; e.g. P. of Ayre (the N. tip of the I. of Man); St. Catherine's P. (I. of Wight); Start P. (Devon).

point-bar deposits Used mainly in U.S.A for alluvium, sand and gravel deposited on the shelving SLIP-OFF SLOPE on the inside of a meander.

polar air-mass An a.-m. which has originated in middle latitudes (40–60°), either over the ocean (*p. maritime*) (*Pm*) or over a continental interior (*p. continental*) (*Pc*). *Note:* Air-masses actually originating near the Poles are unfortunately not called p., but Arctic (*A*) and Antarctic (*AA*), which is a source of possible confusion. The latter terms were coined after p. had become well established.

polar front A f. or frontal zone in the N. Pacific and N. Atlantic Oceans, along which meet P. Maritime and Tropical Maritime air.

polar outbreak The penetration of a cold air-mass from middle and high latitudes into comparatively low latitudes, bringing exceptionally cool, even cold, weather, accompanied by strong winds. This is partic. common when there are no protective transverse barriers of mountains, as in N. America. The cold winds are so distinctive as to be given names: *norther, norte, papagayo, friagem, pampero, southerly burster.*

polar projection A map p. centred on the N. or S. Pole; e.g. the polar cases of the various ZENITHAL P.'s. The example in the diagram is a ZENITHAL EQUIDISTANT. [*f*]

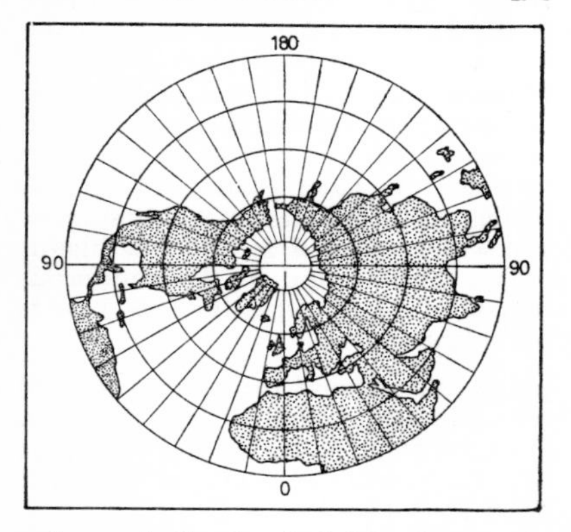

polar wind A wind blowing away from the N. or S. polar regions.

polder (Dutch) A unit of land near, at or below sea-level, reclaimed from the sea, a lake or a river flood-plain by endyking and draining, often kept clear of water by pumping. (Vb. to *empolder.*) The largest scheme ever attempted in the Netherlands is the partial reclamation of the Zuider Zee; about 2227 sq. km. (860 sq. mi.), or 60% of the total area, will be reclaimed in five p.'s, leaving the fresh-water IJssel Meer. [*f opposite*]

pole (i) Geographical p., the N. and S. extremities of the earth's axis. (ii) MAGNETIC P. (iii) CELESTIAL P.

political geography The study of states, their frontiers and boundaries, their inter- and global relations, their contacts and their groupings; the variation of political phenomena from place to place, considered in relation with other features of the earth as the home of Man. 'P. g. is concerned with politically organized areas, their resources and extent, and the reasons for the partic. geographical forms they assume' (N. J. G. Pounds).

polje, pl. **polja** (Serbo-Croat) A large depression in limestone KARST country, usually elliptical in plan and flat floored, often covered with TERRA ROSSA, sometimes with a marsh, or small intermittent or permanent lake. E.g. in Yugoslavia, where most of them are elongated from N.W. to S.E., parallel to the general 'grain' of the country. Some are large; e.g. the Livanjsko p., 64 km. (40 mi.) in length and 403 sq. km. (156 sq. mi.) in area.

pollen analysis See PALYNOLOGY.

Polyconic Projection A CONIC P., on which the parallels are non-concentric circles, each drawn on its own radius $= R. \cot. lat.$ The central meridian is a straight line divided truly, all other meridians are curves. The scale along each parallel is correct, but it increases on the meridians with increasing distance from the central one. Even so, the scale error is less than 1% within 900 km. (560 mi.) of the central meridian. It is easy to draw, using tables giving the co-ordinates of each parallel-meridian intersection. The p. is neither equal area nor conformal, but it is suitable for maps

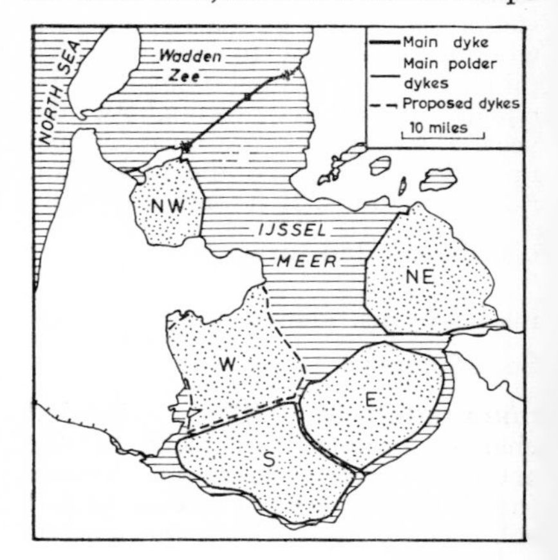

of countries of great longitudinal extent; e.g. Chile. It is also used for large-scale topographical maps in U.S.A., since the scale error within the standard 15 minute quadrangle is very small. The INTERNATIONAL MILLION MAP is drawn on a modified polyconic p., on which the meridians are straight, not curved, with two standard meridians on each sheet, and only the top and bottom parallels are standard. Sheets can be fitted together along N. to S. and E. to W. margins.

polycyclic App. to geomorphological features which have been subjected to a number of partial cycles of erosion. In partic. it is used of a stream whose course reflects base-levelling to more than one former sea-level. Some authorities used *multicyclic*, but p. is gen. preferred.

polygenetic With many origins; e.g. used: (i) of geomorphological features which have undergone several differing processes of erosion; (ii) of mountains formed at different times by different forces (folding, faulting, vulcanicity); (iii) of soils of complex origin; (iv) of pebbles in a conglomerate.

polygon, soil A surface formation which has developed under PERIGLACIAL conditions, often produced by the sorting of fine and coarse weathered materials by freeze-thaw. See PATTERNED GROUND.

Polyhedric Projection A p. used for a large-scale topographical map, whereby a small quadrangle on the spheroid is projected on to a plane trapezoid. Scale is made true either on the central meridian or along the sides.

pond A small area of still water, usually artificial.

ponente A W. wind on the coast of Corsica and Mediterranean France, usually cool, and bringing dry weather.

ponor, pl. **ponore** (Serbo-Croat) A deep vertical shaft in limestone country, leading from the surface into an underground cave system; a stream may fall into it, permanently or intermittently. Syn. with *aven* (Fr.), and in speleological parlance with 'pot', 'pot-hole' or 'swallow-hole'. E.g. Gaping Gill, W. Yorkshire, 111 m. (365 ft.) deep; the Gouffre de Pierre St. Martin in the Pyrenees, 346 m. (1135 ft.) deep.

population The people of any unit area. *P. geography*, involving the study of people, their distribution, density, vital statistics, growth, movement, occupational structure, and groupings in settlements, is an important branch of the subject.

population explosion A sudden and rapid expansion of p. in a partic. area, usually the result of a marked decrease in the DEATH RATE, less usually an increase in the BIRTH RATE. E.g. in Japan, esp. in the Tokyo area.

population potential The possible number of people who can live in a specif. area with a reasonable standard of living, in relation to the available resources in that region.

pororoca (S. Am.) The BORE on the estuary of the R. Amazon.

porosity The nature of a rock with open texture, and usually with coarse grained, widely spaced constituents; e.g. sand, gravel, sandstone, conglomerate, oolitic limestone. The p. of a rock can be stated as a ratio or percentage, in terms of the total volume of pore spaces related to the total volume of that rock. It can be measured by a *porosimeter*, which injects liquid or gas at a specific pressure into a sample of rock. Most porous rocks are PERMEABLE, except clay (where the pore-spaces are minute and are often sealed with water held by surface tension) and unjointed chalk.

porphyry A HYPABYSSAL rock containing large crystals (PHENOCRYSTS) in a fine-grained ground-mass. Hence the textural adj., *porphyritic*.

port A rather loosely used term for a point on a coast where ships can tie-up or anchor, and so load and unload cargoes. This implies the existence of an adjacent settlement, handling facilities, quays and usually docks, and systems of communication inland. Various categories include *terminal p.'s* (New York, Southampton), *p.'s of call* (Colombo), *ferry p.'s* (Dover, Calais), OUTPORTS, ENTREPÔTS, *naval p.'s* (Portsmouth, Toulon), TREATY P.'s (formerly on the coast of China). The largest p.'s in the world, in terms of freight handled, are New York and Rotterdam (approx. 100 million tons per annum). There is a distinction between a p. and a harbour; all p.'s must have a harbour, but the latter can exist without p. facilities.

portage Originating in Canada and U.S.A., signifying the carrying of a boat (usually a canoe) and its contents from one section of navigable water to another, to avoid rapids and falls, or from one lake to another, or even across a divide to another stream.

positive anomaly, of temperature Where the mean t. at a meteorological station is higher than that for all stations in that latitude. See ANOMALY.

positive landform An upstanding l. such as a mountain, a hill and a plateau, in ct. to negative forms.

positive movement, of sea-level A change in the relative level of land and sea, resulting in a net rise of s.-l., in ct. to a NEGATIVE MOVEMENT. This may be the result of: (i) Subsidence of the land as a result of earth-movements or ISOSTATIC loading; (ii) The actual rising of the water-level in the sea on a world-wide and uniform scale (EUSTATIC movement), as in the case of the return of water to the oceans as the Quaternary ice-sheets melted.

possibilism The philosophical doctrine that the environment offers Man a choice of possibilities of which he may or may not take advantage; ct. DETERMINISM, PROBABILISM. This has afforded a fertile field of investigation and writing, including the work of P. Vidal de la Blache, L. Febvre, J. Brunhes, Isaiah Bowman and Carl Sauer. 'There are no necessities, but everywhere possibilities; and man, as master of the possibilities, is the judge of their use' (L. Febvre).

post-glacial App. to all time since the Quaternary ice-age. As, however, this ended at different times in various areas, according to their positions relative to the main centres of ice-dispersal (i.e. is *time-transgressive*), no single date can be ascribed to the beginning of p.-g. time, and indeed glaciation is still present at high latitudes and altitudes. An arbitrary date for E. Anglia would be 15,000 B.C., for S. Scandinavia 8000 B.C., for S. Finland 6500 B.C. In geological terms, the beginning of p.-g. time is the division between the Pleistocene and the Holocene or Recent.

pot Used in a placename as syn. with a CIRQUE, notably in the Mourne Mtns. of Ireland; e.g. P. of Legawherry, cut into the N.-facing slopes of Slieve Commedagh.

potamoplankton Minute organisms, both animal and vegetable, that live in slowly flowing rivers and streams.

potash A gen. term applied to salts of potassium (notably *sylvite* or p. chloride, *carnallite* or p. magnesium chlorate, and *kainite*, a mixture of p. chloride and magnesium sulphate). These are usually found in association with common salt on the site of a salt-lake in past geological times, as in Stassfurt (E. Germany) and Alsace. P. is also recovered from present salt-lakes, as in the Dead Sea and in S.W. U.S.A. Its major uses are as a fertilizer and in the chemical industry gen. C.p. = U.S.A., U.S.S.R., W. Germany, E. Germany, Canada, France.

potential evapotranspiration index An empirical i. devised by C. W. Thornthwaite (1948) as the basis for his second scheme of climatic classification. It is based on the moisture i., relating the amount of precipitation required by e., assuming vegetation to be present, to the actual water available. This is abbreviated to the

P.E.I.; the same initials are sometimes used for the PRECIPITATION EFFICIENCY I. with which it should not be confused.

potential instability, of an air-mass The orig. condition of an a.-m. which when lifted over a relief barrier, or over a mass of cooler air at a FRONT, becomes conditionally unstable. See CONDITIONAL INSTABILITY.

potential model A mathematical construction that measures the force exerted by any defined phenomenon on a point in space by reference to the same phenomenon located at all other points on the spatial domain under study.

pot-hole (i) A h. in the bed of a stream, formed by the grinding effect on the bed-rock of pebbles whirled round by eddies. (ii) The term is used popularly, though incorrectly, to denote vertical cave-systems, esp. by speleologists, hence 'pot-holing'.

pound (i) A field or yard in which stray animals are put until claimed; a fine may be imposed for the animal's return. The p. used to be a common feature of English country life, and it is still in use in places; e.g. in Brockenhurst and Lyndhurst in the New Forest, and even in Southampton. (ii) A British unit of mass: (*a*) *avoirdupois*: 7000 grains; 27·692 cu. in. of water at 4°C.; 453·592 grammes (gm.); 2240 lb. = 1 British long ton. (*b*) *Troy*: 5760 grains; 373·2418 gm.

powder snow Dry, loose s. crystals which have accumulated under conditions of low temperature. As much as 762 mm. (30 ins.) depth of this s. may be equivalent to 25 mm. (1 in.) of rain, as ctd. with wet slushy s. where 100–150 mm. (4–6 ins.) may melt to form 2·5 m. (1 in.) of water. Such p. s. is found gen. in Antarctica, N. Canada and Siberia, and when it occurs in the Alps, etc., it forms an ideal ski-ing medium.

power The *rate* of doing work, as ctd. with energy, the *capacity* for doing work; the basic unit is 1 erg per sec., the practical unit is the watt (10^7 ergs per sec.). 1 horse-p. = 746 watts or 33,000 foot-lb. wt. per minute (i.e. 550 foot-lb. wt. per sec.).

pradoliny (Polish) A large-scale glacial overflow channel; syn. with URSTROMTAL.

prairie An extensive area of grassland, occurring in middle latitudes in the interior of N. America. The use is complicated by the Fr. word *prairie*, lit. a meadow. In its modern world-scale usage, it can be equated with steppe. In its natural form, developed under a rainfall total of 250–500 mm. (10–20 ins.) it presented a continuous cover of tufted grass, coarse and hardy, bluish-green in spring, yellow and straw-like in summer, with bulbous and herbaceous plants. Little of this survives today, since it has been ploughed-up for wheat, or replaced by cultivated grasses. But the term survives as a broad land-use category in Canada and U.S.A.

prairie soil A s. type akin to CHERNOZEM, with a similar profile except that it lacks the deposition of calcium carbonate in the B-HORIZON. It occurs in areas formerly under p. grasses, now growing grain and fodder crops.

Pre-Boreal The climatic phase immediately following the Quaternary glaciation; e.g. in E. Anglia until about 7500 B.C. It was characterized by dry, cold conditions, with a birch-pine flora. At this time sea-level was at least 60 m. below the present, and the British Isles were joined to the continent of Europe.

Pre-Cambrian All geological time (and the associated rocks) before the beginning of the C. period of the Palaeozoic era (i.e. before about 600 million years ago). See ARCHAEAN.

precession of the equinoxes The change which takes place in the rel. positions of the ECLIPTIC and the EQUATOR, the result of the fact that the axis of rotation of the earth describes a slow, slightly conical rotation, caused by gravitational forces acting between the earth and the sun and the earth and the moon; this causes the position of the Celestial N. Pole to appear to sweep round in a complete circle in the heavens in 26,000 years. Similarly, the positions of the EQUINOXES move round the ecliptic once in 26,000 years.

precinct An old word used in various ways, as for the land around a religious foundation. Now used in town-planning and urban geography to denote a specialized and defined unit within a town; e.g. *shopping p.*, *pedestrian p.*

precipice A picturesque term for a high, steep and abrupt face of rock.

precipitation The deposition of moisture on the earth's surface from the atmosphere, including dew, hail, rain, sleet, snow.

precipitation-day A day (24 hours) with at least 0·25 mm. (0·01) of p. In a country with an appreciable part of the p. in the form of snow (as in Canada), this term is more accurate than RAIN-DAY.

precipitation-efficiency (or **p.-effectiveness,** or **p.-evaporation) index** An i. devised by C. W. Thornthwaite for the basis of his first climatic classification system from the formula $i=11{\cdot}5\left(\frac{p}{T-10}\right)^{\frac{10}{9}}$ where p is the monthly mean precipitation in ins. and T is the monthly mean temperature. Five major regions are distinguished on this basis, based on humidity rather than on thermal character. A THERMAL EFFICIENCY I. was used to subdivide these further. The abbr. term P.E.I. should not be confused with POTENTIAL EVAPOTRANSPIRATION I., which is also unfortunately abbr. in the same way.

pressure In climatology, an abbr. for atmospheric p.

pressure gradient See BAROMETRIC G.

pressure-plate anemometer A simple instrument for indicating wind-force, consisting of a wooden base and an upright support, with a metal plate suspended from a knife-edge. The angle from the vertical to which the plate blows is observed, and the velocity is read off from tables. This is not very accurate, esp. when the wind is squally.

pressure release The outward expanding force of p. released within rock masses as a result of 'unloading' (e.g. the removal by denudation of overlying layers), causing the pulling-away of outer layers of the rock. This may occur partic. following deglacierization. It is esp. evident in massive unjointed rocks; e.g. granite. The 'domes' of the Yosemite Valley in California are probably the result of this force, which has caused a succession of curved 'rock-shells', 1 in. to 20 ft. or more in thickness, to pull away.

pressure ridge A r. of floating ice, formed where one floe has been squeezed against another.

pressure system An individual atmospheric circulation s. of high or low p. See ANTICYCLONE, DEPRESSION, RIDGE OF HIGH PRESSURE, COL, SECONDARY DEPRESSION, WEDGE.

pre-urban nucleus A man-made feature in a landscape which existed before the establishment of a city, but around which this subsequently grew. Castles, cathedrals, churches, palaces and bridges have commonly formed p.-u. nuclei.

prevailing wind A w. which blows most frequently from some specific direction; e.g. the S.W. w.'s are p. w.'s over much of W. Britain. This is not to be confused with a DOMINANT W.

Primärrumpf (Germ.) W. Penck's name for an initial, slowly and progressively uprising, flat-topped domed surface, or 'primary peneplain', in ct. to the ENDRUMPF.

Primary First used in the early 19th century for the period of time and associated rocks now referred to as

Pre-Cambrian. Then it was transferred to the Lower Palaeozoic, finally to the whole of the Palaeozoic era. The term has been abandoned in favour of the last by many authorities, though some retain it to conform with the more commonly used Secondary, Tertiary and Quaternary.

primary industry An activity directly concerned with the collecting or utilization of the resources provided by nature; it includes agriculture, fishing, forestry, hunting, mining.

primary product A foodstuff or raw material (minerals, wood, wool) that has not been fabricated, though it may have been processed for consumption. The ct. is with a secondary or manufactured p.

primary (P-) wave A type of compressional vibration produced by an earthquake; this kind of w. is similar to a sound-w., with each particle displaced by the w. in its direction of movement; a 'push-w.'. [*f* EARTHQUAKE]

primate distribution, of city sizes A d. dominated by a single large city. In the d. of city sizes, an exceptionally large gap may exist between the population total of the first and second ranking cities; hence a p. city. Where 2 large cities share this characteristic and dominate all other places, the distribution is bi-p. E.g. Uruguay has a p. d., while Brazil has a bi-p. d.

primeur (Fr.) An early 'out of season' vegetable crop, as grown in France in the Midi or Brittany.

principality A territory over which a prince has jurisdiction, or from which he obtains his title; Monaco and Liechtenstein are now the only ones remaining in Europe. Wales has the title of a p., created in 1301 when Edward became Prince of Wales.

principal meridian The central m. on which a rectangular GRID is based. Used specif. of the systems employed for the U.S. LAND SURVEY, for which there are 32.

prisere The collection of SERAL COMMUNITIES which lead to the development of a CLIMAX COMMUNITY. These may be of several distinctive types, developing from conditions liable to drought (*xeromorphic*), such as bare earth (*lithosere*) or sand-dunes (*psammosere*), or under fresh (*hydrosere*) or salty (*halosere*) water conditions. The emphasis is on the development of a community under conditions which in the first instance are unsuited for vegetation.

prismatic compass A magnetic c., with a small magnet which swings on a central pivot, with a circular card graduated clockwise from N. A peep-sight with a slot and a prism is on one side, on the other is a vertical hair-line in a hinged glass lid. If a sight is taken on an object, when the card has ceased to swing, the angle of the bearing can be read off through the prism.

probabilism A modification of the POSSIBILISM concept of Man's relations with his environment, postulated by O. H. K. Spate; it implies that there are everywhere possibilities, but that some are more probable than others.

probability In the statistical analysis of data (climatic, population, economic productivity), the p. of an event occurring is assessed by dividing the number of occurrences of the event by the total number of cases or trials. P. is normally expressed as a percentage. Thus the occurrence of an absolutely certain event is 100% probable (sometimes written 1·0). If a penny is tossed, the likelihood of its coming down heads is 50%, not exactly so, perhaps, since the coin may not be exactly true, and there is also the outside chance that the coin will land on its edge.

process elements In a MODEL in physical geography, the p. e.'s are the measurable contributions of the various forces involved. E.g. in a model of beach development, the p. e.'s will include the height, period and direction of the DOMINANT WAVES, resulting in the creation of a set of RESPONSE E.'s. All these may be analysed statistically.

process lapse-rate The rate of decrease of temperature of a small parcel of air as it rises, in ct. to ENVIRONMENTAL L.-R. See DRY ADIABATIC L.-R., SATURATED A. L.-R.

producer goods Goods produced which are needed to manufacture other goods; e.g. machine-tools, looms.

profile The outline produced where the plane of a section cuts the surface of the ground, as river p., coast p., dune p. See also COMPOSITE, PROJECTED, SUPERIMPOSED P. [*f*]

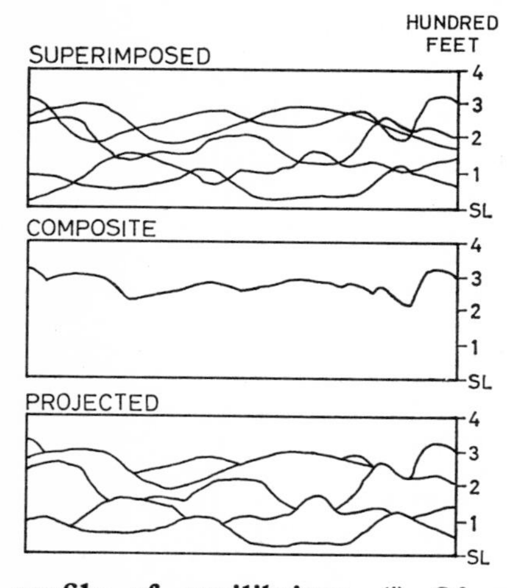

profile of equilibrium (i) Of a *river:* a stream long-p. adjusted to prevailing conditions so that at all points the stream may be said to be in a state of GRADE. This p. is gen. regarded as a flattened parabolic curve, concave upwards. Recent work has questioned this assumption, which is largely based on the concept that the long-p. is the only significant variable in the achievement of grade. (ii) Of a *shore:* A sloping shore where the amount of mud, sand and shingle which accumulates is more or less balanced by the amount removed. This balance is very temporary, and is easily disturbed by exceptionally high tides, strong onshore winds and large destructive waves. After a storm is spent, the processes of accumulation will labour to restore the p. of e., the net result of the average set of conditions obtaining along that stretch of coast.

profitabilism A development of the concept of POSSIBILISM, wherein certain possibilities offer greater prospects of profit than others, which therefore determine the course of action.

proglacial lake An area of water dammed up during a glacial period, usually between the edge of an ice-sheet and some divide; e.g. 'L. Agassiz' in N. America (of which L.'s Winnipeg, Winnipegosis and Manitoba are remnants); 'L. Lapworth' in the W. Midlands of England; 'L. Humber' in the Vale of York; 'L.'s Eskdale, Glaisdale and Wheeldale' on the N. side of the N. York Moors.

progradation An extension outward of land into the sea by deposition and accumulation, usually of river-borne sediment or of material moved along the coast by longshore drift. Ct. RETROGRADATION.

programme map A m. used in town-planning to show the proposed future phasing of areas to be developed or redeveloped.

progressive wave A w. propagated in a channel of theoretically infinite length; its w.-length is defined as the distance between adjoining crests, and its period as the time it takes to move one w.-length. This concept has been applied: (i) to tides (below); and (ii) to wind-generated w.'s in the open ocean.

progressive wave theory, of tides The old theory of t.'s, involving the formation of two tidal w.'s in the S. Ocean, one following (though lagging slightly behind) the moon, the other on the opposite diameter of the earth. From these, branches passed N. into the Atlantic, Indian and Pacific Oceans, and successively into their marginal seas. This has been replaced by the STATIONARY W. theory, though a p. w. does move up a narrow estuary, in extreme cases forming a BORE.

projected profile One of a series of p.'s, usually spaced at equal intervals, plotted on a single diagram, but including only the portions of each not obscured by higher intervening forms. This will give a panoramic effect, with a distant sky-line, a middleground and a foreground, with the effect of an outline landscape drawing showing only summit detail. [*f* PROFILE]

projection See MAP PROJECTION.

promontory A projection, headland or cliff protruding boldly into the sea, usually with rocky cliffs, STACKS and offshore rocks.

proportional dividers An instrument used to copy detail from a map where enlargement or reduction is required. Two bars, pointed at each end, fit diagonally across each other, held by a screw in a sliding groove; the screw is set to the proportion required, according to a scale-line, and screwed down tight. Any distance stepped off with the d.'s at one end of the instrument will give the equivalent distance at the other, increased or decreased proportionally according to the scale-setting.

protected state A territory within the British Commonwealth, over whose foreign affairs the British Government exercises considerable control, but which is internally self-governing; e.g. Tonga; formerly Brunei (now part of Malaysia); and formerly Zanzibar (first an independent republic, united in 1964 with Tanganyika as Tanzania).

protectorate A territory not formally annexed by another country, but in which the latter has, by treaty or grant, some form of political control; e.g. British Somaliland, Aden P., British Solomon Islands.

Proterozoic (i) The younger of the 2 Pre-Cambrian eras, and by some authorities syn. with Algonkian. (ii) By the U.S. Geological Survey, all Pre-Cambrian time and rocks. (iii) By some authorities, the 3rd of the 3 eras of Pre-Cambrian time, succeeding the Eozoic and Archaeozoic.

Protozoic An obsolete term for the Older or Lower Palaeozoic era; i.e. Cambrian, Ordovician and Silurian periods and systems, as ctd. with the also obsolete Deuterozoic (Younger or Upper Palaeozoic).

provenance Partic. used in connection with sedimentary rocks to describe the origin of the constituent materials, or the source area from which the sediments came.

province (derived from the Roman *provincia*) An administrative unit within a country; e.g. Belgium has 9 p.'s, whose boundaries were arbitrarily demarcated after the French occupation of 1795, but broadly correspond to medieval divisions.

proximal map A m. in which value areas are assigned according to their proximity to a datum point, produced by automated cartographic methods.

psammite A rock made of particles of sand 'cemented' together, hence *psammitic*. An ARENACEOUS rock. Ct. PELITE, PSEPHITE.

psephite A coarse, compacted, fragmental rock, with usually rounded particles larger than sand grains; e.g. a conglomerate.

psychrometer A type of HYGROMETER, using wet- and dry-bulb thermometers, with an electrically driven fan which forces a current of air past the wet bulb to ensure max. evaporation. In another form, the *sling p.*, the thermometers are whirled round to ensure max. ventilation. In either case, the aim is to ascertain the RELATIVE HUMIDITY.

Psychozoic Used by some scientists to indicate the era of geological time since man appeared on the earth, from Gk. *psyche* (soul or mind); this is really syn. with the Quaternary.

pteropod ooze A calcareous deep-sea o., formed of the shells of minute conical molluscs (*pteropods*). It is of limited occurrence, being found in relatively small patches only in the N. and S. Atlantic Oceans.

puddingstone A popular name for CONGLOMERATE.

pueblo (Sp.) Applied to: (i) the native Indian inhabitants of S.W. U.S.A. and N. Mexico, who have practised a town-dwelling agricultural economy since before the arrival of Europeans; and (ii) the communal settlements in which they live.

pumice The scum, containing bubbles of steam and gas, on the surface of a LAVA flow, which solidifies to form a spongy or cellular rock. It is so light that it will sometimes float. Its chemical composition is similar to rhyolite, of acid nature, and of fine-grained texture.

puna (S. Am.) A high intermont plateau at about 3600–4800 m. (12–16,000 ft.) between the W. and E. Cordillera of the Peruvian and Bolivian Andes, with a sparse cover of coarse grass and XEROPHYTIC plants. The term is used also: (i) for the vegetation itself; (ii) as a proper name for such a region in Bolivia.

purlieu An outlying part or outskirts, esp. of a forest. It sometimes forms part of the name of an orig. peripheral settlement; e.g. Dibden P. on the edge of the New Forest in Hampshire.

push-moraine Mounds of sand and gravel near the margins of ice-sheet movement, pushed into ridges by the ice when it advanced. The materials are thus folded and contorted by pressure. These are known in Dutch as *stuwwallen*, in Fr. as *moraines de poussée*. They occur in the Netherlands just S. of the IJssel Meer in parts of the Veluwe, Gelderland and Overijssel, and in neighbouring parts of W. Germany. These are the product of the 3rd glacial period (SAALE or RISS), known in the Netherlands as the *Drenthian*. [*f* ESKER]

push-wave (P-wave) See PRIMARY W. [*f* EARTHQUAKE]

puszta (Hungarian) A type of temperate grassland in the plains of Hungary.

puy (Fr.) A small volcanic cone. In Auvergne some are of ash and cinder, some are dome-shaped of acid lava, others have double cones. In all, about 70 cones rise from the plateau-surface at 830–990 m. (2700–3250 ft.); the largest is the P. de Dôme (1465 m., 4806 ft.), of trachyte.

pygmy A member of a racial group of unusually short stature, 137 cm. (4ft. 6 ins.) or less ; e.g. the p.'s of the Congo Basin, Malaya, the Philippines, New Guinea, Ceylon.

pyramid[al] peak A sharp p., formed when 3 or more CIRQUES develop, cutting back into the original upland, with prominent faces and ridges; e.g. the Matterhorn, with its 4 faces and 4 ridges. [*f*]

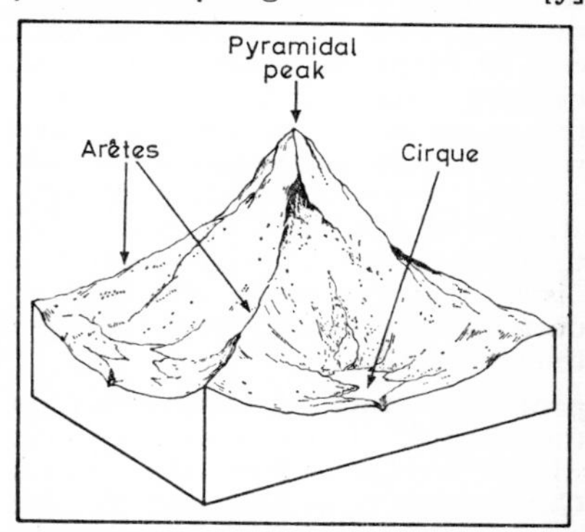

pyrites, pyrite Abbr. for iron p., or iron sulphide, FeS_2, a yellow mineral with a metallic lustre, mined chiefly for its sulphur content. Sometimes called 'fool's gold'. The term also refers to other sulphides; e.g. copper p., but if used alone it denotes iron p.

pyroclast Fragmental volcanic material: lava, cinders, ash and dust, consolidated and compacted; hence adj. *pyroclastic*.

pyrometer An instrument used to measure the temperature of molten LAVA.

pyroxene minerals One of the 3 main groups of ferromagnesian m.'s, including augite, jadeite (the m. of which jade is composed), and many other complex m.'s.

qanat (Persian) An underground channel used to convey water from the foot-hills of Persia to the plains for irrigation purposes.

quadrangle A map of a piece of territory, bounded by parallels and meridians, which fits into a uniform topographical map series. The term is used esp. in U.S.A. for the National Topographic Map Series, ranging from an R.F. of 1:24,000 to 1:1 million. The 15-minute q.'s of the 1:62,500 scale are 17½ ins. in N. to S. dimension, though their width varies from about 15 ins. in the S. to 12 ins. in the N. because of the progressive convergence of the meridians.

quadrant (i) An instrument formerly used for measuring angles and altitudes, consisting of a graduated arc, with an eye-piece. It is now obsolete and has been replaced by the sextant. (ii) The quarter of a circle; an arc subtending 90°.

quadrat A unit area selected in ecology for sampling purposes in order to obtain an accurate statistical description of the vegetation. For sampling ground species, 1 sq. m. might be sufficient, but for woodland a much larger area would be necessary.

quadrature A situation when the sun, earth and moon (or another planet) are at right-angles, with the earth at the apex, which occurs in the case of the moon twice each month. The tide-producing gravitational effects of the sun and moon are then in opposition, and thus the range of the tides is reduced; these are NEAP tides, with low high tides and high low tides. Cf. CONJUNCTION and OPPOSITION.

quadroon A person who has a quarter negro blood, the rest white; i.e. the offspring of a white and a MULATTO parent, esp. in S. America.

quagmire A BOG which shakes under the weight of a man or animal.

qualitative Relating to an object or feature in terms which cannot be expressed quantitatively, but has to be described in subjective terms.

quantification A process dealing with quantities; the use of statistical methods in presenting, explaining and solving a problem in terms of quantifiable concepts and relationships, in handling data, and in producing objective systems of classification. In recent years q. has been applied to virtually every aspect of geographical research. It should be regarded as a supplement to, not a replacement for, any qualitative or descriptive work, and 'high-power' methods should not be applied to 'low-power' data.

quarry A place where stone is excavated from an open surface-working.

quartz The crystalline form of silicon dioxide (SiO_2), with a bright lustre, specific gravity 2·65, and hard enough to scratch glass (No. 7 in the hardness scale). It occurs commonly as clear and transparent, but there are also varieties with different shades of colour; e.g. rose q., smoky q. (or *cairngorm*), amethyst, citrine.

quartzite A rock composed almost entirely of quartz recemented by silica, forming a hard, resistant, impermeable rock. Some q.'s are metamorphic in origin, where sandstones have been recrystallized by heat into a mosaic of quartz grains, others are sedimentary. Q.'s of Cambrian age are found in the Welsh borders, the Midlands (Hartshill quarry), and N.W. Scotland. Much of the main range of the Canadian Rockies between Banff and Jasper (Alberta, B.C.) are of q. In the Ardennes, q.'s are of Lower Palaeozoic age.

Quaternary (i) In Gt. Britain, the 4th of the great eras since Pre-Cambrian times, i.e. post-Pliocene, which started at the onset of the Ice Age. It is divided into 2 periods, the Pleistocene and the Holocene (or Recent). (ii) In U.S.A., the 2nd or younger of the 2 periods in the Cainozoic era, the 1st being the Tertiary.

quebracho forest An evergreen f. of gnarled bushy trees with very hard wood, found in the Gran Chaco of Paraguay–Argentina.

quicksand A thick mass of loose sand and mud impregnated with water, which may swallow up a heavy object such as an animal.

race (i) A large group of people with some basic hereditable physical characteristics in common. The term is used loosely as one of the major divisions of mankind, but all modern populations are the outcome of millennia of race-mixture. (ii) A rapid flow of sea-water through a restricted channel, usually caused by marked tidal differences at either end; e.g. in the Pentland Firth; the R. of Alderney (Channel I.). (iii) A strongly flowing offshore current swirling round a headland or promontory; e.g. off Portland Bill, coast of Dorset. (iv) A narrow channel leading water from a river to the wheel of a water-mill (*head-r.*) and from the mill (*tail-r.*).

Radburn layout A term used in urban studies, derived from the town of Radburn (New Jersey, U.S.A.), where deliberately planned loop-roads and culs-de-sac allow the physical separation of pedestrians and vehicles.

radial drainage A pattern of streams flowing outward down the slopes of a dome- or cone-shaped upland. Volcanic cones afford examples of r. d., their sides scored with the channels of numerous streams. The English Lake District has near r. d. from the W. to E. axis of its elongated dome; the d. goes N.W. via the Derwent, N. via the Calder, N.E. via the Eamont, S. via the Kent, Leven, Crake, Duddon, S.W. via the Esk, Irt and Ehen. This is an example of SUPERIMPOSED D., the pattern of which developed on a cover of new rocks later removed by denudation. [*f opposite*]

radian The angle subtended by an arc of a circle equal in length to its radius $= 57{\cdot}2958^\circ$; π radians $= 180^\circ$.

radiation The process by which energy is propagated or transferred through a medium by means of wave motion. In meteorology this may be partic.: (i) the radiant short-wave energy poured out by the sun into space (see INSOLATION, SOLAR CONSTANT); (ii) the loss of heat into space in the form of long-wave r. from the surface of the earth; the cooling is greatest esp. on a clear night.

radiation fog A shallow layer of white f., formed during settled weather in low-lying areas, esp. in spring and autumn. The surface of the ground, itself rapidly cooled by r. at night, cools the layer of air resting on it, which flows down into hollows by gravity, and is cooled to the dew-point, so causing condensation. The formation of such f. is favoured partic. by the existence of a surface layer of moist air, clear skies permitting max. radiation, and calm air. This is common in fine weather in early summer; as the sun rises, the f. is dissipated. Under cold anticyclonic conditions in late autumn and winter, the r. f. may be thicker and more persistent under an INVERSION layer, and around large towns SMOG may develop.

radioactive decay The progressive break-down of certain unstable 'parent-elements' by the emission of particles from the nuclei of individual atoms, so as to form stable 'daughter' ISOTOPES; e.g. uranium^{238} to lead^{206}, uranium^{235} to lead^{207}, thorium^{232} to

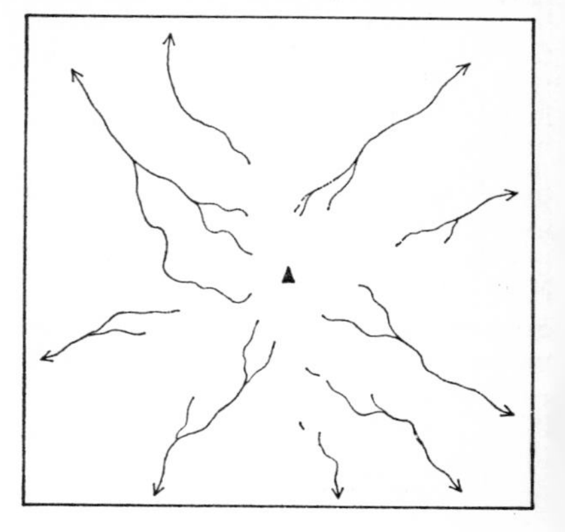

lead^{208}. Hence the age of rocks containing r. elements can be determined; see also LEAD-RATIO, HALF-LIFE. Other radiometric methods of dating include the potassium^{40}/argon^{40} and rubidium^{87}/strontium^{87} ratios. See also RADIO-CARBON DATING.

radiocarbon dating Carbon^{14}, a radioactive ISOTOPE of carbon (atomic weight 14), with a half-life of 5570 years, is created in the upper atmosphere by the bombardment of nitrogen by neutrons liberated by cosmic radiation. The carbon^{14} oxidizes to carbon dioxide, and enters the earth's carbon-cycle. It is absorbed by all living matter; e.g. in the case of plants by photo-synthesis. After the death of an organism, or its burial under sediments, carbon^{14} not only ceases to be assimilated, but the content diminishes at a known rate. Thus the age of a buried piece of wood, a bone in a tomb, peat in a bog, and shells on the ocean-floor can be obtained by finding the proportion of radiocarbon in the total carbon it contains. It can be used to measure ages up to 20,000, sometimes as much as 70,000 years ago, though beyond 30,000 years the accuracy rapidly diminishes. The method was developed by Dr. W. F. Libby, of Chicago in 1947.

radiolarian ooze A siliceous deep-sea o., formed of skeletons of radiolaria, which have a lattice-like structure. It is found in an elongated band in the mid-Pacific Ocean, and in the S.E. of the Indian Ocean.

radiosonde (Fr.) A self-recording and radio-transmitting instrument, carried by a hydrogen balloon to high altitudes, from which meteorological data are sent back by radio-signals.

rag, ragstone A general and rather vague term for a hard rubbly rock outcrop in certain geological formations; e.g. the Coral R. (Jurassic), the Kentish R. (part of the Hythe Beds within the Lower Greensand).

rainbow An arc of multi-coloured light, caused by the refraction and internal reflection of the sun's rays by drops of rain, when the sun is behind the observer and the rain-drops are in front. The light entering each drop is reflected at its far side, and is broken up into the colours of the spectrum. In the primary bow, red is on the outer side, violet on the inner. The angle which the radius of the bow subtends at the observer's eye varies from about 41° for the red end of the spectrum to 43° for the violet. This means that, for an observer at sea-level, a r. can only be seen when the sun's altitude is less than about 42°; the lower its altitude, or the higher the observer, the more bow he sees (from an aircraft he may see a complete circle). The larger the rain-drops, the more vivid the colours; in a fine-dropped fog, the bow is white (see FOG-BOW). A secondary, fainter bow of about 50° angular radius is sometimes to be observed, the result of a double reflection within each rain-drop; the colours are reversed compared with the primary bow, i.e. red on the inner side, violet on the outer.

rain-day A period of 24 hours, commencing at 09.00 hours, with at least 0·01 in. (0·25 mm.) of rainfall in U.K. The U.S. Weather Bureau uses phrases such as a 'day with measurable precipitation', which are probably preferable in that they appear meaningful at first sight, whereas r.-d. may be confusing. Ct. WET-DAY.

rain factor An index devised to express a relationship between precipitation and temperature, so as to give some indication of climatic aridity which might be used in the delimitation of climatic regions; devised by R. Lang.

$$r.f. = \frac{\textit{annual precipitation in mm.}}{\textit{mean annual temperature in °C.}}$$

rainfall When minute droplets of water are condensed from water-vapour in the atmosphere on to nuclei, they may float in the atmosphere as clouds. If the droplets coalesce, they will form larger drops, which, when heavy enough to overcome by gravity an ascending air-current in a cloud, will fall as rain to the surface of the earth. For condensation and precipitation to occur, the ascent of an air-mass is essential; this is brought about in 3 ways, hence there are 3 main types of r.: (i) CONVECTIONAL; (ii) OROGRAPHIC; (iii) CYCLONIC or frontal. R. is measured in a RAIN-GAUGE. Records of r. are given as monthly means, annual means, and in terms of RAIN-DAYS and WET-DAYS. See also INTENSITY OF R., DISPERSION DIAGRAM. *Note:* 1 in. of r. is equivalent to 100·9 tons of water per acre, or 14,460,000 gallons per sq. mi.

rain-forest A dense forest of HYGROPHILOUS trees growing in conditions of heavy, well distributed rainfall; commonly found in: (i) tropical latitudes (e.g. the Amazon Basin, the Congo Basin, Borneo, New Guinea); and (ii) warm temperate latitudes (e.g. S. Brazil, parts of Florida, the coastlands of Natal, central and S. China, S. Japan, E. Australia, the N. Island of New Zealand).

rain-gauge A meteorological instrument, comprising a funnel resting in a collecting vessel, used to measure rainfall. The vessel is emptied periodically into a measuring cylinder. The g. must be carefully sited, if possible twice as far from the nearest building as this is high. Self-recording (or *tipping-bucket*) g.'s are used at many stations.

rain-shadow The markedly drier (or lee) side of a mountain area; see OROGRAPHIC PRECIPITATION. The term introduces some confusion if it is thought that the dryness is simply due to the blocking of or sheltering from air-masses; when heavy rain occurs on the windward side, the moisture content of the air-stream is lower on the leeward. E.g. Alberta is in the r.-s. of the Rockies; the Deccan is in the r.-s. of the W. Ghats; E. Scotland is in the r.-s. of the mountains of W. Scotland.

rain-spell A period of 15 consecutive days, each with at least 0·25 mm. (0·01 in.) of rainfall.

rain-wash The movement of loose surface material down a slope, esp. in semi-arid areas with scant vegetation protection, caused by heavy rainfall and resultant runoff not confined to precise channels.

raised beach A coastline, sometimes backed by a cliff and fronted by a wave-cut platform covered with ancient beach material, standing above present sea-level as a result of a negative movement of sea-level. Such features below about 40–45 m. (130–150 ft.) are called r. b.'s, above this they are known as marine terraces, marine platforms or marine erosion surfaces. In geological writings, the term usually just refers to the beach deposit itself, which may or may not be accompanied by a well-marked morphological feature. Around Britain have been identified: (i) the pre-glacial 3 m. (10 ft.) beach (e.g. Holderness, English Channel, I. of Man, N. Wales); (ii) the 30 m. (100-ft.) beach (e.g. in the W. Isles and W. Scotland opposite Skye); (iii) the 15 m. (50-ft.) beach, found in W. Scotland variously between 14 and 20 m. (45 and 65 ft.) above O. D.; (iv) the 8 m. (25-ft.) beach (e.g. W. Scotland, esp. Arran); (v) r. b.'s a few m. above present high-tide level (e.g. the Gower Peninsula, S. Wales). The order in which these beaches are listed is broadly in decreasing age; no. (iv) is probably of early post-glacial age. Recent work has thrown doubt on the grouping of r. b.'s by height alone because of the factor of isostatic recovery since glacial times and complex sea-level changes. [*f*]

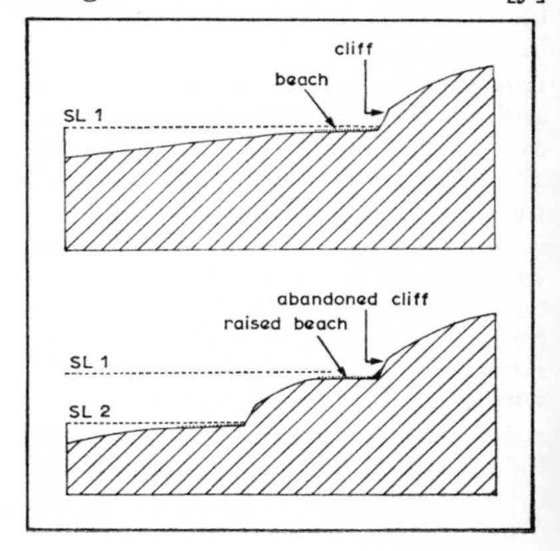

raised bog A thick accumulation of peat, mostly of sphagnum moss, acid in character, forming a lens-shaped sheet in a shallow basin, as much as 5 m. (16 ft.) thick in the centre; e.g. in central Ireland, the source of fuel both for domestic use and for thermal power-stations, R. b. is one of the three main varieties of BOG, the others being *valley bog* and *blanket-bog*.

rake (i) A mining term used in the N. Pennines for a vertical fissure, usually following a joint or a fault, filled with galena, smithsonite and other ores, often linking a horizontal ore body (*flat*) with others above and below. (ii) A sloping terrace on a mountain or rock-face; e.g. Lord's R. on Scafell, Jack's R. on Pavey Ark, Langdale, both in the English Lake District.

ranching The large-scale rearing of cattle on extensive farms, notably in areas which were originally temperate grasslands; e.g. W. U.S.A., W. Canada, Argentina, Uruguay. Once the animals roamed on open 'ranges', now they are kept mainly in enclosures, with alfalfa and other fodder crops grown as feedstuff.

rand (i) A ridge in S. Africa, specif. applied to a rich gold-bearing reef near Johannesburg (Afrikaans). (ii) In U.S.A., the marshy border of a lake.

Randkluft (Germ.) A gap between the surface of a FIRN-field and a surrounding rock-face in a CIRQUE. Ct. BERGSCHRUND, where both sides of the gap are of ice. The r. develops because of melting caused by radiation of heat from the rock-face.

random sample A method of overcoming the problem of coverage of a very large mass of data, or a very large area, in LOCATIONAL research, whatever type of distribution is under investigation. In ct. to *purposive sampling*, where one chooses typical s.'s subjectively, r. sampling is such that every item has an equal chance of selection, and every item selected is quite independent of all others. In a *simple r. s.* the objects of study are listed, each is assigned an index number, and a r. s. of index numbers is obtained by reference to a table of r. numbers, as in the *Cambridge Elementary Statistical Tables*. When a geographer wishes to obtain a r. s. of a spatial variable, either continuous (such as height of land, RELATIVE RELIEF, slope angle, soil features) or discontinuous (such as land use, or any other item expressed as a CHOROPLETH MAP), he may superimpose a GRID over a map of the specif. phenomena, and select a r. series of grid intersections from the table of r. numbers. If in effect the number of items involved is so large as to be virtually infinite, a form of *area sampling* must be used; e.g. a botanist studying a large COMMUNITY of plants may select a number of QUADRATS. If there is a marked spatial clustering of the phenomena, the r. s. method may lead to a biased s., and it may be necessary to make a *stratified r. s.* in which the total phenomena are broken down into classes or 'strata' before the s. is taken. This involves detailed information and considerable prior study; e.g. if occupations are being studied, it may be necessary to group settlements in which those people live according to their size, and then take r. s.'s within the separate groups. Ct. SYSTEMATIC SAMPLING.

Randstad (Dutch) Derived specif. from the so-called 'ring-city', an almost continuous 'conurbation circuit' (incl. Amsterdam, The Hague, Rotterdam, Dordrecht, Utrecht), within which is a less densely populated 'green heart', which the planners would seek to preserve.

range (i) A line of mountains. (ii) An open area, usually unfenced, used for grazing, as in the High Plains of U.S.A. (iii) The difference between the max. and min. of a series of numerical values, esp. of climatic elements, such as seasonal temperature; e.g. Verkhoyansk (Siberia), Jan. mean −50°C. (−59°F.), July mean 15·5°C. (60°F.), hence r. = 65·5°C.

(119°F.) (iv) The limit of habitat of a plant or animal. (v) In the rectangular LAND SURVEY SYSTEM [*f*] of U.S.A., the r. is the N. to S. line of townships between 2 meridians, 6 mi. apart, identified by its distance from the central meridian to which it is parallel. (vi) An area of mineral ore in U.S.A.; e.g. the iron-ore r.'s (Mesabi, Vermilion, Cuyuna, Penokee, Marquette, Menominee) near L. Superior. (vii) The *tidal-r.* between the highest high and lowest low spring tides.

range, of a good The distance over which a g. will be distributed from a CENTRAL PLACE. There are 2 limits to this distance, known as the 'inner' and 'outer' (or 'lower') r.'s. The outer r. is really the ultimate r. of a g., relating to the distance to which a DISPERSED population will be willing to travel for this central g. Beyond this outer r. people will either go to another centre for the g., or will not buy the g. because the gain derived from its purchase will be outweighed by excessive transport costs. The inner r. is defined by the degree of requirement of the THRESHOLD POPULATION for the particular g. Within the inner r., sufficient people must reside to make the distribution of the g. marginally worth-while (i.e. costs are just covered and the g. is therefore not really profitable). If the difference between the inner and outer r.'s of a g. is calculated, then the money spent by the inhabitants of this area represents the profit of the entrepreneur from distribution of the central g.

rank (i) The category of coal, from lignite to anthracite, according to its chemical and physical composition, the result of progressive metamorphism (ii) The rel. position of a town or CENTRAL PLACE in any array of centres plotted by population size or function. See R.-SIZE RULE.

ranking The arrangement of numerical data in descending order according to the specif. attribute under consideration (e.g. land-use, population, productivity, etc.). These ranked data may be used as the basis for a wide range of statistical manipulation, partic. in terms of correlation. The advent, first of punched cards, then of the electronic computer, have both speeded this work and also enabled very large numbers both of cases and variables to be organized.

rank-size rule An empirically derived description of the distribution of town or city sizes in any area that is the antithesis of the PRIMATE city-size distribution. When the population of these towns is ranked in descending order of their populations, according to the series 1, 2, 3, 4 . . . n, the rule specifies that $P_n = P_1 (n)^{-1}$, where P_n is the population size of a town of rank n, and P_1 is the population of the largest city in the area. The r.-s. r. does fit many countries, esp. those that are large and industrialized; e.g. U.S.A., and those that have a long urban history, although its universality is far from complete.

rape (i) One of the former administrative districts of Sussex, 6 in number, as recorded in the *Domesday Book*. (ii) A plant (*Brassica napus*) grown as animal food, and for its seed from which oil is expressed.

rapid(s) An area of broken, fast flowing water in a stream, where the slope of the bed increases (but without a prominent break of slope which might result in a waterfall), or where a gently dipping bar of harder rock outcrops; e.g. the Nile Cataracts; the Istein R.'s on the Rhine below Basle; the r.'s in the Iron Gate of the Danube.

rare earths Widely distributed, though extremely scarce, oxides of elements with atomic numbers from 57 to 71, of atomic weight 138·92 to 175·0, from lanthanum to lutetium. These minerals and their compounds are very difficult to separate.

rating curve A graph showing the DISCHARGE of a river, with the vertical scale the depth of water (as shown by a gauge), the horizontal scale the discharge in CUSECS. If a number of actually measured discharges are plotted and a smooth c. is drawn through them, discharge on a future occasion can be estimated from the single measurement of the depth of water. [*f*]

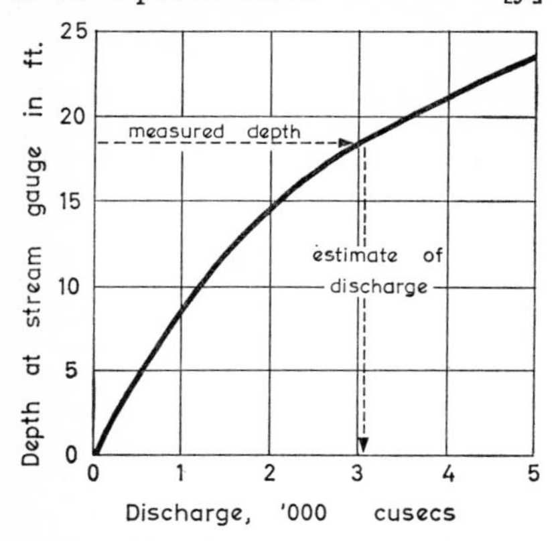

rattan A species of climbing palm in Indonesia with thorny stems, so long that they can be used for ropes and nets, and for weaving baskets.

ravine A narrow steep-sided valley, but larger than a gully or a cleft.

rawinsonde (radar wind-sounding) (Fr.) A hydrogen balloon equipped with self-recording and transmitting meteorological instruments, carrying a radar target to enable its course to be directly followed and plotted, and the apparatus to be recovered after its descent by a parachute. Ct. RADIOSONDE.

raw material A substance intended for processing, fabrication or manufacture; it may be natural (animal, vegetable, mineral) or a product of some other activity (e.g. coal-tar, wood-pulp).

ray-diagram A type of d. used to illustrate dynamic concepts radiating from a point; e.g. the sphere of influence of a town indicated by drawing radial lines from its centre to the villages with which it has service relations; winds (see WIND-ROSE); movements of people or goods; racial affinities. These are often known as *star-graphs*, *star-diagrams*, or *clock-graphs*.

reach (i) A specif. section of a river. (ii) In navigation, a straight section between bends. (iii) In a canal, a section between two locks.

Réaumur scale A thermometric s. in which the melting point of ice is 0°R. and the boiling point of water is 80°*R*. $\frac{R^\circ}{80} = \frac{F^\circ - 32}{180} = \frac{C^\circ}{100}$

Recent The period of time and the system of rocks since the end of the Pleistocene (i.e. the last Ice Age). This varies with the position of any area relative to the main centre of ice-dispersal; for E. Anglia *c.* 15,000 B.C.; for S. Scandinavia *c.* 8000 B.C.; for S. Finland *c.* 6500 B.C. A gen. figure is 10,000 B.C. Syn. with Holocene. R. rocks include alluvium, peat, sand-dunes, shell-beds, coral. *Note:* In the U.S.A. geological hierarchy, R. has the status of an epoch, forming the younger part of the Quaternary period.

recessional moraine A terminal m. which is the result of a brief pause in the retreat, or even of a slight re-advance, of an ice-sheet. Syn. with *stadial m.* E.g. as the ice-sheet of the last glaciation in N. Europe (the WEICHSEL) retreated, it paused at three stages, leaving the *Brandenburg*, *Frankfurt* and *Pomeranian* r. m.'s. [*f* ESKER]

reclamation Although derived from the vb. 'to reclaim', i.e. win back, the wide concept now implies any process by which land can be substantially 'improved' or made available for agriculture: (i) by drainage of temporarily waterlogged land due to seasonal flooding; e.g. in most river flood-plains; (ii) by drainage of marshes; e.g. the Fen District, Pontine Marshes of Italy; (iii) by drainage of lakes or a shallow part of the sea-floor; e.g. in the Netherlands, esp. the Zuider Zee scheme; see POLDER; (iv) by improvement of heathlands; e.g. in the Kempenland of Belgium, E.

Netherlands, W. Germany, W. Denmark; (v) by clearance of scrub-jungle, savanna; e.g. in the Dry Zone of Ceylon; (vi) by clearance of rain-forest for plantations of rubber; e.g. Malaysia, Java, Ghana; (vii) by irrigation; most hot deserts. Some would include only (i) to (iv) in the definition.

reconnaissance mapping A type of exploratory map, produced rapidly yet efficiently, using such techniques as cameras, phototheodolites, car and sun compasses, terrestrial photogrammetry, TELLUROMETERS, air photographs. The essence is speed and rel. cheapness in an area where detailed accurate surveys (which may follow the r. m.) are not yet available, as in deserts, polar areas etc. The map should nevertheless be accurate within certain limits, even if thin in some detail.

rectangular drainage A d. pattern in which tributary junctions are gen. at right-angles, and all streams, major and minor, exhibit sections of approx. the same length. This may be controlled by a rectilinear joint pattern. Ct. TRELLIS D. [*f*]

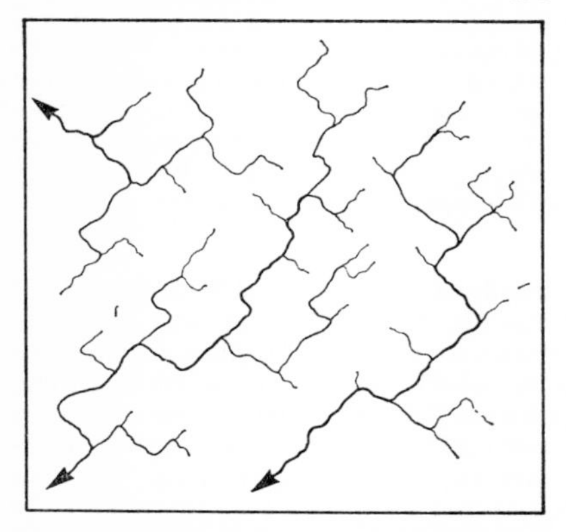

recumbent fold An OVERFOLD which has been forced over into a near horizontal position, with its axial surface nearly horizontal. [*f* OVERFOLD]

Red Clay A fine-grained PELAGIC deposit, consisting of hydrated silicate of alumina coloured by iron oxide, found on the ocean floor (in the ABYSSAL ZONE); it is derived from volcanic and meteoritic dust, material carried by icebergs, and insoluble relics of marine life. This is the most widespread of the pelagic deposits.

red earth A tropical zonal soil, produced by chemical weathering under high temperatures and humidity, with well defined seasonal rainfall. It consists of a loamy mixture of clay and quartz, coloured by iron compounds. This may be as much as 50 ft. thick; e.g. it covers extensive areas of the savanna lands in Brazil, British Guiana, E. Africa, S. Deccan, Ceylon, Burma, Viet-Nam. R. e. is superficially similar to LATERITE, which is more potently leached.

Red Mud A m. of terrigenous origin, found on the continental slope, and stained by ferric oxide.

reef (i) A mass of rock with its surface at or just above low-tide mark. (ii) Specif. a mass of CORAL; see CORAL R., ATOLL, FRINGING R., BARRIER R. (iii) A vein of metal or metal ore, esp. gold; e.g. on the Rand, near Johannesburg, Republic of S. Africa.

reef-flat A platform of coral fragments and coral-sand which has accumulated on the inner side of a coral r., on which dunes may collect and vegetation (esp. palms) establishes itself.

reef knoll A k. made of compacted material that originated as a r., exhumed or exposed by later denudation, so that it now stands out as a rounded or conical k.; e.g. along the S. or downthrow side of the Mid-Craven Fault, between Settle and Appletreewick, Yorkshire (High Hill, Scaleber, Burns Hill, Cowden Pike and at least five others). Further W., in the Bowland Fells, the K. Series of r.-limestones form a number of isolated k.'s rising approx. 600 m. (2000 ft.) above the gen. level (K. Hill, Crow Hill, Worsaw K., Sykes K., Twiston K.).

re-entrant (i) A prominent indentation into a landform, esp. in an escarpment where a transverse valley occurs. (ii) An angular inlet into a coastline.

reference net To facilitate the location of a place on a map, it is covered with squares, lettered along one side, numbered along the other. Thus a place may be in square P4. E.g. Ordnance Survey 1-inch series, 4th (Popular) edition, had 2-inch squares, but obviously this gives only the approx. location of a point within 4 sq. mi., and can refer only to a single sheet. The method is commonly used in the index or gazetteer of an atlas. Ct. GRATICULE, GRID.

reforestation, reafforestation The planting or restocking with trees of a former forested area (which may have been cleared or burnt); this forms an integral part of forestry practice.

reg (Arabic) A stony desert, esp. in Algeria, where sheets of smoothly angular, wind-scoured gravel cover the surface. Strong mineralized solutions drawn to the surface by capillarity evaporate to form a 'cement', which binds the gravel into a hard continuous sheet. See DESERT PAVEMENT.

regelation Pressure within an ice-mass converts ice-grains into molecules of water; these move to points where pressure is less (i.e. downhill) and then re-crystallize and re-freeze thus causing a gradual movement of and within the ice-mass; this is a contributory cause of glacier-flow.

régime (Fr.) (i) The seasonal fluctuation in the volume of a river (also more recently used of a glacier). (ii) *Climatic r.:* the seasonal pattern of climatic changes.

region A unit-area of the earth's surface differentiated by its specific characteristics. The theoretical grounds on which r.'s may be identified and delimited have been the subject of much discussion. There may be single feature r.'s, multiple-feature r.'s, 'total' r.'s (see COMPAGE), GENERIC, SPECIFIC, FORMAL, FUNCTIONAL, NATURAL, NODAL R.'S.

regional accounting The collection of statistics on a r. basis (e.g. for N. Ireland, a Fr. *département*, a Belgian province) in order to evaluate r. accounts (such as gross domestic product by industries and sectors, investment, unemployment), and thereby provide a sound economic base for the preparation of a r. plan and its implementation.

regional analysis A study of the potential for economic or cultural development within a region, by analysing its environmental and human resources and its situation with regard to other areas.

regional development Economic and cultural growth within a specific unit area. This is, to an increasing extent, stimulated, directed or even controlled by direct or indirect Government action. See DEVELOPMENT AREA.

regional geography The geographical study of a unit-area which reveals some degree of identity. In its gen. sense, r. g. is complementary to *systematic g.*; their basic relationship devolves from the fact that whereas systematic g. depends on analysis, r. g. is the product of synthesis, of integration. The region studied may be on any scale, from the Middle E., the Mediterranean Basin or the Monsoon Lands, to the Kempenland, Cumbria or Wirral.

regional hierarchy The ascending scale-pattern of the division of the earth's surface into r. units. J. F. Unstead (1933) used a h. of STOW, TRACT, sub-region, minor region and major region. D. L. Linton (1949) in his delimitation of MORPHOLOGICAL REGIONS, used SITE, STOW, TRACT, section, province, major division and continent. D. Whittlesey (1954) used locality, district, province and realm.

regionalism (i) A regional feeling, identity or group consciousness app. to a region. This can exist within a nation-state, though it may at times threaten its unity, even its existence, if it takes a political or nationalistic trend; e.g. the *Vlaamsch National Verbond*, founded in Belgium in 1935, with the aim of an autonomous

Flanders. Frequently it remains cultural, sometimes linguistic; e.g. the Bretons in Brittany. (ii) Used in the planning sphere to indicate a basic area for regional planning.

regional metamorphism See DYNAMIC M.

regional plan A comprehensive p. for a unit area within a country; this may be an economic planning region or an urban region; e.g. the N. Economic Region in the U.K., the 21 planning regions (*Circonscription d'Action régionale*) in France, the T.V.A. in U.S.A.

regional science An emerging and developing discipline, largely related to LOCATIONAL STUDIES and the social sciences, and making abundant use of such quantitative approaches as *econometrics*. This has revolutionized work by both economic and regional geographers.

register mark A small cross at each corner of a map which is to be printed in more than one colour. The accuracy of printing of each colour is checked by the coincidence of the crosses on the printed map.

regolith A mantle of more or less disintegrated, loose, incoherent rock-waste overlying the bed-rock, together with superficial deposits of alluvium, drift, volcanic ash, loess, wind-blown sand and peat. It also includes the soil layer. The term is derived from Gk. *regos*, a blanket, *lithos*, a stone.

regosol One of the AZONAL group of soils, derived from freshly deposited alluvium, dune-sands and mud-flats. The term is used as one of the two divisions of the Azonal, the other being the stony mountain soils (LITHOSOL).

regur (Indian) A tropical soil, found notably in the N.W. Deccan of India, formed by the weathering of basalt under conditions of heavy rainfall and high temperature; its blackness is due to its titanium content. It is characterized by a low organic content and commonly by a zone of calcium carbonate concretions. In the dry season it forms a black soil which crumbles into dust, or bakes hard if it has not been ploughed; in the wet season it may be plastic or sticky. It is used for cotton-growing, hence is known as the *Black Cotton Soil*. Very similar soils are found in Kenya (also called Black Cotton Soil), Morocco, N. Argentina, and in small areas of the W. Indies.

rejuvenation The revival of erosive activity, esp. by a river, because of: (i) a fall in sea-level; (ii) local movements of land-uplift, both resulting in a change in BASE-LEVEL. These initiate a new cycle, causing features such as knickpoints, terraces and incised meanders. R. also occurs without change in base-level, as when following river CAPTURE or an increase in precipitation, both of which may give increased discharge and therefore greater eroding power to a drainage system. [*f*]

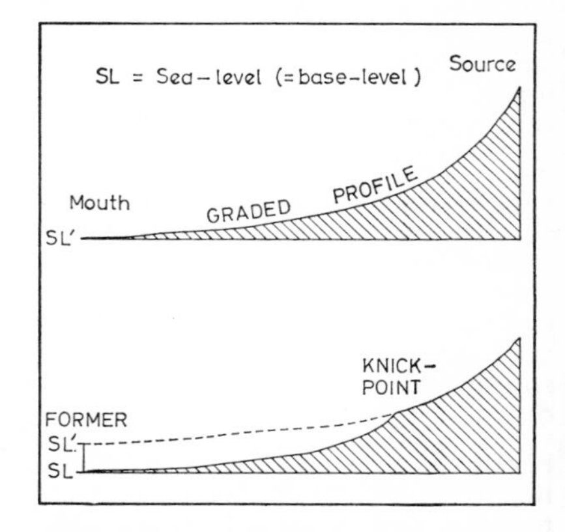

rejuvenation head See KNICK-POINT.

relative age, of rocks The determination of whether one r. is younger than another, thereby producing a tabulated list of the r.'s in order of age. (i) In all beds of sedimentary r.'s, the oldest is at the bottom and each overlying bed is progressively younger (*the law of superposition*), unless inverted by earth movements. (ii) Any sedimentary r. is younger than the fragments of which it is composed. (iii) R.'s containing the same fossil assemblages are similar in age.

(iv) An intruded igneous r. is younger than the r.'s which it cuts across (*the law of cross-cutting relationships*).

relative humidity The water-vapour present in a mass of air expressed as a percentage of the total amount that would be present were the air saturated at that temperature; or the ratio of the air's VAPOUR PRESSURE to the saturation vapour pressure. R. h. varies with: (i) ABSOLUTE HUMIDITY; e.g. saturated air (100%) at 20°C. (68°F.) contains 17·117 gm. per cu. m., but if the air at that temperature contained only 8·262 gm., the r. h. would be $\frac{8 \cdot 262}{17 \cdot 117} \times 100$ = approx. 48%; (ii) *temperature*; if a mass of air saturated (100%) at 4·4°C. (40°F.) is warmed, at 10°C. (50°F.) the r. h. falls to 71%, at 15·5°C. (60°F.) to 51%, at 32°C. (90°F.) to 19%

relative relief The relation of the altitudes of the highest and lowest points of land in any area, sometimes called *local r.*, or *available r.* The difference between the highest and lowest points is the *amplitude of r. r.* Various types of map have been devised to show this, usually dependent on gridding the area on a map, finding a value for the amplitude in each grid-square, and producing an ISOPLETH or dot map to depict the distribution of these values.

relict landscape While in a broad sense all l. features in the present may be said to be r.'s of the past, particular attention may be devoted to 'cultural r.'s', such as deserted village sites, strip LYNCHETS, disused canals, abandoned mines and factories (an aspect of INDUSTRIAL ARCHAEOLOGY), monuments (e.g. Stonehenge), water-meadows, ruins of fortifications. A 'GREEN BELT', a 'NATIONAL PARK', an 'area of outstanding natural beauty' may well by their very nature be preserved or 'fossilized' as r. l.'s. The study of r. l.'s, whether by using the present retrogressively as a key to the past, or by using the past as a key to understanding the present, is essentially part of HISTORICAL GEOGRAPHY.

relict (relic) mountain A surviving upland mass in an area of denudation (MONADNOCK, INSELBERG); an obsolescent term.

relief The physical landscape, the actual configuration of the earth's surface, used in a rather loose sense of differences in altitude and of slope, of inequality of surface, of shapes and forms of the surface. It should not be confused (as it often is) with TOPOGRAPHY, though the American usage of *topographic relief* is permissible.

relief map A m. which shows surface configuration by any of the following methods, some of which may be used in combination: (i) SPOT-HEIGHTS; (ii) CONTOUR-LINES and FORM-LINES; (iii) LAYER-TINTING; (iv) HACHURES; (v) HILL-SHADING (or PLASTIC R. or photo-r.); (vi) cliff- and rock-drawing; (vii) PHYSIOGRAPHIC PICTORIAL symbols.

relief rainfall See OROGRAPHIC PRECIPITATION.

relief model A reconstruction of the surface features of an area by means of a 3-dimensional m., usually with some VERTICAL EXAGGERATION, known in U.S.A. as a *terrain m.* A m. may be made from cardboard (each contour is drawn on card, cut around, and stuck over each other in exact position); plaster; a mixture of sawdust, plaster, paste and a little glue; sheet metal hammered to shape; vinylite and other plastics, pressed over a mould, or by a vacuum-forming process. The m. can be painted and lettered.

relief road A r. designed to remove a proportion of the traffic from some congested route.

rémanie (Fr.) Applied to materials in the earth's crust that have been (lit.) 're-handled'; e.g. the boulders in boulder-clay; pebbles in a conglomerate; masses of country-rock involved in the engulfment or emplacement of a batholith, caught up and solidified within a lava-stream; a fossil that has been removed by natural agencies of denudation from a bed and redeposited in another.

remote sensing The gathering, retrieval and storage of mass data by means of aerial survey, airborne electronic scanning devices, and increasingly by orbital satellites and spacecraft, these sensors operating at considerable distances from the source. While a limited amount of data can be processed manually (e.g. air-photo interpretation), the vast increase in data requires computer techniques for processing and analysis.

rendzina (Polish) An INTRAZONAL soil, dark coloured, with an A-horizon of friable, almost granular, loam, lying on a B-horizon containing chalk or limestone fragments, which in turn rests on the solid rock. It has developed where grassland formerly dominated, notably on the chalk Downs and Wolds. These soils were for centuries under pasture ('downland grazing'), but large areas now grow wheat and barley, others alfalfa. The soil and sub-soil are extremely permeable, and irrigation (in the form of overhead sprinklers) is widely used, esp. on the chalklands of S. England. R.'s are also found in U.S.A. in E. Texas and the 'Black Belt' of Alabama. Some pedologists refer to TERRA ROSSA as 'red r.'

replica line A l. drawn parallel to any specif. l. to indicate a zone of difference from it. E.g. the former Suez Canal Zone was indicated by a r. l. to the E. of it; the Panama Canal Zone is bounded both to E. and W. by r. l.'s.

representative fraction The ratio which the distance between 2 points on a map bears to the corresponding distance on the ground, expressed as a f.

R.F. 1/to	*Miles to 1 inch*	*Inches to 1 mile*	*Km. to 1 cm.*	*Cm. to 1 km.*
million	15·78	0·0634	10·0	0·1
633,600	10	0·1	6·336	0·1578
500,000	7·891	0·127	5·0	0·2
253,440	4·0	0·25	2·534	0·395
250,000	3·945	0·245	2·5	0·4
126,720	2·0	0·5	1·267	0·789
100,000	1·578	0·6336	1·0	1·0
63,360	1·0	1·0	0·6336	1·578
50,000	0·789	1·267	0·5	2·0
25,000	0·395	2·534	0·25	4·0
10,560	0·167	6·0	0·1056	9·468
10,000	0·158	6·336	0·1	10·0
2,500	0·0395	25·34	0·025	40·0
1,250	0·0198	50·69	0·0125	80·0

resection In making a PLANE-TABLE survey, the position of the observing station on the map can be fixed by drawing rays from observed points. The plane-table is orientated by compass before observations are taken. The three rays intersect to form a TRIANGLE OF ERROR, which can be eliminated to fix the position quite accurately.

[*f opposite*]

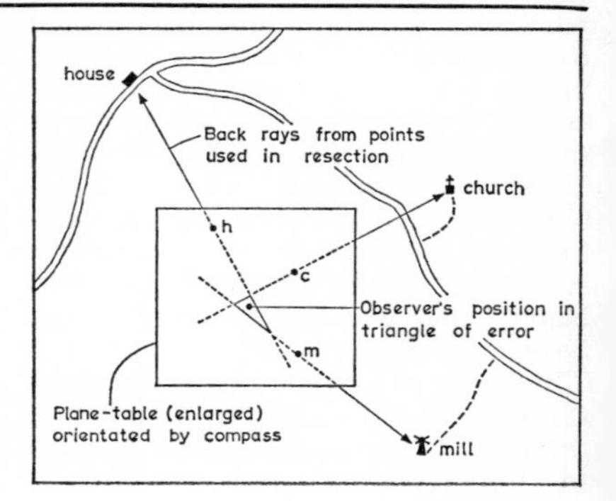

resequent drainage A pattern of d. in which a stream lies more or less along the line of a former longitudinal (synclinal) consequent stream in an area of ancient folding, following a long period of denudation. It is not a SECONDARY CONSEQUENT STREAM, as often defined. E.g. in the Hampshire Basin, the R. Ebble at present occupies a syncline in the Chalk; it may have migrated from its orig. synclinal position to follow a course along either the neighbouring Bower Chalke anticline or the Vale of Wardour anticline, and has now returned to its orig. line. R. streams are common in the Ridge and Valley section of the Appalachians.

resequent fault-line scarp A feature formed when erosion along an OBSEQUENT F.-L. S. produces an escarpment facing in the orig. direction of downthrow.

reservoir An area of water-storage for hydro-electricity production, domestic and industrial consumption, and irrigation, often artificially created by building a dam at a suitable retaining point across a valley. This is esp. necessary when the volume of a river is seasonal, with too much water (causing flooding) at one time of the year (e.g. by snow-melt, or a marked rainy season), and too little at others. A natural lake forms an effective r., though Man usually deepens it and increases its capacity with a dam; e.g. Haweswater and Thirlmere in the English Lake District; Laggan and Sloy in the Scottish Highlands. Some are wholly man-made; e.g. Vyrnwy and Elan in Wales; L. Mead (R. Colorado) and F. D. Roosevelt L. (Columbia R.), behind the Hoover and Grand Coulee Dams respectively; Kariba on the R. Zambezi.

residual deposit The residue of rock disintegration by weathering *in situ*. Hence *r. soil* (sometimes called *sedentary soil*), also formed *in situ*; e.g. Clay-with-Flints on the Chalk, TERRA ROSSA in hollows on limestone.

resistant rock A r. which can withstand weathering and erosion, because of its hardness, compactness, cementation, absence of jointing, and the nature of its chemical composition. This is usually considered in relative terms, as when denudation picks out resistant and unresistant r.'s by differential weathering and erosion.

response elements In a MODEL in physical geography, the r. e.'s are the features produced by the combined operation of a series of PROCESS E.'s. E.g. in a model of beach development, the r. e.'s will be the gen. form of the beach, its gradient, the size and distribution of the material of which it is composed, etc. All these may be analysed statistically.

resurgence The emergence of an underground stream from a cave, usually near the point where an impermeable stratum, underlying a rock such as limestone, intersects the surface. In limestone, water works its way through the mass of the rock, now vertically, now horizontally, finally issuing near its base. A r. is much larger than a spring, involving a considerable stream flowing strongly; e.g. the R. Axe out of Wookey Hole in the Mendips; the R. Aire from under Malham Cove; the Peakshole Water from Peak Cavern at Castleton in Derbyshire; the Echo R. in Kentucky emerging from the vast Mammoth Cave system to join the Green R., hence the Ohio. Syn. with VAUCLUSIAN SPRING.

retrogradation Esp. of shoreline studies, the erosion or cutting back of a beach by wave-action; hence a 'retrograding shore-line'. Ct. PROGRADATION.

reversed drainage See CAPTURE, RIVER.

reverse(d) fault A f. caused by compression, where the older beds on one side of a f.-plane are thrust over the younger beds on the other side; i.e. the HANGING WALL has been raised relative to the FOOT WALL, resulting in crustal shortening. One effect is that landslips are common along a r. f., because of the overhanging strata.

[*f* NORMAL FAULT]

reversing thermometer A mercury t. which has a constriction in its glass tube. When it is reversed or inverted, the mercury column breaks, leaving the reading of the previous temperature. This is used for taking temperatures at depth in the sea.

Rhaetic A series of shales, marls and limestones, sometimes ascribed to the top of the TRIASSIC system, now more usually regarded as transitional between the latter and the Lias at the base of the JURASSIC system.

rheidity In the study of the flowage of a solid, the relationship between its resistance to viscous flow (*viscosity*) and resistance to elastic deformation (*elasticity*). This has important aspects in the folding of rocks and the movement of ice in a glacier. R. obviously involves a time factor. A substance undergoing this deformatory flow is a *rheid*.

rheology From Gk. *rheo*, flow, the study of the flowage of materials, partic. that of plastic solids. See RHEIDITY.

rhinn Used in the W. part of the S. Uplands of Scotland for a rugged ridge; e.g. R.'s of Kells, R.'s of Galloway.

rhombochasm A parallel-sided gap (RIFT-VALLEY) in the sialic crust, floored by dense simatic material (though not basaltic lava), which has probably moved there through rock-flowage (RHEIDITY).

rhumb-line A l. of constant bearing, i.e. it cuts all the meridians at a constant angle; it is syn. with *loxodrome*. It is shown as a straight line on a MERCATOR projection.
[*f* GREAT CIRCLE]

rhymite Syn. with VARVE.

rhyne, rhine A drainage channel in the Somerset Levels, intermediate in size between the small ditches on the one hand and rivers such as the Parrett and Yeo or the larger artificial channels or drains (e.g. the King's Sedgemoor Drain), into which they discharge. These channels also serve as field boundaries.

rhyolite An igneous rock, consisting of alkali-felspars and quartz, commonly with some ferromagnesian minerals; it was extruded, and is fine-crystalled in structure. It may contain porphyritic crystals of quartz, and reveal distinct banding as a result of its flowing as a molten MAGMA (Gk. *rheo* = to flow). R.'s of Ordovician age are found in the English Lake District (e.g. the W. face of the Pillar Rock, Ennerdale), and in Snowdonia; e.g. the upper parts of the Glyders (the huge chaotic blocks on Glyder Fach are r.), and parts of Snowdon itself. Most of Yellowstone National Park, Wyoming, consists of a r. plateau; the rock is here distinctly yellowish, hence the name.

ria (Sp.) A funnel-shaped coastal indentation formed by submergence as a result of a rise in sea-level affecting an area where hills and river valleys meet the coast at right-angles. It decreases in width and depth as it runs inland. The stream which flows into its head, responsible for eroding the orig. valley, is obviously too small for the present size of the inlet. E.g. in N.W. Spain (the R. de Vigo, de la Coruña, del Ferrol); S.W. Ireland (Dingle Bay, Kenmare R., Bantry Bay); W. Brittany (Rade de Brest, Baie de Douarnenez). The term is often used more widely for the submergence of any land margin which is dissected more or less transversely to the coastline; e.g. the S. coast of Devon and Cornwall (the Tamar estuary, with Plymouth Sound, and the Fal estuary, with Carrick Roads).
[*f*]

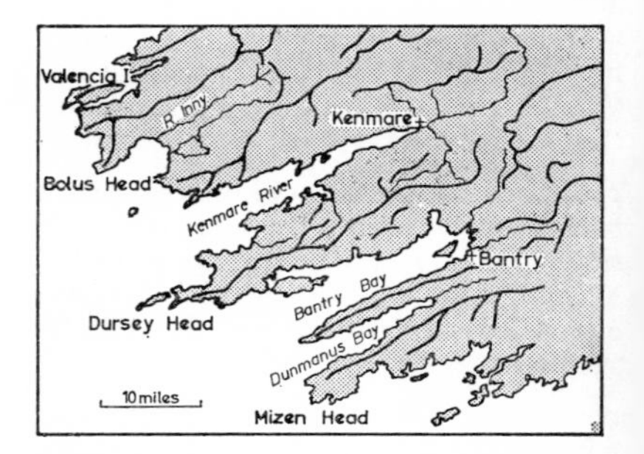

ribbon development The building of houses along each side of roads radiating from a town, esp. in Britain in the inter-war years.

rice A cereal (*Oriza sativa*) grown mainly in monsoon and equatorial climates, but also as far N. as N. Italy, S. France, S. Spain and Arkansas in U.S.A. It provides more food per acre and feeds more people than any other cereal. The yield has been raised still further by the development of new hybrids; e.g. *Japonica* in Japan. C.p. = China, India, Japan, Pakistan, Indonesia, Thailand.

Richter Scale The currently used s. of earthquake magnitude, which has replaced the Rossi-Forel and Modified Mercalli S.'s of earthquake intensity. The R. S. was devised and published in 1935 by C. F. Richter, seismologist at the California Institute of Technology. The s. is based simply on instrumental records, making allowance for the distance from the EPICENTRE. The s. assigns the largest numbers to the largest earthquakes, from 0 to over 8·0; 8·9 has been recorded on 3 occasions. It cts. with the Mercalli S. which is based on observed effects on buildings and people.

ridge A long narrow upland, with steep sides, but the term has no very specif. application.

ridge, of high pressure An elongated region of high atmospheric p. between two areas of low pressure, rather wider than a WEDGE. It brings fine, though short-lived, weather, often within a general period of rather rainy conditions. 'The borrowed day, too good to last'.

ridge and furrow The remarkably persistent pattern of r.'s and f.'s often visible in an area of permanent grassland formerly under the plough.

ridge and valley A type of relief characterized by a close pattern of nearly parallel r.'s and v.'s; the type-region is the R. and V. region in the Appalachians (E. U.S.A.), lying between the Allegheny-Cumberland plateau on the W. and the Blue Ridge on the E. The r.'s consist of resistant sandstones, quartzites and conglomerates, the v.'s of weaker shales and limestones.

Riding One of the three administrative divisions of Yorkshire: N., W. and E.

Ried (Germ.) A marshy flood-plain, specif. of the Rhine, consisting of marshland and backwaters, with damp pasture, clumps of willows, alders and poplars, often flooded in spring and early summer. Though much has been drained, there are still considerable remaining areas in the section of the Rhine valley between Basle and Strasbourg.

rift-valley A narrow trough between parallel FAULTS, with THROWS in opposite directions, so forming a long steep-sided, flat-floored v. There may be a series of step-faults on either side of the trough, or the sides may be clean-cut as a result of the downthrow along a single major fault on either side. Its exact origin is a matter of argument: (i) *Tension* in the crust may have pulled the 2 sides apart, leaving the centre to subside. (ii) *Compression* from either side may have thrust the masses on either side higher than the central block, which may also have been forced down. (iii) There may have been a gentle upbending of the strata, so that a gaping crack developed along the crest of the swell. E.g. the line of the Jordan V. (the floor of the Dead Sea is 750 m., 2500 ft., below the surface of the Mediterranean Sea), Gulf of Akaba, Red Sea, Abyssinia, E. Africa to the Zambezi, length 4800 km. (3000 mi.); the middle Rhine r.-v. between the Vosges and the Black Forest; the Midland V. of Scotland between 2 boundary faults. Ct. GRABEN. [*f*]

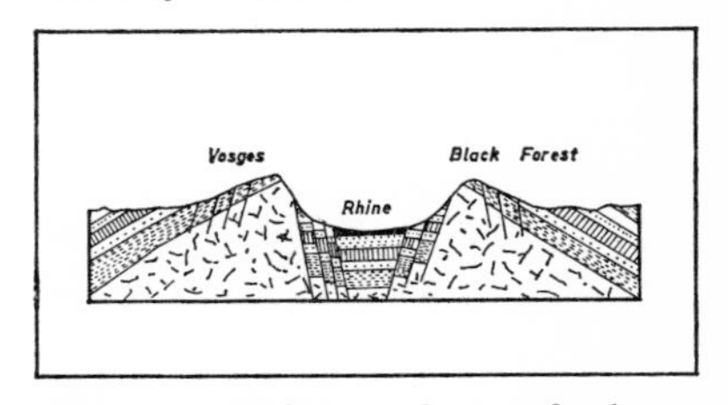

right ascension One of the 2 references (the other being DECLINATION) which enable a heavenly body to be exactly located on the Celestial Sphere, equivalent to longitude for terrestrial reference. For all heavenly bodies the r. a. and declination are tabulated in the *Nautical Almanac*. The lines of r. a., known as *hour-circles*, pass through the Celestial Poles, and cut the Celestial Equator and all parallels of declination at right-angles. R. a. is reckoned E.ward, starting from the point of the spring EQUINOX on the Celestial Equator, known as the 'First Point of Aries' (♈); this is analogous to the 0° Greenwich meridian. It was so called because when this point was first chosen 2000 years ago, it was situated in the constellation of Aries. Since then, the point has moved along the Celestial Equator (see PRECESSION OF THE EQUINOXES) into the next constellation (Pisces), though it is still termed the First Point of Aries. The r. a. is therefore the arc of the Celestial Equator intercepted between the First Point of Aries and the hour-circle through the body, measured in time E.ward from 0 to 24 hours. [*f*]

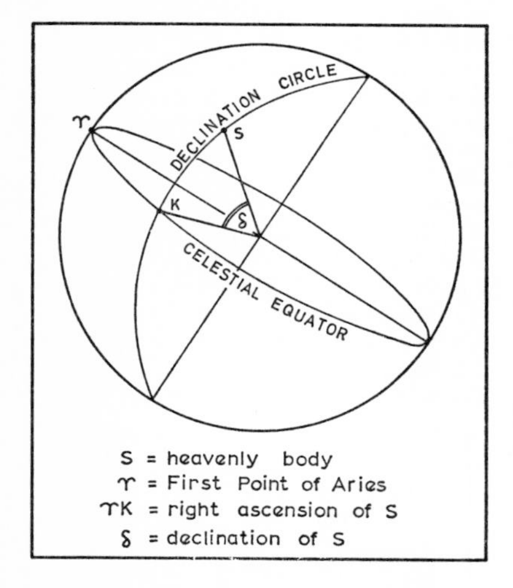

rill erosion The uneven removal of surface soil by the formation of numerous small-scale r.-channels, a process known in U.S.A. as *rilling*. If this continues, the r.'s will coalesce into larger gullies (*concentrated wash*).

rimaye (Fr.) The Fr. form of the Germ. BERGSCHRUND.

rime An accumulation of white opaque granular ice-particles, formed when a fog composed of super-cooled droplets is driven by a slight wind against objects the temperatures of which are below freezing, such as telegraph poles and wires, trees and rock buttresses. When the particles coalesce, they are popularly known as 'frost-feathers'.

ring-city See RANDSTAD.

ring-dyke A dyke in a zone surrounding, in more or less arcuate form, a circular or dome-shaped igneous intrusion. The MAGMA forming the intrusion seems to have exerted pressure upward and outward, forming fractures; these are filled with the magma, which solidifies as d.'s. If the fractures are vertical in section, r.-d.'s are formed. E.g. in Mull, Skye, Arran, Ardnamurchan and N. Ireland. Ct. CONE SHEET.

ring road A r. which encircles a built-up area, as either an 'inner r. r.' designed to serve the town centre itself, or an 'outer r. r.' designed as a bypass.

rip Turbulence and agitation in the sea, caused by: (i) the meeting of two tidal streams; (ii) a tidal stream suddenly entering shallow water; (iii) the return of water piled up on the shore by strong waves, esp. when they break obliquely across the line of a longshore current. Hence ***r.-tide***, ***r.-current***. [*f*]

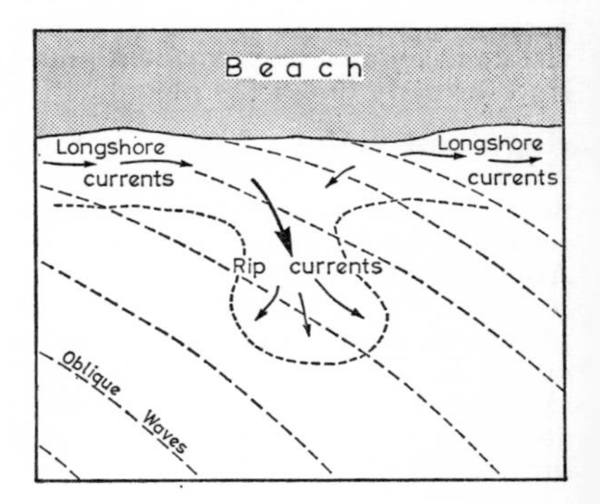

riparian App. to the banks of a river, as 'r. rights' or 'r. states'.

ripple marks The familiar patterns on sand or mud, made by wind, waves or tidal currents, which have been preserved on a surface of deposition, now in the form of compacted rock (sandstone, shale), covered and preserved by sediments, and later 'exhumed' by the removal of the sediments to reveal the orig. patterns.

rise An elongated, gently sloping elevation rising from the sea-floor, though its crest lies far from the surface; e.g. the Dolphin Rise in the N. Atlantic Ocean.

Riss The 3rd of the four main glacial periods of the Quaternary glaciation, identified and named by A. Penck and E. Brückner in 1909 in the Alpine Foreland, where they recognized distinctive fluvioglacial gravels on the High Terrace, associated with moraines and drumlins nearer the main Alpine ridges. The R. is equated with the *Saale* in N. Europe, and with the *Illinoian* in N. America. The R. was probably the period of max. glaciation. [*f* MINDEL]

river A gen. term for water flowing in a definite channel towards the sea, a lake, a desert basin, a main r., a marsh. The longest r.'s are:

	km.	*mi.*
Nile	6649	4132
Amazon	6276	3900
Mississippi–Missouri	6111	3860
Ob	5570	3461
Yangtse	5520	3430
Hwangho	4672	2903
Congo	4667	2900
Amur	4509	2802
Irtysh	4421	2747
Lena	4269	2653
Mackenzie	4240	2635
Mekong	4184	2600
Yenisey	4132	2566
Parana	3942	2450
Volga	3690	2293
St Lawrence	3057	1900
Rio Grande	3034	1885

For the main features associated with r.'s, whose names are prefaced with r. (e.g. *r.-cliff*, *r.-terrace*), see under specif. term.

river profile The outline of the shape of a r. valley, which may be: (i) *longitudinal*, from source to mouth, the *long-p.;* (ii) *transverse*, across the valley at right-angles to the r., the *cross-p.* See also THALWEG and PROFILE. [*f* KNICKPOINT]

riviera (It.) An area along a coast, esp. where popular tourist resorts have developed; e.g. the Italian R., French R. It is usually characterized by extensive sandy beaches.

road metal Broken angular fragments of stone, usually hard, which are quarried and used for road surfacing; e.g. Penmaenmawr granite (N. Wales); Mount Sorrel granite (Leicestershire); basaltic rocks of the Great Whin Sill (Northumberland).

roadstead, roads An anchorage outside a harbour, off a coast, or in an estuary or bay, with some degree of protection against wind and heavy seas; e.g. Cowes Roads, Carrick Roads. The Fr. form is *rade*; e.g. Rade de Brest.

Roaring Forties The uninterrupted ocean S. of latitude 40°S., where the N.W.–W. (or 'Brave West') winds blow with great strength and constancy; an area of gales, stormy seas, overcast skies, and damp, raw weather associated with a constant W. to E. procession of low pressure systems.

robber economy (from Germ. *Raubwirtschaft*) The removal or extraction by Man of various resources offered by the earth. Although the term is used in a wide sense to include mining and quarrying, it implies more specif. the rapid and ruthless destruction of resources (which with care could be maintained) for immediate profit, with no thought of the future; e.g. forests, fish, whales, seals, soil fertility.

roche mountonnée (Fr.) A glacially-moulded mass of rock, with a smooth, gently sloping, rounded upstream side (the result of abrasion), and a steep, rough, irregular downstream side (the result of plucking). It was orig. so called because of its fancied resemblance to wigs smoothed down with mutton grease, common at the end of the 18th century. They are common features in most glaciated areas; e.g. on the sides of the Pass of Llanberis, N. Wales. [*f*]

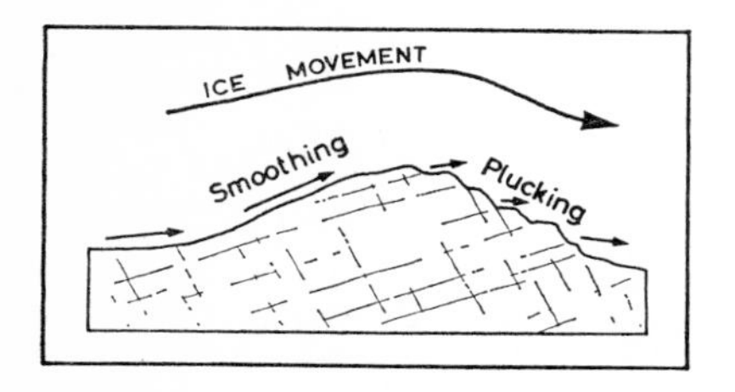

rock An aggregate of mineral particles, forming part of the lithosphere. In popular use, it is something hard, consolidated, compact and massive, but in the correct sense the term includes sand, gravel, mud, shells, coral and clay. R.'s may be classified: (i) *by mode of formation*, including (*a*) IGNEOUS; (*b*) SEDIMENTARY; (*c*) METAMORPHIC; and (ii) *by age*, i.e. a tabulation according to their relative ages, so forming a convenient time-chart or chronology. See STRATIGRAPHY. The term is used as a place-name where a r. is a striking feature; e.g. R. of Gibraltar, Spider R. (Arizona), Ship R. (New Mexico).

rock-drumlin A smooth elongated hummock of boulder-clay, but with a mass of rock as its 'core', round which the material is plastered.

rock-fall A free fall of massive rocks or individual boulders down a steep mountain side. Ct. ROCK-SLIDE.

rock-flour The fine r.-material produced by the grinding effect of glaciers. A stream flowing from a glacier-snout is usually milky white because of its suspended load of r.-f. So too are lakes into which glacial streams flow; e.g. the S.E. end of L. Geneva; L. Louise in the Canadian Rockies.

rock-salt Also known as *halite.* A clear, white, grey, yellow or brown mass of sodium chloride, crystallized in cubic form, sometimes with calcium and magnesium sulphates and chlorides. It was formed in Europe esp. during the desert conditions of the Permian and Triassic periods, on the dried-up beds of inland seas and lakes; e.g. near Stassfurt (E. Germany); near Cracow (Poland); in Cheshire; and very widely in U.S.A. It forms with brine the main source of commercial and industrial salt. See SALT, COMMON.

rock-slide A mass of r. which slides *em masse* down a gentle hill-side over a bedding-plane or a fault-plane; e.g. in the Madison R. valley, S.E. Montana, on 18 Aug., 1959, when a vast r.-s. ('triggered-off' by an earthquake) blocked the valley, so ponding up a lake. A section of the Dorivarz mountain slid into the valley of the Zeravshan R. on 26 April, 1964, in Soviet Uzbekistan, the result of an earthquake, and ponded up a lake whose waters threatened Samarkand; engineers lowered the natural dam and partly diverted the river.

roddon A raised bank in the Fen District of E. Anglia, probably the LEVEE of a former river built up by deposition of silt. It was emphasized by the compaction, oxidation and general lowering of the peat-lands between the rivers. Thus a r. may stand above the level of the general land surface, whereas once it formed the levee of a stream.

roller A popular word for an ocean swell, the result of a long FETCH, which forms immense breakers on coasts exposed to it, even in calm weather; e.g. on the Atlantic islands (St. Helena, Ascension, Tristan da Cunha).

rond See BROAD.

roof pendant A mass of country rock which projects downwards into the top of an igneous intrusion; e.g. older rocks penetrating a BATHOLITH, and completely surrounded by igneous rocks.

root A term used in a tectonic sense in connection with NAPPE structures in highly folded mountain ranges, where the 'core' of the recumbent fold appears to turn steeply downwards. Hence in the Alps the 'r. zone' of the great nappes of the Bernese Oberland and the Pennine Alps lies in N. Italy, each individual vertical nappe r. being separated by metamorphosed sediments.

ropy lava See PAHOEHOE.

Rossi-Forel Scale A s. of intensity of earthquake shocks, devised in 1878 by M. S. de Rossi and F. A. Forel, and used until 1931, when the modified MERCALLI SCALE was introduced. In its turn the latter has been superseded for many purposes by the RICHTER SCALE.

rotational slip A downhill movement *en masse* on a s.-plane of solid material, such as rock or ice, which seems to pivot about a point. The mass is left with a marked back-slope facing uphill. E.g. the Warren, Folkestone. [*f*]

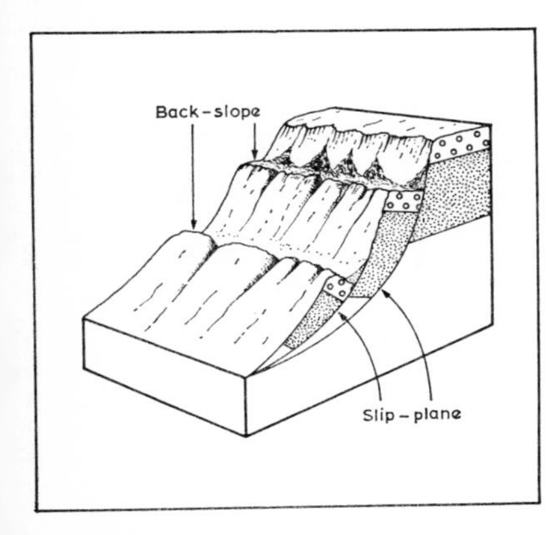

rotation of crops The cultivation on a piece of land of a different c. annually over a period of three, four or more years, to avoid the exhaustion of partic. minerals in the soil and the increase of specif. pests, and by growing certain legumes to enable atmospheric nitrogen to be fixed in their root-nodules, hence in the soil. The earliest form of r. was simply grain one year, grass the next, but by the early 18th century the 'Norfolk r.' of wheat—roots—barley—legumes (clover or beans) had been evolved.

rotation of the earth The movement of the e. on its polar axis from W. to E.; this results in an apparent daily r. of the sun, moon and stars from E. to W. The period of r., as measured in terms of two successive transits of a meridian by a partic. star, is 23 hours, 56 minutes, 4·09 seconds; this is a SIDEREAL DAY. The average interval of time between two successive transits of the sun across the meridian (i.e. the period of r. of the earth relative to the sun) is 24 hours; see MEAN SOLAR TIME, EQUATION OF TIME. At the equator, the velocity of the e.'s rotation is about 1050 mi. per hour, at 60°N. and S. about 525 mi. per hour, at the Poles zero. To find the velocity per hour in any latitude, divide the length of any parallel ($2\pi R. \cos \theta$) by 24.

rubber, natural A substance derived from latex, the milky fluid which issues from cuts in the bark of certain species of trees, the most valuable being *Hevea brasiliensis*, originating in the Amazon basin, now grown on plantations. C.p.=Malaysia, Indonesia, Thailand, Ceylon, Nigeria.

rubber, synthetic A substitute for natural rubber, manufactured mainly from hydrocarbons derived from petroleum by the process of polymerization. The U.S.A. and Germany were pioneers in this field, and now the former manufactures 80% of the free world's supplies. Production totals about two-thirds of that of natural rubber.

rudaceous rocks A category of sedimentary r.'s of coarse texture, both compacted (BRECCIA, CONGLOMERATE, TILLITE) and unconsolidated (SCREE, GRAVEL, BOULDER-CLAY).

run Used in the E. part of U.S.A. for a small stream and its valley; e.g. Bull R., Virginia.

rundale (Irish) (i) The joint occupation and cultivation of a number of separate non-contiguous holdings or patches of land in Ireland. Some of this land is commonly redistributed among the joint holders at regular intervals. Syn. with *runrig* in Scotland. (ii) An area of upland pasture grazed in common by the flocks of a number of individuals.

runnel See SWALE.

running mean A statistical procedure which endeavours to smooth out irregularities in a time-series of values (e.g. of annual precipitation, population, agricultural output), thus producing a moving average, which may be plotted for each time-interval on a graph. E.g. in a series $x_1 \ldots xn$, on a 5-year moving average, the value for year 3 would be $\frac{x_1+x_2+x_3+x_4+x_5}{5}$, while the value for year 4 would be

$$\frac{x_2+x_3+x_4+x_5+x_6}{5}$$

runoff The surface discharge of water from rainfall or snow-melt down a slope, also including the surface discharge of underground streams. The proportion of the precipitation that becomes r. depends on: (i) the slope; (ii) the nature of the rock and soil; (iii) the presence or absence of a vegetation cover; (iv) the rate of evaporation. On a steeply inclined bare rock surface, r. will comprise 100% of precipitation; on a nearly level surface of sand, r. may be virtually nil.

rural population In popular use, people living in the country, as opposed to the urban population (living in the towns). Specif., people living in an administrative unit of a certain defined size or p., according to the country. In France, U.S.A. and Japan, people in communities with populations below 2000, 2500 and 10,000 respectively are classified as rural. In England, people living in a Rural District are so classified.

rural-urban continuum A conceptual dichotomous model used in sociological studies, in which a community under consideration is placed at an appropriate point on an imaginary line joining two 'polar' types, truly rural and wholly urban, by a study of its sociological characteristics. The validity of this concept has been challenged.

rural-urban fringe A zone of transition around a town in which u. functions, uses and activities are mixed with agricultural ones. This zone separates the area exclusively devoted to u. uses from that exclusively devoted to agricultural uses.

rurban Introduced into the literature of urban geography, app. to the indeterminate area between town and country, the 'rural-urban continuum'; i.e. the RURAL-URBAN FRINGE.

rye A very hardy cereal (*Secale cereale*), which will resist adverse climate, disease and pests, and will grow in poor sandy soils. It was formerly consumed widely by peasant-farmers ('black bread'), but now has largely been replaced by wheat; it is still grown for fodder and straw, and to some extent for food. C.p. = U.S.S.R., Poland, W. Germany, E. Germany, Turkey, Czechoslovakia.

ryot An Indian tenant-cultivator.

Saale The early part of the 3rd glacial period in N. Europe, corresponding to the Riss in the Alps and the Illinoian in N. America, occurring about 150,000 years ago. The S. ice-sheet extended further S. than those of other glacial periods; its. S. limit is represented by a discontinuous series of low sand-hills extending from near Utrecht in the Netherlands to near Krefeld in Germany. It is known as the *Drenthian* glaciation in the Netherlands.

saddle A broad flat col in a ridge between two mountain summits; e.g. the Saddle between Dodd and Red Pike, Buttermere, English Lake District.

saddle-reef A lens-shaped mass of ore-bearing rock between the beds near the axis of an ANTICLINE.

saeter, seter (Norwegian) (i) An upland pasture in Norway. (ii) A farm high up in the mountains of Norway used only in summer, following snow-melt.

sagebrush A scrub vegetation in semi-desert areas, of greyish heath-like shrubs (*Artemisia* spp.), esp. in the Great Basin of Utah, Arizona, Nevada and Mexico.

St Elmo's Fire A small electrical brush discharge, seen mainly at night playing around the masts and spars of ships, with the appearance of luminous flames. It is esp. common in the DOLDRUMS. The same type of brush discharge is experienced in high mountains, usually making a hissing sound and tingling a climber's skin; this is rarely visible (though the writer has seen a friend's head outlined luminously). It is a sign of stormy weather and is usually associated with the passage of a FRONT.

sakiyeh (Arabic) A cumbersome irrigation device for lifting water from a river on to land at a higher level. It consists of a dipper at the end of a beam on a pivot, which is dipped into the river, swung round, and the water tipped on to the land, using an animal-powered cog-wheel system.

salar A basin of inland drainage in S.W. U.S.A., usually containing a salt-lake or a salt-flat; syn. with PLAYA.

salina (Sp.) A PLAYA with a high salt content; a term used in the deserts of S.W. America.

saline soils A group of INTRAZONAL s.'s characterized by a considerable proportion of salts, esp. of sodium. They occur widely wherever there is strong evaporation, both in the hot deserts and in the cool temperate continental interiors with high summer temperatures. Strong s. solutions rise by capillarity, and the salts form a greyish surface-crust, below which lies a granular salt-impregnated horizon. It is syn. with *solonchak*. E.g. in the Great Basin of Utah, esp. around the Great Salt Lake; around the Caspian Sea; in the Tarim Basin. Many of the salt-crusts are the result of the shrinkage and drying-out of former extensive salt-lakes in basins of inland drainage; s. s.'s occur around their margins where there is enough silt and clay to call it a s. Such a s. is of little use for irrigation, unless it can be desalinized; e.g. some areas around the Caspian Sea now grow cotton. Constant irrigation can cause an increase in the salinity of s., necessitating expensive systems of drains through which water can be periodically flushed.

salinity In sea-water, the proportion of dissolved salts in pure water, stated in parts per thousand by mass. The mean figure for the seas as a whole is 34·5 parts per thousand, written 34·5‰. In terms of actual salts, this is made up of sodium chloride (23), magnesium chloride (5), sodium sulphate (4), calcium chloride (1), potassium chloride (0·7); minor ingredients include salts of bromine, strontium, carbon, boron, silicon, phosphorus, fluorine and many others, including 'trace elements' of great importance to marine plants and animals. S. can also be expressed in terms of the actual elements; on an average, sea-water contains 18·98 grammes per kilogramme of chlorine (about 55% of the total salt content), 10·56 of sodium (31%), 1·27 of magnesium, 0·88 of sulphur, 0·40 of calcium, and 0·38 of potassium. The proportions of the constituents of sea-water remain very constant from place to place, despite changes in total s. The standard method of determining s. is by precipitating the halides by adding silver nitrate. Surface s. varies according to temperature (causing evaporation and concentration), supplies of additional fresh water from rivers, rainfall, melting ice and snow (therefore causing dilution), and the degree of mixing by surface and sub-surface currents. In the open ocean, differences are small, ranging from about 37‰ near the tropics, 35‰ near the equator and in mid-temperate latitudes, to 34‰ towards the Poles. In partially or wholly enclosed seas the variation is much greater; the Red Sea is about 40‰ (high summer temperatures and evaporation, few inflowing rivers), the Baltic off Bornholm only about 8‰ (low evaporation, numerous inflowing rivers).

salt, common (sodium chloride) (*NaCl*) This occurs in solution as brine, as solid sheets and crusts around the margins of s.-lakes (e.g. the Dead Sea, Great Salt Lake of Utah, Aral Sea), in s.-domes, and as deposits esp. in the Permo-Triassic rocks (e.g. in Cheshire, Stassfurt in E. Germany, Texas and neighbouring U.S. states). See also ROCK-S. C.p. = U.S.A., China, U.S.S.R., U.K., W. Germany, India, France, Canada.

saltation The process by which solid material moves along the bed of a stream in a series of hops. The term has also been used for similar movement of sand grains in deserts.

salt-dome (sometimes called a **salt-plug**) A roughly circular mass of solid s., varying from 90 m. (100 yds.) to 1·6 km. (1 mi.) in diameter, but extending vertically to great depths, sometimes as much as 13 km. (8 mi.). It was forced up by slow flowage from more deeply buried deposits of rock-salt. It is commonly associated with deposits of gypsum, anhydrite and petroleum, and usually appears to be crowned with a limestone cap-rock.

salt-flat A horizontal stretch of s.-crust, representing the bed of a former s.-lake, temporarily or permanently dried up; e.g. the Bonneville S.-f.'s W. of Salt Lake City, where motor speed-trials are held as the surface is so level and firm; around L. Eyre in Australia.

salting The slightly higher areas of a SALT-MARSH (though still sometimes inundated by tides), where grass is present and there is little bare mud.

salt-lake A highly saline lake, located in a basin of inland drainage, in an area with high temperatures and high evaporation rates. Water entering brings in some saline material, which is left when the water evaporates; e.g. Great Salt Lake (Utah) (220‰); Dead Sea (238‰); L. Van in Asia Minor (330‰).

salt-lick A natural occurrence of s. on the ground (e.g. on a large ranch), where animals go to lick. On an English farm, a block of s. may be placed in the fields by a farmer for the same purpose.

salt-marsh A coastal marsh found along a low-lying shore, usually enclosed by a shingle-bar or a sand-spit, or in the sheltered part of an estuary. Fine silt and mud are deposited by the tides in backwaters, often added to by alluvium brought down by rivers. Vegetation gradually spreads, and helps the process of accretion: eel-grass (*Zostera*), marsh samphire (*Salicornia*) and rice-grass (*Spartina townsendii*). These plants form increasingly dense communities, which help to trap silt; first hummocks of vegetation, then more continuous areas develop, and the whole surface is raised naturally, while the tidal waters flow in increasingly restricted channels. Gradually other plants establish themselves, and the marsh may turn into a SALTING, esp. if Man helps the process by dyking or building wicker-work fences. E.g. along the Norfolk coast; the marshes of the rivers of S. Suffolk and Essex; Romney Marsh behind Dungeness (now reclaimed);

the Solway marshes; and the extensive areas between the Frisian I'.s and the mainland coasts of the Netherlands, W. Germany and Denmark (known as *Watten* or *Wadden*).

salt-pan A small basin containing a s.-lake, surrounded by and lined with a solid deposit of s.; e.g. the SHOTTS on the plateaus in the Atlas Mtns.

samun, samoon (Persian) A warm, dry descending wind in Persia, of the same nature as the FÖHN.

sand Small particles, mainly of quartz, with a diameter of between 0·02 and 2·0 mm. The general category is sometimes subdivided into: (i) *fine* (0·02 to 0·2 mm.); (ii) *medium* (0·2 to 0·5 mm.); (iii) *coarse* (0·5 to 1·0 mm.); (iv) *very coarse* (1·0 to 2·0 mm.). The term is also applied to soils which consist of more than 90% s.

sandbank An accumulation of sand in the sea or river, usually exposed at low water.

sandr, sandur (Icelandic) An OUTWASH PLAIN of sand.

sandstone A sedimentary rock, consisting mainly of grains of quartz, often with felspar, mica and a number of other minerals, consolidated, cemented and compacted into a rock. S. can be classified according to the 'cementing' material which binds together the individual grains into: (i) *calcareous*; (ii) *siliceous*; (iii) *ferruginous*; and (iv) *dolomitic* types. The colour varies from dark brown or red through yellow to grey and white, mainly due to the iron content and its degree of oxidation or hydration; some s.'s have a greenish shade as the result of the presence of glauconite or reduced iron compounds. The sands were laid down before compaction in: (*a*) shallow seas; (*b*) estuaries and deltas; (*c*) along low-lying coasts; (*d*) in hot deserts. The s.'s are commonly laminated and sometimes show FALSE-BEDDING, though some are FREE-STONES. S. is widely spread in space and through the geological record; e.g. Torridon S. (Pre-Cambrian) in Wester Ross; Bala S. (Ordovician) of Caer Caradoc in Shropshire; Old Red S. (Devonian) of Herefordshire, S. Wales, central and N.W. Scotland; various sandstones in the Millstone Grit (Upper Carboniferous); Coal Measure S. (Upper Carboniferous) in most coalfields; New Red S. (Permo-Triassic) of N.W. England (St. Bees S.), the S. part of the I. of Arran; Bunter and Keuper S.'s (Triassic) of the Midlands, Cheshire, Merseyside; the Hastings S., Lower and Upper Greensand (Cretaceous) of the Weald; various rocks of Eocene age (Thanet S., Bagshot S.) in the London and Hampshire Basins; various younger Tertiary sands.

sandstorm A storm in a desert or semi-arid area, in which the wind carries clouds of sand, usually near the surface and rarely above 15 m. The erosional effect of sand in a storm may be very potent, akin to a 'sand-blast'.

Sangamon interglacial The 3rd i. period in N. America, corresponding to the Riss-Würm in the European Alpine Foreland.

Sanson-Flamsteed Projection (also known as the **Sinusoidal P.**) An equal area p., on which the equator is taken as the standard parallel and drawn as a straight line to scale $=2\pi R$, and the central meridian is also drawn as a straight line to scale $=\pi R$; each is divided truly (i.e. by 36 for 10° intervals). Parallels are drawn as straight lines through the points of division on the central meridian; each is made its true length $=2\pi R \,.\, \cos\theta$, and divided truly (for 10° intervals $=\dfrac{2\pi R \,.\, \cos\theta}{36}$). The meridians are sine-curves drawn through the corresponding points on each parallel. It can be used for a world-map, but extensive 'shearing' occurs in high latitudes because the meridians are very oblique to the parallels. It is useful for maps of the S. continents, esp. when they fall about the central meridian, since shape is good. The p. may be 'INTER-

RUPTED', and it is used in the GOODE'S HOMOLOSINE P. for the section in tropical latitudes, the rest being drawn on MOLLWEIDE. The S.-F. P. is in fact a partic. case of BONNE, with the equator as the standard parallel. [*f*]

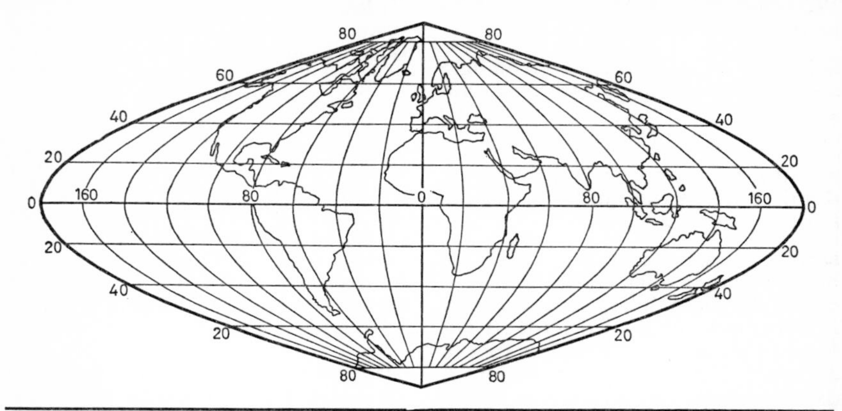

Santa Ana (Sp.) A hot, dry wind blowing from the N. and N.E., descending from the Sierra Nevada across the deserts of S. California; it is a wind of FÖHN type, though it is often laden with dust. It can do much damage to fruit orchards through the desiccation and withering of buds or blossom.

sapping (i) Syn. with glacial PLUCKING. (ii) See SPRING-SAPPING.

saprolite A residual weathered cover, the result of the rotting of the bedrock *in situ*. This is partic. evident in warm humid climates, where hydrolysis is effective, and here it may be as much as 90 m. thick. LATERITE and REGUR are both soils of saprolitic origin.

sapropel, hence adj. **sapropelic** A sludge or ooze, which collects in freshwater swamps and in shallow sea-basins, rich in organic matter. The decomposition and putrefaction of the organic content is believed to be the basis of origin of petroleum and natural gas, esp. when the organic matter is derived from fatty algae. The word is derived from the Gk. *sapros* (rotten) and *pelagos* (sea).

saprophyte A plant (such as a fungus) that lives on decaying vegetable matter. Ct. EPIPHYTE.

sarn (Welsh) A kind of natural 'causeway' (though there are claims that it is actually of artificial origin), notably in Cardigan Bay where there are at least 5. Formed of loose stones and boulders, sections are exposed at very low ebb-tide; e.g. about 14·5 km. (9 mi.) of the total length of 34 km. (21 mi.) of S. Badrig. It is obviously the result of a rise of sea-level, possibly inundating a low watershed of boulder-clay, washing away the finer materials, later developing like a storm-beach as more material accumulates.

sarsen A large block of siliceous sandstone, found mainly on the chalk Downs, the remnants of a now vanished cover of rocks of Tertiary age, probably Eocene. The precise constituents of a s. vary from place to place. They may be either relatively homogeneous or conglomeritic. They are often concentrated in isolated valleys in England, esp. in Wiltshire. They have been used widely as building stone, as gate-posts, and in prehistoric times in the construction of megalithic monuments; e.g. Stonehenge (partly) and Avebury. In N. France they have been even more economically important in an area otherwise devoid of building stone; e.g. in the construction of *pavé* roads. They are often called 'greywethers', because of their likeness to sheep when seen from a distance.

sastruga, pl. **sastrugi** (Russian) A sharp ridge on a snow-field or ice-field, caused by the scouring and furrowing erosion of wind laden with ice-particles. These are partic. evident to polar travellers, since they interfere with the smooth running of a sledge, partic. if the journey lies transverse to the general pattern (a result of the direction of the prevailing winds).

satellite town A t. which is associated with a major city; many of the world's large t.'s have several smaller ones clustering around them, and in part dependent on them. The s. t. lies within the functional orbit of the larger city and is dominated by it. E.g. Charleroi in Belgium had a population in 1961 of 25,605, but with its satellite towns of Jumet, Gilly, Montignies, Marcinelle and Marchienne-au-Pont, its total was 300,000.

saturated (or **wet** or **moist**) **adiabatic lapse-rate** The rate of decrease in air temperature by dynamic cooling, the result of expansion within a vertically ascending s. 'parcel' of air. This rate is less than the DRY A. L.-R. because of the release of latent heat. The actual rate varies with the amount of water-vapour condensed, which in turn varies with the amount of water-vapour present, which depends on the temperature. It can lie anywhere between about 0·4°C. and 0·9°C. per 100 m. Thus an air-mass of about 27°C. (80°F.) may contain so much water-vapour, and therefore releases so much latent heat, that the s. a. l.-r. may be as low as 0·4°C. per 100 m. (2°F. per 1000 ft.). In a very cold air-mass there may be so little water-vapour that the s. a. l.-r. differs little from the dry a. l.-r.

saturation, saturated (i) The state of the atmosphere when it can hold no further water-vapour; i.e. as many molecules of water enter the air as leave. The amount of water-vapour that can be held by the atmosphere varies with its temperature and pressure. Thus if s. air is cooled, a certain amount of water-vapour may condense. However, for condensation to occur nuclei are required, unless the water-vapour molecules coagulate under very high VAPOUR PRESSURE. If nuclei are not present, the atmosphere may become SUPERSATURATED at temperatures below the theoretical DEW-POINT. S. must always be considered with respect to the physical condition of given 'parcels' of air. (ii) A rock which is holding the max. amount of water in the interstices of its mass is said to be in a state of s.; hence the *zone of s.* See WATER-TABLE. (iii) The term is used in another sense by petrologists to indicate minerals which can crystallize out from magma in the presence of an excess of silica; e.g. felspar, mica, amphiboles and many more. If a rock is over-saturated, the excess of silica occurs in the form of free quartz.

saturation deficit The amount of water-vapour required to bring non-saturated air at a given temperature and pressure to the point of saturation. Syn. with *vapour-pressure d.*

savanna, savannah, savana Open tropical grassland, with tall grass, and scattered trees and bushes usually of a XEROPHILOUS character. In some areas the grass forms a discontinuous cover, separated by bare ground; it may be only 0·3 m. high, or as much as 4 m. (e.g. elephant grass, a stiff yellow straw crowned with silvery spikes). Patches of low thorny scrub are common. Clumps of trees—various palms, baobab, ceiba, euphorbia and acacia—grow in hollows where ground-water approaches the surface. Many of the trees are wedge- or umbrella-shaped as a result of the strong winds. As the desert border is approached, the grass becomes shorter, more tufted, with more bare sand between clumps; equatorward the trees become taller and more numerous, merging into light forest and then rain-forest. The vegetation pattern is the result partly of a marked precipitation régime, partly of soil conditions, partly of extensive fires; during the dry winter, the grass is parched and the trees bare, but with the rains in summer there is a short-lived luxuriance. The s. covers a large area in Africa, in S. America on the Guiana Plateau (*llanos*) and on the Brazilian Plateau (*campos*) to N. and S. of the rain-forest, and in the N. of Australia.

scabland An American term for a type of landscape in parts of the Columbia-Snake Plateau in the N.W. of U.S.A., where the basalt surface was extensively eroded by glacial floodwater, and the bare rock has been exposed or covered with angular debris derived from its own disintegration; there is a thin soil cover, and scanty vegetation. Ct. the Badlands (S. Dakota), in areas of eroded sedimentary strata.

scale (i) The proportion between a length on a map and the corresponding length on the ground. It may be: (*a*) expressed in words; (*b*) shown as a divided line; and (*c*) given as a REPRESENTATIVE FRACTION. (ii) A place-name element used in the N. of England, esp. in Cumberland, derived from the Middle English *schele*, a shepherd's summer shed, or Old Norse *skala;* e.g. Winscale, Portinscale, S. Hill. The same element is in Scafell, Scawdale. It is syn. with the Scottish SHIELING.

scalogram technique A method of quantitative analysis in REGIONAL SCIENCE, esp. where statistical information is limited, providing '. . . a means of ordering ranked data in such a way that a single unidimensional scale is produced, along which effective measurement is possible' (D. Timms). A number of observations are reduced to a single dimension, producing a 'scale score' which summarizes the attributes of an area.

scar A rock-face, partic. in N. England, where it refers to a limestone cliff, in some cases outcropping across country for some distance; e.g. Attermire S., Langcliffe S., Gordale S. One of the massive resistant bands in the Carboniferous Limestone is known as the Great S. Limestone.

scarp, scarp-slope, scarp-face The steep slope of a CUESTA [*f*]; ESCARPMENT is preferable.

scarp-foot spring A s. that breaks out at or near the foot of an ESCARPMENT, esp. where chalk lies on clay (S. Downs), or limestone and sandstone lie on clay (Cotswolds).

scarp retreat See PARALLEL RETREAT OF SLOPES.

scarth A bare rock-face, esp. in the English Lake District; e.g. S. Gap above the Buttermere valley.

scatter-diagram A d. giving a graphic indication of the amount of correlation between 2 sets of statistical data, one set plotted as ordinates, the others as abscissae. If when the values are plotted they tend to be grouped along a diagonal line, some degree of correlation is manifest.

Schattenseite (Germ.) The shady side (N.-facing in the N. hemisphere) of a deep valley; syn. with UBAC. It is characterized by thick growth of conifers in the Alps, in ct. to settlements, terrace cultivation and pasture on the S.-facing sunny side. Ct. SONNENSEITE.

schist A medium-grained rock which has been affected by regional metamorphism, causing re-crystallization; it usually has a foliated, sometimes a wavy, texture. Flakes of such 'platy' minerals as mica are usually visible. This texture is quite independent of the bedding-planes of the original rock. There are varieties of s.; e.g. quartz-s., metamorphosed from sandstone; hornblende- and biotite-s. from basalt and gabbro; mica-s. (the Fr. *schistes lustrés*) from phyllite.

Schuppenstruktur (Germ.) The Germ. name for IMBRICATE STRUCTURE [*f*], though used quite commonly in English geological literature.

scirocco See SIROCCO.

sclerophyll An evergreen tree or shrub with small hard leaves, hence *sclerophyllous*. Found mainly in lands with a Mediterranean climate, with long hot, dry summers, during which the leathery leaves resist TRANSPIRATION. E.g. OLIVE, cork oak (*Quercus suber*), Holm oak (*Q. ilex*), Aleppo pine (*Pinus halepensis*), lavender (*Lavandula latifolia*).

scoria (i) A coarse clinkery mass of volcanic rock, of a slaggy nature. It is usually basic in composition, fine-grained but cellular in texture, hence *scoriaceous* as a result of the former abundant gas-bubbles and steam-blisters. It results from the rapid cooling of the surface of a lava stream which contained much gas and steam. (ii) The accumulation of similar clinkery material which has been blown out of a volcano as PYROCLASTS.

'Scotch mist' A fine drizzling rain, occurring among the British hills, when clouds lie at or near to the ground.

scour (i) The powerful erosive effect of a tidal current, removing deposits on the sea-bed close inshore. The importance of tidal s. in forming gentle submarine erosion surfaces is becoming increasingly apparent. It is now known that a tidal current of 4½ knots is capable of scouring shingle at a depth of 59 m. off Hurst Castle Spit in Hampshire, while 'sand waves' have been observed 164 m. down on the CONTINENTAL SHELF off S.W. England. A tidal s. may be felt partic. in the mouth of a bottle-necked estuary; e.g. the Mersey. (ii) The powerful and concentrated erosive effect of a river current; esp. on the outside curve of a bend. Concrete training-walls are built to direct the s. so as to assist in the regularization of a shipping channel in a river; e.g. along the Rhine below Basle; the Danube.

scree Slopes of angular rock-debris on a mountainside, of all sizes, lying at an angle of rest of *c.* 35°, which remains remarkably uniform. The material is mainly formed as the result of frost action, hence it occurs most strikingly at the foot of steep rock buttresses, on which frost weathering is potent; sometimes a distinct s.-cone tapers outwards from the base of a buttress or the foot of a rocky gully. In U.S.A. the word TALUS is also used, sometimes syn., though by other authorities s. is defined as all loose material lying on a hill slope, while talus accumulates specif. at the base of cliffs. Some refer to s. as the ingredient, talus as the whole slope or feature.

scribing In cartography, a modern method of drawing a map before printing. A sheet of glass or transparent plastic is coated with an opaque medium, into which is cut all the detail to be shown, using a 'scriber' or 'graver', of which various types are used; some have sapphire tips or tungsten-steel points. S. is done either as a negative, so that when laid on the sensitized metal printing plate a positive is produced, which by offset printing will produce correct-reading copies, or as a positive.

scrub A vegetation association in a semi-arid climate, or on poor sandy or stony soils, characterized by stunted trees, bushes and brushwood. The s. may be: (i) tropical and semi-desert type (MULGA, SPINIFEX, CHAÑARAL, ACACIA); (ii) warm temperate type (MAQUIS, CHAPARRAL, GARIGUE, MALLEE, BRIGALOW, SAGEBRUSH). The term is also used loosely of any rough vegetation on heathland. The plants are mainly XEROPHILOUS in character, including cacti, thorny aromatic shrubs, small gnarled evergreens, saltbushes, mesquite, creosote and sharp spiny grasses.

scud Fractostratus; tattered, ragged masses of cloud, driven along by the wind beneath the main cloud-layers of nimbostratus; gen. it is a bad-weather cloud, and a sign of wind and storm.

sea (i) A gen. term for the salt waters of the earth's surface, as in 'land and s.'. (ii) A proper name for any specif. area of water, usually on the margins of continents, in ct. to an ocean; e.g. North S., Mediterranean S., China S. A few large inland bodies of water are also so-called; e.g. Dead S., Caspian S., Aral S.

sea breeze A local b. blowing from the s. during the afternoon towards a low pressure area (produced by heating and convectional uplift) over the land, esp. in equatorial latitudes. It is felt for only short distances inland, but it considerably ameliorates the stagnant, humid heat and freshens the air. It blows only during periods of calm, settled weather when not masked by the Trades. Ct. LAND B. [*f*]

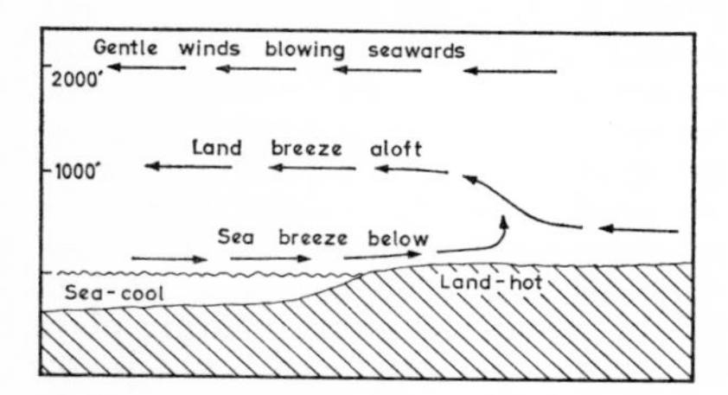

sea-level See MEAN S.-L.

sea-marsh See SALT-M.

sea-mile See MILE.

seamount An isolated peak rising from the floor of an ocean, the summit of which is 900–1800 m. (sometimes as much as 3000 m.) below the water surface. They are esp. numerous in the Pacific Ocean, where about 10,000 (including GUYOTS) are estimated. (A s. has a prominent crest, whereas a guyot is flat-topped.) Both are usually volcanoes, formed by sub-oceanic eruptions. See SUBMARINE RIDGE [*f*].

seascarp An ESCARPMENT on the floor of the ocean, the result of a large-scale faulting movement, notably off the W. coast of S. America; e.g. Mendocino, Murray, Clarion and Clipperton S.'s.

season One of the distinct periods into which the year may be divided, partic. in terms of the duration of daylight and of climatic conditions, as a result of changes in the duration and intensity of solar radiation. If the s.'s are defined astronomically, spring is from Mar. 21 (when the apparent sun is at the First Point of Aries; see ZODIAC) to June 22 (at the First Point of Cancer); summer is from June 22 to September 23 (at the First Point of Libra); autumn is from September 23 to December 22 (at the First Point of Capricorn); and winter from December 22 to March 21. Because the earth's orbit is elliptical and the orbital rate of movement changes, the s.'s are of unequal length. For the N. hemisphere, spring = 93 days; summer = 94 days; autumn = 90 days; winter = 89 days. In Europe a 4-fold division, essentially reflecting the life cycle of cultivated plants, is usual: dormancy (winter), sowing (spring), growth (summer), harvest (autumn). These periods can seldom be precisely defined in terms of climatic factors, and the main contrast is between the extremes of summer and winter in terms of daylight hours and temperatures, usually taking the months of July and January as indicative. The low latitude s.'s may be differently defined; e.g. in India by rainfall and drought.

seat-earth Given by miners and others to the layer of material on which coal-seams rest; this is a fireclay, and is used for making fire-resistant bricks. The roots of the swamp-forests from which coal has been derived grew in this layer, and their fossil remains are commonly found.

seaway A SHIP-CANAL; an inland waterway which can take sea-going ships; e.g. the St. Lawrence S. 220 km. (140 mi.) long, with a min. depth of 8 m. (27 ft.).

Secant Conic Projection An inaccurate name sometimes given to the C. P. WITH 2 STANDARD PARALLELS. Strictly a secant is a straight line cutting the circumference of a circle at 2 points, and a true S. C. P. would have its 2 standard parallels separated by this secant distance, whereas the conic with 2 standard parallels actually has its 2 standard parallels separated by the arcuate distance. [*f, page 311*]

second (i) 1/60 of a minute of arc in the measurement of latitude, longitude, and angles gen. (ii) 1/60 of a minute, 1/3600 of an hour, in the measurement of time.

Secondary A term used for the second of the eras of time and corresponding group of rocks. It is syn. with Mesozoic, which has gradually superseded it in gen. use.

secondary consequent stream A tributary to a SUBSEQUENT STREAM, flowing parallel to the main CONSEQUENT STREAM. A. s. c. s. is gen. thought of as being initiated after the formation of subsequent streams, but in a direction consistent with that of the original consequent proper. In areas of folded rocks, it is also a stream draining the flanks of an ANTICLINE, leading into a synclinal depression, which is followed by a *primary* or *longitudinal c.*

[*f* SUBSEQUENT STREAM]

secondary depression A small area of low atmospheric pressure on the margins of a main d. It may be just a bulge in the isobars, or it may be an individual system with closed circular isobars. It may be more intense, with lower pressure and associated with more stormy, rainy weather than the primary one. The s. d. seems to travel around the primary, in the N. hemisphere in an anti-clockwise direction.

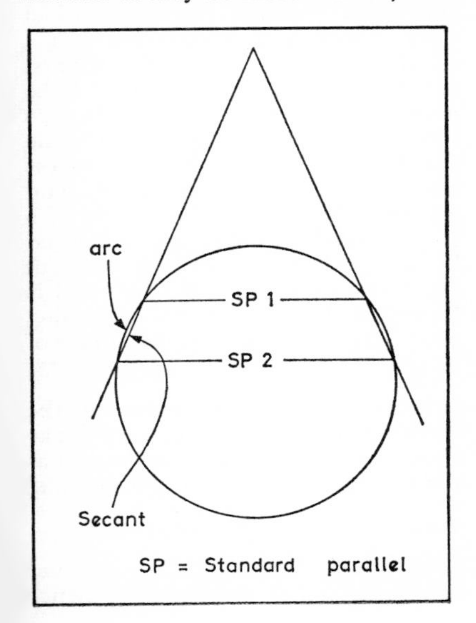

secondary industry The working up or fabricating of materials derived from PRIMARY I. into manufactured articles, including building and public works construction.

section (i) A vertical cut through soil and/or rock, either natural or artificial, or a representation of such a cut. The term is used only when the details of the underlying strata are shown; the surface outline alone is properly known as a PROFILE (though SOIL PROFILE seems to contradict this). Geological s.'s may be: (*a*) *diagrammatic* (small-scale and generalized); (*b*) *semi-diagrammatic* (with an accurately plotted surface profile, but with diagrammatic representation of the strata); and (*c*) *accurate*, constructed with no vertical exaggeration which would falsify the dip of the strata. (ii) In U.S.A. a unit, 1 mi. sq., representing 1/36 of a township in the LAND SURVEY SYSTEM.

secular An adj. implying a long period of time; i.e. a s. change is a very slow, virtually imperceptible one.

sedentary agriculture A. as practised in one place by a settled farmer; strictly, virtually all a. is now s., but the term was used of primitive agriculturalists in Africa who farmed the same land indefinitely, in ct. to SHIFTING A.

sedentary soil A s. formed *in situ* from the underlying parent rock, in ct. to one derived from transported parent material. This term is not much used now by pedologists.

sediment Deposited particles or grains of rocks. Sometimes the term is extended to include all residual and detrital material laid down by rivers, wind, ice and the sea. Hence *sedimentation*, the process of deposition of s. In its broadest sense, sedimentation includes a consideration of the rocks from which the s. is derived, the processes involved in the transportation of the fragments, and the actual settling of the material, whether: (i) in the deep calm water of an ocean basin or a GEOSYNCLINE; (ii) in a shallow marginal sea; (iii) under deltaic conditions; (iv) on a river FLOOD-PLAIN; (v) in a lake; or (vi) in a desert.

sedimentary rock A r. consisting of sediments, laid down in layers and/or 'cemented'. The main types are: (i) *mechanically formed* (CLASTIC): (*a*) arenaceous (sand, sandstone, conglomerate, grit); (*b*) argillaceous (mud, clay, mudstone, shale); (*c*) rudaceous (breccia, conglomerate, tillite, scree, gravel, boulder-clay); (ii) *organically formed:* (*a*) calcareous (coral limestone, crinoidal limestone, shelly limestone); (*b*) ferruginous (ironstone); (*c*) siliceous (diatomaceous earth); (*d*) carbonaceous (peat, brown-coal, lignite, cannel coal, bituminous coal, anthracite); (iii) *chemically formed:* (*a*) carbonates (travertine, dolomite); (*b*) silicates (sinter, flint, chert); (*c*) ironstone (limonite, haematite, siderite); (iv) *formed by desiccation: evaporites*; (*a*) sulphates (anhydrite, gypsum); (*b*) chlorides (rock-salt).

sediment discharge rating In hydrology, the ratio between the total discharge of a stream and the discharge of s. carried by it.

'seeding' of clouds The stimulation of CONDENSATION and then PRECIPITATION by dropping from aircraft particles of 'dry ice' (solid carbon dioxide), silver iodine, volcanic dust, etc. into supercooled c.'s. These can sometimes be seeded from the ground by coke-burning braziers from which silver iodine suffuses upwards. The disadvantage is that conditions must in any case be suitable for precipitation before s. is effective as a 'triggering' agent, and the proportion of successful attempts is not high, though much research is proceeding. S. of c.'s seems most practicable on high plains adjacent to mountain ranges (e.g. in the Midwest of U.S.A.), where orographic ascent causes SUPERCOOLING.

seepage (i) The slow sinking of surface water into the soil and subsoil. (ii) The oozing-out of water along a fault or joint-plane, though not moving strongly enough for a spring. (iii) A 'show' of mineral oil at the surface, often a guide to prospectors. *Seep* is also used as a vb., and in U.S.A. as a noun ('a seep').

segregation (i) In an ecological sense, the spatial separation of people or institutions into distinct areas. Whereas ecological concentration or centralization implies the separation of functions that operate in relation to the whole area, s. stresses the separation of one urban function from all other functions. The opposite situation, the breakdown of s., is referred to as *ecological desegregation*. (ii) See APARTHEID.

seiche A short-term standing-wave oscillation in the surface of a lake, the result of changes in atmospheric pressure and wind direction. The oscillation may be periodic, determined by the physical characteristics of the enclosing basin. Occas. a s. may be caused by seismic forces; e.g. an earth tremor.

seif-dune (Arabic) A longitudinal dune, forming a steep-sided ridge, often many mi. in length, aligned across the desert in the direction of the prevailing wind, which sweeps through the depressions between the parallel lines. The sand supply is less than in the case of BARKHANS and extensive 'sand-seas'. The origin of a s.-d. is not clear, but it may be formed by the coalescence of lines of small crescentic dunes, so that the wind is funnelled between them, sweeping away their 'tails', and leaving a pair of ridges. [*f*, *page 313*]

seismic focus The point in the earth's crust at which an earthquake shock originates; this is also known as the s. origin. [*f* EARTHQUAKE]

seismic wave See TSUNAMI.

seismology The scientific study and interpretation of earthquakes; hence *seismologist; seismograph* (a pendulum-based instrument for recording seismic waves); *seismogram* (the record thus obtained); *seismometer* (a delicate instrument for receiving seismic impulses).

selective logging The felling of mature trees at intervals during the growth of a mixed forest, so as to make the max. use of it. This is opposed to complete clearance ('clean l.') at one time. S. l. is not desirable for such trees as Douglas fir, which thrive best in stands; when they reach maturity, complete clearance is carried out.

selva (Portuguese) The TROPICAL RAIN-FOREST in the Amazon Basin, and thus used gen. for any area of such vegetation.

semi-artesian well A w. driven through the overlying impermeable beds in a structural basin, in which the water rises under hydrostatic pressure, but does not reach the surface; e.g. at Bovington Camp, Dorset, a bore was driven through the Tertiary beds into the Chalk to a depth of 221 m., and water rose to within 28 m. of the surface of the ground.

semi-desert, semi-arid climate A transition zone of climate and vegetation between savanna grassland and true desert, and further from the equator between the true desert and the Mediterranean vegetation. It also occurs in continental interiors; e.g. in parts of the Great Basin and the Colorado Plateau of U.S.A. The s.-d. is characterized by scrubby XEROPHILOUS bushes, coarse grass and bare patches of sand.

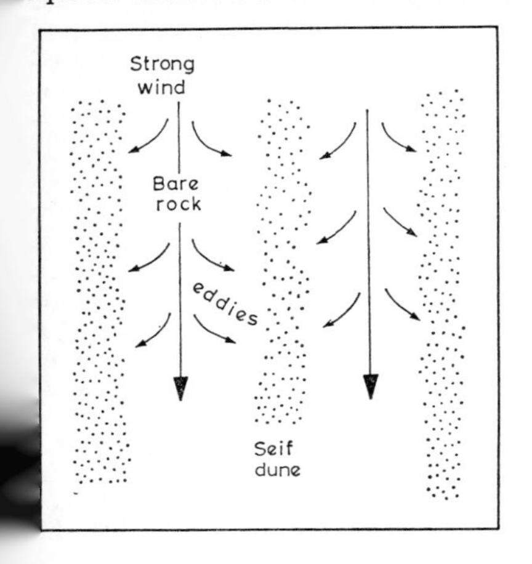

semi-diurnal tide Two tidal rises and two falls occurring during each lunar day at intervals of approx. 12 hours, 25 minutes; each high t. attains more or less the same level, as does each low t. [*f* TIDE]

semi-natural vegetation A plant-association that is 'wild' in that it has not been planted by Man, yet owes its characteristics to his direct or indirect influence. Much of what was formerly thought to be natural v. has been shown to be s.-n.; e.g. if rain-forest is cleared and then neglected, it will be rapidly covered by secondary forest. SAVANNA may be the result of the destruction of forest by SHIFTING CULTIVATION, HEATH the result of the clearance of light deciduous forest.

send, of waves See SWASH.

senile river A r. in the 'old-age' stage of its cycle of development. This concept is derived from W. M. Davis's CYCLE OF EROSION.

sensible temperature The sensations of heat and cold felt by the human body, involving not only temperature, but also humidity and wind. A wet bulb thermometer reading gives an accurate indication; if it reads much above 24°C. (75°F.), sustained manual labour is difficult. Damp raw cold is much more trying than dry cold; very low t.'s can be endured when the air is dry and there is no wind. Dry heat mitigates high t.'s by accelerating evaporation from the skin surface.

sequential landform In some classifications of l.'s, these are divided into INITIAL and s.; the latter include those where modifications by subsequent denudation are so pronounced that only vestiges remain of the initial structural forms.

sequent occupance D. Whittlesey's term for a succession of 'pictures in time' of the changing geography of an area. E.g. in parts of U.S.A.: (i) the Indian hunter and French trader (before 1840); (ii) the pioneer trapper and frontier farmer (1840–80); (iii) the stock farmer and sportsman fowler (1880–1900); (iv) the corn belt farmer and river resorter (since 1910). (Alfred H. Meyer, of the Kanakee Marsh in N. Indiana.)

sérac (Fr.) A pinnacle in an ice-fall on a GLACIER, the result of intersecting CREVASSES.

seral community A plant COMMUNITY which forms a stage in the development of a CLIMAX COMMUNITY, and which is itself only temporary, affording conditions more favourable for colonization by successively more demanding communities.

sere A developmental sequence of a plant COMMUNITY; it may be a *hydrosere* (in water), a *xerosere* (originating under dry conditions), a *lithosere* (on a bare rock surface), or a *psammosere* (on sand).

sericulture The rearing of silk-worms and the production of raw silk, as in Japan, China, Syria, N. Italy and the S. of France.

series A stratigraphical unit, forming a division of a system, and corresponding in the time-scale to an EPOCH; e.g. Carboniferous Limestone is a s. within the Carboniferous; the Chalk is a s. within the Cretaceous.

serir (Arabic) The stony desert of Libya and Egypt, where sheets of angular gravel cover the surface; it is syn. with REG in Algeria.

serozem (Russian) See GREY EARTH.

serpentine A common rock-forming mineral of complex chemical composition, containing hydrous magnesium silicate (*chrysotile*), related to asbestos. It is of variegated shades of green, sometimes with streaks of red, and often with attractive markings resembling the skin of a serpent, with a curious greasy or soapy feel. It sometimes occurs in large masses, as near the Lizard Peninsula, Cornwall, where it is worked for ornamental purposes.

servicing industries A category of occupations involving communications, transport, trade, finance, the professions, administration and personal service.

set (i) The direction of a wind, tide or current. (ii) A clearly definable collection of things or items, known as the *elements* or *members* of that particular s., listed individually (finite) or defined collectively (infinite). The use of s.'s is an integral part of modern mathematics and is increasingly used in geography.

set theory The use of SETS forms a basic aspect of modern mathematics, forming a common vocabulary for the different branches of the subject. In recent years s.'s have been increasingly used in various forms of geographical generalization, esp. in examining analogous attributes and situations. E.g. s. t. has been used in an attempt to integrate some of the alternative approaches to GEOGRAPHY [*f*], which can be regarded as lying variously within the earth sciences, the social sciences and the geometrical sciences (α, β and γ respectively), each being viewed as a s. Overlap and intersection of 2 sets and 3 sets indicate the complexity of the definition of geography, its approaches and constituent material. Thus the interesection of the α and γ s.'s (written as $\alpha \cap \gamma$) indicates geomorphometry and surveying; $\alpha \cap \beta$ = human ecology; and $\beta \cap \gamma$ = location analysis.

settlement (i) Any form of human habitation, usually implying more than one house, though some would include a single isolated building; s. may be rural or urban. (ii) The opening up, colonizing and 'settling' of a hitherto unpopulated or thinly populated land, esp. of immigrants to a 'new' country.

sextant An instrument used for measuring angles, esp. in navigation for obtaining the altitude of the sun or a star. The observer looks through the eye-piece, and can see both the horizon and the mirrored image of the sun at the same time, then reads off the angle.

shade temperature Air t. as indicated by a thermometer sheltered from the sun's rays, from radiation from neighbouring objects, and from precipitation. This sheltering may be achieved by using a STEVENSON SCREEN. This t. is the one quoted in climatic records and textbooks, unless 'sun t.' is specif. used.

shaduf, shadouf, shadoof (Arabic) A device for raising water from a well, consisting of a long pole with a weight at one end and a bucket at the other, mounted and hinged on a vertical pole in the ground; man-power is used; ct. SAKIYEH. The bucket is dipped into the well, and it may be lifted, swung round, and emptied into a trough or channel with very little effort by virtue of the counterpoised weight. This device is used in many parts of the world.

shakehole A steep-sided hole or depression in the ground, found commonly in limestone, sometimes in Millstone Grit, usually the result of the collapse of an overlying cave-roof, leaving a mass of debris in the bottom of the hole. It is usually dry, as ct. a SWALLOW-HOLE, which has a stream descending it. The more accepted scientific term is a *collapse doline*, or in U.S.A. a *collapse sink*.

shake-wave An *S*-wave in an earthquake shock, similar to a light-wave, where each particle is displaced by the wave at right-angles to its direction of movement. It is also known as a *transverse wave*. [*f* EARTHQUAKE]

shale A sedimentary rock formed from deposits of fine mud, compacted by the compression of the overlying rocks, and characteristically laminated in thin layers which easily split apart. The lamination is gen. that of the bedding. The name is used widely in geological proper names; e.g. Ludlow S. and Wenlock S. of Silurian age; Yoredale S. of Carboniferous age; Alum S. of the Upper Lias of Jurassic age.

shamal (Arabic) A strong, dry, dust-laden N.W. wind, blowing across the plains of Iraq in summer, producing a hazy sky.

share-cropping A type of agricultural tenancy, still found in France (*métayage*), whereby the owner of the land supplies seed and fertilizer, and takes part of the produce.

shatter-belt A zone in which the rock has been broken through earth-movements into angular fragments (FAULT-BRECCIA) along the line of a FAULT (or faults) which is (are) not clean-cut. It forms a distinct line of weakness on which weathering and erosion can concentrate.

shearing The cracking and breaking of a rock at a fault or thrust-plane through compression or tension, with resultant slipping; hence *shear-fault*, *shear-plane*. In effect, the volume of the rock remains the same, but its form is altered as the 2 adjacent portions slide past each other. Sometimes actual crushing and shattering may take place along the line of s.

sheet erosion, sheet wash SOIL E. in which the soil is gradually removed by surface runoff in a thin s. over a large area, as it moves down gentle slopes. It is a slow process, far less spectacular than other forms of soil erosion, but cumulatively it is very serious.

sheetflood, sheetflow Water flowing down a slope in thin continuous sheets, rather than concentrated into individual channels. Such flow occurs before RUNOFF is sufficient to promote concentrated flow, and also after torrential rainfall, when the existing rills cannot carry the increased runoff.

sheeting The process of the splitting-off of shells of rock in a massive, little jointed rock, probably the result of the outward expanding force of PRESSURE RELEASE, causing curved rock-shells to pull away from the mass. This causes a doming or rounding effect; e.g. the granite domes of the Yosemite Valley, California.

sheet lightning The effects of a flash of l. in a cloud discharge or cloud-to-cloud discharge, where the cloud causes the illumination to be diffused in a sheet-like appearance.

shelf A ledge or platform, esp. CONTINENTAL S. Hence *shelf-sea*, syn. with EPICONTINENTAL SEA.

shelf-ice A floating ice-mass formed by the coalescence of glaciers along the margins of Antarctica, ending in an ice-cliff; it is sometimes called *barrier ice*.

shell-sand A s. which consists of comminuted fragments of shell; e.g. along the W. coast of Skye at the heads of the sea-lochs; the MACHAIR of the Outer Hebrides, esp. on S. Uist and Tiree; on the beaches of Brittany.

shield A rigid mass of Pre-Cambrian rocks, the 'nucleus' of a continent, which has remained relatively stable since an early period in the earth's history. It may have undergone gentle warping, but otherwise it has been little disturbed. These are: (i) the Baltic S. (Fennoscandia); (ii) the Siberian Platform (Angaraland); (iii) the Chinese Platform; (iv) the Deccan; (v) Arabia; (vi) the continent of Africa; (vii) W. Australia; (viii) the Canadian (Laurentian) S.; (ix) the Guiana Plateau; (x) the Brazilian Plateau; (xi) Antarctica.

shield volcano A volcanic cone of basic lava, with a small angle of slope and a large basal diameter; e.g. Mauna Loa, the largest in the world, with a base 480 km. (300 mi.) in diameter on the ocean floor, its total height from the ocean floor to the summit about 9750 m. (32,000 ft.), the height of its summit above sea-level 4168 m. (13,675 ft.), 113 km. (70 mi.) in diameter at sea-level, and with a broad shallow crater 16 km. (10 mi.) in circumference. The angle near its cone is about 10°, but near its base only about 2°. Mauna Loa is still active, and large-scale lava flows frequently occur.

shieling (Scottish) A summer pasture, with rough temporary dwellings for the herdsmen, in the Highlands and islands of W. Scotland. Syn. with SCALE in English, and used as a place-name element as 'shield'; e.g. N. Shields (orig. *Nortscheles*).

shift The max. or total displacement of rocks on opposite sides of a FAULT, far enough from it to be outside the actual dislocated zone. In a fault which also involves a bending or flexing of the strata, the actual movement of the rocks, or SLIP, along the fault-line will be only part of the total displacement of the rocks.

shifting cultivation C. by a nomadic people for a few years until the soil becomes exhausted; the group then moves and clears a fresh piece of land, usually by burning the vegetation and digging in the ashes, leaving the old area to become overgrown. Frequently the abandoned, unprotected soil may be subject to rapid SOIL EROSION. It is sometimes possible for the site to be reoccupied again later, when the soil has had time to recover its fertility. E.g. in E. Africa; S.E. Asia. This form of c. was extremely common (though now less so), not only in primitive societies, but also in the cultivation of some cash crops by methods of ROBBER ECONOMY.

shifting rule A concept whereby analogy with a river moving as a result of SEDIMENTATION during a period of 'CAPACITY strain', a road may be moved as a result of its overcrowding by traffic, and a place is thus by-passed. The same concept has been applied to the movement of shopping centres.

shingle A mass of water-worn, rounded stones of various sizes; usually the term is limited to a beach-deposit.

ship-canal An artificial waterway large enough to accommodate ocean-going vessels; e.g. the Manchester S. C. (57 km. long, 37 m. wide, 9 m. deep) (35·5 mi., 120 ft., 28–30 ft.); Suez (160 km., 60 m., 10 m.) (100 mi., 197 ft., 34 ft.); Panama (81 km., 90 m.,

14 m.) (50·5 mi., 300 ft., 45 ft.); and Kiel (98 km., 45 m., 14 m.) (61 mi., 150 ft., 45 ft.). See also SEAWAY.

shire An Old English administrative district, derived from the Anglo-Saxon *scire*; e.g. *Wirecestre Scire* (Worcestershire), as listed in *Domesday Book*. Many modern counties, corresponding broadly to the s.'s, have the suffix -s.; e.g. Hampshire, Lancashire. Some s.'s, however, are not now counties or administrative units at all; e.g. Hallamshire, Hexhamshire. 'The S.'s' refer specif. to the English Midlands' fox-hunting counties.

shoal (i) A shallow area in a sea, river, lake. (ii) The bank of sand, mud or pebbles responsible for that shallow area. (iii) As a vb., esp. in navigation, to shallow gradually.

shore The area between the lowest low-water spring-tide line and the highest point reached by storm-waves. It is also used in nautical terms simply to refer to the land as seen from the sea. [*f* COAST]

shore-face terrace The area of deposition at the outer edge of a marine ABRASION PLATFORM, where wave-worn material is deposited in deep, calm water.

'short time-scale' A t.-s. of the PLEISTOCENE glaciation, in which its onset is put as recently as 600,000 years ago, by others at about 1 million years ago. Ct. 'LONG TIME-SCALE'. This dating is largely based on the astronomical theory of periodical perturbations in the orbit of the earth to explain climatic fluctuations (F. E. Zeuner), not on the radioactive dating of Pleistocene deposits.

shott (Arabic) (i) A shallow fluctuating salt-lake in a hot desert. (ii) A hollow or depression in which such a salt-lake lies; e.g. S. el Jerid, S. Melrir, on either side of the Algerian-Tunisian border.

shoulder (i) A bench on the side of a valley. This is partic. developed on the sides of a valley which has been deepened by the erosive action of a glacier; the s. occurs at the marked change of slope between the steep-sided inner valley and the more gentle upper slopes which were above the level of glaciation. See also ALP. (ii) A short rounded spur on a mountain-side.

S.I. The *Système Internationale d'Unités*, a rationalized and much simplified metric system coming into international use; 23 countries, including the U.K., have already agreed the system (1970), which may well become universal. Nearly all the quantities required are based on 7 units: (*a*) *length*: METRE; (*b*) *mass*: KILOGRAMME; (*c*) time: SECOND; (*d*) *electric current*: ampère (A); (*e*) *temperature*: the degree KELVIN (°K); (*f*) *luminous intensity*: the candela (cd); and (*g*) *pressure*: NEWTONS per sq. m. (N/m^2). In addition, there will be a number of coherent derivatives, and some other units in common use (e.g. the nautical mile and the knot for the U.K., and the °Celsius) are also included. The metre and the kilogramme are basic units, as opposed to the centimetre and the gramme; multiples and sub-multiples of these will be used, as follows: 10^{12} (prefix *tera*, T); 10^9 (*giga*, G); 10^6 (1 million) (*mega*, M); 10^3 (*kilo*, k); *10^2 (*hecto*, h); *10^1 (*deca*, da); *10^{-1} (*deci*, d); *10^{-2} (*centi*, c); 10^{-3} (*milli*, m); 10^{-6} (*micro*, u); 10^{-9} (*nano*, n); 10^{-12} (*pico*, p); 10^{-15} (*femto*, f); 10^{-18} (*atto*, a). Preference is expressed for multiples or sub-multiples separated by the factor 1000; it follows that the values indicated by * will be non-preferred; i.e. 1 cm. will not be used, but that unit of length will be expressed as either 10 mm. or 0·01 m.

sial, adj. **sialic** The surface granitic rocks of the continental crust, composed largely of (*si*)lica and (*al*)umina, from the initial 2 letters of each of which the name is derived. It has a density of 2·65 to 2·70. Ct. SIMA. [*f* ISOSTASY]

sidereal day The interval of time between 2 successive transits of a star over the same meridian, i.e. 1

rotation of the earth relative to the stars, which takes 23 hours, 56 minutes, 4·0996 seconds of solar time. A s. d. is thus nearly 4 mins. shorter than a MEAN SOLAR D., and the stars appear to rise nearly 4 mins. earlier every night.

sidereal year The period of time in which the earth makes a complete revolution of the sun with reference to the stars = 365·2564 mean solar days, or 365 days, 6 hours, 9 minutes, 9·54 seconds.

siderite (i) Ferrous carbonate ($FeCO_3$), an ore of iron, found in the Coal Measures and in the Jurassic limestones of England; alt. *chalybite*. (ii) An 'iron METEORITE', consisting of an iron-nickel alloy.

sierra (Sp.) A high range of mountains with jagged peaks projecting like the teeth of a saw; e.g. S. da Guadarrama, S. Nevada. The name is used widely in Spain, and in S.W. U.S.A., Mexico, and Latin America.

sieve-map A series of maps, drawn on transparent paper, each of which depicts the distribution of some factor; if superimposed, the s. m.'s will 'sieve out' suitable or unsuitable areas. The term was coined by E. G. R. Taylor in 1938 in connection with a study of industrial locations.

silage Compressed, part-fermented fodder; it consists of layers of grass, clover and alfalfa, compressed in a s.-pit, usually with the addition of molasses. It can be kept and cut out in blocks as animal feed. Also *ensilage* in U.S.A., where the whole MAIZE plant, cut while green, may be used.

silcrete A hard silicified sandstone; e.g. a SARSEN.

silicon An element (atomic no. 14), the second most abundant (to oxygen) in the crust, comprising about 28% by volume. It combines with oxygen to form s. dioxide (SiO_2), known as *silica* or *quartz*, and with various other oxides as a large group of rock-forming silicates, including the felspars, hornblende, the micas, olivine, augite. Hence *siliceous*, containing *Si.*; vb. *to silicify*, make siliceous; *silicification*, replacement by silica. Cf. PETRIFACTION.

sill (i) A near-horizontal or tabular sheet of igneous rock which has solidified from MAGMA intruded between bedding-planes; i.e. it is concordant with the structure. E.g. the Great Whin S. in N. England, from which the term was derived, extends over about 3900 sq. km. (1500 sq. mi.) from the Northumberland coast to the W. edge of the Pennines, varying in thickness from 1 m.–75 m. (2–3 ft. to 240 ft.). It consists of dolerite. Other examples are the Salisbury Crags near Edinburgh; Drumadoon in S. Arran; the Palisades [*f*] along the W. bank of the Hudson R., New Jersey State, U.S.A. (ii) A submarine ridge separating one ocean basin from another, or from a sea; e.g. in the Straits of Gibraltar; in the Straits of Bab-el-Mandab, over which the water is 365 m. (1200 ft.) deep, ct. over 2100 m. (7000 ft.) deep on either side in the Red Sea and the Indian Ocean. (iii) A submarine s. or threshold near the mouth of a FJORD, usually of rock, though sometimes with a capping of morainic material. [*f*]

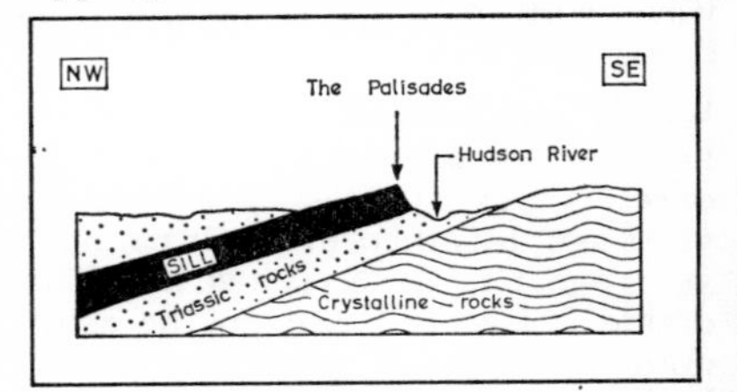

silt Material laid down in water, strictly with particles of diameter 0·002 to 0·02 mm.; they are coarser than those of clay, finer than those of sand. The term is used of *s.-soil*, containing over 80% s.

Silurian The 3rd of the geological periods of the Palaeozoic era, and the associated systems of rocks, succeeding the Ordovician and preceding the DEVONIAN. It is dated *c.* 440–400 million years ago. The British Geological Survey included the Ordovician rocks in the S. until the end of last century, and this is still the practice in continental Europe. Rocks of this age occur in Shropshire, central and S. Wales, the S. part of the English Lake District, the S. Uplands of Scotland, N.E. Ireland. The accepted series are: (i) *Llandovery;* (ii) *Wenlock;* (iii) *Ludlow;* they consist of shales, sandstones, flags, limestones.

silver A precious metal, often occurring in native form in association with others such as gold, and as a mineral, silver sulphide or *argentite* (Ag_2S), found among most lead, zinc and copper ores. Silver is used in alloy coins, in ornaments, and in photography. C.p. = U.S.A., Mexico, Peru, Canada, U.S.S.R., Australia.

silver thaw A descriptive name used in U.S.A. for GLAZED FROST.

silviculture The cultivation of forest trees for timber, as by the Forestry Commission, or by private owners.

sima, adj. **simatic** Dense (2·9–3·3) rocks of (*si*)lica and (*ma*)gnesia underlying the less dense continental SIAL masses, and forming the floors of much of the oceans. [*f* ISOSTASY]

similar triangles A method of reduction or enlargement of some narrow strip on a map, such as a section of road, railway or river. [*f*]

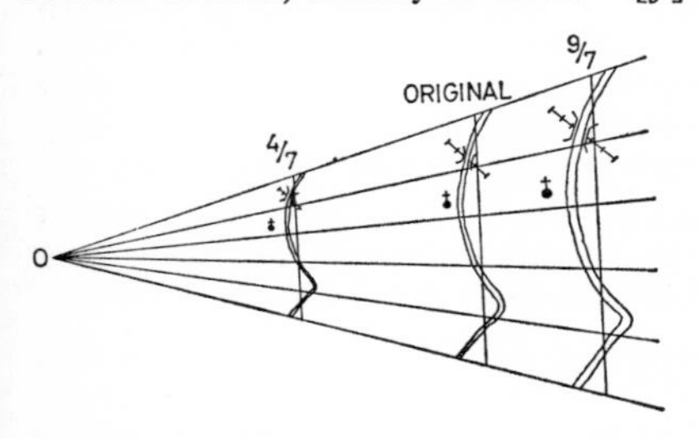

simoom, simoon (Arabic) A very hot, swirling, intensely dry, sand-laden wind, experienced in summer in the N. Sahara. It is usually associated with a low pressure system passing to the N.

Simple Conic Projection See CONIC P.

Simple Cylindrical Projection See EQUIRECTANGULAR P.

simulation The operation of a MODEL by manipulation of its constituent elements, subjectively by a scientific worker, objectively by a computer; the s. may be projected indefinitely into the future.

simulation technique An approach to research in which experiments and natural analogies are employed to confirm theories, esp. in human geography. E.g. the solving of problems in spatial equilibrium by using an electric current analogue, in human migration by heat flowing through a sheet of copper, in CENTRAL PLACE THEORY by floating magnets.

singularity In climatology, the tendency of a type of weather to recur with reasonable regularity each year around the same date.

sinkhole, sink (i) A hole, shaft or funnel-shaped hollow in limestone or chalk country, corresponding to a PONOR in the karst terminology. It is the result of: (*a*) solution under the surface soil, hence *solution-s.*; (*b*) a collapse of rock above a large cavern system, hence *collapse-s.*; in Gt. Britain this is a SHAKEHOLE. (ii) A hollow down which surface water disappears; i.e. syn. with SWALLOW-HOLE. But it seems preferable to regard a s. gen. as a depression, dry or down which water merely seeps, and to restrict the term swallow-hole to a vertical shaft with a waterfall. In the broad sense, there are said to be 60,000 s.'s on the plateau of Kentucky, U.S.A. (iii) A volcanic s., in a large SHIELD VOLCANO or lava dome; as molten lava is removed from beneath, subsidence forms a large depression, several hundreds of m. deep, and 1·5 km. or more across.

sinter A hard, crusty deposit of silica formed around a GEYSER or hot spring. S. alone signifies *siliceous s.*, or *geyserite*. Sometimes the term *calcareous s.* (or *calc-sinter*) is used, but this is strictly TRAVERTINE.

Sinusoidal Projection See SANSON-FLAMSTEED P. [*f*].

siphon A passage in a cave-system, in section an inverted U, which will allow water to flow up through the passage whenever the head of water in the cave behind rises above the level of the top of the s. passage. If the cave below the s. opens out on to the surface, this may form an *intermittent spring*. The term s. is often inaccurately used to signify a SUMP OR TRAP. [*f* INTERMITTENT SPRING]

sirocco, scirocco (It.) A hot S. or S.E. wind, sometimes oppressively humid, sometimes dry, which blows from the Sahara across N. Africa, Sicily and S. Italy. It blows in advance of depressions moving E. through the Mediterranean, and though dry when leaving the desert, it may be very humid on reaching S. Italy.

site (i) The position of a town in respect of the actual area it occupies. Ct. SITUATION. (ii) The smallest unit in D. L. Linton's hierarchy of MORPHOLOGICAL REGIONS.

situation The location of a town in relation to its wider surroundings. Ct. SITE.

Sixth Power Law A l. relating to the ability of a stream to move particles of a certain size, formulated in 1842 by W. Hopkins. This ability, or COMPETENCE, is said to be proportional to the 6th power of its velocity; i.e. if a stream flows twice as rapidly, it can carry particles 64 times ($=2^6$) as heavy, if 3 times as rapidly it can carry them 729 ($=3^6$) times as heavy. But it must be remembered that other factors modify this, partic. grain size. A critical velocity exists for each grain size at which it can be moved, though once in motion it may continue to be transported. Some fine-grained substances, e.g. mud, are cohesive and difficult to move, though once in motion they continue easily in suspension.

skeletal soil See AZONAL S.

skerry (derived from Swedish *skär*) A low hummocky island, often in a series offshore and parallel to the main trend of the coast. It consists sometimes of solid rock, sometimes of morainic material; e.g. off W. Norway and S.E. Sweden; Sule S. and Stack S. in the S. Bank, 64 km. (40 mi.) W. of the Orkneys. See *f* FIARD.

skerry-guard An inaccurate translation of Swedish and Norwegian terms as a line of skerries, forming a 'breakwater'. Correctly it refers to the calm water between the line of skerries and the mainland; e.g. the 'inner-lead' along the W. coast of Norway.

skiagraphy (i) The art of drawing features in 3 dimensions with the aid of shading. (ii) (*Astronomy*) The method of finding time by the shadow of the sun.

sky cover The amount of s. covered or obscured by clouds. This may be considered as the amount completely hidden (*opaque s. c.*), or the amount which is just obscured but not completely hidden (*total s. c.*). S. c. is measured in Britain on a scale of 0 (cloudless) to 8 (sky entirely covered). It is also measured in 10ths of the sky covered in accordance with international practice.

slack (i) A depression between lines of coastal sand-dunes. (ii) A period of stand-still about high and low water, when the tide is neither flowing nor ebbing. (iii) A quiet part of a stream where the current is slight, as on the inside of a bend. (iv) Small coal, now usually burnt in a thermal-electricity generator as its most efficient use.

slag The waste after-products of the smelting of metal, esp. iron in a blast-furnace. See BASIC S.

slaking The crumbling of clay-rich sedimentary rocks into small pencil-like fragments when exposed to air, as a result of their alternately absorbing and giving up moisture. The term is also used in a more general sense for any crumbling and disintegration of earth on drying-out after being saturated with water.

slash An area of swampy ground, esp. in the S. and S.E. of the U.S.A. Much of the Gulf Coast lowlands has been planted, esp. with s. pine (*Pinus caribaea*), from which gum and turpentine are obtained. Other trees found in similar locations are bald cypress, longleaf pine, pond pine.

slate A fine-grained metamorphic rock, formed from deposits of mud (clay-slate) which were subject to pressure as a result of earth-movements; it splits easily into thin layers. It can also be formed from deposits of fine-grained volcanic ash similarly compacted; e.g. the Buttermere S.'s in the Borrowdale Volcanic Series, English Lake District. The term is occas. used for any easily split rock, though in strict geological usage only for rock which splits along lines of CLEAVAGE, and not along the bedding-planes (as is the case with shales). It is used as a proper name in many stratigraphical formations; e.g. Llanberis S., N. Wales (Cambrian); Skiddaw S., English Lake District (Ordovician); Delabole S. (Devonian).

sleet (i) (British definition): a mixture of snow and rain, or partially melted falling snow. (ii) (U.S. definition): raindrops which have frozen and then partially melted.

slickenside A polished, sometimes finely fluted or striated, surface-plane of a FAULT, caused by friction and sometimes fusion by heat, as a result of movement along the divisional planes of the fault; e.g. in the Bunter Sandstone near Frodsham, Cheshire; in the Chalk to the W. of Lulworth Cove, Dorset.

slide (i) A mass of rock or earth that has fallen through gravity down a hillside, with a speed constant throughout the mass. (ii) The mark left on the hillside by such a fall. See ROCK-SLIDE.

sling psychrometer A WET- AND DRY-BULB THERMOMETER mounted on a frame which is fixed on a handle, such that the frame whirls round the handle when operated. This ventilates the wet-bulb thermometer, and readings of the temperatures enable the RELATIVE HUMIDITY of the atmosphere to be read off from tables.

slip (i) A form of landslide involving a mass of material, usually saturated with water, and moving *en bloc* as a constituent mass which commonly stays intact. (ii) In a FAULT, the actual relative movement along the fault-plane, which may be only part of the total displacement or SHIFT of the rocks. The component of the movement may be in the direction of the STRIKE of the fault-plane (hence *strike-s.*), or in the direction of the DIP of the fault-plane (hence *dip-s.*).

slip-face The sheltered leeward side of a sand-dune, steeper than the windward side. Eddy motion helps to maintain a slightly concave slope. The sand constantly blows up the windward side, and down the s.-f., hence the dune advances, unless its movement is checked by vegetation, some prominent obstacle, or water.

slip-off slope A low, gently sloping spur projecting from the opposite (convex) side of a meander to the river-cliff in a valley. The stream is thought to have migrated outward down the s.-o. s., cutting into the opposite bank. [*f* MEANDER]

slope length The length of a s., measured on its surface from the higher to the lower point, and not by its projection on to a plane (as on a map), which is the HORIZONTAL EQUIVALENT. [*f*]

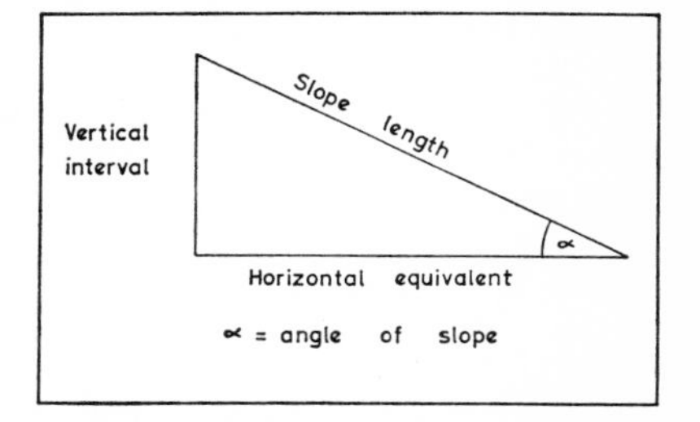

slope-profile line A straight line of traverse down a hill-slope, orthogonal to rectilinear contours. See PROFILE.

slope retreat The progressive back-wearing of the s. profile during the course of erosion. Three basic concepts have been advanced. W. M. Davis suggested that s.'s would progressively decline in angle. W. Penck proposed the idea of s. replacement, a form of parallel r. in which s. facets are replaced by others of different angle. L. C. King has maintained that PARALLEL R. OF S.'s as a whole takes place, and that they maintain their form almost until the end of the cycle.

slough (i) A somewhat fanciful, dramatic or poetic name for an area of marsh, bog, mud or waterlogged ground; e.g. 'the S. of Despond'. (ii) A term used in U.S.A. for a backwater.

sludging The flow of a thawed surface sheet of mud over the still frozen ground beneath; syn. with SOLIFLUCTION. It is common in early summer in high latitudes and at high altitudes.

slump, slumping A type of mass-movement involving an actual shearing of the rocks, a tearing away of a mass of material (a *s.-block*), usually with a distinct rotational movement on a curved concave-up plane, leaving a fresh scar on a hill-side. This is partic. common when more massive rocks overlie clay or a weak shale, along a sea-cliff or an escarpment; e.g. along the S. coast of the I. of Wight, and the Kentish coast between Folkestone and Dover, where Chalk overlies Gault Clay; along the Devon coast between Exmouth and Lyme Regis; along the Cotswold scarp between Gloucester and Cheltenham, where Jurassic limestones slump down over the clay. [*f* EARTH-FLOW]

slurry A flow of very wet, highly mobile mud.

small circle Any c. on the surface of the earth, the plane of which does not pass through its centre (i.e. it is not a GREAT C.). All lines of latitude, with the exception of the equator, are s. c.'s.

smallholding A small unit of agricultural land, farmed intensively, usually by a man (a smallholder) and his family without paid labour, including market-gardening, horticulture, glass-house cultivation, sometimes with poultry and pigs. In Britain the term is legally defined as a unit under 50 acres, or below a specif. value of rent. Such agricultural units are common in Belgium, Netherlands and France.

smelting The extraction of metal from its ore by means of heat, usually by the reduction of the oxide of the metal in a furnace.

smog Thick yellow RADIATION FOG over a town, where large quantities of soot act as nuclei for condensation, and sulphur dioxide adds to the acrid flavour. The term was coined in 1905, and has recently become well known as a result of some very bad s.'s, such as those of London in the winter of 1952. These are said to have resulted in 4000 deaths from bronchitis and pneumonia. This state of affairs has led to campaigns for smoke abatement, the creation of smokeless zones, and the use of smokeless fuels. Improvements can be effected, as shown by the case of Pittsburgh, formerly one of the dirtiest towns in the U.S.A., which achieved a reduction of atmospheric pollution of over 90% in six years.

snout The end of a valley-glacier, usually with a cave from which flows a melt-water stream. Sometimes the s. is obscured by masses of morainic material.

snow The type of precipitation formed when water-vapour condenses at a temperature below freezing-point, so passing directly from the gaseous to the solid stage, and forming minute spicules of ice. These unite into crystals which are either flat hexagonal plates or hexagonal prisms, revealing infinite variations in their patterns. These crystals aggregate into *s.-flakes*. Where the lower atmosphere is sufficiently cool, they will reach the ground without melting. S. may be very dry and powdery under low temperature conditions, as in Antarctica; there about 760 mm. of fresh s. may be equivalent to 25 mm. of rain. It may be wet and compact, so that only 100–150 mm. of s. will melt to form 25 mm. of rain. For records in Britain, s. is melted and added to the precipitation total. A tall cylinder is placed on a rain-gauge to collect the s., but it may become choked, and readings are then difficult.

snow-avalanche A fall of a mass of s. down a hill-side; it may be: (i) *wind-slab*, where the s. is crusted and compacted; (ii) a *dry s.-a.*, usually of new s. in winter; and (iii) a *wet s.-a.*, caused by a sudden spring thaw.

snow-bridge A mass of s., more or less compacted, which spans a CREVASSE or BERGSCHRUND.

snowdrift A bank of s. drifted by the wind which has accumulated against obstacles, sometimes to great depths; this may block roads and railway-lines, unless protected by barriers, sheds or s.-fences.

snow-field An area of permanent s. which has accumulated in a basin-shaped hollow among the mountains or on a plateau; e.g. the Ewig Schneefeld near the Jungfrau in the Bernese Oberland.

snow-line The lowest edge of a more or less continuous s.-cover. (i) The *permanent s.-l.* is the level at which wastage of s. by melting in summer fails to remove winter accumulation. This varies with latitude, altitude and aspect. At the Poles it lies at sea-level, in S. Greenland at 600 m. (2000 ft.), in Norway at 1200–1500 m. (4–5000 ft.), in the Alps about 2700 m. (9000 ft.), in E. Africa at 4900 m. (16,000 ft.). (ii) The *winter s.-l.* fluctuates from year to year, but is markedly lower than (i). There is no permanent s. in Gt. Britain, though in winter in the Highlands of Scotland s. lies above 900 m. (3000 ft.) for about 80 days.

snow-patch erosion See NIVATION.

social geography This is often used simply as the equivalent of HUMAN G., or in U.S.A. as 'cultural g.'; but usually it implies studies of population, urban and rural settlements, and social activities as distinct from political and economic ones.

soffoni, soffione, suffione (It.) See FUMAROLE.

softwood Timber obtained from coniferous forests, used for construction work, and for making pulp and cellulose. It is lighter, softer and has a more open texture than a hardwood.

soil The thin surface-layer on the earth, comprising mineral particles formed by the break-down of rocks, decayed organic material, living organisms, s.-water and a s. atmosphere. It has a physical structure, and various chemical constituents. S.'s may be classified into a number of major groups: (i) ZONAL, occurring in broad latitudinal zones or belts; (ii) INTRAZONAL, resulting from special kinds of parent rock (e.g. limestone), the presence of salt (solonchaks), the presence of water (peat-s.'s); (iii) AZONAL, *skeletal* or *immature s.'s*, new materials on which the soil-forming processes have not had sufficient time to work. See under individual s.-types.

soil climate The temperature and moisture conditions of the s.

soil erosion The removal of s. by the forces of nature more rapidly than the various s.-forming processes can replace it, partic. as a result of Man's ill-judged activities (over-grazing, burning and clearance of the vegetation cover, unprotected fallow, DRY-FARMING, up-and-down slope ploughing). This causes a loss of agricultural land which Man can ill afford. The chief types of s. e. are: (i) *wind e.* or DEFLATION; e.g. the 'Dust-bowl' of S.W. U.S.A.; (ii) SHEET or SHEET-FLOOD E.; (iii) RILL E.; (iv) GULLY E. S. e. can be checked by the maintenance of an effective vegetation cover, CONTOUR-PLOUGHING, ROTATION OF CROPS, terracing, composting, the planting of COVER CROPS, the creation of wind-breaks (trees, fences), pipe drainage to prevent gullying, the damming of gullies or the filling of them with brushwood.

soil profile A vertical section of the s., from the surface down to bed-rock, showing the individual HORIZONS, often of different colours and textures, sometimes grading into each other, sometimes changing sharply. Each main s.-type has its own p. A scientist may remove a complete p. in a s.-box for laboratory examination. See also various s.-types. [*f*]

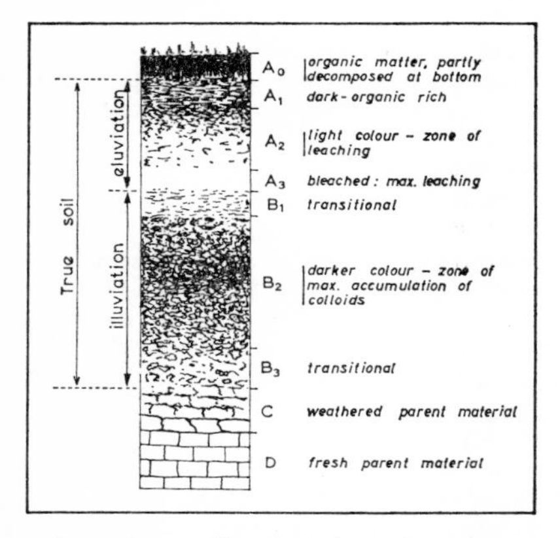

soke A small administrative region in Gt. Britain, now surviving only as a proper name in the S. of Peterborough which lies outside the administration of Northamptonshire.

solano (Sp.) An E. or S.E. wind, hot and oppressive, which sometimes brings rain in the summer to the coast of S.E. Spain.

solar constant The rate per unit area at which s. radiation is received at the outer limit of the atmosphere; this is equal to approx. 2 gramme-calories per sq. cm. per minute. It is thought to vary slightly, but observational problems make such calculations very difficult.

solar day The interval between 2 successive transits of the sun across a meridian; since the earth's orbit is an ellipse and is inclined to the equator, this interval varies slightly (see EQUATION OF TIME), and therefore a *mean s. d.* of 24 hours is used. Ct. LUNAR DAY.

solarimeter An instrument for measuring the intensity of solar radiation.

solar system The 9 PLANETS and the ASTEROIDS revolving in very nearly circular orbits around the sun in nearly the same plane.

solar year The period of time taken by the earth to make 1 complete orbit around the sun = 365·2422 MEAN S. DAYS.

sole (i) The ice-base of a glacier. (ii) The lowest THRUST-PLANE in an area of such intense compressional crustal movements that overthrusting has occurred; e.g. the lowest of the Caledonian thrust-planes in the N.W. Highlands of Scotland, which is actually called the S.

solfatara (It.) A vent quietly emitting sulphurous gases, usually associated with the approaching extinction of volcanic activity. It is so called after a small volcano of the same name in the Phlegraean Fields, near Naples. Ct. FUMAROLE.

solid geology The g. of the rocks underlying the layers of superficial deposits (broadly known as DRIFT). The British Geological Survey produces for many areas maps both of s. g. (as if the drift were removed) and of drift g., in which case the solid rocks are often not shown. The maps are not entirely consistent, as usually the more extensive and thicker areas of alluvium are shown as an integral part of the geological picture even on a s. g. map, otherwise it would be unrealistic.

solifluction, solifluxion The downhill viscous flow of surface deposits saturated with water, esp. when released by thaw, over the still frozen ground beneath. Formerly the term was used as syn. with soil creep, that is, downward movement of material whether aided by water saturation or not. Hence *s. slope, s. sheet, s. lobe, s. stripe.*

solonchak (Russian) An INTRAZONAL SOIL in which soluble salts are present in considerable quantity.

These are of widespread occurrence wherever there is a sufficient degree of evaporation, both in the hot deserts and in the cooler continental interiors where summer heat allows seasonal evaporation. The strong salt solutions rise by capillarity and form a greyish surface crust, below which is a granular salt-impregnated HORIZON. E.g. in the Great Basin of Utah, the Jordan Valley and around the Caspian S.

solonetz (Russian) A saline soil in an area with appreciable rainfall, so that some of the salt in the surface layer is leached out, to be concentrated in a lower horizon.

solstice One of the two solstitial points on the plane of the ECLIPTIC, where the overhead midday sun is at its furthest declination (angular distance) from the equator (approx. 23½°N. and 23½°S.). The sun reaches the N. s. (at the Tropic of Cancer) about 21 June, the S. s. (at the Tropic of Capricorn) about 22 Dec. These are the *summer* and *winter s.'s* respectively, as far as the N. hemisphere is concerned. [*f*]

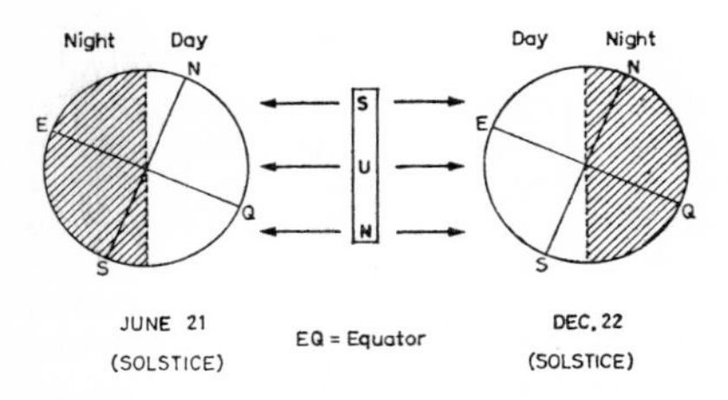

solution The state of a substance (the *solute*) being dissolved in a liquid (the *solvent*). This process can act as a form of weathering; e.g. the s. of rock-salt in water. The term is also used to include the process by which rain-water (acidulated with carbon dioxide derived from its passage through the atmosphere) acts as a dilute acid upon calcareous rocks (chalk and limestone). The $CaCO_3$ is slowly dissolved and removed in the form of calcium bicarbonate, $Ca(HCO_3)_2$. It is more correct to refer to this as CARBONATION-SOLUTION. Rivers carry a vast load in s.; the Mississippi transports to the sea each year 136 million tons of matter in s. (ct. 340 millions in suspension and 40 by SALTATION). It is estimated that each sq. mi. of the earth's surface loses 50 tons of material annually by s.

solution collapse, subsidence In limestone country, the sub-surface CARBONATION-SOLUTION of the rock along joints, fissures and underground water-courses may lead to subsidence, so producing surface hollows.

solution pipe A cylindrical or cone-shaped hole, filled with debris, occurring partic. in chalk country. These vary in size from quite small features to vertical shafts as much as 120 m. deep. They are the result of CARBONATION-SOLUTION concentrated along fissures and joints. A p. is commonly filled with material from overlying Tertiary rocks or superficial deposits such as gravel and sand. E.g. in the E. of the plateau of Hesbaye, Belgium, are numerous p.'s known as '*aard pijpen*' or '*orgel pijpen*', in diameter from 0·3 to 2·0 m., in depth from 1 m. to 20 m.

Sonnenseite (Germ.) The sunny slope in a valley; syn. with ADRET.

sop A vertical mass of iron-ore, usually occurring in the Carboniferous Limestone, e.g. in W. Cumberland. It was deposited from mineralized solutions derived from the former overlying Triassic sandstones, which percolated along lines of weakness in the limestone.

sorghum See MILLET.

sotch (Fr.) A term in the Grands Causses (Central Massif) for a DOLINE.

sough A level passage in a mine which has been driven to assist drainage; syn. with adit.

sound (i) An area of water connecting 2 larger areas, usually too wide to be called a strait; e.g. in W. Scotland, as S. of Jura, Sleat. (ii) A lagoon along the S. and S.E. coast of U.S.A.; e.g. Pamlico S., N. Carolina. (iii) A term applied to an inlet of the sea; e.g. Plymouth S. (Devon); Marlborough S. (New Zealand). (iv) 'The S'. (proper name) separates Zealand (Denmark) from Sweden.

sounding (i) The depth of water in the sea or a lake, formerly obtained by the use of 'the lead', a fine hemp line with a weight attached, then by a wire attached to a s.-machine, now by sonic or ultrasonic echo-s. By the last method, vibrations are transmitted down through the water to the sea-floor, to return in the form of an echo which is electrically recorded. (ii) The actual depth of water thus obtained, normally expressed as being 'below chart datum', usually the lowest possible tide-level, i.e. the worst navigational conditions. Port authorities (e.g. the Mersey Docks and Harbour Board, the Southampton Harbour Board) sometimes have their own Datum, to which all s.'s in waters under their control are related. S.'s are marked on charts as 'spot-depths', by the British Admiralty in fthms., and submarine contours are interpolated.

source The point at which a river rises, or begins to flow as an identity: (i) at a spring; e.g. R. Churn (a Thames headstream) at Seven Springs in the Cotswolds; (ii) from a lake; e.g. R. Mississippi; (iii) from a glacier; e.g. R. Rhône; (iv) as a resurgence from a cavern; e.g. R. Axe from the Wookey Hole in the Mendips; (v) from a marsh, bog or swamp; e.g. Goredale Beck, Malham, Yorkshire. Many streams just begin as tiny rills or trickles on a hillside, or as a seepage, and gradually grow into a distinctive flow of water.

Southeast Trades The TRADE WINDS of the S. hemisphere.

Southerly Burster A strong, dry wind, bringing unusually low temperatures to New South Wales, Australia. This is an example of a 'polar outbreak', whereby cold polar air-masses 'burst' into normally milder areas, drawn N. behind a depression.

sovkhoz (Russian) A state farm in the U.S.S.R., which uses hired (though directed) labour.

soya, soy bean The seed from the pods of a leguminous plant (*Soja hispida*), of immense value both to peasant-farmers (for s. flour, s. oil, animal fodder) and to modern industry (used for plastics, adhesives, paint, margarine, cooking-fat). The seeds are rich in protein and have a 15% content of vegetable oil. C.p. = U.S.A., China, Indonesia, Brazil, Japan, U.S.S.R.

spa A watering place, with medicinal springs; e.g. Spa in S.E. Belgium; Leamington S. in Warwickshire.

spalling Used in U.S.A. as syn. with EXFOLIATION.

spate A sudden flood or downflush of water on a river, caused by intensive rain or by sudden snow-melt higher up its valley.

spatter cone A large mass of lava ejected violently from a volcano, which 'spatters' and congeals as it hits the ground to form a small c., 3–6 m. high; e.g. around Sunset Crater, Arizona, U.S.A.

special area See DEVELOPMENT A.

specific gravity The ratio of the mass of a body to the mass of an equal volume of water at the temperature of its max. density (i.e. 3·945°C.). S. g. is a rel. quantity, DENSITY is an absolute quantity, though numerically they are the same.

specific heat The amount of heat (in calories) required to raise the temperature of 1 gm. of a partic. substance through 1°C. 1 calorie is required to raise 1 gm. of water from 0° to 1°C.

specific humidity The ratio of the weight of the water-vapour in a 'parcel' of the atmosphere to the total weight of the air (including the water-vapour), stated in gm. of water-vapour per kilogram of air; e.g. very cold dry air may have a s. h. of only 0·2, while very humid warm air near the equator may have a s. h. of 15·0 to 18·0. Ct. MIXING RATIO.

specific region A unit-area with a definite name and identity, in ct. to one of a given type (i.e. a GENERIC R.). E.g. Cumbria, the Kempenland. Ct. the Mediterranean Basin, a s. r., with other genetic areas with Mediterranean characteristics of climate and vegetation (central Chile, California, S.W. of S. Africa, S.W. Australia).

speleology, spelaeology The science of cave-exploration. In U.S.A. it is known as 'spelunking', its exponents as 'spelunkers'.

sphenochasm A triangular gap in the sea separating 2 continental blocks bounded by converging faults; e.g. the Arabian S., the Gulf of California, the S. of Japan, the Ligurian S. in the W. Mediterranean.

sphere A solid figure, any point on the surface of which is equidistant from its centre.

Area of surface $=4\pi r^2$; volume $=\frac{4}{3}\pi r^3$.

sphere of influence (i) In a politico-economic sense, an area in which a foreign country has special interests, rights and privileges; e.g. the Russian s. of i. over E. Europe. (ii) An area depending on an urban centre for services, or with which it has special relations (commuting, distributing trades, delivery areas, hospital services, local newspaper circulation). Ct. UMLAND.

spherical triangle A t. on the surface of a sphere, bounded by the arcs of 3 GREAT CIRCLES. The solving of a s. t. by means of s. trigonometry is a basic operation in geodetic surveying.

spheroid A nearly spherical body generated by the rotation of a sphere about its axis.

spheroidal weathering The swelling or expansion of the outer shells of a rock by penetration of water, so forcing them to pull successively away and loosen. It is a form of chemical w. similar to the larger scale mechanical w. (or EXFOLIATION). It is partic. noticeable in basalt and granite.

spiegeleisen An alloy of iron, carbon and manganese added to the molten metal in a Bessemer steel-converter.

spillway See OVERFLOW CHANNEL.

spine, volcanic A solidified mass of viscous lava, either forced out by extrusion and congealed, or hardened in the pipe and exposed by denudation; e.g. the now destroyed s. of Mont Pelée, formed in 1902; the Santiagito s. in Guatemala, formed 1922–4; the Devil's Tower, Wyoming.

spinifex A plant with tufted, sharp-pointed, spiny leaves, growing in the deserts of central Australia in large clumps separated by bare sand; sometimes it is called *porcupine grass*.

spit A long narrow accumulation of sand and shingle, with one end attached to the land, the other projecting into the sea or across the mouth of an estuary; e.g. Calshot S. at the S. end of Southampton Water; Hurst Castle S. at the W. end of the Solent; Blakeney, Norfolk. [*f*]

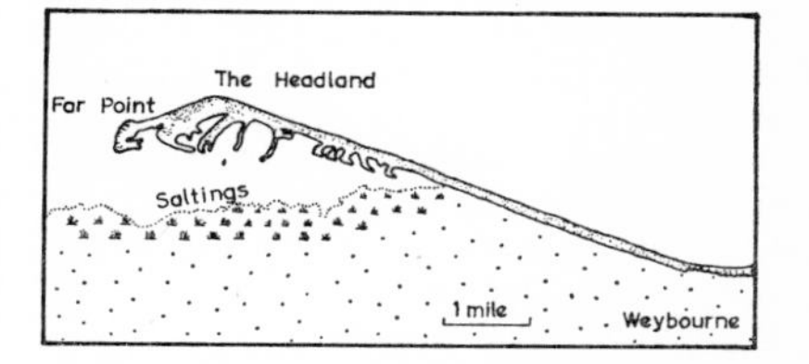

splaying crevasse A c. which starts as a longitudinal feature and splays outwards towards the side of a glacier.

spoil Debris or waste material from mines or quarries; hence *s.-banks, s.-dumps.* Its disposal is often a major problem, and many industrial landscapes are marred by its accumulation.

spot-height A precise point, the height of which above a given datum is measured and indicated on a map, though not necessarily on the ground; ct. BENCH-MARK, which is a s. h. actually indicated on the ground.

spring (i) A natural flow of water from the earth's surface, gen. issuing strongly, though at times it may just ooze or seep out. Its occurrence is related to the nature and relationship of the rocks, esp. of permeable and impermeable strata, together with the profile of the surface relief. A s. occurs where the WATER-TABLE intersects the surface. It may be a permanent, INTERMITTENT, SCARP-FOOT, DIP-SLOPE, FAULT, JOINT or VAUCLUSIAN s. (ii) The season between winter and summer; in the N. hemisphere it occurs astronomically between the s. or vernal EQUINOX (about 21 Mar.) and the summer SOLSTICE (about 21 June); popularly the period March, April, May.

spring-line A line of s.'s indicating the level where the WATER-TABLE intersects the surface, as at the foot of an escarpment. Villages along such a line are often known as *s.-l. villages;* e.g. along the N. edge of the S. Downs. While this availability of water is doubtless very important in settlement establishment, it is probably not the only factor involved, for villages may be so situated as to take advantage of 2 types of land.

spring-sapping Erosion around the issue-point of a strongly flowing stream, which removes material from around it, thus creating a small amphitheatre which is cut progressively back into the slope. This is thought to have been an important agency in the development of an escarpment DRY VALLEY.

spring tide A twice-monthly t. of considerable amplitude (i.e. a high high t. and a low low t.), occurring when the moon, sun and earth are in the same straight line (SYZYGY), either in conjunction (new moon) or opposition (full moon), so that the gravitational effects are complementary. Ct. NEAP T.

sprinkler irrigation A type of i., using a system of pipes with nozzles at intervals, so that a strip across a field can be readily watered, then the pipes are moved to the next section. This simulates rainfall. It is cheap to install, and does not necessitate the expensive work of levelling, grading, ditching and banking required by ground-water i. It also needs much less labour to operate. It has long been used in Washington, Oregon and in S. British Columbia, and is making rapid headway in parts of tropical Africa.

spur A prominent projection of land from a mountain or a ridge. See INTERLOCKING S., TRUNCATED S.

squall A sudden, violent, short-lived wind, usually accompanied by a heavy shower of rain or hail. In U.S.A. a s. implies a wind-speed of 16 knots for at least 2 minutes. See LINE-SQUALL.

stable equilibrium A state of the atmosphere where the actual LAPSE-RATE of a column of air is less than the DRY ADIABATIC LAPSE-RATE, or in which temperature increases with altitude (see INVERSION), If the air is displaced upwards, it will be colder and therefore denser than the surrounding air, and will tend to sink back to its orig. level. Rising motion cannot be sustained, and the atmosphere is in NEUTRAL EQUILIBRIUM. Conditions in an ANTICYCLONE and in subsiding air are usually s. Ct. UNSTABLE E.

stac An isolated mass of hard igneous rock, the result of marine erosion, in the St. Kilda group lying W. of the Outer Hebrides. Some have been worn down to or just below sea-level, others are upstanding, such as S. an Arnim (191 m., 627 ft., the highest in Britain), S. Lee (160 m., 524 ft.).

stack A steep pillar of rock rising from the sea, formerly part of the land but isolated by wave action. It forms part of a cycle of marine erosion, in the sequence cave—arch—s.—stump—reef; e.g. the Needles (at the W. end of the I. of Wight), Old Harry (I. of Purbeck), both made of chalk; the Old Man of Hoy (Orkneys) of Old Red Sandstone; Sule S. (on the Skerry Bank, W. of Orkney) of gneiss, 40 m. (130 ft.) high. See also STAC. [*f, page 329*]

stadial moraine See RECESSIONAL M.

stage (i) In its restricted geomorphological sense, the point of development reached in the course of a CYCLE OF EROSION. Characteristic s.'s are thought to be recognizable, viz. 'youth', 'maturity', 'old age'. Recent critics of Davisian geomorphology have urged that s. should not be given this restricted meaning of a phase in a progressive and inevitable evolution of landforms in a cyclic manner, but a wider, more realistic implication of changes in the causal relationships between structure and process with the passage of time. (ii) A subdivision of a SERIES in stratigraphy, corresponding to a unit such as a FORMATION, or in a mineral zone, or a zone of a partic. fauna, and corresponding in time to an AGE. (iii) The division of the Pleistocene; i.e. glacial s.'s. (iv) In U.S.A., the height of the surface of a stream, i.e. the depth of the water; see RATING CURVE [*f*].

stage-discharge curve Syn. with RATING C. [*f*].

staith (staithe) A shipping berth, landing-stage or wharf which is equipped with rails from which waggons may discharge coal into vessels; e.g. along the lower Tyne.

stalactite A pendant mass of calcite hanging vertically from the roof of a cave, deposited from drops of water containing calcium bicarbonate which have seeped through crevices and joints. The calcite is deposited partly because of evaporation, partly because some of the carbon dioxide in the water escapes and so part of its dissolved calcium bicarbonate is changed back into calcium carbonate. In the Ingleborough Cave, near Clapham, Yorkshire, the rate of growth of a s. was found to be 7·483 mm. (0·2946 in.) per annum, or about 760 mm. (30 ins.) per century, much more rapid than is gen. believed.

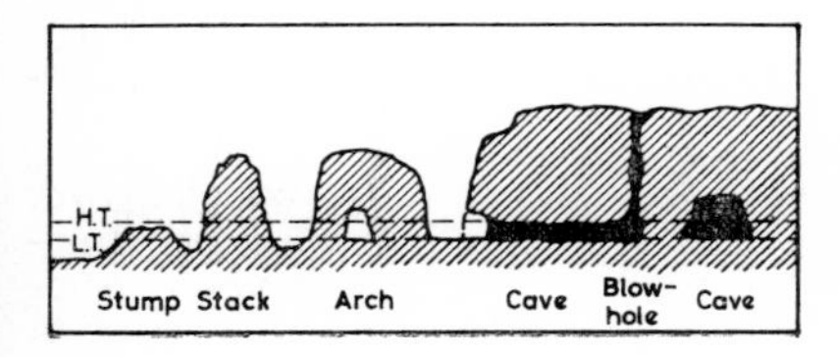

stalagmite A similar mass to a STALACTITE, usually more stumpy, growing from the floor of a cave, deposited from drips from the roof. This forms immediately below a stalactite, so that they may ultimately join to form a single pillar or column. In the Aven Armand, in the Grands Causses, France, these pillars are 23–25 m. (75–80 ft.) high and 1 m. or more in diameter.

stand A group of similar trees growing together; e.g. a s. of Douglas fir, or of teak. This makes for economic forestry practice. In tropical rain-forests, the timber species are scattered, and s.'s rarely occur, but they are much more usual in mid-temperate forests.

Standard Metropolitan Statistical Area (U.S.A.) A term replacing Standard Metropolitan District from 1961. It is defined by the U.S. Bureau of the Census by ref. to several criteria dealing with the size, m. character and functional integration of a.'s. Some of the more important criteria relate to the a. containing at least one central city of at least 50,000 inhabitants (or several contiguous urban areas with this total), together with adjacent counties if these have 75% of their labour force in non-agricultural activities, and over half of the population lives in contiguous minor civil divisions with a density of 150 persons per sq. mi. Moreover, if 15% of the workers living in the county work in the central city a., or 25% of the central city a. workers live in the county, then this county will be considered to be part of the S.M.S.A.

standard parallel A partic. p. of latitude, selected for calculating and drawing a specif. projection (e.g. the conical group), or for forming the horizontal axis of a grid-system.

Standard Time The mean time of a meridian centrally located over a country, and used for the whole area, instead of the inconvenience of APPARENT (LOCAL) TIME. In a large country each zone, of 15° longitude, has its own *Zone S. T.*; e.g. U.S.A. has Eastern S. T., Central S. T., Mountain S. T. and Pacific S. T., related to the 75°, 90°, 105° and 120°W. meridians, and respectively 5, 6, 7 and 8 hours slow on Greenwich. [*f*]

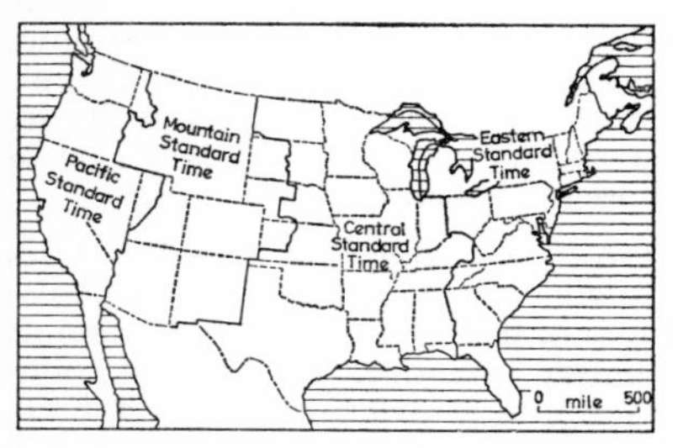

standing wave Syn. with an OSCILLATORY or stationary w. in air or water.

staple (i) A basic item of food (e.g. wheat), or a dominant article of commerce (e.g. wool). (ii) The length of a textile fibre; e.g. long-s. cotton, in which the fibres are 38 to 63·5 mm. (1·5 to 2·5 ins.) long (such as Sea Island cotton), in ct. to short-s. Indian cotton of 13 mm. (0·5 in.) s.

star-dune A sand-dune shaped like a pyramid, often quite high (up to 90 m.) and apparently fairly permanent in position, with radiating ridges of sand from a central point.

state A group of people occupying a specif. territory, organized under a government. It may refer gen. to a single country, or to a unit of regional or local government within a country; e.g. U.S.A. (with 50 individual s.'s), *Estados Unidos de Brazil* (20 s.'s), *Estados Unidos Mexicanos* (32 s.'s).

static lapse-rate Syn. with ENVIRONMENTAL L.-R. Ct. DYNAMIC L.-R.

static rejuvenation The r. of a river caused by the re-stimulation of erosive activity by an increase in precipitation due to climatic change, or to an increase in volume due to CAPTURE. Ct. DYNAMIC R.

stationary wave Syn. with OSCILLATORY W. in the modern concept of tide-formation.

statute mile See MILE.

Staublawine (Germ.) An avalanche of fine powder-snow.

'steady state' Syn. with DYNAMIC EQUILIBRIUM.

steam-coal A hard shiny c., intermediate in carbon content between BITUMINOUS coal and ANTHRACITE. It has been widely used for steam-raising, esp. in locomotives and ships, but its use is rapidly decreasing in face of oil fuel.

steam-fog A f. formed when cold air passes over the surface of much warmer water, from which moisture condenses into minute droplets, so that the water appears to 'steam'. In very low temperatures the moisture is converted directly into ice-particles, hence ICE-FOG.

steel An alloy of iron containing between about 0·5 and 1·5% of carbon (*cementite*, or carbide of iron, Fe_3C). S. is much harder and less brittle than cast-iron. Iron is converted into s. by: (i) the Bessemer process; (ii) the open-hearth process; (iii) the electric furnace. If the iron ore contains phosphorus, the resulting s. is 'cold-short' and liable to fracture; this is overcome by lining the converter with dolomite bricks, i.e. Basic Bessemer and Basic Open-hearth, so producing 'basic' or 'mild' steel; ct. acid steel produced from non-phosphoric ores in converters lined with silica bricks. Most modern steels are alloyed with small amounts of other metals for hardness, cutting power, and resistance to corrosion (cobalt, tungsten, nickel, chromium, vanadium). C.p. = U.S.A., U.S.S.R., Japan, W. Germany, U.K., France, China, Italy, Poland, Czechoslovakia, Canada, Belgium.

steering Used in meteorology for the directional effect of some atmospheric influence on another atmospheric phenomenon; e.g. the s. effect of high altitude STREAMLINES or of temperature differences on the movement of surface depressions.

step-fault A series of parallel f.'s, each with a greater throw in the same direction, so producing a 'stepped slope'; e.g. the W. slopes of the Vosges and the E. slopes of the Black Forest, bounding the Rhine rift-valley. [*f* RIFT-VALLEY]

steppe (Russian) The mid-latitude grassland extending across Eurasia from the Ukraine to Manchuria, usually regarded as syn. with prairie. Some writers, however, regard s. as being somewhat drier than prairie.

Stereographic Projection A ZENITHAL P. in which the meridians and parallels are projected on to a tangent plane at a point on the opposite extremity of the diameter. Both parallels and meridians are circles. The p. is CONFORMAL. Exaggeration increases outward symmetrically from the central point. It has one unique property: all circles on the globe appear as circles on the p. It is used for maps of the world in two hemispheres, for star maps, and for maps used in geophysics (since problems in spherical trigonometry on it are easy to solve). It can be constructed in the polar, equatorial [*f*] and oblique cases. As for all zenithal p.'s, Great Circles passing through the centre are straight lines; it can be used therefore for aeronautical maps centred on an airport, and the polar case is used for high latitude navigation. Graphically the polar case is easy to construct. The meridian circle is drawn to scale, with a line tangential to the Pole, and the angles are projected on to the tangent line from the opposite side of the polar diameter; where these lines intersect the tangent line will be the radii of the individual parallels. Meridians are straight lines radiating from the centre (360° divided according to the meridian interval). Trigonometrically, the radius of each parallel $= 2\,R\,.\,tan\,\frac{1}{2}\,(90° - \text{latitude})$.
[*f opposite*]

stereoscope An instrument with lenses or eye-pieces the approx. distance apart of human eyes (about 57 mm., 2·25 ins.), through which a pair of photographs of the same piece of landscape, taken from slightly different angles, are viewed. This gives a 3-dimensional effect. The simplest instrument comprises 2 lenses with folding legs; more complex patterns have binocular viewers, and large stereogrammetric plotters are used for accurate map-making from air photographs (hence *stereoplotting*).

Stevenson Screen A white-painted wooden box on legs, 1 m. above the ground, with louvred sides for ventilation, in which meteorological instruments are placed, so as to give shade readings unaffected by direct sunshine and strong winds.

stillstand A period of time during which the level of the sea relative to that of the land remains virtually undisturbed.

stochastic (Gk.) Introduced by J. Neyman and E. L. Scott (1957) for growth of any kind in which chance plays a major part, esp. in the growth of population or settlement. A s. locational model can be developed using probability matrices.

stock A small intrusive mass, sometimes defined as having an upper surface extent of not more than 100 sq. km.; i.e., a small BATHOLITH. When more or less circular, it is known as a BOSS; e.g. the Cheviot, in the Cheviot Hills.

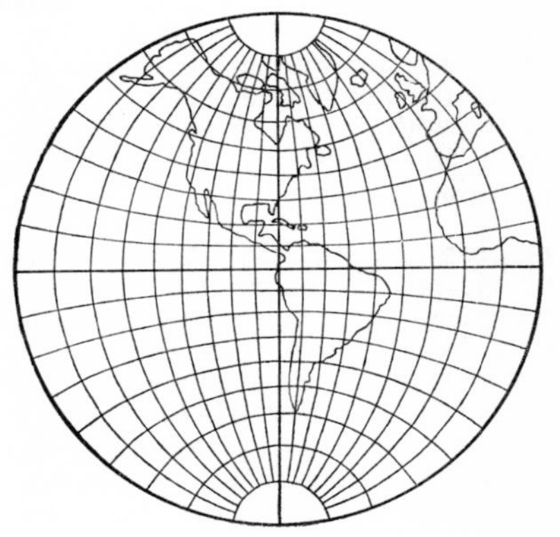

stock-grazing index The average number of s.-units per unit of area for each administrative division for which statistics are available. The U.S. Department of Agriculture uses a basis of 1 unit = 1 horse, 1 mule, 1 cow, 5 pigs, 7 sheep and 7 goats. J. C. Dunn drew a map of N. Northumberland, in which s.-units of cattle and sheep per 100 acres of each parish were computed, and mapped according to a CHOROPLETH grading. S.-units were here defined as: cows and heifers in milk, cows in calf, 1; heifers in calf, $\frac{3}{4}$; other cattle over 2 years, $\frac{7}{8}$; other cattle, 1–2 years, $\frac{2}{3}$; calves, $\frac{1}{8}$; lowland ewes with lambs, $\frac{1}{4}$; lowland yearling sheep, $\frac{1}{5}$; hill ewes, with lamb, $\frac{1}{7}$; hill hoggs and others, $\frac{1}{8}$.

Stone Age The name given by archaeologists to the period lasting approximately from the beginning of the Pleistocene period to the BRONZE AGE, during which time Early Man used stone implements or *artifacts*. The period has been divided into the *Eolithic* ('dawn' s. a.), *Palaeolithic* ('old' s. a.), *Mesolithic* ('middle' s. a.) and *Neolithic* ('new' s. a.). It ended about 4000 years ago in W. Europe.

stone-polygon, -stripe See PATTERNED GROUND.

'stop-and-go determinism' Introduced by T. Griffith Taylor, envisaging that the environment offers certain possibilities, but the rate at which man takes advantage of, or pursues, these opportunities may vary, with periods of acceleration (perhaps the result of a sudden technological advance) alternating with periods of slower progress or even temporary stagnation or recession.

stoping, magmatic A process by which MAGMA penetrates the overlying country-rock by shattering it, so that blocks sink into the magma (ENGULFMENT), thus allowing it to ascend. This was once held to account for large-scale m. penetration, but various objections, mainly geophysical, lead to the belief that the penetration is small-scale, confined to the upper parts of BATHOLITHS and around RING-DYKES.

storm (i) Any severe disturbance of the atmosphere, together with its effects, esp. at sea. There are many types: thunder-, rain-, snow-, hail-, sand-, dust-. (ii) The term for Force 11 (64–75 m.p.h.) on the BEAUFORT WIND SCALE. Hence there are many compound words: s.-wave, s.-tide, S.-SURGE, S.-BEACH.

storm-beach An accumulation of coarse material (shingle, cobbles, boulders), formed during a period of exceptionally powerful storm-waves, usually above the foreshore, and well above the level reached by normal high spring-tides.

storm-surge A rapid rise of the level of the sea well above the predicted tidal heights, whereby the water is 'piled up' against the coast by strong onshore winds. This may result in the breaching of coastal defences and widespread flooding. E.g. during the night of 31 Jan.–1 Feb. 1953, s.-s.'s developed along the coasts of E. England and the Netherlands, following an intense atmospheric depression crossing the N. part of the N. Sea. This produced gale-force winds from the N. and N.N.W., and caused a 'surge' or piling-up of water in the N. Sea basin, which narrows to the S. The vertical rise in the water-level above predicted high water was 2–2·5 m. between the Wash and the Straits of Dover, and as much as 3–4 m. along the coast of the Netherlands. Widespread flooding, damage (due also to high seas), and loss of life occurred.

storthe A dialect word (probably of Norse origin), for a small wood, notably in E. Yorkshire.

stoss (Germ.) The side of a prominent crag facing the direction from which the movement of a glacier or ice-sheet occurred, striated and roughened by the ice. The opposite side is the LEE; hence *s. and lee relief*. See also ROCHE MOUTONNÉE [*f*].

stow (i) The unit of smallest order in the hierarchy of regional divisions devised by J. F. Unstead. (ii) The second order in D. L. Linton's system of MORPHOLOGICAL REGIONS.

strait, straits A narrow stretch of water linking two areas of sea; S. of Gibraltar, Dover, Magellan.

strand A fanciful name for the shore.

strandflat (Norwegian) A wave-cut platform off the coast of W. Norway, which now stands above sea-level as a result of ISOSTATIC recovery. It exceeds 48 km. (30 mi.) in width. Sub-aerial weathering (mainly frost action) has been so potent that the cliffs have been rapidly worn back, and the powerful waves have quickly removed the debris. The word is also used as syn. with any wave-cut platform.

strath (Scottish) A long steep-sided flat-floored valley, though wider than a GLEN. It is commonly used as a prefix; e.g. Strathclyde, Strathmore, Strathpeffer, though not always; e.g. S. Spey. Occas. used in N. England; e.g. Langstrath, Borrowdale, Cumb.

stratification The accumulation of sedimentary rocks in layers or strata; hence *stratified*, vb. to *stratify*. Some authorities use the term rather more widely to indicate any rocks occurring in layers, incl. igneous intrusions (SILLS), but this is not advisable. *S. index:* the number of beds in a single formation, multiplied by 100, and divided by the total thickness of ft. in the formation; e.g. in a formation 50 ft. thick with 7 distinct beds, s. index =14.

stratiform A term covering all sheet clouds, incl. STRATUS, CIRRO-STRATUS, ALTOSTRATUS, STRATO-CUMULUS, NIMBOSTRATUS.

stratigraphy A branch of geology dealing with the order and succession of rock-strata, their occurrence, sequence, lithology, composition, fossils and correlation; the basis of HISTORICAL GEOLOGY.

stratocumulus (*Sc*) A uniform heavy cloud, consisting of dark grey globular masses within a continuous sheet, occurring up to 2400 m. (8000 ft.), though usually lower.

stratosphere The layer of the atmosphere above the TROPOPAUSE, extending up to the IONOSPHERE at about 90 km. (55 mi.). Its base height averages about 18 km. (11 mi.) at the Equator, 9 km. (5·5 mi.) at 50°N. and S., and 6 km. (4 mi.) at the Poles; this varies slightly with the season (it is rather higher in summer) and with gen. atmospheric conditions. The s. contains very little dust or water-vapour. At the base of the s. temperatures over the Equator vary during the year only from about −79°C. (−110° F.) to −90°C. (−130°F.), though over the polar regions the seasonal difference is much more marked (from about −40°C. (−40°F.) in summer to −79°C. (−110°F.) in winter). In the s. the progressive decrease of temperature with altitude ceases, and until recently temperatures were regarded as being virtually constant. Now it is believed that there is a rise in temperature to about 15°–20°C. (60°–70°F.) at about 30 miles, a 'hot layer' caused by the concentration of OZONE. Near the upper limit of the s., at about 80 km. (50 mi.), occur the NOCTILUCENT CLOUDS. Temperatures in the upper part of the s. are about −80°C. (−112°F.).

strato-volcano Used in U.S.A. for a COMPOSITE VOLCANIC CONE.

stratum A layer or bed of sedimentary rock (pl. *strata*). Usually the syn. bed is used in the singular, strata in the pl., but there is no fixed practice. Some geologists, esp. in U.S.A., use the term s. in a wider sense, so that it may be made up of a number of beds.

stratus (*St*) A gray, uniform cloud-sheet, often very persistent, occurring in a thin sheet at any height up to about 2400 m. (8000 ft.). It is not usually responsible for precipitation, other than a fine drizzle.

stream A body of flowing water, covering all scales from a small rill to a large river. Hence the term is used to denote all the characters of, processes and landforms resulting from s.'s.

streamline The direction of movement of all 'parcels' of air within an area, measured at a single moment in time. This gives an instantaneous overall picture of air movement. Ct. TRAJECTORY, which follows a single 'parcel' of air over a period of time.

stream order A classification of s.'s according to their position in the drainage net. The scheme usually adopted is that of A. N. Strahler, modified from that of R. E. Horton. The s. network can be thought of as a branching tree: those branches terminating at an outer point are designated *1st order s.*'s. The junction of 2 of these will produce a *2nd order s.*, 2 *2nd* order s.'s will combine to form a *3rd order s.*, and so on until the 'root', or s. mouth, is reached. See BIFURCATION RATIO. [*f*]

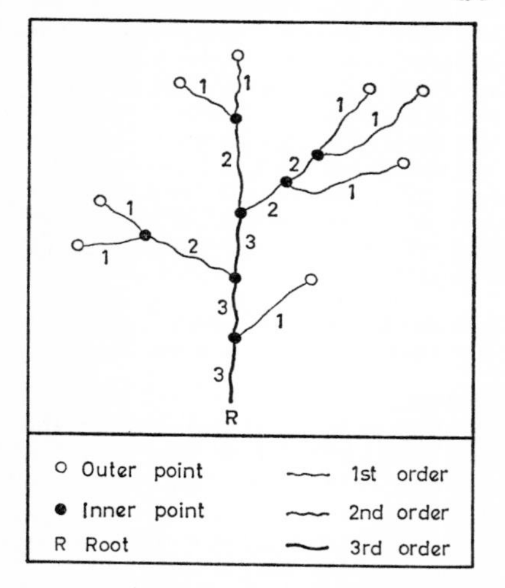

street village A rural settlement built along a single road, from the Germ. *Strassendorf*, where farmhouses are in 2 rows facing each other across the s.

stress A force per unit area, obtained by dividing the total force by the area to which it is applied. If 2 forces act away from each other the stress is a *tension;* if they act towards each other it is a *compression;* if they act parallel to each other it is a *shear*. When rocks are exposed to stress, they suffer strain.

striation, striae Scratches, grooves and scorings on ice-worn rocks, caused by the dragging of rocks frozen into the base of a glacier by its movement over the valley floor. These afford an indication of the direction of ice-movement.

strike The direction along a rock stratum at right-angles to the TRUE DIP. S. is used widely as an adjective in connection with features developed along it; e.g. s.-fault, -valley, -joint, -stream. [*f*]

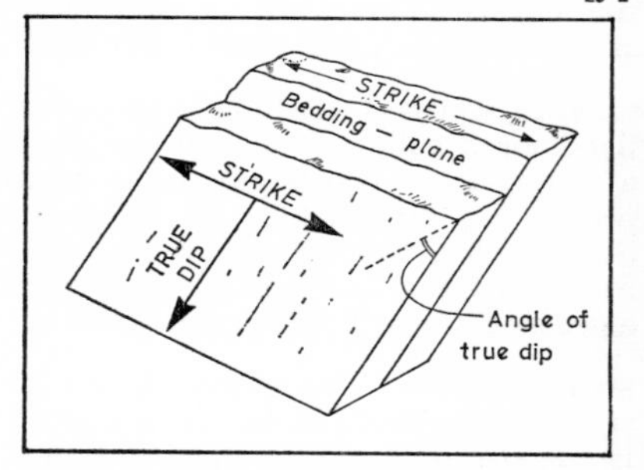

strike-fault A f. whose direction is parallel to the strike of the strata it affects.

strike-slip fault See TEAR-F.

strike-valley A v. developed in rocks parallel to the strike of the strata; the v. of a subsequent stream in a TRELLIS DRAINAGE PATTERN. E.g. in the Front Ranges of the Rockies, between Denver and Boulder (Colorado), where the rocks dip E., s.-v.'s trending from N. to S. have developed along the s. of the outcrop of the weaker shales between the more resistant rocks (which form HOGBACKS).

string town Used descriptively for a long, drawn-out urban settlement. LINEAR TOWN is preferred.

strip cultivation (i) A large field cultivated in long s.'s, each worked by an individual tenant or owner; this was the basis of the medieval 2- and 3-field system in England. It also results from '*parcellement*' and '*morcellement*' in France. (ii) The alternation of narrow s.'s of arable and pasture to check soil erosion, as in parts of the American Mid-West.

striped ground See PATTERNED G.

strip-mining An American term for OPENCAST MINING or quarrying.

stripped surface The denudation of the overlying rocks until a resistant stratum is reached, leaving a more or less smooth s.

strombolian eruption So called after Stromboli, a volcano in the Lipari I. to the N. of Sicily. The crater contains molten lava, and gases can readily escape, with small-scale intermittent e.'s at frequent intervals, but with no great accumulation of pressure which might cause a more violent e. A cloud constantly hangs over Stromboli, coloured pink by reflected light at night; hence its name 'lighthouse of the Mediterranean'.

structure The arrangement and disposition of the rocks in the earth's crust. In some cases, this implies more than just arrangement, including the nature of the rocks and even the initial forms prior to erosion. It is in this sense that the term is used in W. M. Davis's classic trilogy for the explanatory description of landscape: 's., process and stage'. However, it is better as a rule to restrict the term to the actual ATTITUDE of the rocks. Structural is used widely as an adj.: -basin, -feature, -terrace, -trap.

structure plan Under the 1968 Town and Country Planning Act, a Written Statement with supporting maps prepared by a local planning authority which formulates policy and gen. proposals for the development of the physical environment, and management of traffic. These proposals have to be related to those of neighbouring areas.

subaerial Occurring or forming on the surface of the earth, as distinct from *subaqueous* or *submarine*.

subarctic (i) Used with reference to latitudes immediately S. of the Arctic Circle, and to the phenomena occurring there. (ii) A group of climatic types defined by W. Köppen, with the mean temperature of the warmest month above 10°C. (50°F.) and with the mean temperature of the coldest month under −3°C. (26·6°F.); evaporation and precipitation are both low. They are denoted by letter *D*, the subdivisions being *Dfc*, *Dfd*, *Dwc*, *Dwd*. (iii) An alternative name for the PRE-BOREAL climatic phase.

Sub-Atlantic The climatic phase which has lasted since 500 B.C., following the SUB-BOREAL, and is still in being. Conditions became milder and moister, and an alder-oak-elm-birch-beech flora gradually spread, with the beech gen. dominant.

Sub-Boreal The period from about 3000 to 500 B.C., following the ATLANTIC climatic STAGE, with a renewal of the cooler climatic conditions of before 5500 B.C., though rather drier. The dominant tree of the Atlantic phase (oak) gradually gave way to pine.

subclimax A persistent SERAL community which has been permanently prevented from reaching a state of CLIMAX VEGETATION by factors such as bad drainage, poor soil, or exposure to strong winds; e.g. in the high heathlands in W. Europe, such as the Kempenland.

sub-consequent stream See SECONDARY CONSEQUENT S.

subglacial moraine Material carried by a glacier, frozen into the ice at or near its base. [*f* MORAINE]

subhumid A climate which is transitional in its precipitation received, lying between the areas with a well distributed rainfall and those with near arid conditions. The vegetation consists of tall grass; i.e. it is too dry for much of the year for tree growth, but not so dry as to be desert or scrub. It is a matter both of seasonal distribution and total of precipitation.

sublimation The conversion of a solid to a vapour, or vice versa, without an intervening liquid stage. Specif. the formation of ice-crystals directly from water-vapour when condensation takes place at temperatures below freezing-point.

sublittoral The zone which extends from low-tide level to the edge of the CONTINENTAL SHELF, syn. with the shallow water zone. On it are deposited sands, muds and coral fragments.

submarine canyon A steep-sided trench on the floor of the sea which crosses the CONTINENTAL SHELF, sometimes continuing across the continental slope into deep water. The cause of such features is something of a mystery; they may be due to tidal scouring, to the erosive activity of s. turbidity currents, to former river erosion when sea-level was greatly lower than at present, to faulting, or to the sapping action of powerful s. springs which burst out on the sea-floor. Some of these s. c.'s are obviously the submerged valleys of rivers continuing across the continental shelf, but others cross the entire shelf and slope into deep water, and a few occur only near the outward margins, thus forming deep gorges cut into the slope. E.g. the Fosse de Cap Breton in the Bay of Biscay [*f below*]; off the mouth of the R. Congo; off the R. Hudson. [*f*]

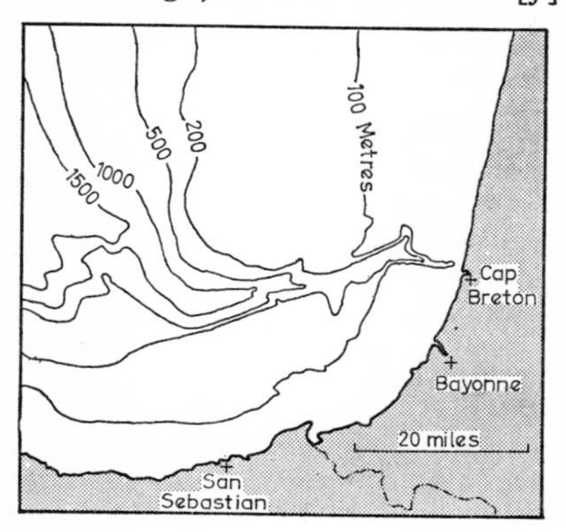

submarine ridge A long narrow area of land rising appreciably from the floor of an ocean; e.g. the Mid-Atlantic R. in the Atlantic, S-shaped, parallel to the general trend of the coastlines of the continents on either side, covered by about 3100 m. (1700 fthms.) of water (known as the Dolphin Rise in the N., the Challenger Rise in the S.). The Walvis R. and the Rio Grande R. are transverse ridges in the S. Atlantic. [*f*]

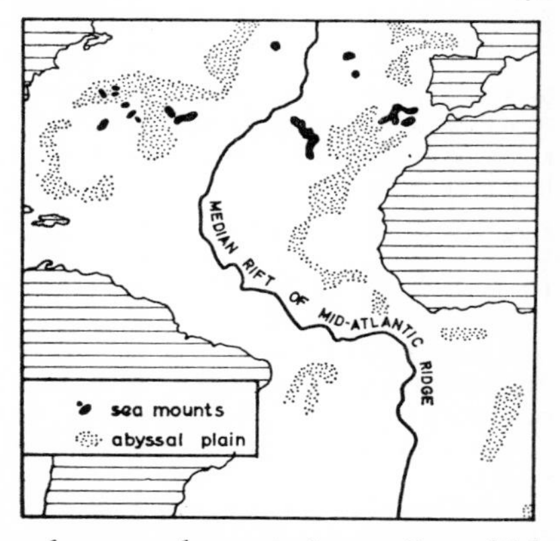

submerged coast A coastline which owes its form largely to a relative rise of sea-level, so covering the margins of the former land-surface. The chief categories are: (i) *s. upland coast*: (*a*) RIA; (*b*) FIORD; (*c*) DALMATIAN or longitudinal; (ii) *s. lowland coast*: (*a*) broad shallow estuaries, with marsh and mud-flats uncovered at low tide, creeks and winding shallow inlets; e.g. S. Suffolk; Strangford Lough (N. Ireland) and Boston Harbour (Mass., U.S.A.), where a DRUMLIN coast is submerged; (*b*) FIARD; (*c*) FÖHRDE; (*d*) BODDEN. See also S. FOREST.

submerged forest A layer of peat in which tree-stumps and roots are embedded, occurring between tide-levels or even below present low tide, indicative of a positive change (a relative rise) of sea-level; e.g. revealed by the excavations for dock basins at Southampton and Barry (Glamorgan). Peat, some stumps and even trunks can be seen at low tide along the coast near Formby (Lancs.), Wirral, near Harlech, and at Borth on the Dovey estuary. At Pentuan in Cornwall a layer of sediment containing oak stumps and roots lies 20 m. (65 ft.) below present sea-level.

subsequent boundary An international b. which has been defined and demarcated after a geographical region has developed some identity, so that the political b. cuts across and divides up this region. E.g. the Middle Danube Basin.

subsequent stream A tributary to a CONSEQUENT S., which has developed its valley, mainly by headward erosion, along an outcrop of weak rocks; if outcrops occur more of less at right-angles to the consequent slope, the s. s. will join the consequent river at right-angles. See TRELLIS DRAINAGE. [*f*]

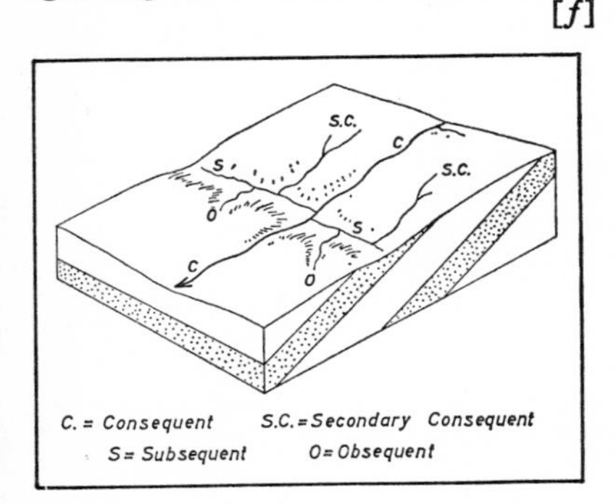

subsere A SERAL community which has developed to a certain level, but which has been prevented by extraneous factors from developing into CLIMAX VEGETATION. Only if the delaying factors are regarded as temporary is the community known as a s.; otherwise the appropriate term is SUB-CLIMAX.

subsidence (i) The depression of part of the earth's surface relative to its surroundings. This may be: (*a*) a large-scale crustal movement; e.g. a RIFT-VALLEY, a warped basin; (*b*) a small-scale local movement; e.g. the result of collapsed tunnels and passages in a coalfield area; the result of solution of salt deposits (e.g. in Cheshire; see MERE, FLASH); the collapse in a limestone area of the roof of a cave-system formed by carbonation-solution, leaving a hollow or gorge on the surface; e.g. Gordale Scar near Malham (Yorkshire); Cladagh valley, near Enniskillen (N. Ireland). (ii) Used in climatology for the slow settling of large masses of air, as in an ANTICYCLONE.

subsistence crop A c. consumed by the grower and his family, usually as a basic item of their diet; e.g. rice in S.E. Asia; yams, millet and bananas in W. Africa; potatoes in Ireland. The ct. is with a CASH C., grown for sale. Hence *s. agriculture*.

subsoil A transitional layer of partially decomposed rock underlying the top-soil and overlying the bed-rock. It forms the C-HORIZON in the SOIL PROFILE.

subtopia A term (derived from *surburb* and *utopia*) implying the ugliness of the suburban sprawl around a town, in spite of the utopian ideals of town planners.

subtropical App. to the latitudinal zones between the tropic of Cancer and about 40°N., and the tropic of Capricorn and about 40°S. It is used in respect of: (i) belts of atmospheric high pressure; (ii) climatic regions with no cold season; i.e. with no month below about 6°C. (43°F.). The term is also used gen. of lands, climates, vegetation outside the tropics and merging into the warm temperate zone.

suburb The outer residential parts of a town; hence 'residential s.', 'dormitory s'. The word is of much older origin, formerly meaning the territory immediately outside the walls of a town or city. Hence *suburban*, *suburbia*. The *Index of Suburbanization* (devised by J. E. Rickert) is the percentage of the rural non-farm population in a partic. rural territory.

succession (i) The order of beds of rock in time. (ii) A gradual sequence of changes or phases in vegetation over a period of time, even if the climate remains unaltered; hence *plant s*. This will proceed until some situation of equilibrium is attained, and a CLIMAX COMMUNITY is established.

sudd (Arabic) A compact mass of floating vegetation on the White Nile in the Sudan, which blocks the main channel, forcing it to split up into a maze of backwaters and shallow lakes.

sugar limestone A crumbly soft material formed by the THERMAL METAMORPHISM of Carboniferous L.; e.g. by an INTRUSION, as in the N. Pennines above and below the Great Whin Sill.

sulphur An element obtained: (i) from iron pyrites (FeS_2), as in Spain; (ii) from deposits of natural s., sometimes in the neighbourhood of volcanic regions, also in vast beds in Texas and Louisiana associated with salt-domes, and in Sicily associated with Tertiary limestones and sandstones; (iii) in recent years from natural gas and oil (e.g. in France from gas at Lacq and oil at Parentis). C.p. = U.S.A., Mexico, France, Japan, Italy, Spain.

sumatra A line-squall wind (or 'linear disturbance') in the Malacca Straits, usually accompanied by thunderstorms, occurring very suddenly, gen. at night, and esp. during the period of the S.W. monsoon.

summer (i) The warmest season of the year, in ct. to winter. (ii) In the N. hemisphere, the period June, July and August; in the S. hemisphere, the period December, January and February. (iii) In astronomical usage, the period between the summer solstice (about 21 June) and the autumn equinox (about 22 Sept.) in the N. hemisphere, and between about 21 Dec. and 21 March in the S. hemisphere.

summit The highest point or level of a mountain, railway, road, canal.

summit plane A plane passing through a series of ACCORDANT SUMMITS indicating the existence of either: (i) a former PENEPLAIN; or (ii) a regional balance in down-wasting, so that divides are reduced in level together (the GIPFELFLUR hypothesis).

sump A deep pool in a cave-system below the WATER-TABLE, the exit of which lies below the level of the water.

sun An incandescent, almost spherical body, the centre of the solar system, with a diameter of about 1,392,000 km. (865,000 mi.), a mean surface temperature of about 5700°C., and a very high internal temperature. It rotates once in 24·5 days at its equator and the whole solar system moves through space at about 18·5 km. (11·5 mi.) per second. Its mean distance from the earth is 150 million km. (92·9 million mi.). See APHELION, PERIHELION.

sunrise, sunset colours At these times of day, the light-waves of solar radiation have had a long passage through the atmosphere because of their low angle to the earth's surface. When these light-waves encounter obstacles (such as air particles, water-vapour, dust, smoke or volcanic ash), the constituents are scattered, esp. at the short-wave (blue-violet) end of the spectrum, which results in the colours at the long-wave (red) end of the spectrum becoming increasingly dominant. 'Red sky at night shepherd's delight; red sky at morning shepherd's warning.' Cf. ALPINE GLOW.

sunshine An important climatic element, the duration of which is partly a function of latitude (i.e. the total hours of daylight and therefore of possible s.), partly a function of daytime cloudiness. It is measured by a CAMPBELL-STOKES RECORDER, and tables of data can be given in terms of duration in hours per day, month or year, or as a percentage of the possible amount. Lines of equal mean duration of s. (plotted for various stations) are *isohels*. The sunniest parts of the earth are the hot deserts; at Helwan, Egypt, the mean s. for the year is 3668 hours, or 82% of the possible for the latitude. The lands around the Mediterranean Sea have about 90% of the possible in summer, though they are somewhat cloudier in winter. The equatorial and cool temperate latitudes have much less s.; Valentia in S.W. Ireland has a mean of 1·3 hours of s. per day, or 17%. The sunniest place in Gt. Britain is claimed to be Shanklin, I. of Wight.

supercooling This occurs when water remains liquid below 0°C. (32°F.). In clouds well below freezing-point, water droplets frequently remain unfrozen if undisturbed. Freezing may ultimately occur when these droplets come into contact with flying aircraft, causing 'icing' (which can be extremely dangerous unless provided for), and can freeze on telegraph wires. S. may also lead to the growth of very large hail-stones.

superficial deposit Materials lying more or less loosely on the earth's surface, formed independently of the rocks below and usually transported and deposited there by natural agencies: sand and loess blown by the wind, glacial drift, river-borne alluvium and gravels. Peat is also included, although it has developed *in situ*. A soil which has developed through weathering of the underlying rock is not regarded as a s. d.

superglacial stream A s. of melt-water flowing during a summer day on the surface of a glacier in a deeply cut runnel, until it vanishes down a crevasse. See MOULIN.

superimposed drainage, superimposition A pattern of d. which orig. developed on a cover of rocks now removed. This pattern seems to bear no relation to the existing surface rocks; e.g. the radial d. of the English Lake District; the rivers of E. Glamorgan which cross at right-angles the varied Devonian and Carboniferous rocks of the S. Wales coalfield basin and its margins; the Meuse, which crosses the Ardennes transverse to the gen. trend of the structure; the rivers of the Hampshire Basin which flow gen. S. across the E. to W. folds of mid-Tertiary age. [*f opposite*]

superimposed profile The construction of a diagram on which a series of p.'s spaced at regular intervals across a map of a piece of country is drawn. This may bring out such features as ACCORDANT SUMMITS and erosion platforms. [*f* PROFILE]

Superposition, Law of In sedimentary rocks (and also extrusive igneous rocks) an upper bed of rock is younger than a lower one, providing that earth-movements have not reversed the order.

supersaturation, supersaturated The state of a body of air with a relative humidity of more than 100%, with water-vapour sufficient to produce condensation, but in which this has not actually occurred. For condensation to take place, nuclei such as particles of dust, smoke, salt, pollen, even negative ions, are usually necessary, but not always available. Thus s. is a fairly common atmospheric phenomenon.

supranational An organization which is concerned with the related activities of a number of nation states; e.g. O.E.C.D., E.E.C., E.F.T.A.

surazo A cold wind experienced in S. Brazil, the result of an ANTICYCLONE in winter.

surf A mass of broken foaming water, formed when a large powerful wave breaks on a reef or on a steep shore.

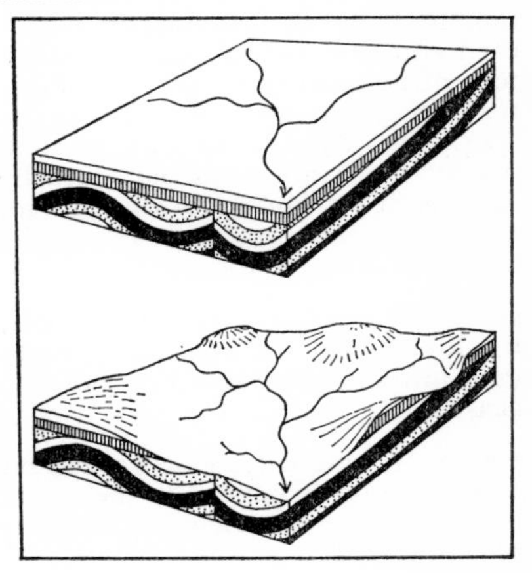

surge, glacier In a g., the downstream passage of a bulge of ice at a considerably greater speed than the normal g. flow. R. P. Sharp describes a s. that has been watched for 15 years moving down one of the g.'s on the flanks of Mount Rainier in the Cascade Mtns., U.S.A.

survey (i) The measuring and recording of lines and angles so as to make an accurate map of part of the earth's surface; this has become a very exact process. Hence *surveying, surveyor*. See GEODETIC s. The various operations include BASE-LINE measurement, TRIANGULATION and LEVELLING. Simple but effective methods of local s. include compass traversing and plane-table s.'s. (ii) To examine, record and depict certain features of the earth's surface, in cartographic, diagrammatic or written form; e.g. an industrial s., a land-use s., a regional s., a geological s., a CADASTRAL S.

suspension The holding-up of small particles of matter, transported by moving water, by turbulent upward eddies. The resistance of the water through which a particle is sinking by gravity helps to keep the finest material in s. for a long time; e.g. if a jar of muddy water is left, it takes many hours for most material to sink to the bottom even though the water is still; coarse sand, however, sinks very rapidly. The R. Mississippi carries each year to the sea about 340 million tons of material in s., ct. 136 million tons in SOLUTION, 40 millions by SALTATION. Approx. 80 tons of solid matter in s. are removed annually from each sq. km. (300 tons per sq. mi.) of the earth's surface. Hence the *suspended load* of a stream.

swag A shallow water-filled hole, resulting from mining subsidence; used in the Midlands of England. Cf. FLASH.

swale A long narrow depression on a beach, broadly parallel to the coastline, separating two ridges of shingle; syn. with LOW.

swallow-hole, swallet (i) A deep, vertical shaft, usually in limestone, down which a surface stream disappears as a waterfall; e.g. Gaping Gill, Ingleborough, Yorkshire, down which the Fell Beck pours as a waterfall 111 m. (365 ft.) high; Hunt Pot, Pen-y-ghent, Yorkshire. Known also as a PONOR in KARST terminology. (ii) A hole in the bed of a stream down which water disappears; e.g. in the Mole Valley near Dorking, Surrey; near N. Mimms, Hertfordshire.

swamp A gen. term applied to a permanently waterlogged area and its associated vegetation, commonly reeds; ct. MARSH, only temporarily inundated; and BOG, where the vegetation is partially decayed. A s. is intermediate between a wholly aquatic habitat and a marsh.

swamp-forest A waterlogged area in which certain trees can grow; e.g. the cypress swamps of S.E. U.S.A.; the tropical fresh-water swamps of E. Sumatra; MANGROVE (salt-water) swamps.

swash The mass of broken foaming water which rushes bodily up a beach as a wave breaks; syn. with *send*.
[*f* LONGSHORE DRIFT]

S-wave See SHAKE-W.
[*f* EARTHQUAKE]

sweep zone The max. vertical difference in surface profile of a beach during the period of observation.

swell (i) The regular undulating movement of waves in the open ocean, not breaking, and with an appreciable distance between crests; there may be as much as 12–15 m. (40–50 ft.) vertically between the crest and trough, and the longest s. measured was 1128 m. (3700 ft.) horizontally between 2 successive crests. (ii) A long, gently sloping elevation rising from the sea-floor, though its summit is still far from the surface; e.g. the Hawaiian S., 1000 km. (600 mi.) from W. to E. and 3000 km. (1900 mi.) from N. to S.

swing, of pressure belts The seasonal migration of the atmospheric pressure-belts, N.-ward in the N. summer, S.-ward in the N. winter, following the overhead sun and the THERMAL EQUATOR, i.e. the zone of greatest solar heating and low pressure.

sword-dune A type of sand-d., syn. with SEIF-D.

syenite A group of coarse-grained igneous rocks, composed mainly of alkali-felspar. One type resembles granite (though without quartz, or with the mica replaced by hornblende). It is usually classed as intermediate in chemical composition, and occurs in plutonic form.

syke, sike A small stream in the N. Pennines, specif. in Teesdale.

Symap Abbr. for the Synagraphic Mapping System, developed during recent years at the Laboratory for Computergraphics, Harvard, Mass., U.S.A. The production of THEMATIC MAPS by converting computer tabulation to graphic output on a co-ordinate basis, utilizing a grid and a standard line printer. The programme can produce contour maps, choropleth maps (where mean values are applied to an administrative unit), and PROXIMAL MAPS (where value areas are assigned according to their proximity to a datum point). See COMPUTERGRAPHICS.

symbiosis In an ecological sense, a form of adaptation in which each unit located in an area is nominally independent, but derives benefits because of its proximity and adaptation to other units. E.g. many shops in a CENTRAL BUSINESS DISTRICT are unlike each other, but each adds to the combined importance of the centre and so attracts trade for the benefit of all.

synagraphic mapping A technique of composing spatially distributed data of wide diversity into a map, graph or other visual display (*symap*), by means of a computer programme; e.g. as carried out in the Laboratory for Computergraphics, Harvard, U.S.A.

synclinal valley A v. formed by a downfold; e.g. the *vaux* (sing. *val*) of the Jura Mtns. S. v.'s are not common when much erosion has taken place, because of the development of longitudinal streams, the destruction of the anticline, and thus the formation of synclinal peaks.

syncline A downfold in the earth's crust, with strata dipping inward towards a central axis, caused by compressive forces. The s. may be broad and shallow; e.g. the London Basin, the Paris Basin; while in strongly folded districts the LIMBS of the s. may be much steeper; e.g. S. limb of the Hampshire Basin. Ct. ANTICLINE. [*f*]

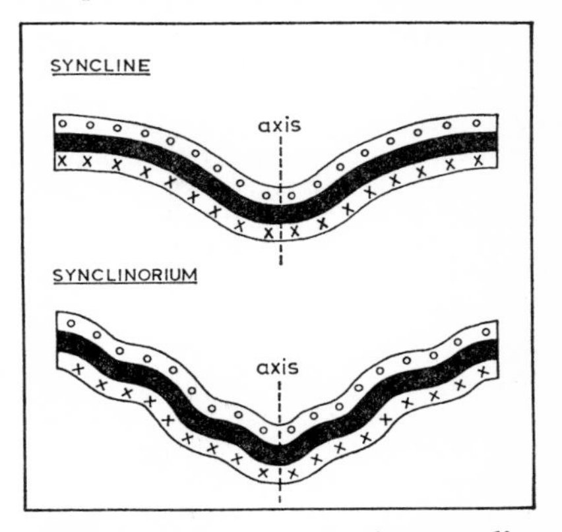

synclinorium A complex syncline upon which minor upfolds and downfolds are superimposed. [*f* SYNCLINE]

synoptic chart A c. depicting meteorological conditions (isobars, winds and other features) at a moment of time. The construction of a s. c. is essential in weather forecasting. In Britain and U.S.A. such c.'s are published daily.

synoptic climatology The overall study on a regional basis of the condition of the atmosphere at a moment of time, primarily in connection with weather forecasting. Ct. DYNAMIC C.

synoptic map (i) A cartographic technique of presenting all the various factors involved in any problem on a single m., or by a series of overlays. E.g. for choosing the site of a NEW TOWN, various m.'s of relief, slope, drainage, soil, accessibility, presence of minerals, liability to undue frost, fog and flooding, landscape value, etc. may be superimposed to give a 'cumulative picture'. Cf. SIEVE MAP. (ii) Used specif. for climatic charts, showing ISOBARS, temperatures, wind-force and direction, etc.

system (i) A division of the succession of stratified rocks, equivalent to a period in geological time; e.g. Devonian, one of the six (seven in U.S.A.) divisions of the Palaeozoic. (ii) See GENERAL SYSTEMS THEORY.

systematic geography An approach to g. by way of its various contributory aspects (physical, human, economic, historical g., etc.), each organized and presented in a s. manner; it was once known as 'general g.', from the '*geographia generalis*' of B. Keckermann (1617) and B. Varenius (1622-50). S. g. depends essentially on analysis, its complement, REGIONAL G., on synthesis.

systematic sampling (i) A method by which the sample is taken at regular intervals, using a map, e.g. at every grid intersection. (ii) By listing the features and sampling at regular intervals; e.g. every 10th item in the list. Ct. RANDOM S.

systems analysis A search for generalizations based on the whole rather than on individual parts; a consideration of a set of objects and the functional and structural relationships and organizations linking these objects. It is not a replacement for analytical methods, but an alternative additional line of scientific enquiry. See GENERAL SYSTEMS THEORY, CLOSED SYSTEM, OPEN SYSTEM.

syzygy When the sun, moon and earth are in the same line, either in conjunction or opposition.

tableland A level-topped upland, esp. where its edges are well defined; e.g. in S. Africa, terminating in the Drakensberg Mtns. Used loosely on various scales; e.g. for a MESA in Arizona, and for the Chinese T., the plateau of S. China.

tabular iceberg A large horizontal floating ice-mass in the S. Ocean, which has broken off from the edge of the Antarctic ice-barrier. These are often of great size; over 60 mi. length has been recorded.

tacheometer A THEODOLITE adapted for the measurement of distance, with which both the horizontal and vertical position of a point can be established by instrumental observations. Hence *tacheometry* or *tachymetry*, the making of a contained plan.

taconite A low-grade iron ore, consisting of ferruginous chert, which contains unleached haematite, magnetite, siderite and hydrous iron silicates. It has an iron content of 25% or less, and until recently was not economically workable. It exists in vast quantities in the L. Superior region; geologically it represents the unleached portion of the 'iron ranges' (haematite), which have hitherto supplied most of American needs. In recent years, methods of concentration into pellets for shipment have led to a rapidly increasing use of the Superior t. deposits in the steel industry of Ohio and Pennsylvania.

tactual map A m. produced for the visually handicapped, in which features are shown in raised-line symbols, with either Braille or large type names.

tafoni (Corsican patois) Hollowed out blocks of granite, as in the massif of Agriates in N. Corsica. Nightly dew-fall and daily heating lead to the capillary rise of mineral solutions, forming a surface 'varnish', or hard crust, behind which the rock-core gradually disintegrates as the 'cements' are removed. Some granite blocks may become virtually empty shells.

taiga (Russian) A needle-leaf evergreen forest, extending across the N. continents where they are at their broadest, from Scandinavia to the Pacific coast of the U.S.S.R., and from Alaska to Labrador and Newfoundland. In W. Europe they extend S. to 60° N., in E. Asia to about 50°N. The trees have to withstand very cold winters, short cool summers, a light summer rainfall, and winter precipitation in the form of snow. They grow slowly; their hard needle-shaped leaves reduce transpiration, and their conical structure helps their stability against wind and snow. Their timber is 'soft'. There are not many varieties of tree: pine (*Pinus*), spruce (*Picea*), fir (*Abies*), with larch (*Larix*), silver birch (*Betula pendula*), aspen (*Populus tremuloides*). In parts the t. becomes boggy, as the MUSKEG in Canada. *Note:* Some authorities differentiate between the main coniferous belt (BOREAL FOREST), and the thinner more open t. on the TUNDRA margins.

tail See CRAG-AND-TAIL.

tail-dune A sand-d. on the lee-side of an obstacle, tapering gradually away, varying in length from 3 m. (10 ft.) to 0·8 km. (0·5 mi.). [*f* DUNE]

talc Hydrous magnesium silicate, sometimes called *soapstone* because of its greasy or soapy feeling. It can be scratched with the finger-nail, and is given the number 1 on the Mohs' hardness scale. Most of the commercial production is marketed in ground-up form, to be consumed by the paint, cosmetics, ceramic and paper industries, and gen. as a 'filler'. It occurs widely among weathered magnesium-rich basic igneous rocks such as peridotite, or in association with dolomite and marble. U.S.A. produces nearly half the world output. The largest individual quarry, yielding 85,000 tons a year, is at Trimouns in the French Pyrenees, where it occurs in beds within masses of mica-schist.

talus By most authorities, t. is syn. with SCREE. Some would make a distinction, regarding t. as a land-feature, scree as the ingredient (broken rock-debris) of which it is formed; others refer to the t.-slope as specif. at the base of a cliff.

talwind A wind which blows up-valley, in ct. to a BERG WIND.

tank An artificial pool or small lake in India and Ceylon, used to store water for irrigation.

tanker Used gen. of any transportable container for liquids, by road, rail or sea. If the term is unqualified, it denotes a sea-going oil-tanker; these have increased rapidly in size in recent years (*super-t.'s*). The first 100,000 deadweight tonnage t.'s had a length of 296 m. (970 ft.) and beam of 40 m. (135 ft.) (cf. *Queen Elizabeth I*, 36 m. (118 ft. 6 ins.)). The largest afloat (1968) is of 312,000 d.w.t. (Japanese), 346 m. (1135 ft.) length, 53 m. (175 ft.) beam, 24 m. (79 ft.) draught. T.'s now comprise more than half the world's total.

tarn A small lake among the mountains, usually on the floor of a CIRQUE-basin, specif. in the English Lake District; e.g. Sprinkling T., Red T., Bleaberry T.

tar-sand See OIL-S.

tautochrone 'A line joining together points of varying condition or value referring to a partic. moment or period of time; e.g. soil temperatures at varying depths at 16.00 hrs. on April 1st.' (J. A. Taylor).

taxonomy The scientific classification of features according to gen. principles and laws.

tea The leaves of an evergreen shrub which thrives in a tropical monsoon climate. After picking, they are sun-dried (withered), rolled, allowed to ferment, roasted, rolled again, dried and packed for shipment as 'Indian' tea. In 'China' tea the fermentation process is omitted, leaving a greenish leaf. C.p. = India, Ceylon, China, Indonesia, Japan, U.S.S.R.

tear-fault A FAULT in which the displacement of the rocks is mainly horizontal along its line; usually regarded as syn. with STRIKE-SLIP FAULT. [*f*]

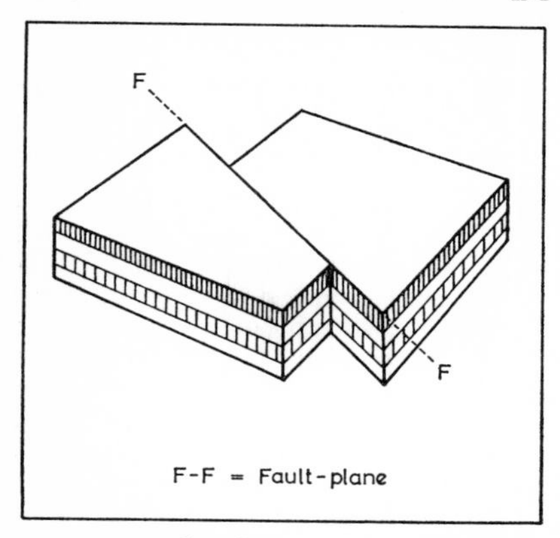

tectogenesis The processes which collectively create an OROGENIC zone, incl. FOLDING, FAULTING, OVERTHRUSTING and VULCANICITY.

tectonic From a Gk. word, *tekton*, 'builder', app. to all internal forces which build-up or form the features of the crust, incl. both DIASTROPHISM and VULCANICITY. Hence the term is applied to the features which result; e.g. *t. valley*.

tektite A small rounded stony mass, black to green in colour, with a glassy appearance; probably of meteoritic origin.

teleological An approach to a study of the world involving the doctrine of final causes, implying that developments are the result of the purpose or design served by them; a divine causality.

tellurometer A modern surveying instrument which can measure distance up to 64 km. (40 mi.) to a very high degree of accuracy, by sending short-wave radio signals from the observing station to the other point.

temperate Introduced in classical times for one of the 3 temperature zones, the others being TORRID and FRIGID. In climatic terms, t. is usually thought of as 'moderate', without any extremes, whereas in actual fact climates within the t. zone may vary considerably and be far from moderate. Thus the expression MID-LATITUDE is often preferred. T. is commonly used as a qualifying adj.; e.g. t. forest, t. grassland, t. low-pressure belt. W. Köppen included the name t. in 5 of his climatic types: *Cfa* (T. rainy, with hot summers); *Cfb* (T. rainy, with warm summers); *Cfe* (T. rainy, with cool short summers); *Csa* (T. rainy, with dry hot summers); *Csb* (T. rainy, with dry warm summers).

temperature As a climatic element, the degree of sensible heat or cold within the atmosphere. It is measured on various thermometric scales (see CENTIGRADE, CELSIUS, FAHRENHEIT, RÉAUMUR, ABSOLUTE). The highest shade t. ever recorded in the world was 58·0°C. (136·4°F.), at Azizia in Tripoli and at St. Louis Potosi, in Britain 38°C. (100·5°F.) at Tonbridge (Kent). The lowest shade t. recorded is −89°C. (−127°F.) in the Antarctic continent, in Britain −27·2°C. (−17°F.) at Braemar.

temperature-humidity index An i. constructed from various formulae, which affords an indication of the physiological effects of the weather on human comfort. One such i., used by the U.S. Weather Bureau, is $0{\cdot}4 \times (\textit{dry bulb temperature} + \textit{wet bulb temperature}) + 15$. On this scale, 60–65 represents ideal conditions, while 80 is so uncomfortable that offices and factories may be closed.

temporary base-level A b.-l. in the course of a stream, such as a lake, or a resistant stratum, or (for a tributary) the level of the main stream which it joins. A dam is a man-made t. b.-l. These are impermanent, in ct. to the ocean, the ultimate b.-l. (though even that may change as a result of relative movements of land and sea); a lake may be filled up or its outlet downcut, a hard bar may retreat upstream.

tension A stress caused by 2 forces moving in opposite directions, in ct. to COMPRESSION. In the earth's crust,

t. involves an extension of the strata, and results in JOINTS and NORMAL FAULTS.

tepee butte A residual conical hill found in Arizona, esp. in the Painted Desert. The edge of a plateau, consisting of mostly rather soft, horizontal strata of sandstone, shale, bentonite and clay, has been rapidly eroded, but a layer of more resistant sandstone near the top forms a capping which helps to preserve isolated hills, with softer material resting on the hard stratum giving them a conical appearance. These are sometimes called similarly '*tent-hills*' in Australia.

tephigram A thermodynamic diagram on which are plotted temperature and dew-point data at different isobaric levels.

tephra Used more commonly in U.S.A. for volcanic dust, ash, cinders, scoria, pumice and bombs, ejected during an eruption in solid form.

tephrochronology A geological time-scale based on the dating of volcanic ash layers found around volcanoes and widespread as buried layers in soil horizons. Introduced by S. Thorarinsson, and used with success in Iceland where volcanic eruptions have been numerous.

Terai A zone of marshy jungle along the edge of the Himalayan foot-hills, at a height of about 1500 m. (5000 ft.).

terlough In W. Ireland, a shallow depression, with a sink-hole, containing water when the WATER-TABLE rises to the surface.

terminal moraine A crescentic mound of boulder-clay deposited beyond the snout of a glacier or at the edge of an ice-sheet; syn. with *end-m.* A large t. m. indicates a prolonged stand-still of the ice; e.g. in N.W. Europe, a clearly defined line of t. m., representing the max. advance of the ice-sheet during the 4th (*Weichsel*) glacial period, can be traced through Denmark, E. Germany and Poland as the Baltic Heights or Baltic End Moraine. At several points these ridges exceed 300 m. in height. The t. m.'s of former valley-glaciers often form natural dams, thus creating long narrow lakes; e.g. the English Lakes; the Italian Lakes.

[*f* MORAINE, ESKER]

terminal velocity The rate of fall attained by a particle moving through a fluid; it represents a balance between gravitational force and the resistance of the fluid (related to the size of the particle). This is an important factor in the transportation of solid matter in SUSPENSION in a stream current, where TURBULENCE is also involved, and in the development of rain-drops and HAIL.

terrace A shelf or bench, relatively flat and horizontal, sometimes slightly inclined. A *river t.* lies along the side of a valley; it represents either: (*a*) the remains of a former FLOOD-PLAIN at a higher level dissected by renewed downcutting by a rejuvenated river, with a t. at each side (*paired* t.); or (*b*) the formation left by freely swinging MEANDERS cutting into a former flood-plain (*meander* t.). River-t.'s are carefully identified (often named) and are correlated by their height, sequence and the deposits (esp. of gravels) on them. E.g. the Thames in London has three well-known pairs: (*a*) the present Flood-Plain T.; (*b*) the Taplow T.; (*c*) the Boyn Hill T., as well as a number of intermediate minor pairs. Near Oxford are: (*a*) the Summertown-Radley T. (on which the city stands); (*b*) the Wolvercote T.; (*c*) the Hanborough T., which is 30 m. above the present river-level. See also KAME-T. [*f*], ALLUVIAL T. [*f*], MEANDER T. [*f*]

terrace cultivation The c. of a hill-slope by constructing a series of horizontal steps, usually where the relief is steep and level land is scarce. The soil is retained by building earth banks or dry-stone walls. E.g. vineyards along the valleys of the Rhine and Moselle; paddy-fields in S.E. Asia; tea-plantations in Ceylon; in the Mediterranean lands gen. Irrigation water can move down by gravity from step to step; the steps are well-drained, and SOIL EROSION is reduced.

terrace-gravel A deposit of g. on a river-t. The orig. material deposited by a river consists of alluvium and g.; the former is more easily removed when renewed erosion takes place, so that most older, higher terraces are covered with t.-g.'s, which are often named and correlated.

terracette A small terrace, from a few cm. to 0·3 m. or so high, across the face of a slope on which SOIL-CREEP is prevalent. These are sometimes called 'sheep-tracks', though it is not likely that the features have been caused by animals treading out paths, though they may be used in this way once formed.

terrain The physical character of an area, its configuration (as in '*t. studies*' and '*t. intelligence*').

terrain model See RELIEF MODEL.

terrain-type map Partic. in U.S.A., a m. on which the land-surface is divided into a number of categories (or '*t.-types*'), and shaded distinctively. Sometimes quantitative methods are used, in order to attain some degree of objectivity, and t.-t.'s such as 'nearly flat plains', 'rolling and irregular plains', 'partially dissected tablelands' and 'hills' are numerically defined. The surface of a map is gridded with squares, for each of which a value is obtained, using such criteria as max. difference in elevation, slope and proportion of near-level land. Boundaries are drawn to include all areas within the same quantitatively determined categories.

terra rossa (It.) A reddish residual clay-soil, rich in iron hydroxides, the indissoluble residue of chemical weathering, which accumulates in depressions in limestone country, notably under semi-arid or summer drought (Mediterranean) conditions, as in Yugoslavia, S. Italy, Malta. They also develop, less markedly, on limestones in Britain.

terrestrial App. gen. to the land or earth; e.g. t. radiation, -deposits, -sphere, -magnetism.

terrigenous deposits Inorganic d.'s of sand and shingle derived from the denudation of the land, and laid down in the littoral zone of the sea-floor. One of the 13 categories of marine d.'s. Ct. NERITIC, PELAGIC.

territorial waters The coastal w.'s over which a bordering state has jurisdiction; under International Law this was orig. defined as a distance of 5 km. (3 mi.), but attempts have been made by various countries to extend this. Some, notably Iceland, have already extended the width of their fishing limits to 20 km. (12 mi.). The question is becoming more acute with the recent development of offshore prospecting for oil and gas.

territory (i) Land belonging to a sovereign state. (ii) In international law, an area dependent on a sovereign state, a 'TRUST' T. or 'TRUSTEESHIP T'., supervised on behalf of the United Nations. (iii) Any large area of land. (iv) In U.S.A., Canada, Australia, areas not included in any individual state, nor admitted within the Union or Federal Government, and with a separate yet dependent organization; e.g. Yukon T. (Canada); Northern T. (Australia); Oklahoma T. (U.S.A.) before it became a state.

Tertiary A division of geological time, gen. recognized as lasting from 70 million to about 1 million years ago, but about whose use and definition there is considerable confusion. (i) In Britain, it is regarded by many as one of the 4 eras, with its corresponding groups of rocks, and with its 4 divisions (periods) of Eocene, Oligocene, Miocene and Pliocene. (ii) In U.S.A. it is regarded as a PERIOD, the earlier part of the Cainozoic era (the Quaternary being later), together with the corresponding system of rocks; this is divided into 5 epochs and the corresponding rock series: Paleocene, Eocene, Oligocene, Miocene, Pliocene. *Note:* This practice is followed in A. Holmes's revised time-scale (1959).

(iii) Some European and American workers regard it as an era (with its corresponding rock group, but divide it into 2 periods (systems): Palaeogene and Neogene; this puts the Pleistocene and Holocene (or Recent) into the Neogene, and avoids the word Quaternary.

tertiary industry One of the 3 major categories of i., incl. transport, communications, finance, trade, domestic service, personal service, the professions and public administration. Cf. PRIMARY, SECONDARY I.

Tethys The GEOSYNCLINE of mid-Tertiary times, which extended from W. to E. across what is now the Old World, between Laurasia and Gondwanaland.

Tetrahedral Theory Formulated by Lowthian Green in 1875. He envisaged the earth as a 4-faced figure, contained by 4 triangles, with its apex at the S. Pole, its base at the N. Pole (representing the Arctic Ocean). This pattern corresponds with the greater part of the land-masses in the N. hemisphere, of the oceans in the S., the triangular tapering nature of the continents, and the antipodal arrangement of land and sea. A spheroidal tetrahedron would fit the concept of a shrinking globe. But there are serious geophysical objections, and the theory is no longer considered tenable.

texture, of soil The physical quality of a s., depending on the sizes of its individual constituent particles. It may be: (i) *coarse-grained*, with a gritty feel; i.e. a sandy s., with particles between 0·02 and 2·0 mm. in diameter; (ii) *fine-grained*, with a sticky feel; i.e. a clay s., with particles less than 0·002 mm. in diameter; (iii) *intermediate*, with a silky feel; i.e. a silt s., with particles from 0·02 mm. to 0·002 mm. in diameter; (iv) a mixture of particles of many sizes; i.e. a loamy s.

thalweg, talweg (Germ.) Lit. a 'valley-way', but used to denote the longitudinal profile of a river.

thaw The physical change from snow and ice to water when the temperature rises above freezing-point, and hence is applied gen. to periods during which this happens; i.e. 'the t.'.

thematic map In ct. to a TOPOGRAPHIC m., a t. m. presents and stresses a partic. distribution, theme, topic or aspect under discussion.

theodolite A precise optical surveying instrument used for measuring angles, with both horizontal and vertical graduated arcs, a telescopic sight and a spirit-level, mounted on a tripod. Some t.'s are so accurate that with a vernier or micrometer microscope they enable fractions of a second of arc to be read.

thermal (noun) A vertically rising current or updraught of air in the atmosphere. On a sunny day, the sun's rays heat the earth's surface, parts of which (e.g. bare sand, concrete) will heat more rapidly than others (e.g. woodland or pasture), causing conductional heating and absolute instability, and therefore the vertical currents will rise more rapidly. Both birds and glider-pilots look for t.'s to assist their ascent. A t. may cause condensation, the formation of CUMULUS clouds, and possibly heavy convection rainfall and a thunderstorm. T. is also used as an adj., app. to temperature gen.

thermal depression A small-scale, though sometimes intense, low pressure system, the result of local heating and convectional rising of air, responsible for thunder and heavy rainstorms, and in hot deserts for certain storms such as DUST-DEVILS, the SIMOOM of the Sahara, and the KARABURAN of the Tarim Basin.

thermal electricity E. produced by means of steam-turbines, which create mechanical energy; this is converted into electrical energy in a dynamo. The heat for steam-raising is produced from coal, lignite, brown-coal, peat, gas, oil, a nuclear reactor. One advantage is that sub-standard unportable fuels can be consumed to produce more efficient and transportable power in the form of e.

thermal equator A line drawn round the earth joining the point on each meridian with the highest average temperature. If this is done for each month, this line will move N. and S. with the apparent motion of the sun, though for the most part it lies N. of the equator because of the larger land-masses and therefore the greater heating in summer. An annual t. e. is sometimes defined, esp. in U.S.A., as the line connecting places of the highest mean annual temperatures for their longitudes. Syn. with *heat e.*

thermal fracture The cracking or fissuring of rocks as a result of sudden temperature change. This is partic. likely: (i) where the rock is heterogeneous; i.e. containing various minerals with differing coefficients of expansion, so setting up strains; (ii) on the face of a steep crag which receives and loses the sun's rays rapidly; (iii) where there is a rapid drop of temperature after sundown. *Note:* Doubt has recently been cast on the efficacy of t. f., or indeed whether it happens at all in the absence of moisture.

thermal (or **temperature**) **efficiency index** An empirical i. devised by C. W. Thornthwaite, in conjunction with a PRECIPITATION EFFICIENCY I., as the basis for his first scheme of climatic classification. T. e. is defined for each month as $\frac{T-32}{4}$, where T = mean monthly temperature in °F.

thermal metamorphism M. caused by a rise in temperature, usually as a result of the INTRUSION of a mass of molten igneous rock at a high temperature. It can produce a fusion or recrystallization of the minerals or grains in a rock; e.g. a coarse-grained sandstone changes into quartzite; limestone into marble.

thermal spring See HOT S.

thermograph A self-recording THERMOMETER, which contains a bimetallic strip in the shape of a coil; one end is fixed, the other actuates a pen which traces a continuous record on to a chart fixed to a rotating drum.

thermokarst The formation of irregular depressions in the ground as the result of the melting of ground-ice.

thermomeion An area with a high negative temperature ANOMALY.

thermometer An instrument for measuring temperature. One type consists of a glass tube graduated with a thermometric scale (see CENTIGRADE, FAHRENHEIT, RÉAUMUR, ABSOLUTE) in degrees, containing a column of mercury or alcohol, rising from a bulb of the same liquid, which expands or contracts with temperature change. Other types include metals which expand or contract at a known extent with temperature, or which possess varying resistance to the passage of electricity with temperature change.

thermopleion A region with a high positive temperature ANOMALY.

thermosphere Syn. with IONOSPHERE.

tholoid A steep-sided volcanic dome which has grown up inside the CRATER or CALDERA of a volcano as the result of the slow extrusion of viscous acid lava. E.g. a t. grew within the caldera of Bezymianny, Kamchatka, U.S.S.R. in 1956. A t. developed in the crater of Mt. Pelée in 1902, through which a SPINE was later forced.

thorn forest Dense thickets of thorny scrub, growing in semi-arid climates, such as in N.E. Brazil. See CAATINGA.

threshold In urban geography, the conditions that any GOOD requires for entry into a CENTRAL PLACE SYSTEM. Before any good is offered for sale at a central place, all the costs involved in its production must be covered. This min. number of sales of any good represents the t. requirements for that good.

threshold population The min. number of people required to support any CENTRAL FUNCTION. It relates to the number of people who are needed in any area before it will be profitable for any GOOD or service to be distributed from a point within it. The term is simplified to relate to the number of people, although it really relates to that part of the expenditure of each person used to purchase each partic. good.

throw, of a fault The vertical change of level of strata or rocks as a result of their displacement by faulting. The blocks of rock on either side of the fault-line are referred to as upthrow or downthrow, relating to their relative displacement. The amount of throw may vary from a fraction of an inch to thousands of feet.
[*f* FAULT]

thrust A compressional force upon strata in a low-angle or a near-horizontal plane, the cause of an extreme RECUMBENT FOLD or of a very low-angled REVERSE FAULT. See T.-FAULT, T.-PLANE.

thrust-fault A REVERSE F. of very low angle, in which the upper beds have been pushed far forward over the lower beds.

thrust-plane The p. or surface of movement in a REVERSE FAULT, over which the upper strata are thrust, usually inclined at a very low angle; e.g. four main t.-p.'s can be distinguished in the N.W. Highlands of Scotland: the Glencoul, Ben More, Moine and Sole (all of CALEDONIAN age); in the HERCYNIAN folding in the N. Ardennes of Belgium is a major t.-p., the Grande Faille du Midi, over which rocks from the S. have been driven N. towards and partly over the Namur syncline; in Glacier National Park, Montana, the Pre-Cambrian 'Belt' Series has been driven E., as a result of the LARAMIDE folding, over the Lewis T.-p., so that these rocks now lie over Cretaceous strata.
[*f* OVERTHRUST]

thunder The result of the expending of electrical energy in a flash of lightning, which heats the gases in the atmosphere, causing their expansion and creating sound-waves.

thunderhead Graphic name, used esp. in U.S.A., for the top of a towering CUMULONIMBUS cloud.

Thünen's (von) Rings A concept put forward by J. H. von Thünen (1875) concerning the development of concentric rings of land-use in an isolated state with a single dominant market centre, an early example of a simple MODEL. His assumptions and empirical constants were based on his own estate near Rostock in Mecklenburg. Outside the urban-industrial trade-centre, he postulated 7 r.'s or zones in a broad open plain: (i) intensive agriculture; (ii) forest (for fuel, timber, etc.); (iii) a, b, c, progressively more extensive agriculture; (iv) ranching; and (v) waste.

thunderstorm A storm in which intense heating causes convectional uprising of air, extreme conditions of INSTABILITY, CUMULONIMBUS clouds, heavy rain and/or hail, lightning and thunder. The exact mechanism of a t. is still far from understood. *Frontal t.*'s are associated with the passage of a COLD FRONT. Other t.'s accompany OROGRAPHIC PRECIPITATION, esp. in the tropics where very warm and moist air-masses rise sharply over steep mountain ranges.

tidal current A movement of water set up in areas affected by the rise and fall of the tides. A distinction is sometimes made between the normal movement in and out of an estuary (T. STREAM), and an hydraulic t. c. set up by differences of water-level at either end of a strait due to differing tidal régimes. The latter is the stricter, more limited, usage; e.g. in the Menai Straits high tide occurs at different times at either end, resulting in a powerful t. c. flowing through the Straits. The same phenomenon takes place in the Pentland Firth, in the N. of Scotland. In the Seymour Narrows, between Vancouver I. and the mainland of British Columbia, there can be as much as 4 m. (13 ft.) difference in water-level at either end, with resulting strong t. c.'s.

tidal datum A d. obtained from a long period of t. records. It may serve as: (i) a d. for national geodetic and topographical survey; e.g. the ORDNANCE D. (O. D.) of Gt. Britain; (ii) a d. for a port authority, or for an Admiralty chart, by indicating the lowest possible depth of water, i.e. the worst navigational conditions (CHART D.), or a basis for the port tide-tables (*port d.*).

tidal flat An area of sand or mud uncovered at low tide. See also MARSH.

tidal (tide-water) glacier A valley-g. which reaches the sea, and there discharges bergs or floes; e.g. along the E. and W. coasts of Greenland.

tidal range The difference in the height of the water at low and high tide; this varies constantly, but a mean is given for most ports; e.g. about 3·7 m. (12 ft.) at Southampton, 5 m. (17 ft.) at Sheerness in the Thames estuary, 7 m. (23 ft.) at London Bridge, 9 m. (30 ft.) at Liverpool, 13 m. (44 ft.) at Avonmouth. In the open ocean, the r. may be less than 1m. The t. r. is at its max. at SPRING TIDES.

tidal stream The normal movement of water in and out of an estuary or other inlet, as a result of the alternate high and low water stages, known as the flood-t. and the ebb-t. The former may be responsible for a BORE. This powerful movement of water, apart from its obvious importance to navigation, may have marked geomorphological effects; see SCOUR. Most flood-tides seem to flow strongly for about 3 hours before high tide, and ebb for about 3 hours after, with a period of slack water about high and low tide But there are anomalies; e.g. the YOUNG FLOOD STAND at Southampton. A t. s. should not be confused with a T. CURRENT.

tidal wave An inaccurate term for a TSUNAMI.

tide The periodic rise and fall in the level of the water in the oceans and seas, the result of the gravitational attraction of the sun and moon. The strength of this attraction varies directly as the masses of the sun and moon, and inversely as the square of their distances. The sun's mass is 26 million times that of the moon, but it is 380 times farther away; its t.-producing force is 4/9 that of the moon. See also SPRING T., NEAP T., PERIGEAN T., APOGEAN T., OSCILLATORY WAVE THEORY OF T.'S. [*f*]

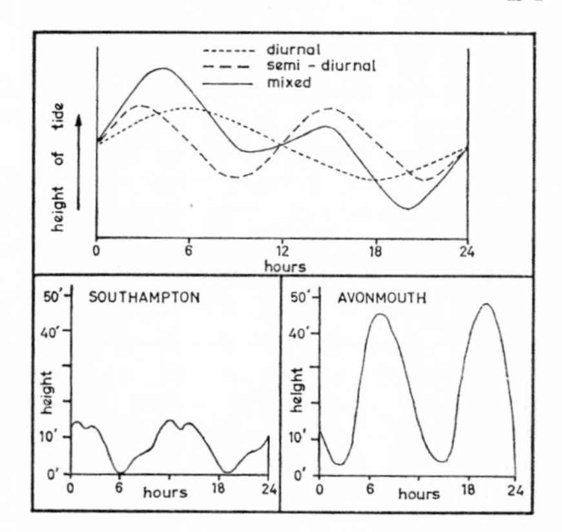

tide-mill A mill situated near the mouth of a coastal inlet. At high t. the sluices are closed, thus providing a head of water which can be let out gradually to drive the mill-wheel during the ebb-t.; e.g. at Ashlett Creek, on the W. shore of Southampton Water, now disused. The same principle is used for the tidal hydro-electricity station under construction near St. Malo, France, and contemplated for the Bay of Fundy, N. America.

Tierra (Sp.) An altitudinal climax zone, distinguished specif. in Mexico and central America: (i) *T. Caliente*, the hot tropical coastlands from sea-level to about 900 m.; (ii) *T. Templada*, the zone between about 900–1800 m., with monthly means of 18°–24°C. (65°–75°F.); (iii) *T. Fria*, the zone above 1800 m., with monthly means of 12°–18°C. (55°–65°F.); (iv) *T. Helada*, the zone of high peaks and permanently frozen or snow-covered terrain.

till An unsorted, heterogeneous mixture of rocks, clays and sands, carried within, upon or below a glacier or ice-sheet and deposited directly on melting, without any subsequent water transport. It is more usually called BOULDER-CLAY in Gt. Britain, though t. is perhaps a better word as it does not imply any precise constitution. It is increasingly used as a proper name in Britain; e.g. the Cromer T., Lowestoft T., Gipping T., which have replaced former boulder-clay names. T. is one of the two main types of DRIFT, the other being *stratified drift*, or FLUVIOGLACIAL deposits. *T. plain*: an extensive area covered with a fairly level sheet of t.; e.g. much of Illinois, Iowa and Indiana, though later stream erosion has produced modifications.

tillage The cultivation of land, particularly plough-land.

tillite A compact and very ancient BOULDER-CLAY, deposited during one of the ice-ages earlier than that of the Quaternary. See DWYKA T.

tilt-block A crustal b. left upstanding between prominent fault-lines, but bodily tilted, so giving contrasting steepness of its bounding slopes; e.g. the Basin-and-Range country in the Great Basin of Utah; the Sierra Nevada, California and the Grand Tetons, Wyoming; the Central Massif of France (with the Cévennes forming its steep S.E. margin); the Alston Block of N. England (with the steep Cross Fell Edge on the W.).

timberline (in Gt. Britain, **tree-line**) The height on a mountain-side, or the latitudinal limits, at which tree growth ceases. It is rarely an actual line, since it varies with species, latitude and climate, and locally with shelter, aspect and slope. Occasional stunted trees may survive beyond the gen. limit. The t. in S.W. U.S.A. is at about 3500 m. (11,500 ft.), up to which grow aspen and Engelmann spruce, and above which is a zone of Arctic flora or TUNDRA. In this area the Douglas fir grows to about 2900 m. (9500 ft.), the pinyon pine to about 2100 m. (7000 ft.). Some authorities in U.S.A. define: (i) a *dry t.*, a lower one, separating the forested slopes above from the semi-arid scrub below; and (ii) a *cold t.*, an upper one, separating forested slopes from the Tundra.

time-distance curve, of an earthquake A c. produced on a graph (vertical axis, time; horizontal axis, distance), showing the travel of an earthquake from its EPICENTRE. The time intervals between the arrival of the various waves increase with distance, and the curves help to determine the distance from the epicentre and the time of the earthquake. If curves from several stations are available, the position of the earthquake can be accurately determined.

time signal A world-wide radio-signal transmitted, e.g from Greenwich, at frequent intervals, indicating an exact moment of time for the regulation of chronometers and thus for the calculation of longitude.

time-transgressive Used of the climatic sequence in post-glacial times; the S. boundary of each stage varies with latitude.

time zone A longitudinal division of 15°, or less if a country is small, within which the MEAN T. of a meridian near the centre of the z. is adopted as standard for the whole, hence STANDARD T. [*f*].

tin A metal resistant to corrosion by air, water and most acids, derived from its ore, *cassiterite* (tin oxide) (SnO_2), which occurs either as 'alluvial' (or 'placer') deposits which have weathered out from metallic veins in granite (e.g. in Cornwall, Malaya), or directly from lodes (as in Bolivia). Its main use in the past was to alloy with copper to make bronze, now for coating tinplate and in some alloys. C.p. = Malaysia, Bolivia, Thailand, China, Indonesia, Nigeria.

tind (Norwegian) Syn. with HORN, and used thus occas. in English. A common mountain name in Norway;

e.g. Glittertind (2484 m., 8150 ft.) in the Jotunheim.

tipping-bucket rain-gauge A type of self-recording R.-G.

Tissot's indicatrix An application of the law of deformation, developed by M. A. Tissot in 1881, which enables the amount of angular deformation, or of areal exaggeration or reduction, to be determined for any point on a map projection.

tjaele (Swedish) Permanently frozen ground under PERIGLACIAL conditions, containing lenses of ice within the solidly frozen soil. English form = *taele*.

toadstone Sheets of dark-coloured basaltic rock occurring among the Carboniferous Limestone of Derbyshire. The t. is impermeable, and thus may create a PERCHED WATER-TABLE among the limestone, of great importance in the siting of villages. The t. also seems to be associated with mineral veins near its margins, though it contains no metallic ore itself. Hence one theory as to its name: *todstein* (Germ. for dead or worthless stone), the other being that its dark roughish surface resembles a toad's skin.

tobacco An annual plant, producing leaves which are cut, dried, cured, processed and blended for manufacture into pipe-t., cigarettes and cigars. It grows best in a subtropical or warm temperate climate. It provides a major source of revenue (through taxation) in most countries. C.p. = U.S.A., China, India, U.S.S.R., Brazil, Japan, Turkey.

tombolo (It.) A bar of sand or shingle, linking an island with the mainland; e.g. Chesil Beach joining the I. of Portland to the mainland beyond Weymouth; the island of Monte Argentario, on the W. coast of Italy between Leghorn and Rome, is linked to the mainland by 2 bars, T. della Gianetta, T. di Feniglio. [*f opposite*]

ton A measure of mass: (i) *avoirdupois* (*long t.*) = 2240 lb. (ii) *metric* = 2204·6223 lb., or 0·98421 of an avoirdupois t. (= 1000 kilogrammes) (iii) *American* or *short t.* = 2000 lb. (20 cwt. of 100 lb. each) = 0·907 of a metric t.

ton-kilometre The product of each load in tons multiplied by the distance which it travels in kilometres along a road, railway or waterway. This gives a reasonable impression of 'work done', for it makes adjustments as between long- and short-distance journeys which statistics of absolute tonnages carried would not reveal. It may, however, give a misleading impression of relative importance and activity as between one route and another, since the longer route may record a higher figure of t.-k.'s, yet may be no more active. E.g. in 1948 the R. Meuse in Belgium had 138 million t.-k.'s, the Albert Canal, Section I, 69 millions. But the Meuse is 113 km. in length, that section of the Albert Canal only 28 km. Therefore a derived average figure of t.-k. per km. may be used; on this basis, the Albert Canal $= \frac{69}{28} = 2{\cdot}43$ million t.-k. per km., while R. Meuse $= \frac{138}{113} = 1{\cdot}22$ million t.-k. per km.; i.e. traffic is twice as dense on the Albert Canal.

topographic science The collection of measurable data concerning the distribution of features on the earth's surface, and of the phenomena related to them; i.e. the whole field of surveying, photogrammetry, cartography and certain aspects of geodesy.

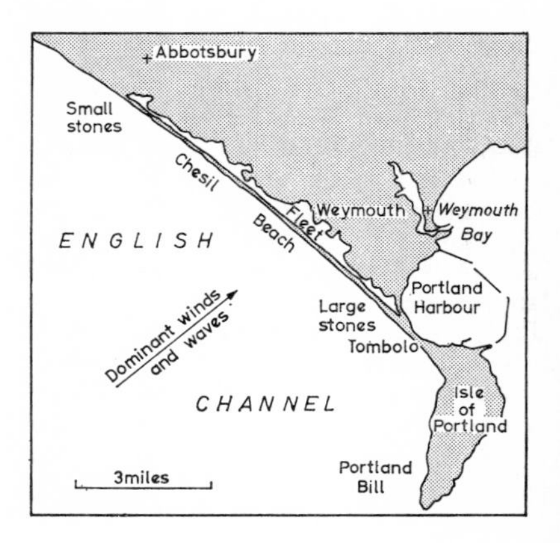

topography Derived from the Gk. *topos*, a place. The description (or representation on a map) of the surface features of any area, including not only landforms, but all other objects and aspects both of natural or human origin. Hence *topographic, topographical*, as applied to a map of fairly large scale (e.g. 1:63,360) showing t. The term is commonly but incorrectly used for relief features alone, even as syn. for relief; this should be strictly avoided. Some American authorities use the term *topographic map* as indicating a large-scale map of a small area, as distinct from a *chorographic map* of scale between 1:500,000 and 1:5,000,000, and a *global map*, on a smaller scale still.

topology A branch of geometrical mathematics concerned with order, contiguity and rel. position, rather than with actual distance and orientation. In much geographical research, topological relationships are expressed in terms of networks (*graph theory*). T. is sometimes referred to as 'the rubber sheet geometry', since a pattern on such a sheet can be deformed yet points on it remain in the same order or relationship.

topological map In ct. to a topographical m., which retains a familiar scale and orientation, a t. m., while retaining *contiguity of relationships* (e.g. boundaries, positions of towns), uses other criteria (such as e.g. density of population, area, gross income, communications system) to determine scale, i.e. is subject to a *t. transformation*. Such a diagrammatic m. may bring out novel relationships and patterns. E.g. a t. m. is seen in every London Underground ('tube') train, showing the correct sequence of stations though not to scale.

toponymy The study of place-names.

topset beds The fine material deposited on the surface of a delta, continuous with the landward alluvial plain. [*f* BOTTOMSET BEDS]

topsoil Used by a gardener or farmer for the top 'spit' or layer of mature soil; that part of the soil which is cultivated.

tor An isolated exposure of much jointed rock, e.g. granite, standing as a prominent castellated mass above the gen. surface of a plateau, notably in Cornwall and Devon; e.g. Haytor, Dartmoor, at 454 m. (1490 ft.). The origin of t.'s has been the subject of much controversy. D. L. Linton suggested that they result from sub-surface (rather than subaerial) rotting of granite, through the action of acidulated rain-water penetrating along the joints into the body of the granitic mass. The pattern of the t. is controlled by the joints, which leave between them broadly rectangular 'core-stones'. Where the jointing is widely spaced, massive core-stones remain; where the jointing is close, there is more shattering and removal, forming depressions between the t.'s. This may have taken place in pre- or interglacial times. Then followed a period of exhumation, when the overlying weathered material and the fine-grained products of the rock-decay were removed by melt-water or SOLIFLUCTION, thus revealing the 'blockpile' character of the t. J. Palmer believes that PERIGLACIAL processes account for the formation of t.'s. [*f*]

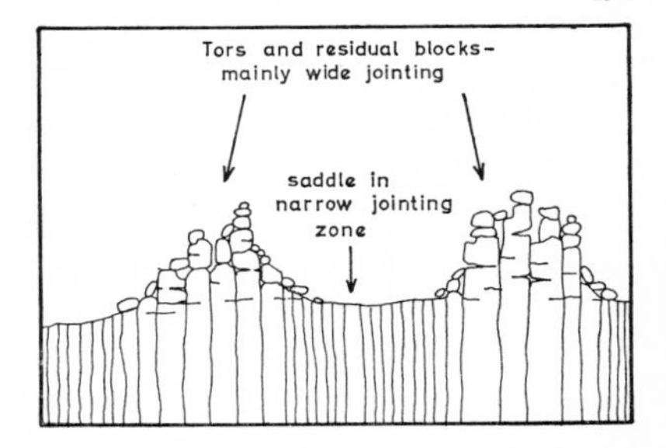

tornado (i) A counter-clockwise whirling storm (sometimes called a '*twister*' in U.S.A.), formed around an intensely low pressure system, with winds of great velocity (300 km.p.h. plus), and often a dark funnel-shaped cloud. It occurs partic. in the Mississippi basin in spring and early summer. It is associated with a trough of low pressure, where cool air from the N. and warm, damp air from the Gulf of Mexico come into frontal contact (commonly along a SQUALL-LINE). Local heating contributes to a vortical uprush of air. It is very short-lived, lasting for only an hour or two, and is usually only a hundred metres or less across. Its destructive effects are limited in extent, but it can cut a swathe across a town. Not only are buildings destroyed by the winds, but the very low pressure at the centre causes them to collapse outwards. T. warnings are put out by radio from a centre in Kansas. (ii) A name given to a squall associated with a thunderstorm and torrential rain on the coast of W. Africa.

torrent A rapidly flowing stream in an upland area; hence *t. tract*, or mountain tract, of a stream.

torrid Introduced in classical times for the warmest of the 3 latitudinal temperature zones they designated, meaning 'burning' or 'hot'. Hence *t. zone*, the others being temperate and frigid.

Torridonian, Torridon Sandstone The upper of the 2 main Pre-Cambrian divisions of rocks in N.W. Scotland and the Hebrides. It is a stratified deposit, often horizontally bedded or with a gentle dip, including red felspathic grits and sandstones. It is characterized by much false-bedding. The constituent grains are mainly rounded, and it appears to have been laid down in water. It rests unconformably on the denuded surface of the Lewisian Gneiss, the other main Pre-Cambrian division.

tourism The study of the geographical aspects of the tourist industry; e.g. G. W. S. Robinson's 'Touristic Map of Europe'.

town A compact settlement larger than a village, with a community pursuing an urban way of life. In Gt. Britain the term has no administrative connotation; an urban district, borough or county borough may all be t.'s, with some form of municipal administration. Many have specialized functions: e.g. market-t., seaside-t., mining-t., bridge-t., fortress-t., colonial-t., fishing-t., garrison-t., railway-t., resort-t., port-t., spa-t., dormitory-t., walled-t. Others have locational features: gap-t., riverside-t., estuary-head-t. Note the special usage of 'ghost-t.' (now uninhabited), 'shanty-t.' (a group of shacks and hutments in some African and American suburbs), 'chinatown', DOWNTOWN, 'twin-t.'s' (Minneapolis–St. Paul, Hastings and St. Leonards, Brighton and Hove, Reigate and Redhill), NEW T.

town map A m. forming part of the specif. DEVELOPMENT PLAN, showing on a scale of 6 in. to 1 mi. the future uses proposed for land within a town. Such maps were prepared under the Town and Country Planning Acts of 1947 and 1962, and will be superseded by STRUCTURE PLANS and LOCAL PLANS showing respectively less and more detail under the Town and Country Planning Act (1968).

townscape The whole urban landscape, the form and pattern of buildings, streets and open spaces. In an effort to evaluate this, a scale has been suggested by K. D. Fines of: (i) slums and derelict areas; (ii) modern industrial and commercial areas; (iii) modern suburbs; (iv) towns of architectural and historical interest; (v) 'classic' towns (e.g. Florence, Venice, Edinburgh).

township (i) A CROFTING district in Scotland, with a number of individual crofts, each with its own land, together with common grazing. (ii) In U.S.A. it is used in two senses in connection with the LAND SURVEY SYSTEM [*f*]: (*a*) as a 6-mi. square, the basic unit in the survey, known as a *congressional t.*; (*b*) the N. component in locating and identifying the t.'s; i.e. t.'s run from S. to N. parallel to the central meridian.

trace element A minute quantity of an e., such as boron, manganese and iodine, which is usually present in the soil; any deficiency may have adverse effects upon the development of organic cell-structure in plants and animals. Its presence or absence can be detected, but quantitative determination demands refined techniques.

trachographic map Used by E. Raisz for a type of m. showing ruggedness by means of perspective symbols, using 2 major elements: relative relief and average slope.

trachyte An extrusive, fine-grained igneous rock, composed mainly of alkali-felspar, with crystals of orthoclase, plagioclase and hornblende; it has a distinctive flow-structure. In composition it is regarded as intermediate. Phonolite is a variety of t. Many of the puys of Auvergne, France, are of t.; e.g. the Puy de Dôme; and there are outcrops in the S. Uplands of Scotland.

tract (i) A unit of the 2nd smallest order of magnitude in the hierarchy of regional divisions, devised by J. F. Unstead in 1933. (ii) The 3rd order of unit in D. L. Linton's system of MORPHOLOGICAL REGIONS. (iii) The term is used as a Census area unit in Canada and U.S.A.

traction load, of a river Syn. with the BED-LOAD of a r.

Trade Winds Winds blowing from the Sub-tropical High-Pressure 'cells' towards the Equatorial Low from a N.E. and a S.E. direction in the N. and the S. hemispheres respectively, hence N. E. T. and S. E. T. winds. The name comes from the phrase 'to blow trade' (Lat., *trado*), i.e. in a constant direction, and has nothing to do with commerce. The w.'s are noted for their constancy of force and direction, esp. over the E. side of the oceans, though there are interferences by pressure disturbances (esp. E. waves) near the W. sides of the oceans.

trading estate A comprehensively planned industrial e. designed primarily to diversify employment opportunities in areas of high unemployment or unbalanced industrial structure (as where there is undue dependence on a major basic industry). Facilities offered include services, standard designed factories, and centralized information, publicity and administrative services. E.g. Trafford Park (Manchester), Slough (Buckinghamshire), Team Valley (Durham), Lillyhall (Workington, Cumberland).

traffic flow diagram See FLOW-LINE MAP.

train A series of uniform ocean waves passing through deep water.

trajectory The path followed by a single 'parcel' of air moving during a period of time. Ct. STREAMLINE.

tramontana (It., Sp.) A cold dry N. or N.E. wind in the W. Mediterranean basin; the name is applied commonly to any wind blowing down from the mountains, as in Italy and Spain. It occurs in Corsica in winter behind a depression, commonly bringing snow and bitter weather, and sometimes affects the Balearic I's.

transcurrent fault Syn. with TEAR-FAULT.

transgression An extension of the sea over a former land-area as a result of a positive movement of sea-level; this may be caused either by a EUSTATIC rise of sea-level, or by an actual sinking of the land. E.g. the Flandrian T. of post-glacial times, which created the S. part of the N. Sea, the Straits of Dover and the English Channel (a eustatic rise).

transgressive intrusion Syn. with DISCORDANT I.

transhumance The seasonal movement of men and animals to fresh pastures. There are three main categories: (i) *Alpine* or *Mountain*, a movement from the valley floors to the high summer pastures (ALPS or SAETERS) for the summer, returning in autumn to the valleys, as in Switzerland and Norway; (ii) *Mediterranean*, a movement from the drought and heat of the lowlands in summer into the mountains, as in Spain; (iii) in semi-arid grassland margins, a movement of *nomadic pastoralists* near the borders of the deserts according to fluctuations in rainfall and therefore pasture, usually following set seasonal tracks.

transit In surveying and astronomy, the apparent passage of a heavenly body over the MERIDIAN.

transit trade Freight traffic which passes from one country to another across a third; cf. international t. which passes from one country directly to another. E.g. a vast volume of freight enters the Netherlands via Rotterdam (incl. Europoort, 'the gateway of Europe'), and moves up the Rhine valley to W. Germany, Switzerland, etc. Luxembourg derives considerable revenue from t. rail-freight across its territory.

transliteration The rendering of a geographical name from one alphabet to another; this is an important problem in map compilation. Akin to this is the rendering of a name in an official language in ct. to the conventional English form; e.g. *Suomi* (Finland); *Bruxelles* (Brussels); *Lac Léman* (L. Geneva); *Firenze* (Florence); *Athenai* (Athens); *Livorno* (Leghorn).

transpiration The loss of water-vapour from a plant through the minute pores (*stomata*) which cover the leaf-surface. See also EVAPOTRANSPIRATION.

transport (of sediment, etc.) In the whole process of denudation, t. is that phase which is concerned with the actual movement of material by some natural agent: rivers, glaciers, ice-sheets, the wind, waves, tides and currents. It does not include mass-movement by gravity. The material transported is the LOAD, which itself acts as an eroding agent (abrasion), and suffers progressive diminution by impact both against other parts of the load and the surface over which it is being transported (attrition).

transverse coast See DISCORDANT C.

Transverse Mercator Projection A case of the MERCATOR P. in which the cylinder is tangential to the globe not along the equator, as in the normal case, but along a meridian; i.e. it has been turned transversely through 90°. The central meridian is divided truly. The p. is used mainly for maps of small areas with the main dimensions from N. to S. It is used for all British Ordnance Survey maps, and as a basis of the NATIONAL GRID. It is also used for the American U. T. M. GRID. The scale error increases away from the central meridian. Hence on the T. M. P. for Gt. Britain the lines of correct scale are transferred 200 km. E. and W. of the 2° W. central meridian, resulting in a negative error between them, a positive error outside them. The overall error is thus spread out, and in effect is halved. On this p. the LOXODROMES are curved lines, not straight as in the normal case. It is also known as the *Gauss Conformal P.*

transverse valley A v. which breaks across a ridge at right-angles; e.g. a *cluse* in the Jura Mtns.

trap (i) In a cave-system part of which is below the WATER-TABLE, a t. is formed where the roof of the cave dips down below the water, but rises again above the water-level some distance further on. This involves diving, sometimes necessitating 'frog-man's' apparatus, and is one of the main hazards (and thrills) of speleology. (ii) A gen. name applied to dark-coloured basaltic rocks, notably occurring in DYKES and SILLS. (iii) A structural arrangement in the rocks favourable for the accumulation of oil and or natural gas.

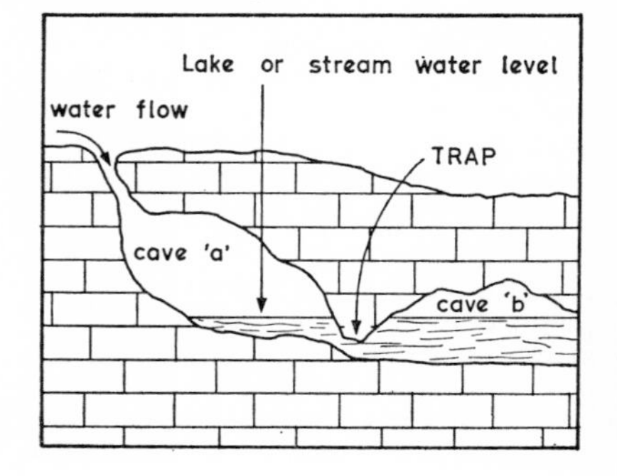

traverse A surveyed line. The easiest method is by using a magnetic compass, hence a *compass t.*, entering the observations in a field-book. In such a survey method, the error is cumulative, and at the end may be considerable; this error is distributed proportionally along the t. A series of such lines (or 'legs') can be linked together in a circuit to make a *closed t.*, from which a map of the whole area can be made.

travertine A crusty deposit of calcium carbonate formed from a strong solution around a hot spring; sometimes it is called *calc-sinter*, or calcareous tufa; e.g. the Mammoth Terraces, Yellowstone National Park, Wyoming, which are magnificent cascades of delicately coloured t.

treaty port A p., sometimes even an inland city, open by treaty to representatives of a foreign power for trade and residence, with special privileges. After the Opium War in China (1840–2), the Chinese opened five t. p.'s, followed by a large number of others; at one time there were as many as 56. These included: (for Gt. Britain) Amoy, Canton, Hankow, Tientsin, Shanghai; (for Japan) Amoy, Foochow, Hankow, Hangchow, Tientsin and Newchwang; (for France) Canton, Hankow, Shanghai and Tientsin; (for Belgium and Italy) Tientsin. A t. p. was a concession where foreigners were able to live, manage their own affairs, levy taxes, have their own legal courts (i.e. with extra-territorial jurisdiction), own warehouses, and other privileges. The rights of t. p.'s were given up in 1943.

tree-line The l. or zone marking the limit of tree growth, both altitudinally on a mountain and latitudinally as regards distance from the Poles. It is known in U.S.A. as the TIMBERLINE.

trellis(ed) drainage A rectilinear pattern of d. with CONSEQUENT, SUBSEQUENT, OBSEQUENT and SECONDARY CONSEQUENT streams, usually occurring in scarplands where outcrops of alternately more and less resistant rocks occur at right-angles to the initial slope, and where adjustment to structure has occurred. [*f*]

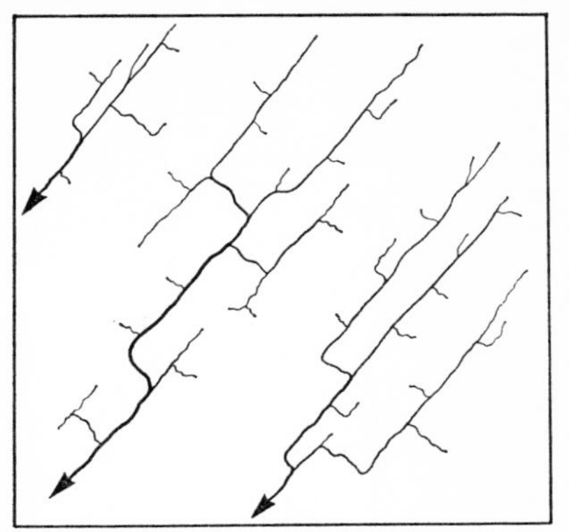

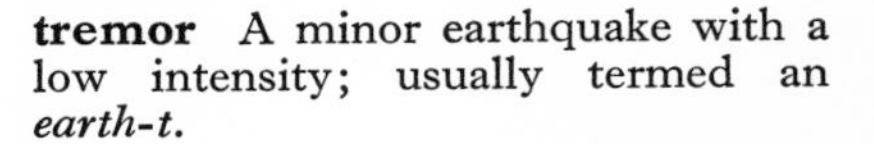

tremor A minor earthquake with a low intensity; usually termed an *earth-t.*

trench (i) An elongated trough or DEEP in the ocean floor; e.g. the Mariana T. off Guam in the W. Pacific. (ii) A U-shaped valley in the mountains.

trend line The gen. tendency or pattern of the structural l.'s in an area (e.g. folds, faults, dykes); akin to GRAIN.

trend surface map A m. which isolates, measures and depicts quantitative components in any geographical pattern. The 2 methods which can be used are: (i) *filter mapping*, and (ii) *nested sampling*. In (i), where complete data are available, the area is covered with a grid, and the pattern is expressed as a ratio or value for each grid-sq. (e.g. forested/non-forested); isopleths are then interpolated. This gives a response s., resulting from both regional and local factors, which can be filtered to show local anomalies. (ii) Can be used when information is incomplete, as in a reconnaissance or exploratory survey, when a few equal-sized regions are selected at random, and then broken down, also at random, into smaller units. Values for each level can be determined using an appropriate type of variance analysis. Thus by either methods (i) or (ii), the variability of areal patterns can be broken down and sampled. T. s. mapping can be used partic. effectively to indicate the possible existence of a dissected surface and to reconstruct former EROSION SURFACES.

triangle of error In finding the position of a point on a plan by RESECTION [*f*], using a plane-table, if three intersecting back-rays are drawn from known objects, a t. of e. will result. (i) If the observer's position is within the t. formed by the 3 known points, the position on the plan will lie within the t. of e. To fix this more exactly, the true position will be vertically away from each ray a distance proportional to its total length. (ii) If the observer's position is outside the t. formed by the 3 known points, the true position will be either to the right or to the left of all 3 rays, at a vertical distance from each proportional to the length of each. In each case, estimate the true position by eye, realign the plane-table, and check; the t. of e. should have been eliminated, or at least be much smaller, in which case the operation is repeated.

triangular diagram The plotting of 3 related or associated aspects of some feature or item on t. graph-paper, ascribing a max. value for each aspect to each apex of the triangle; e.g. 3 aspects of climate (pressure, temperature, humidity), population age (young, middle-aged, old), sediment (sand, clay, silt), hill-slope analysis (% of slope-angle observations falling within summit, midslope and basal portions, as used by A. F. Pitty). [*f*]

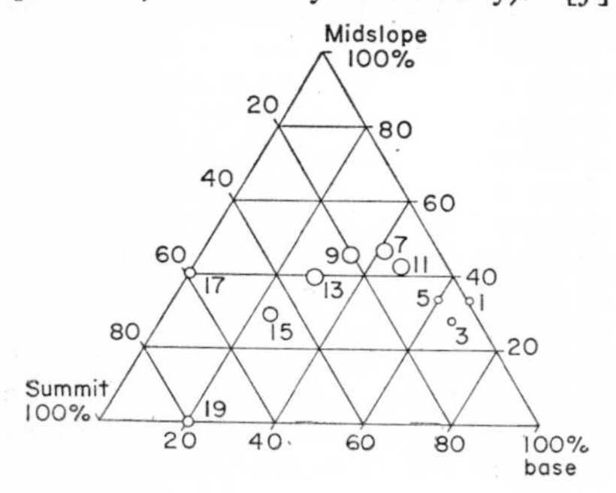

triangulation A system of triangles developed by angular measurement with a theodolite from a BASE-LINE, thus affording an accurate geodetic framework for a topographical survey. The principles of t. are lost in antiquity, but the 1st national t. was made by the Cassini family in France in the 18th century. The current O. S. maps in Britain are based on a new t., commenced in the 1930s. The t. may be *primary*, *secondary* and *tertiary*, according to the size of the triangles and the standards of accuracy.

Triassic The 1st of the geological periods of the Mesozoic era, and the system of rocks laid down during that time, which lasted from about 225 million to 180 million years ago. The system was named by F. A. von Alberti in 1834 from its 3-fold division: (i) *Bunter* (sandstones and pebble beds); (ii) *Muschelkalk* (shelly limestone); (iii) *Keuper* (red sandstones and marls, with beds of rock-salt and gypsum). In W. Europe *Muschelkalk* is quite widespread (e.g. in central Germany), but there is no trace in Britain. The rest of the T. is strikingly developed in the Bristol Channel area, the W. Midlands, S.W. and N.E. Cumberland, and in small areas of N. Ireland. The T. rocks are difficult to distinguish from the underlying Permian system, and commonly the term New Red rocks is used for the Permo-Trias. See RHAETIC.

tributary A stream or river which joins a larger one.

triennial fallow A pattern of arable farming, probably brought to Britain by the Romans, which persisted until the introduction of modern ROTATION OF CROPS; it consisted simply of 2 years' cereals, followed by 1 year off.

'trigger action' In an AIR-MASS, any process which initiates the development of CONDITIONAL INSTABILITY; e.g. mechanical uplift (a wind), heating of the surface air by CONDUCTION followed by CONVECTION, increase in the water vapour content of the atmosphere, and the rising of a warm air-mass over a cold air-mass along the WARM FRONT in a DEPRESSION.

trimetrogon photography A method of rapid reconnaissance mapping, using 3 airborne cameras (pointing vertically and at 60° to port and starboard), exposed simultaneously along the flight-lines.

tropic One of the 2 parallels of latitude of approx. 23½° N. and S. (see CANCER, CAPRICORN). In the pl., broadly the zone between these 2 parallels, hence *tropical*, app. to this zone, as -forest, -grassland, -climates.

tropical air-mass An a.-m. which has originated within the sub-tropical ANTICYCLONE belt, either over the ocean (*tropical maritime*) (*Tm*, or *mT* in U.S.A.), or over a continental interior (*tropical continental*) (*Tc* or *cT*).

tropical climates A group of climatic types occurring within the tropics, but of varying definition. (i) W. Köppen defines t. c.'s as having an average temperature for each month above 18°C. (64·4°F.), thus no winter season, with a considerable precipitation, mostly of a convectional nature with a marked summer max. Köppen includes three main types: *Af.* T. Rain-forest; *Am.* T. Monsoon; *AW.* T. with dry winter, hence SAVANNA. (ii) A. A. Miller's classification includes (*a*) T. Marine; (*b*) T. Marine (Monsoon); (*c*) T. Continental (summer rain); (*d*) T. Continental (Monsoon).

tropopause The discontinuity plane between the TROPOSPHERE and the STRATOSPHERE, marked by an abrupt change in the lapse-rate; it varies in height slightly with the seasons, but lies at about 18 km. (11 mi.) above the equator, 9 km. (5·5 mi.) at latitude 50°, and 6 km. (4 mi.) at the Poles. Recent research shows that the t. is not a single plane, but a series of overlapping planes.

tropophyte A plant, not XEROPHYTIC, which possesses various adaptations (e.g. leaf-shedding) to enable it to survive a period of seasonal adversity (cold or drought). It behaves as a XEROPHYTE at one period of the year, as a HYDROPHYTE at another.

troposphere The lower part of the atmosphere, from the surface of the earth to the TROPOPAUSE.

trough (i) An elongated TRENCH or DEEP in the ocean floor. (ii) A U-shaped valley, as a glacial t.; e.g. Lauterbrunnen, Switzerland; Yosemite, California. These t.'s are of varied size and form, and have been usefully classified by D. L. Linton into four types: (*a*) *Alpine*, where the t.'s occupy pre-glacial valley systems; (*b*) *Icelandic*, where plateau-accumulation of ice rather than high ground has led to discharge into peripheral valleys; (*c*) 5 sorts of *composite* types; and (*d*) *intrusive* t.'s, which have been formed by lowland ice intruding into and through uplands. (iii) A depression between the crests of two successive waves in the sea. (iv) A narrow elongated area of low atmospheric pressure between 2 areas of higher pressure.

trough-end A steep rock-wall forming the abrupt head of a glaciated valley; e.g. Warnscale Bottom, the head of the Buttermere valley, English Lake District; the t.-e. at the head of the Rottal, under the S.W. face of the Jungfrau, Switzerland.

truck-farming In U.S.A. the specialized cultivation of vegetables and fruit, grown further away from markets than market-gardening, and involving transport.

true dip The max. d. of a stratum, as ct. with APPARENT D. [*f* STRIKE]

true north The direction of the Geographical N. Pole along the meridian through the observer. Ct. MAGNETIC N., GRID N.

truncated spur A s. which formerly projected into a pre-glacial valley, partially planed off by a glacier which moved down the valley; e.g. the t. s.'s of Saddleback (Blencathra), a mountain lying E.N.E. of Keswick in the English Lake District. [*f*]

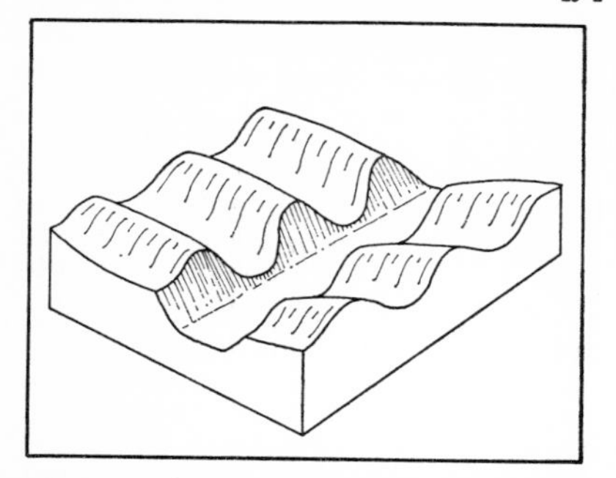

trust territory, trusteeship An area administered by another country, appointed by and responsible to the United Nations, with the intention of enabling that area ultimately to attain self-government; e.g. New Guinea (t. t. of Australia); W. Samoa (of New Zealand).

tsunami (Japanese) A large-scale seismic sea-wave (incorrectly known as a tidal wave), caused by an earthquake shock in the ocean-floor. It may travel for long distances over the sea, then surge over the land margins, where it may cause considerable damage. In the open ocean the wave height may only be a foot or so, but the wave-length may be 80–190 km. (50–120 mi.), travelling at 640–960 km.p.h. (400–600 m.p.h.). On entering shallow water the wave steepens rapidly up to 15 m. (50 ft.) or more. The eruption of Krakatoa in 1883 caused a t. to travel right round to the other side of the world. Another t. off Japan in 1896 drowned 27,000 people and destroyed nearly 11,000 houses.

tufa (It.) A porous, cellular or spongy type of TRAVERTINE, deposited around the point of issue of a spring of calcareous water.

tuff A rock formed from a compacted or cemented mass of fine volcanic ash and dust, with particles less than 4 mm. in diameter. Ct. BRECCIA. See also WELDED TUFF.

tuffisite Fragmentary material produced by explosive volcanic activity, but deposited in the PIPE, rather than over the surface as TUFF. E.g. in the Swabian Jura, W. Germany.

tumulus See BARROW.

tundra (Lapp) A zone between the N. latitudinal limit of trees and the polar regions of perpetual snow and ice, experiencing a climate with a brief summer above freezing-point (W. Köppen's *ET* type), in which the mean temperature of the warmest month is below 10°C. (50°F.), but above 0°C. (32°F.). It is characterized by PERMAFROST, a vegetation cover of dwarf shrubs, berried plants such as cloudberry in Norway and cranberry in Canada, herbaceous perennials, lichens (e.g. *Cladonia*) and mosses, with no trees other than a few stunted Arctic willow and birch, and with widespread marshland in summer. The pattern of vegetation in detail is very varied, in response to slight differences in slope, aspect and drainage.

tungsten A metal mostly obtained from the minerals wolframite and scheelite, commonly found in VEINS cutting through metamorphosed sedimentary rocks. T. has a very high melting-point and high tensile strength, and is used for high-speed tool-steels, and wherever a super-hard steel is required. *T.-carbide* is probably the hardest known cutting agent. T. is also used for filaments in electric light bulbs. C.p. = China, U.S.S.R., U.S.A., S. Korea, N. Korea, Bolivia, Australia.

turbary The right of digging and removing peat or turf.

turbidity The churning up of sediment by water, forming a dense, heavy flow, the sediment remaining in suspension, esp. in the sea. This applies partic. to bottom currents on the ocean floors, which are believed to have important erosional effects on the continental slope, and also to currents formed when rapidly flowing, sediment laden, river waters enter a lake. See SUBMARINE CANYON.

turbulence (i) An irregular eddying flow, in ct. to a smooth laminar flow, partic. used in a meteorological context for the flow and mixing of air by this means. Though it is easy enough to say what it is in gen. terms, the mathematical definition and explanation of the phenomenon is a difficult problem. Instability due to the uneven heating of the earth's surface may cause the rapid rise of 'pockets' of air. This may make an aircraft distinctly uncomfortable, as it abruptly rises and falls; see THERMAL. T. can usually be recognized by the presence of CUMULONIMBUS clouds and thunderstorms, but '*clear-air*' t. may occur, esp. in a JET-STREAM, which is difficult for a pilot to avoid. (ii) In a stream t. contributes to the transport of material in suspension. (iii) T. is a marked feature of an ocean DRIFT, which has an overall direction of water movement, but contains irregular speeds and directions within it.

turnpike A gate across a toll-road, opened only when the required toll is paid. Toll-roads in Britain constructed in the late 18th century became characterized as t. roads. The term has been revived in U.S.A.; e.g. the Pennsylvania T.

twilight The reflection of the sun's light before it has appeared above the horizon in the morning and after it has disappeared below it in the evening. Its duration depends on the angle made by the sun's path across the horizon, i.e. on date and latitude. (i) *Astronomical t.* extends in the morning from when the sun's centre is 18° below the horizon until dawn, and in the evening from sunset until the sun's centre is 18° below the horizon, indicating theoretical perfect darkness. In high latitudes at times around the SOLSTICE the sun's centre is never as much as 18° below, and thus twilight lasts from sunset to sunrise. (ii) *Civil t.* extends from and to 6° respectively. Lighting-up time for vehicles is approx. civil t. (iii) *Nautical t.* extends from and to 12° respectively.

'twilight area' An a. of a town where, because of ageing buildings, obsolete layout and inadequate facilities, environmental standards are deteriorating, even though the buildings themselves may not be unfit for use.

tychoplankton Minute organisms, animal and vegetable, that have been transported from the margins of lakes and ponds into those bodies of water by currents.

typology The study or classification of types. Urban t. should not be confused with URBAN HIERARCHY, which refers to specif. functional characteristics.

typhoon A small, intense, vortical tropical storm in the China Sea and along the margins of the W. Pacific Ocean, accompanied by winds of terrific force (160 km.p.h., 100 m.p.h. plus), torrential rain and thunderstorms. Cf. CYCLONE, HURRICANE.

ubac (Fr.) A hill-slope, esp. in the Alps, which faces N., and so receives minimum light and warmth; cf. Germ. *Schattenseite*. It is usually left under forest, while settlements, terraced agriculture, and meadows (ALPS) occur on the S.-facing slopes. Ct. ADRET.

Uinta structure A type of structure named after the Uinta Mtns. in N.E. Utah, where a broad flattened anticlinal flexure was upraised in late Cretaceous times, and then extensively denuded, so that Pre-Cambrian rocks (sandstones and quartzites) are exposed over most of its surface. At the end of the Eocene, the worn-down mass was again bodily uplifted, with major faults flanking it to S. and N. The Uinta Mtns. experienced further uplift during late Pliocene-Pleistocene times; their highest point rises to 4114 m. (13,499 ft.) in King's Peak.

ulotrichous The quality of having 'woolly' hair; hence in some classifications of mankind (e.g. by A. C. Haddon) 1 main category is of the

Ulotrichi, with 2 divisions: *U. Orientales* and *U. Africani*. Ct. CYMOTRICHOUS, LEIOTRICHOUS.

ultrabasic rock An igneous r. containing less than 45% silica, and more than 55% basic oxides, mainly ferromagnesian silicates, metallic oxides and sulphides; e.g. peridotite, consisting mainly of olivine.

ultraviolet rays That part of the solar RADIATION which lies just beyond the blue end of the visible spectrum, with a wave-length range from 4×10^{-5} cm. to 5×10^{-7} cm. Much of the u. light is absorbed by ozone molecules in the upper atmosphere, but some reaches the earth's surface; the intensity is most marked in high mountains.

Umland (Germ.) Formerly included gen. within HINTERLAND, now used specif. as an 'urban hinterland': the sphere of influence, economic and cultural, of a town. The term was used in this sense by E. Van Cleef in 1941, though much earlier by Germ. geographers in the general sense of 'surroundings'.

unconformity A break or gap in the continuity of the stratigraphical sequence, where the overlying rocks have been deposited on a surface produced by a long period of denudation. Strictly, an overlying stratum does not conform (i.e. is *unconformable*) to the dip and strike of the underlying strata. See also NONCONFORMITY, DISCONFORMITY, ANGULAR U., which are various types of u. *[f opposite]*

underclay See SEAT-EARTH.

undercliff A mass of material at the base of a cliff, the result of falls due to weathering; this is partic. pronounced where chalk lies over clay. E.g. the U. along the coast of the I. of Wight to the S.W. of Ventnor; along the Devon coast between Seaton and Lyme Regis.

undercutting Partic.: (i) the erosive action of a river current as it impinges against its bank on the outside of a bend; hence undercut slopes; (ii) the erosion by sand-laden wind near the base of a rock in the desert; (iii) of coastal cliffs; e.g. the Dorset coast, where removal of the Lower Portlandian Sands has undercut the overlying Purbeckian Beds.

underdeveloped land An area where the natural resources are not developed or used to the best account, usually through lack of capital, technological backwardness, and absence of stimulus. The term is less invidious than 'backward', although it is often used too gen., as where 'undeveloped' would be more appropriate. The modern concept of the term was embodied in President Harry S. Truman's 'Point Four' of his Inaugural Address in 1949; hence 'Point Four Aid'.

underfit App. to a stream; syn. with MISFIT.

underground stream In areas of rock characterized by joints and fissures, water flow below ground may be concentrated in a distinct channel. The courses of such u. s.'s, from SINKHOLE to RESURGENCE, have been traced in a number of cases, sometimes by actual exploration, sometimes by putting fluorescein (green colouring) into the water. These are frequently far more complicated than one would suspect from the surface; e.g. the R. Aire at Malham, Yorkshire. N. Casteret proved that the Garonne

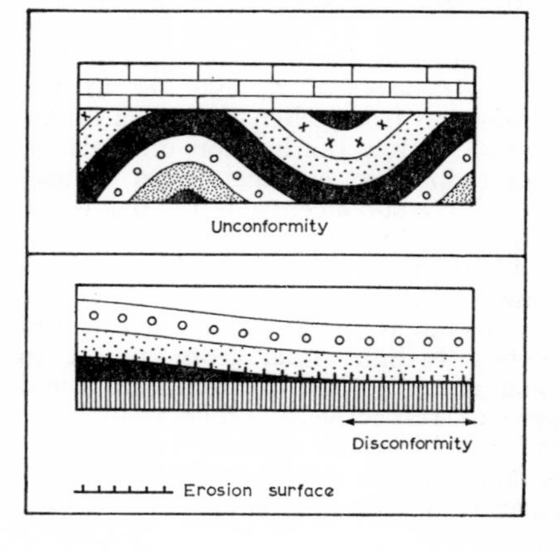

rises in Spain, on the S. side of the main Pyrenean watershed, which it penetrates as an u. s.

undertow An undercurrent flowing down the beach near the bottom of the water, the result of a back-flow of water piled up on the beach by a breaking wave.

unequal slopes, Law of A principle which states that where the opposing s.'s of a ridge are steep and gentle respectively, the former will be eroded more rapidly, and the ridge-line will therefore move back on that side. This process is involved in the recession of an ESCARPMENT and the development of a COL.

uniclinal Used by some authorities for beds of rock dipping uniformly and evenly in one direction; hence *u. structure*. Others use 'monoclinal', but this is inadvisable, as the word MONOCLINE has specif. implications.

uniclinal shift If the cross-profile of a river valley is asymmetrical, esp. one which lies along the line of the STRIKE in an area of gently dipping rocks, the course of the river may tend to migrate down the DIP, and the escarpment forming the valley-side on the down-dip side will move sideways in a similar manner. [*f*]

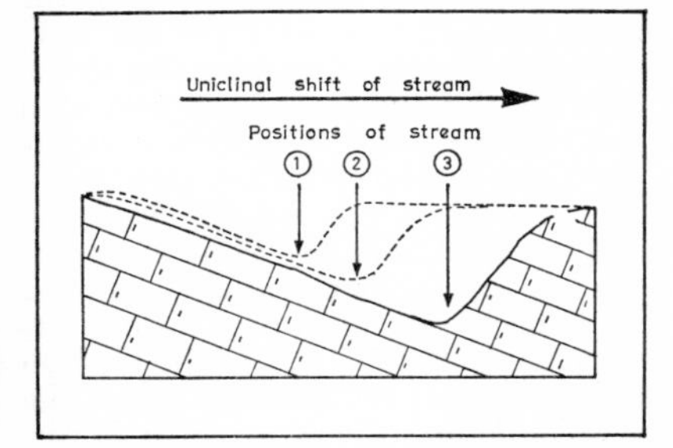

uniformitarianism An important geological principle, established by J. Hutton in 1785, and formulated by C. Lyell in 1830, that processes and natural laws which existed in geological time are sensibly those that may be observed today. This is contrasted with CATASTROPHISM, which holds most physical phenomena to be the result of short-lived and exceptional events.

unloading See PRESSURE RELEASE.

unroofed anticline See BREACHED A.

unitary area Defined in the report of the Royal Commission on Local Government in England (the Radcliffe-Maud Report, 1969) as the future primary local government unit, responsible for all services, and covering both a town and its surrounding a. As far as possible, the aim is to make a single authority responsible for the entire sphere of influence of each major English town.

unstable equilibrium The state of the atmosphere where the actual LAPSE-RATE of an air-mass is greater than the DRY ADIABATIC LAPSE-RATE (i.e. it is warmer and therefore lighter than its surroundings), and so will continue to rise. A warm, very damp air-mass may rise to great heights and cause very u. atmospheric conditions, building up large CUMULUS clouds, possibly causing heavy rainfall, hail and even thunderstorms. Vertical ascent will cease when it reaches the same temperature as the surrounding air, and it is then in NEUTRAL (or indifferent) E. Cf. CONDITIONAL INSTABILITY, and ct. STABLE E.

upland A gen. term for higher land, in contrast to lowland, with no specif. connotation.

upland plain A not very satisfactory term for a comparatively level area of land at some altitude; a high-lying planation surface.

U. P. S. Grid (Universal Polar Stereographic Grid) The U.S. military g. for areas between 80°N. and the N. Pole, and 80°S. and the S. Pole, complementary to the U.T.M. G. It is superimposed on a Polar STEREOGRAPHIC PROJECTION bounded by the parallel 80°, and divided into 100-km. squares, with their sides parallel to a line joining 180° to 0° longitude (central meridian), crossed at right-angles by that joining 90°E. and W. The origin of the g. is the

Pole, the centre of the projection; the FALSE ORIGIN is transferred 2 million m. W. and 2 million m. S. in the N. Zone, 2 million m. W. and 2 million m. N. in the S. Zone.

urban App. to a town or city; hence a systematic branch of the subject, *u. geography*, the geography of towns, their situations, functions, patterns and lay-out, origins and evolution.

urban conservation area An a. with special architectural or historic interest, defined under the Civic Amenities Act (1967) with the object of preserving or enhancing its character and appearance. Within such an a., partic. strict planning controls apply.

urban fence Used in planning to denote a line enclosing an area fundamentally devoted to u. occupation and activity, outside which lies land primarily devoted to rural and agricultural interests.

urban field The sphere of influence of a town; the district around, to which it is functionally linked.

urban form The spatial arrangement of zones within an u. area, seen essentially from the point of view of accessibility. In modern u. planning, this commonly involves evaluation of linear, concentric and grid f.'s.

urban hierarchy The classification of towns according to size, or to their stages of development.

urban mesh The locational network of u. places, a concept derived orig. from W. Christaller's theory of CENTRAL PLACES. The restrictive assumptions of the theory resulted in a uniform hexagonal network or m. of central places, though the incorporation of other, less uniform, variables into the theory has modified such a rigid geometrical pattern.

urban morphology The systematic study of the form, shape and plan of a town.

urban renewal The planned r. of the obsolete fabric of a town, though retaining the same essential character and use matrix, and also the continuity of the community.

urban sprawl An unplanned, sporadic spread of building, often the result of the outward expansion of a town and the coalescence of adjoining units. See CONURBATION.

Urlandschaft (Germ.) A past landscape, as used in HISTORICAL GEOGRAPHY.

Urstromtal, pl. **-täler** (Germ.) Lit. an 'ancient stream-valley'. A broad shallow trough or depression, eroded by a melt-water stream flowing along the front of the continental ice-sheets which lay at various stages across N.W. Europe. As the ice-sheets retreated, so a series of U. was successively formed; 5 main lines can be traced across Germany and Poland. The post-glacial rivers (Elbe, Oder, Vistula) flow in a general direction from S.E. to N.W. across the N. European Plain to the North Sea. They occupy various sections of the U., then continue sections of their N.W. courses, which helps to explain the frequent 'elbows' in their courses. Some of the U. are dry, others have facilitated the construction of sections of Germany's '*Mittelland*' Canal.

U-shaped valley A glaciated v., with a flat floor and steep sides, the result of glacial erosion on not only the floor but also the sides (up to the level of the surface of the ice) of a pre-glacial river v. The v. is straightened, projecting spurs are planed off, high tributary v.'s are left hanging, and rock-steps in the floor are common. The v. ends abruptly at its head in a steep wall, known as a TROUGH-END. E.g. the radial v.'s of the English Lake District; the Scottish glens; the v.'s containing the Norwegian fjords. Two of the most spectacular U-shaped v.'s are Lauterbrunnen in Switzerland and Yosemite in California. [*f* ALP]

U. T. M. Grid (Universal Transverse Mercator Grid) The standard g. for all U.S. military maps throughout the world. It is drawn on the TRANSVERSE MERCATOR PROJECTION between 80°N. and 80°S. (see U. P. S. G.).

The world is divided into 60 grid-zones, each 6° of longitude (with $\frac{1}{2}$° overlap into adjacent zones on each side). Each zone is numbered, 1 to 60, beginning at 180°W. and moving E.-ward. Each latitudinal division of 8° is lettered from C (80°S. to 72°S.) to X (72°N. to 80°N.). Hence a quadrilateral is known by a figure and letter; e.g. 4Q is between longitude 162°W. and 156°W., 16°N. and 24°N. A 2nd degree of reference is given in respect of squares of 100 km., indicated by 2 other letters. The 3rd degree of reference is within each grid-zone, with 6 figures; i.e. a full world reference could be 4QXC264839. A km. g. is drawn within each zone, based on the central meridian and the equator. The FALSE ORIGIN for each zone is 500 km. W. of the central meridian on the equator for the N. hemisphere, 10,000 km. S. of this for the S. hemisphere.

uvala (Serbo-Croat) A hollow in limestone country, larger than a DOLINE (and commonly formed by the coalescence of several of these), but smaller than a POLJE. Its floor is more uneven than that of a polje.

vadose water 'Wandering' w. moving through permeable rock above the WATER-TABLE. Ct. GROUND W.

val, pl. **vaux** (Fr.) A valley; specif. a longitudinal valley in the Jura Mtns. [*f* CLUSE]

vale A somewhat poetical name for a valley, used commonly in place-names; e.g. V. of York, Evesham, Lorton (English Lake District), Clwyd (N. Wales).

valley An elongated depression sloping towards the sea or an inland drainage basin, usually though not always occupied by a river. See LONGITUDINAL, TRANSVERSE, RIFT-, U-SHAPED, HANGING, and V-SHAPED V.'s.

valley-glacier A tongue of ice moving downward and outward from a FIRN-field, along a pre-existing v.; it is sometimes called an alpine or mountain g. See GLACIER.

valley train A line of fluvioglacial outwash material deposited in a v. by melt-water from a glacier-snout.

valley wind See ANABATIC, KATABATIC W.

valloni (It.) A long narrow gulf or channel in the Adriatic Sea between the islands of the Dalmatian coast of Yugoslavia; syn. with *canali*. They were formed by a rise of sea-level in an area of mountains and valleys parallel to the coast, thus turning the valleys into gulfs of the sea.

vanadium A metal widely dispersed as a mineral-ore in many igneous and sedimentary rocks, though in small quantities. Most is used in steel-alloys; it helps to make the metal less prone to fatigue under stress. C.p. = U.S.A., S. Africa, S.W. Africa, Finland.

vapour-pressure The p. exerted by the water-v. present in the atmosphere. Ct. the v.-p. in England (7–15 mb.) and near the equator (30 mb.). The max. v.-p. at any temperature obviously occurs when the air is saturated. *V.-tension* is syn., though obsolescent.

vapour trail See CONDENSATION T.

variation, magnetic See MAGNETIC DECLINATION.

Variscan App. to the mountain-building movements of late Palaeozoic (Carbo-Permian) times. Sometimes the name is given in central Europe to the equivalent of the ARMORICAN in the W.; sometimes it is regarded as syn. with HERCYNIAN and includes all the European ranges of this age, hence *Variscides*. The name was first used in 1888 by E. Suess, after the Germanic tribe of the *Varisci*. It seems preferable to regard the V. uplands as the E. representatives of the Hercynian, the Armorican as the W. ones.

varve A distinctive banded layer of silt and sand, deposited annually in lakes ponded near the margins of ice-sheets; the coarser material, lighter in colour, settles first during summer melting, the finer darker deposits in winter. Each band of light and dark material represents one v. Thus it is possible to count the number of v.'s and so find the number of years involved in the formation of the v. deposit; as v. characteristics are fairly distinctive and recognizable in relative thickness for each individual year's sedimentation, correlations over quite wide areas are possible, affording a contribution to the glacial chronology. The concept was worked out by the Baron G. de Geer in Sweden, and first published in 1910.

Vauclusian spring The resurgence or reappearance of an underground stream, called after the Fontaine de Vaucluse in the lower Rhône valley. It occurs commonly in limestone country, where water wears subterranean ramifications, finally issuing from the limestone at its base. E.g. the R. Axe, issuing from Wookey Hole, Mendips, Somerset.

vector analysis A statistical method of analysing changes in the amount and direction of movement. A v. is a quantity which has direction as well as magnitude (e.g. a wind, a tidal current), and can be denoted by a line drawn in stages from an original to a subsequent position. E.g. a diagram of a tidal current may comprise a continuous series of lines (a *v. traverse*), each representing the hourly observation drawn on a correct bearing, and each proportional in length to the mean velocity during that hour. Another simple form is a WIND-ROSE. V. a. is also used in locational studies, as in manufacturing industry (e.g. determining the optimum location for a plant in terms of transport costs).

veering, of wind (i) A change of direction of the w. in a clockwise direction; e.g. from N.E., to E., to S.E. Ct. BACKING. (ii) In U.S.A. the term is used as above in the N. hemisphere, but in the reverse (i.e. anti-clockwise) in the S. This is not gen. usage, which implies a change in a clockwise direction in both hemispheres.

vega (Arabic-Sp.) An irrigated lowland in Granada, Spain. The name is sometimes restricted to an area which produces a single crop each year. Ct. HUERTA.

vegetable oil The o. which can be extracted, usually simply by compression, from the fruits and seeds of various plants. Ct. mineral o., animal o. The main v. o. yielders are olive, coconut, palm (copra), o.-palm, ground-nut, soya, cotton-seed, linseed (from flax), tung, castor, rape, sunflower, sesamum.

vegetation The living mantle of plants (*flora*) which covers much of the land-surface, forming an important aspect of the physical environment. Hence v. regions, NATURAL V., SEMI-NATURAL V., cultivated v., v. types. V. is controlled by various groups of factors: (i) *climatic* (temperature, moisture, wind, light); (ii) *edaphic* (soil conditions); (iii) *physiographic* (relief, aspect and drainage); (iv) *biotic* (the effects of organisms); and (*v*) *human* (burning, felling of trees, drainage, irrigation, grazing by domestic animals).

vein A fissure or a crack in a mass of rock containing minerals deposited in crystalline form, frequently associated with metallic ores. Syn. with LODE, the miners' term.

veld (Afrikaans) An area of open grassland on the S. African plateau, of great variety, including the *High* (above 1500 m., 5000 ft.), *Middle* (1500–900 m., 5000–3000 ft.) and *Low v.* (below 900 m., 3000 ft.), *Grass v.*, *Bush v.*, and *Sand v.* (semi-arid).

Venn diagram A conventional V. d. is used as a SET THEORY MODEL; it depicts visually the inter-relationship of SETS. It may display the overlap of 2 sets in 2 dimensions, or it may be extended into 3 spatial dimensions, or into time. E.g. the overlap of service areas of 2 or more towns; a definition of geography as the overlap of 3 sets; see SET THEORY. The diagram is under SET THEORY. [*f*]

vent An opening in the surface of the earth's crust, through which material is forced during a volcanic eruption.

ventifact A boulder, stone or pebble, worn, polished and faceted by wind-blown sand in the desert. See DREIKANTER, EINKANTER.

verglas A thin layer of clear, hard, smooth ice which forms over rocks, the result of a hard frost following rain or snow-melt; this adds considerably to climbing difficulties.

vernier An auxiliary scale on a ruler, barometer, theodolite, etc., which enables readings to be taken to an additional significant figure. E.g. if a scale is graduated in 10ths of an inch, the v. s. is a length of 9/10 of an in., divided into 10 equal parts. The v. is slid until its zero is opposite the quantity to be measured. By observing which division of the v. s. is exactly opposite a division on the main scale, the second decimal place can be read off.

vertical exaggeration (V. E.) A deliberate increase in the v. scale of a SECTION, in comparison with the horizontal scale, in order to make the section clearly perceptible. E.g. if the horizontal scale is 1 in. to 1 mi. (5280 ft.) and the vertical scale is 1 in. to 1000 ft., the v. e. is 5·28 times.

vertical interval The difference in v. height between 2 points; see GRADIENT. A 1° rise in a HORIZONTAL EQUIVALENT of 100 ft. involves a v. i. of 1·74 ft. [*f* SLOPE-LENGTH]

vertical photograph An aerial p. taken from as v. a position as possible; the ct. is with an *oblique p.*

vertical temperature gradient See LAPSE-RATE.

vesicular, vesicule (noun) Applied to the texture of a rock containing many small cavities (vesicules), the results of the presence of bubbles of steam or gas in molten rock (LAVA) as it cooled. See AMYGDALE.

village A collection of houses in rural surroundings, smaller than a town, larger than a hamlet, and without a municipal form of government. Usually v.'s were founded as agricultural settlements, but they may not be so today, even though retaining the name; e.g. Dulwich V. in London; Greenwich V. in New York City. Most v.'s in England are administered by a rural district council and by a parish council, under the overall administration of the county in which they are situated.

virga The trailing shreds of cloud under the low dark line of a passing FRONT.

virgation A bunching and divergence of fold-ranges from a central 'knot'; e.g. the Pamirs in central Asia, Pasco in the Andes.

viscous App. to a fluid, specif. LAVA with a high melting-point, a stiff, pasty consistency and rich in silica; it solidifies rapidly, does not flow far, and builds up a high, steep-sided cone.

visibility The distance which an observer can see, depending on: (i) his height above sea-level (see HORIZON), with which is involved the curvature of the earth's surface; (ii) the amount of DEAD GROUND; (iii) the clarity of the atmosphere; see FOG, MIST, HAZE; (iv) the time of day or night.

viticulture The cultivation of the vine.

vitrain Thin horizontal bands of strata in bright glossy coal (from Fr. *vitre*, glass), fracturing easily at right-angles to the bedding.

volcano Popularly a conical hill or mountain, built up by the ejection of material from a vent; better, a volcanic peak. The adj. *volcanic* is correctly used to signify all types of extrusive igneous activity, as opposed to PLUTONIC; the noun *volcanicity* is also sometimes used. The world's highest extinct v. is Aconcagua (6959 m., 22,834 ft.) in the Andes; the highest active v. is Guayatiri (6060 m., 19,882 ft.), also in the Andes, which last erupted in 1959. See under specif. volcanic feature: AGGLOMERATE, ASH, BOMB, CALDERA, CINDERS, CRATER, DOME, DUST, ERUPTION, PIPE, PLUG, SPINE, THOLOID, TUFF.

volume A measure of bulk or space. In Gt. Britain, U.S.A., etc., different scales are used for solids and liquids.

Solids

1728 cu. ins.	= 1 cu. ft.
1 cu. in.	= 16·387 c.c.
27 cu. ft.	= 1 cu. yd.

Liquids

4 gills	= 1 pint (0·5682 litre)
2 pints	= 1 quart
4 quarts	= 1 Imperial gallon (4·546 litres)
2 gallons	= 1 peck
4 pecks	= 1 bushel
8 bushels	= 1 quarter

Metric

1000 cu. mm.	= 1 cu. cm. (c.c.)
1000 c.c.	= very nearly 1 litre

Conversions

1 cu. m.	=	35·315 cu. ft.
1 litre	=	0·22 gall.
1000 litres	=	1 cu. m.

Voralp (Germ.) The lower pastures of an Alpine valley, above its floor but below the ALP proper; occupied by transhumant flocks and herds as a stage in their seasonal movement.

V-shaped valley Gen. a valley eroded by a river, as opposed to most U-shaped glaciated valleys. The angle of the V depends on: (i) the resistance of the rocks both to river erosion and to the weathering of the containing slopes; (ii) the stage of the cycle of river erosion, whether young (a steep V), or mature (a wider, more open V). In old age the V is replaced by a broad, almost level valley, bounded by low bluffs which may be far back from the river itself.

vulcanicity, vulcanism The processes by which solid, liquid or gaseous materials are forced into the earth's crust and/or escape on to the surface. This includes igneous activity gen., besides that popularly associated with volcanoes.

wadi (Arabic) A steep-sided rocky ravine in a desert or semi-desert area, usually streamless, but sometimes containing a torrent for a short time after heavy rain. E.g. in Arabia.

wake-dune A sand-d. which may occur on the lee side of a larger d., trailing away in the direction of wind movement. [*f* DUNE]

Wallace's Line A l. between S.E. Asia and Australia, demarcating the distinctive Asiatic and Australian flora and fauna. [*f*]

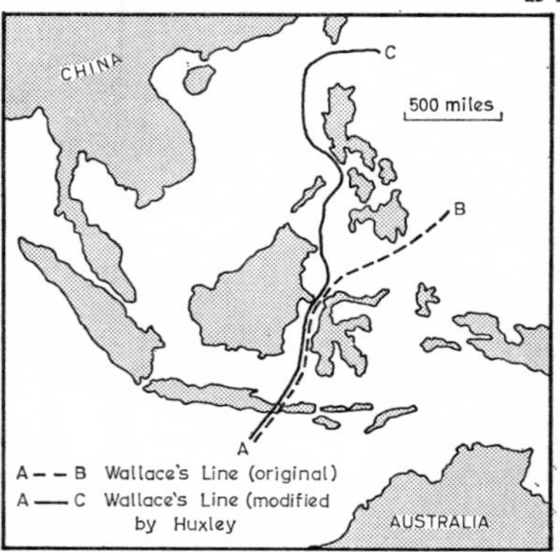

wall-sided glacier A valley-g. which flows out on to a plain, so forming a steep-sided tongue of ice unconfined by a valley.

waning slope Used by W. Penck and A. Wood for the gentle s. of finer material (sometimes called the '*wash slope*'), accumulated at the foot of a CONSTANT S. These low-angle s.'s are said to become more and more dominant as erosion proceeds, and the final peneplain stage may be regarded as consisting of coalescing w. s.'s of progressively gentler angles. [*f* FREE FACE]

warm front The boundary-zone at the front of the w. sector of a DEPRESSION, where a mass of w. air is overriding and rising above the cold air which it is overtaking. The frontal surface is at a very low angle, of only $\frac{1}{2}°–1°$. Ahead of the w. f., a broad belt of continuous rain falls from a heavily overcast sky, while the wind BACKS before the w. f. arrives, and then veers. As the w. f. approaches, a distinctive series of thickening cloud-forms can be seen: cirrus, cirrostratus, altostratus, stratus, and then nimbostratus as the w. f. passes. [*f* DEPRESSION]

'warm glacier' A g. where the ice-mass is at or near 0°C., mainly warmed by the percolation of melt-water produced by conduction heating at the surface. As the water passes down through the mass, it is cooled by contact with the ice particles, giving up latent heat as it refreezes, and so raising the internal temperature of the entire mass. Thus a 'w. g.' may approach 0°C. throughout its mass in summer, though in winter superficial cooling will produce a very cold crust. These physical facts play an important part in g. motion. Called a 'temperate g.' by H. W. Ahlman. Ct. 'COLD GLACIER'.

warm occlusion An o. where the overtaking cold air is not as cold as the air-mass in front. [*f* OCCLUSION]

warm sector A bulge or 'bay' of w. air in the S. part of a depression (in the N. hemisphere), its E. edge being a W. FRONT, its W. edge a COLD FRONT. Within the w. s. over Britain, the air-flow is between S. to S.W., with a complete cloud-cover, and continuous drizzling rain, which gradually becomes heavier and concentrated in showers as the cold front approaches. [*f* DEPRESSION]

warp, warpland Sediment deposited in a tidal estuary; e.g. on the shores of the Solway Firth, Cumberland.

warping (i) A gentle deformation of the crust over a considerable area. Though it may involve a vertical movement of only a few ft., it may have important results on the surface of an uplifted peneplain, or along a coast. The Central Massif of France shows several warped surfaces intersecting each other at low angles. There has been w. in the Gt. Lakes region of N. America during uplift. One type of w. is the result of ISOSTATIC depression (*down-w.*) or uplift (*up-w.*), the latter esp. after the melting of an ice-sheet; e.g. in Scandinavia. (ii) The deposition by flooding of a layer of mud and alluvium over the low-lying land adjacent to an estuary. Hence WARP, WARPLAND.

warren An area of waste land, often consisting of sand-dunes, in former times reserved for breeding game, esp. rabbits. E.g. Dawlish W., near the mouth of the R. Exe, Devon.

Warthe A stage of the Quaternary glaciation in N.W. Europe, the moraines of which can be traced in the Fläming (to the S.W. of Berlin) and in Poland (esp. Upper Silesia). Some authorities contend that the W. was an early stage of the 4th (Weichsel) main glaciation, but the consensus of opinion is that it represents a temporary halt and slight readvance during the retreat of the ice-sheets towards the end of the 3rd (Saale) glaciation.

wash (i) Fine material moved down a slope, esp. where there is little vegetation to fix and hold it; sometimes called *downwash*. (ii) The movement of water up a beach after a wave breaks; ct. BACKWASH. (iii) An area of tidal sand- and mud-banks. (iv) A term applied in the S.W. of U.S.A. to a shallow, dry, stream-channel in the desert.

washboard moraine A landscape in an area of former ice-sheet stagnation and melting, where a series of closely parallel m.'s (mainly of sandy TILL) were laid down; e.g. in W. Alberta.

washland Embanked low-lying lands bordering a river or estuary, deliberately allowed to flood so as to cope with high water-level in the river; e.g. the land between the Old and New Bedford Levels in the Fen District. W. can be used for grazing at some times of the year.

washout (i) The result of a sudden concentrated downpour of rain, causing extensive scouring, the sweeping away of bridges, and the undermining of the river-banks. (ii) A gap in a coal-seam, usually filled with sandstone; it represents concentrated stream erosion during or after the formation of the coal-seam; a stream similar to a distributary in a delta formed a channel, which was later filled with sand.

wastage (i) The loss of ice in a glacier or ice-sheet; a better term than shrinkage. It is sometimes regarded as syn. with ABLATION. (ii) A gen. term for the denudation of the earth's surface.

waste mantle See REGOLITH.

waterfall A steep fall of river water, where its course is markedly and suddenly interrupted. This may be the result of: (i) a transverse bar of resistant rock across the river's course, interrupting its progress to a graded profile; e.g. the Nile cataracts, Niagara, Kaieteur Falls (Guyana), Gibbon Falls (Yellowstone National Park, Wyoming); (ii) a sharp well-defined edge to a plateau; e.g. Aughrabies Falls in S.W. Africa, where the Orange R. crosses the edge of the African plateau; (iii) faulting, later forming a FAULT-LINE SCARP; e.g. Gordale Scar, Malham, Yorkshire; Zambezi Falls, due to faulting in part; (iv) the presence of a deep glaciated valley, with HANGING tributary valleys; e.g. Yosemite Falls, California; the falls (e.g. Staubbach) of the Lauterbrunnen valley, Switzerland; (v) along the edge of a cliffed coast; e.g. Litter Water on the Devon coast near Hartland.

Some Major Falls

	m.	*(ft.)*
Angel F., Venezuela	979	3212
Sutherland Fall, New Zealand	580	1904
Ribbon F., Yosemite, U.S.A.	491	1612
Upper Yosemite F., U.S.A.	436	1430
Uitshi, Guyana	366	1200
Staubbach, Switzerland	264	866
Vettisfoss, Norway	260	852
Kaieteur, Guyana	251	822
Victoria, Africa	110	360

Note: (*a*) The Yosemite Falls include the Upper Falls (436 m., 1430 ft.), the intermediate cascades (248 m., 815 ft.), and the Lower Falls (98 m., 320 ft.); total (782 m., 2565 ft.) (*b*) The Victoria Falls on the Zambezi R. are 1372 m. (4500 ft.) in width. (*c*) Niagara Falls descend 51 m. (168 ft.), with 2 distinct falls, the Canadian or Horseshoe Falls with a frontage of 853 m. (2800 ft.), separated by Goat I. from the American Falls with a frontage of 323 m. (1060 ft.); these falls are receding upstream at an average rate of 0·67 m. (2·2 ft.) a year. [*f*]

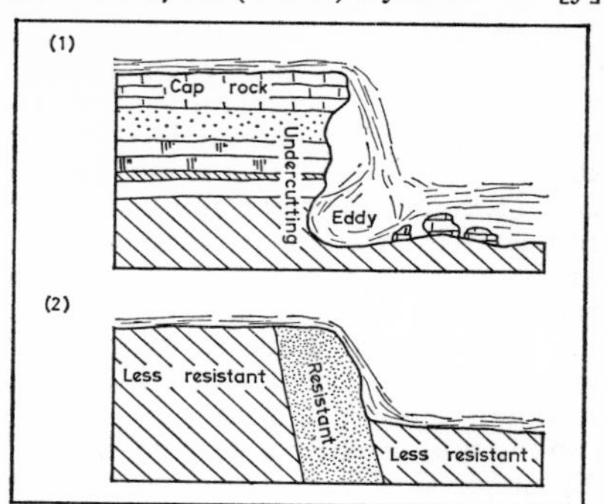

water-gap A low-level valley across a ridge, through which flows a river; e.g. the Wey g., near Guildford, the Mole g. near Dorking, through the N. Downs; the Ouse g. near Lewes through the S. Downs; the Goring g., where the R. Thames flows between the Berkshire Downs and the Chilterns.

water hemisphere Half of the globe, more or less centred on New Zealand, which contains only 1/7 of the land surface, as opposed to the land h.

waterhole A hole containing water, found esp. in the hot deserts or savanna lands, sometimes in the bed of an intermittent stream, used by animals and men.

water-meadow An area of m. on the flood-plain of a river, which is naturally or artificially flooded, so stimulating grass growth. It is found in chalk country adjacent to a stream, where flooding early in the year produces an early crop of grass. Some w.-m.'s are intersected with channels which can be flooded from a river; e.g. along the Itchen flood-plain S. of Winchester.

water-power Orig. used of the p. generated through the turning of a mill-wheel by running water; now syn. with HYDRO-ELECTRICITY.

watershed (i) The line separating headstreams which flow to different river systems; it may be sharply defined (the crest of a ridge), or indeterminate (in a low undulating area). In U.S.A. this is equivalent to a

DIVIDE. The term *water-parting* is used in both Britain and U.S.A. (ii) In America, the whole 'gathering-ground' of a single river-system, equivalent to a DRAINAGE BASIN. The term is thus used with 2 quite different meanings. [*f*]

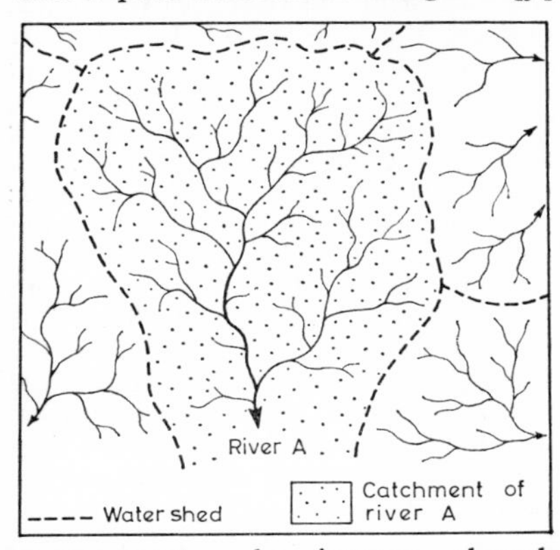

waterspout An intense, though small-scale, rapidly moving, low pressure system, similar to a TORNADO, but over the sea. From the low-lying base of a CUMULONIMBUS cloud, a whirling cone or funnel of cloud elongates until it touches the surface of the sea. The water-drops are derived both from condensation caused by cooling within its vortex and from water picked up from the agitated surface of the sea.

water-table The upper surface of the zone of saturation in permeable rocks; this level varies seasonally with the amount of percolation. Where it intersects the land surface, springs, seepages, marshes or lakes may occur.

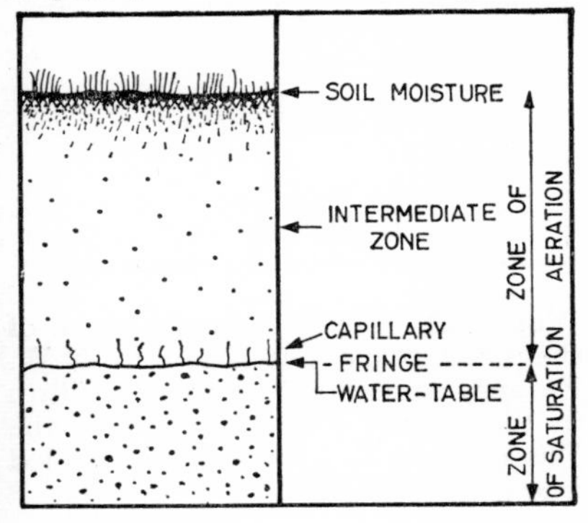

The w.-t. in a permeable AQUIFER roughly follows the surface profile of the ground, with its gradients somewhat flattened out. The slope of the w.-t. is inversely proportional to the permeability of the aquifer. The situation becomes more complicated, however, in rocks where flow is concentrated along faults and fissures, when no simple w.-t. may exist. See also PERCHED W.-T. [*f previous column*]

waterway A stretch of inland water, either natural (a river) or artificial (a canal), which can be used for transport. See SHIP-CANAL, SEAWAY.

Watten (Germ.) Tidal marshes lying between the mainland coasts of Denmark, W. Germany and the Netherlands and the protective line of the offshore N. and E. Frisian Islands. At low tide a maze of creeks and channels separates the sheets of mud. Syn. with *wadden* in Netherlands.

wave, ocean An oscillation of water particles. In the open o., the particles describe a circle in a vertical plane as the w. passes; each moves slightly forward on the crest, and returns almost to its orig. position in the trough. The oscillation is caused by the friction of wind upon the surface of the water. Its size is determined by the wind speed and duration, and by the length of FETCH. The nature of the w. is defined by its: (*a*) *height*, the distance from crest to trough, which may be as much as 12–15 m. (40–50 ft.), the record instrumentally measured being 21 m. (70 ft.) in 1961; (*b*) *length*, the distance between two successive crests (the longest measured is 1128 m., 3700 ft.); (*c*) *period*, the time taken for the w.-form to move the distance of one w.-length; (*d*) *velocity*, the speed of the forward movement of an individual crest; (*e*) *steepness*, the ratio of its height to its length, which increases as the w. enters shallow water until it reaches 1:7, when it breaks (hence a BREAKER); (*f*) *energy*, dependent on all the preceding features, including fetch; the average pressure exerted by a large w. in winter on an exposed coast is nearly 11 tons per sq. m. (a ton per sq. ft.)

(e.g. on the coast of W. Ireland), and during a storm this may be 3 times as great. A w. is an erosive agent through: (i) hydraulic action; (ii) corrasion; (iii) attrition; (iv) solution. The greatest depth at which sediment is disturbed on the sea-floor is the *w.-base*. Deposition is carried out by constructive w.'s, transport by w.'s breaking obliquely, hence LONGSHORE DRIFT. See also SWASH, BACKWASH, DOMINANT W.'S.

wave-built terrace A feature of marine deposition beyond the W.-CUT BENCH [*f*].

wave-cut bench A feature of marine erosion at the base of sea-cliffs; this can develop into a w.-c. platform. [*f*]

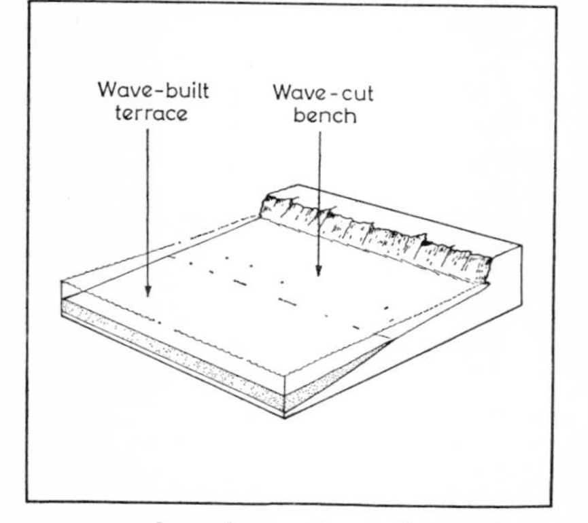

wave refraction A tendency for a w.-front to be turned from its orig. direction as it approaches the coast. It may be retarded by the shallowing of the water. Oblique w.'s tend to turn parallel to the coast, while coastal indentations produce curved w.-fronts and a concentration of energy on headlands. [*f opposite*]

waxing slope The upper convexity on a hill-slope, as defined by W. Penck and A. Wood, thought to result from the weathering of an initial break of slope from 2 sides at once. [*f* FREE FACE]

weather The condition of the atmosphere at any place at a specif. time, or for a short time, with respect to the various elements (temperature, sunshine, wind, clouds, fog, precipitation). This is an hour to hour, or day to day, condition. Ct. CLIMATE.

weathering The disintegration and decay of rock, so producing *in situ* a mantle of waste, depending on: (*a*) the nature of the rock; (*b*) the relief; and (*c*) the potency of the climatic elements in operation. W. may be: (i) *mechanical* or *physical*: frost action, temperature change; (ii) *chemical*: SOLUTION, CARBONATION, HYDROLYSIS, OXIDATION, HYDRATION; (iii) *biological*: the presence of moss and lichen, tree-roots, worms, moles, rabbits; this is not strictly w., but it assists, both mechanically and physically. The work of wind and rain (except for the latter providing lubrication of material and water which may freeze in cracks) are not included in w., since they involve transport of material, and are part of EROSION.

weather type A generalized type of synoptic pressure pattern, with its associated set of characteristic w. conditions. See ANALOGUE.

wedge, of high pressure A region of high atmospheric p. between 2 depressions, narrower than a RIDGE, bringing a brief spell of fine weather.

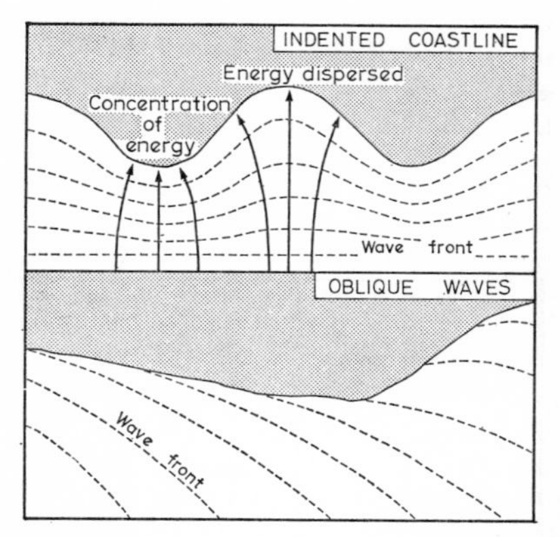

Weichsel The final main phase of the Quaternary Glaciation in N.W. Europe, corresponding to the Würm in the Alpine Foreland, and to the Wisconsin in N. America. It is divided into 3 main stages:

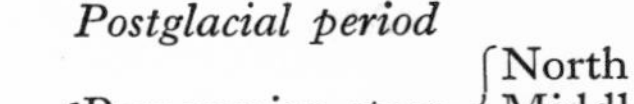

Postglacial period

Weichsel glaciation
- Pomeranian stage
  - North
  - Middle
  - South
- Frankfurt stage
- Brandenburg stage

weir A dam across a river for raising and maintaining the level of the water, so controlling the current, keeping a navigable depth, or providing a head for a mill-wheel.

welded tuff A mass of hot volcanic ash that has fused to form such compact rocks as RHYOLITE.

Wentworth scale A geometric s. of factor 2 of the size of particles of sediment, published by C. K. Wentworth in 1922. It ranges from particles of clay (0·004 mm. diameter), silt, sand, granule, pebble, cobble to a boulder (exceeding 256 mm. in diameter).

Westerlies The air-flow from the Sub-tropical High Pressure 'cell' to the Temperate Low Pressure zone between 35°N. and 65°N., and 35°S. and 65°S., blowing from the S.W. in the N. hemisphere, from the N.W. in the S. hemisphere. The W. strengthen with altitude, and are locally concentrated into JET STREAMS.

wet adiabatic lapse-rate See SATURATED A. L.-R.

wet-bulb temperature The t. recorded on a thermometer which has its bulb surrounded by a moist muslin bag, thus lowering the temperature by loss of latent heat through evaporation. With the aid of a dry-bulb thermometer and a set of tables, RELATIVE HUMIDITY can be ascertained.

wet-day In British meteorology, a day (24 hours, commencing at 09.00 hours) with at least 1·0 mm. (0·04 in.) of rainfall.

wet spell A duration of at least 15 successive WET-DAYS.

wetted perimeter In hydrological studies, the actual length of the line of cross-sectional contact between the water in a river and its bed or channel.

whaleback (i) A rounded elongated mass of rock, commonly granite, esp. shaped by moving ice; e.g. in the Canadian Shield; Finland. (ii) A smooth elongated mass of sand in a hot desert.

wheat A widely cultivated cereal, the world output of which is second only to rice. It may be sown in autumn and harvested in summer (*winter w.*), or sown in spring and harvested in late summer or autumn (*spring w.*). It is grown extensively in the former Temperate Grasslands (prairies, steppes, etc.) and intensively in such countries as Denmark, Belgium, France and the U.K. C.p. = U.S.S.R., U.S.A., China, Canada, Australia, France, India.

whinstone A quarryman's term in N. England for dolerite; hence the Great Whin Sill.

whirlpool (i) A violent circular eddy in the sea, produced by a powerful tidal current flowing through an irregular channel, or by the meeting of 2 currents; e.g. the Maelstrom in the Lofoten I.'s. (ii) A similar phenomenon produced at the base of a large waterfall; e.g. Niagara Falls. Another w. has developed 5 km. (3 mi.) down the gorge of the Niagara R. where it bends at right-angles; the current thus impinges against the N. side of the gorge, swirls violently round, and flows off N.E. This w. is apparently cut into the S. end of the drift-filled and abandoned interglacial gorge.

whirlwind A rapidly rotating column of air, produced by local heating and convectional uprising. See CYCLONE, DUST-DEVIL, TORNADO, WATERSPOUT.

'white area' An a., normally of open agricultural land, shown without notation in a DEVELOPMENT PLAN. In such an a. it is intended that the existing uses shall remain for the most part undisturbed.

'white coal' A fanciful name for hydro-electricity, derived from the Fr. term *houille blanche*.

'white man's grave' The name formerly applied to W. Africa, because of the widespread prevalence of diseases affecting Europeans. With modern sanitation, better water supply, pest control, inoculation and drugs, the term is no longer merited.

white-out Under BLIZZARD conditions with a total snow-cover, it is extremely difficult to find one's direction; the impression is of being swathed in a white opacity.

wildcat, wildcatting Used in U.S.A. in respect of attempts to locate oil or minerals, involving some risky, chancy or speculative elements.

wilderness Used in CONSERVATION, indicating an area left untouched in a natural state, with no human control or interference. Ct. NATURE RESERVE.

williwaw A sailor's term for a sudden squall, esp. in the ROARING FORTIES.

willy-willy An intense tropical storm originating off the coast of N.W. Australia, sometimes crossing on to the land; a type of tropical CYCLONE.

wilting point A measure of soil moisture used by botanists and agriculturalists. It indicates the amount of water in the soil, below which plants will be unable to obtain further supplies, and will therefore wilt.

wind A horizontal current of air, varying from 'light air' to 'hurricane'. W.'s can have a vertical movement, but this seldom happens at the earth's surface. See BEAUFORT SCALE.

wind-break A shelter, either natural (e.g. a line of trees or a thick hedge) or artificial (a screen), which 'breaks' or interrupts the force of the w. This is commonly used by horticulturalists in exposed areas; e.g. in the lower Rhône valley, where the MISTRAL can do great damage. Groups of trees are grown around isolated and exposed farmhouses.

wind-chill index An i. of physiological significance in cold climates, obtained from a formula involving temperature and wind-force.

wind-gap A g. or notch in a ridge of hills, without a river flowing through, usually (though not always) at a somewhat higher level than a WATER-G. These are partic. identified as COLS in escarpments through which former consequent streams were thought to flow before CAPTURE or 'beheading' by scarp retreat and by scarp-foot streams. E.g. Clayton, Pyecombe and Saddlescombe cols in the S. Downs. [*f* CAPTURE, RIVER]

'window' See FENSTER [*f*].

wind-rose A diagram with radiating rays drawn proportional in length to the mean percentage frequency of w.'s from each cardinal direction. The rays may be subdivided to show the frequency of various w. strengths associated with the partic. direction, each direction of ray being of different width [as *f*, p. 375]. An octagonal w.-r. can be constructed, with each side representing one of the 8 cardinal directions, and the 12 monthly frequencies of w. from each of these directions are plotted as columns. Calms may be given as a percentage in the centre of the circle or octagon. [*f, page 375*]

wind-shadow A 'dead air space' in the lee of an obstacle in the path of the w., though it is rarely calm there, but the scene of eddying. This is an important fact in the creation of a sand-dune and in determining its shape.

wind-slab A sheet of snow which is hard-packed by the w., requiring great care by a ski-er, since it is liable to avalanche *en masse*.

Winkel's Tripel Projection A development from LAMBERT'S ZENITHAL EQUAL AREA P., used by J. Bartholomew for world maps of climatology, vegetation and population because of its distributional merits.

winter (i) In gen. terms, the coldest season of the year, in ct. to summer. (ii) In the N. hemisphere the period Dec., Jan., Feb., in the S. hemisphere the period June, July, August. (iii) In astronomical terms, the period between the winter solstice (about 22 Dec.) and the spring equinox (21 March) in the N. hemisphere, and between about 21 June and 22 Sept. in the S. hemisphere.

winterbourne A stream breaking out in the floor of a dry valley only after a prolonged period of rainfall. This forms a common place-name element in chalk country; e.g. Winterborne Abbas, -Strickland, -Zelstone and many more in Dorset. Syn. with BOURNE.

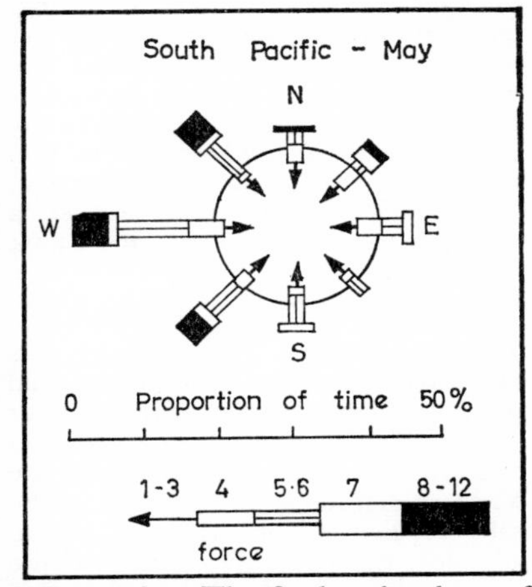

Wisconsin The final main phase of the QUATERNARY glaciation in N. America, corresponding to the Würm in the Alpine Foreland and to the Weichsel in N.W. Europe. The chronology of the W. is complex, but it has been closely studied because of its wide distribution, as indicated by the extensive deposits of boulder-clay and moraine, esp. in the area S.W. of the Great Lakes, and between Lakes Michigan, Huron and Erie. The area covered in N. America was greater than that of all the other glacial phases combined. Various stages and interstadial intervals have been suggested:

Wisconsin glaciation
- Mankato glaciation
- Valders glaciation
  - Two Creeks interstadial
- Cary glaciation
- Tazewell glaciation
  - Peorian interstadial
  - Iowan glaciation (formerly regarded as a separate stage)

Sangamon (Third) interglacial

The glacial relief is very varied; it includes numerous well-defined terminal and recessional moraines, KAME-AND-KETTLE features (esp. in Wisconsin State), and thick clay-TILLS.

wold An open rolling chalk upland, used more specif. as an element in a proper name; e.g. Lincoln W.'s, Yorkshire W.'s. The term is sometimes used for hills of other rocks; e.g. Cotswolds (Jurassic).

wolfram An ore of tungsten, ferrous tungstate ($FeWO_4$), sometimes called *wolframite*.

wool A fibre derived from a variety of animals, mainly sheep, with a small amount (*mohair*) from goats (Angora, Kashmir), camels, llamas, alpacas, vicuñas. W. now forms under 10% of all fibres used in the textile industry, but it is still very important for the manufacture of clothing for temperate and cool climates. C.p. = Australia, U.S.S.R., New Zealand, Argentina, S. Africa, U.S.A.

wrench-fault A nearly vertical STRIKE-SLIP FAULT, with the horizontal displacement on a large scale; e.g. San Andreas Fault, California; the Alpine Fault of the S. Island of New Zealand.

Würm The name given by A. Penck and E. Brückner to the last of the series of 4 periods of fluvio-glacial deposition during the Quaternary glaciation; see GÜNZ, MINDEL, RISS. This concept was subsequently widely applied to glacial periods in general, though the latest research indicates that the pattern of glacial and interglacial periods is not a simple 4-fold one. The W. is equivalent to the Weichsel in N.W. Europe, and probably to the Wisconsin in N. America. The W. is divided into 3 stages, known as W. I, II, III.

[*f* MINDEL]

xenocryst A crystal in an igneous rock which is foreign to the rock in which it now occurs.

xenolith A metamorphosed piece of foreign rock occurring near the margin of a BATHOLITH within the solidified granite, usually derived from the invaded country rock.

xerophyte, xerophilous, xerophytic One of a category of plants adapted to withstand dry conditions, seasonal or perennial, in various ways: long roots; small, hard, glossy leaves; thick bark; thorns; various water storage devices. Ct. TROPOPHYTE.

xerosere (alt. **xerarch**) Applied in a botanical context to a plant succession (SERE), developing under markedly dry conditions. It may comprise either a rock sere (*lithosere*) or a sand sere (*psammosere*).

xerothermic index An i. of the relationship of drought to plant-growth.

yard The British standard unit of length. 1 yd. = 0·914399 m. 1 m. = 1·09361 yds.

yardang A sharp keel-like crest or ridge of rock, separated from a parallel neighbour by a shallow groove or furrow. It is the result of differential erosion in a desert by the scouring effect of sand-laden winds. The ridges may be as much as 6m. (20 ft.) high and 37 m. (120 ft.) wide.

yazoo A DEFERRED JUNCTION [*f*] of a tributary, called after the R. Yazoo which joins the lower Mississippi.

year A measure of time related to the revolution of a heavenly body around the sun, specif. the earth. The exact concept differs: (i) *Sidereal Y.*: the time taken by the earth to make 1 complete revolution in its orbit with reference to the stars, 365·2564 mean solar days, or 365 days, 6 hours, 9 minutes, 9·54 seconds. (ii) *Tropical Y.*: the average time taken by the earth to make 1 complete revolution in its orbit with reference to the vernal equinox, as indicated by the First Point of Aries (see RIGHT ASCENSION), at present 365·2422 solar days, or 365 days, 5 hours, 48 minutes, 45·51 seconds; this diminishes by about 5 seconds in a millennium. It is also known as the *Equinoctial*, *Astronomical*, *Natural* or *Solar Y.* (iii) *Civil Y.* (or *Gregorian Calendar Y.*): a period of 365 mean solar days of 24 hours; to compensate for the extra 0·2422 solar days, every 4th is a Leap Y. with 366 days (one extra in Feb.). This is still not quite right (0·01321478 days too long, or approx. 0·04 days every 4 years, and 1 day per century of 25 leap-years), so the last year of a century is only a leap y. when the first 2 figures are divisible by 4; i.e. 1600 A.D. and 2000 A.D. are, but 1700, 1800, 1900 are not; (iv) *Anomalistic Y.*: the time between 2 successive PERIHELIONS, 365·25964 solar days.

'yellow ground' See KIMBERLITE.

Younger Drift The younger tills deposited after the last interglacial period, which in part overlie the older glacial tills which have been weathered and eroded (OLDER D.). In Europe the Y. D. is defined as the product of the 4th (Weichsel or Würm) glacial advance, the Older D. as that of the three earlier ones. In Britain the same distinction applies; the S. limit of the Y. D. can be traced from the coast of N. Norfolk through the Vale of York, the W. Midlands and S. Wales to the mouth of the Shannon in S.W. Ireland. The Older D. and Y. D. are of course much subdivided.

young flood stand A period of slack-water interrupting the normal tidal rise, though not the ebb; e.g. in Southampton Water. This occurs about 1½ hours after low water (near mean tide level), and lasts for nearly 2 hours before the tidal rise is resumed. This is the result of a double entrance to the Solent, which is not in phase with the tidal rise and fall outside the I. of Wight. A hydraulic gradient prevails at certain times between the Needles and Spithead entrances; at spring high water, the heights of the tide there are 1·07 m. and 2·0 m. respectively. This causes a flow from

E. to W., the result of this 'head'. The reverse applies at spring low water when the level at the Needles is 0·9 m. above that at Spithead. These interruptions to the normal tidal flow produce a pause in the normal tidal rise until the hydraulic 'head' has been wiped out. Added to the stand at high water, this gives 7 hours of slack in each 24.

'young' mountains Fold-mountains created during the last great period of folding (e.g. the Alps), by contrast with earlier ones (the 'old' f. m.'s of HERCYNIAN and CALEDONIAN age).

youth The 1st stage in the cycle of landform development; the orig. structure is still the dominant feature of the relief. Slopes are steep, gradients irregular, and denudation processes rapid (e.g. a 'young' river). Recent criticism of the indiscriminate use of this evocative terminology has been put forward; using the human life cycle as a simile can be carried too far. It cannot be assumed that the youth stage is necessarily the most active in terms of erosion, nor that 'old age' streams are the most sluggish. See CYCLE OF EROSION.

zambo An Indian-negro racial mixture in Latin America.

zawn A narrow rocky inlet in a cliffed coast, specif. in Cornwall.

zenith The point in the heavens (i.e. on the CELESTIAL SPHERE) vertically above the observer.

zenithal (or **azimuthal**) **projection** A class of p. in which the globe is projected on to a plane touching at the Pole (*polar z.*), at the equator (*equatorial z.*), or anywhere between (*oblique z.*). All bearings are true from the centre of the projection. See GNOMONIC, Z. EQUAL AREA, Z. EQUIDISTANT, STEREOGRAPHIC, ORTHOGRAPHIC, LAMBERT'S Z. EQUAL AREA P.'S.

Zenithal Equal Area Projection (also called AZIMUTHAL EQUAL AREA) When drawn as the polar case (with the Pole at the centre), it has concentric parallels, with the area between each the same as on the globe to scale, of radius $= \sqrt{2R\ (R - R \sin latitude)}$, where R is the scale of the globe. Meridians are straight lines radiating out from the Pole. In the other cases, both meridians and parallels are curves. It forms a useful p. for representing a single hemisphere. The oblique and equatorial cases are most conveniently constructed from tables.

Zenithal (Azimuthal) Equidistant Projection The only p. in which all points are the true distance and true direction from its centre. In the polar case, the meridians are radii from the centre of the p., spaced at the desired angular interval. Parallels are concentric circles, their true scale distances apart; i.e. $2\pi R.\frac{x}{360}$, where x is the parallel interval in degrees. The equatorial case is rarely used. The oblique case is used esp. for a map centred on an important city or an airport, since every place in the world is in the correct direction and at the correct distance, although shape becomes very distorted away from the centre. The emblem of the United Nations Organization is drawn on this p. [*f* POLAR PROJECTION]

Zeuge (pl. **Zeugen**) (Germ.) A tabular mass of resistant rock up to 30 m. high, standing out from softer underlying rocks because of its protective capping. It is produced by differential erosion in a desert through the scouring effect of sand-laden winds. [*f*]

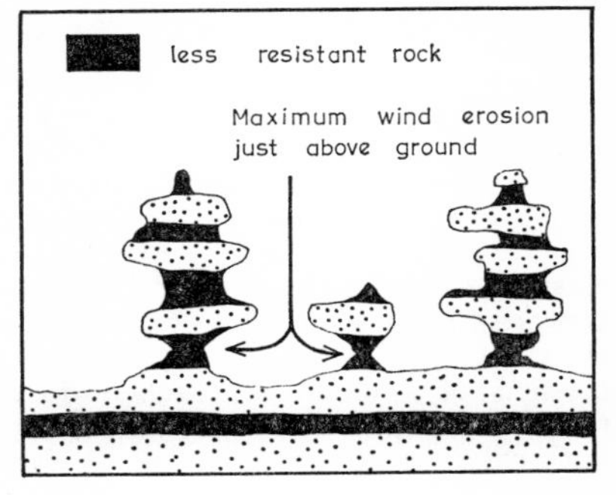

zinc A metal mainly obtained from *zincblende* (*ZnS*) (the mineral *sphalerite*), and *calamine* ($ZnCO_3$), an ore which often occurs in association with lead and silver, now commonly known in Britain as SMITHSONITE. Commercially, calamine includes silicates of zinc as well as the carbonate. It is used in alloys (with copper to form brass), as a thin coating over iron and steel (galvanizing), and in its oxide form as a pigment. C.p. = Canada, U.S.A., U.S.S.R., Australia, Peru, Japan, Mexico.

zincblende A major ore of ZINC, the mineral *sphalerite*, *ZnS*.

Zodiac An imaginary belt or zone in the heavens, within which are the apparent courses of the sun and the planets (except Venus and Pluto), bounded by 2 lines parallel to the ECLIPTIC, and 8° away on either side. In astrology, the Z. is divided into 12 equal portions of 30°, their names derived from constellations which at the time of the Greeks lay in that partic. portion of the sky, but now, owing to the PRECESSION OF THE EQUINOXES (about 50 seconds a year), lie some distance (about 30°) to the W. The Signs begin from the First Point of Aries (see RIGHT ASCENSION), where the plane of the ecliptic intersects the CELESTIAL EQUATOR at the time of vernal equinox, about 21 March. The Signs are Aries, Taurus, Gemini, Cancer, Leo, Virgo, Libra, Scorpio, Sagittarius, Capricornus, Aquarius and Pisces. Each Sign now lies in the constellation next to the W.; e.g. the First Point of Aries is in Pisces, which is astrologically awkward.

zodiacal light A cone of faint l., extending upward from a base below the horizon, best seen after sunset in spring and before sunrise in autumn (in the N. hemisphere), esp. on a clear moonless night. It seems to slant upward to the left in the evenings, to the right in the mornings. It probably consists of a cloud of rarefied dust particles or gas molecules, extending outward from the sun. In the tropics the z. l. sometimes appears as a band right round the sky.

zonal flow The movement of air along the parallels of latitude, in ct. to MERIDIONAL F.

zonal soil A s.-type largely resulting from the climatic factors which contribute to the s.-forming processes. The 2 main groups are the PEDOCALS and the PEDALFERS.

zonda A warm, humid, sultry wind in the Argentine, blowing from the N. in front of a low-pressure system. The name is also given to a FÖHN-type wind blowing down the E. slopes of the Andes in the same country.

zone (i) Used gen., even vaguely, for various belts (esp. of climate and soil) of a latitudinal character, or for any region defined between specif. limits with special characters (e.g. plants, animals). (ii) In geology, a thickness of rock with a partic. fossil-assemblage. (iii) In town planning, an area designated or allocated for a specific purpose ('to z.', 'zoned', 'zoning'). The word z. is used in conjunction with many other terms, esp. within rocks or the soil: z. of aeration, capillarity, discharge, eluviation, illuviation, flow, fracture, saturation.

zone standard time The mean time within a longitudinal zone of 15°; see STANDARD TIME [*f*].

zoogeography The study of the geographical distribution of wild animals.

zooplankton A collective name for the minute marine animal life floating in shallow seas, esp. where warm and cold currents meet, and where cold water upwells. The main groups of z. are *Radiolaria*, *Foraminifera* and *Copepoda*. Ct. PHYTOPLANKTON, similar vegetable organisms, though gen. much smaller.

Zusammenhang (Germ.) The stressing by Germ. geographers, esp. in the 19th century, of the causal interdependence of man and the landscape, in some ways syn. with Fr. *connexité*.